D0942869

A HISTORY OF ASTRONOMY

A. PANNEKOEK

A HISTORY OF ASTRONOMY

ILLUSTRATED

DOVER PUBLICATIONS, INC.
NEW YORK

Published in Canada by General Publishing Company, Ltd., 30
Lesmill Road, Don Mills, Toronto, Ontario.

This Dover edition, first published in 1989, is an unabridged
and unaltered republication of the work first published by George
Allen & Unwin Ltd., London, in 1961. It is reprinted by special
arrangement with Unwin Hyman Ltd., 15-17 Broadwick Street,
London W1V 1FP. The original Dutch edition, *De Groei van ons
Wereldbeeld*, was first published in 1951 by Wereld-Bibliotheek,
Amsterdam.

Manufactured in the United States of America
Dover Publications, Inc., 31 East 2nd Street, Mineola, N.Y.
11501.

Library of Congress Cataloging-in-Publication Data

Pannekoek, Anton, 1873–1960.
    [Groei van ons wereldbeeld. English]
    A history of astronomy / A. Pannekoek.
        p.   cm.
    Translation of: De groei van ons wereldbeeld.
    Bibliography: p.
    Includes index.
    ISBN 0-486-65994-1
    1. Astronomy—History. I. Title.
QB15P28313 1989
520'.9—dc20   89-7794
                                       CIP

CONTENTS

PART THREE: ASTRONOMY SURVEYING
THE UNIVERSE

ILLUSTRATIONS

INTRODUCTION

ASTRONOMICAL science originated in a much earlier period of human history than the other natural sciences. In the remote past, when practical knowledge in daily life and work had not yet led to a systematic study of physics and chemistry, astronomy was already a highly-developed science. This antiquity determines the special place which astronomy has occupied in the history of human culture. The other realms of knowledge developed into sciences only in later centuries and this development took place mainly within the walls of universities and laboratories, where the noise of political and social strife seldom penetrated. Astronomy, on the contrary, had already manifested itself in the ancient world as a system of theoretical knowledge that enabled man to prophesy even the terrifying eclipses and had become a factor in his spiritual strife.

This history is associated with the process of the growth of mankind since the rise of civilization and, to a great extent, belongs to times in which society and the individual, labour and rite, science and religion, still formed undivided entities. In the ancient world and in the following centuries astronomical doctrine was an essential element in the world concepts, at once religious and philosophical, which reflected social life. When the modern physicist looks back at his earliest predecessors, he finds men like himself, with similar though more primitive views on experiment and conclusion, on cause and effect. When the astronomer looks back at his predecessors, he finds Babylonian priests and magicians, Greek philosophers, Mohammedan princes, medieval monks, Renaissance nobles and clerics—until in the scholars of the seventeenth century he meets with modern citizens of his own kind. To all these men astronomy was not a limited branch of specialist science but a world system interwoven with the whole of their concept of life. Not the traditional tasks of a professional guild but the deepest problems of humanity inspired their work.

The history of astronomy is the growth of man's concept of his world. He always instinctively felt that the heavens above were the source and essence of his life in a deeper sense than the earth beneath. Light and warmth came from heaven. There the sun and the other celestial luminaries described their orbits; there dwelt the gods who ruled over his destiny and wrote their messages in the stars. The heavens were near and the stars played their part in the life of man. The study of the stars was the unfolding of this higher world, the noblest object that human thinking and spiritual effort could find.

This study, continued through many centuries, and even in antiquity, taught two things: the periodic recurrence of celestial phenomena and the vastness of the universe. Within the all-encompassing celestial sphere with its stars, the earth, though for man the centre and chief object, was only a small dark globe. Other world bodies—sun, moon and planets, some of them of larger size—circulated around her. This was the world concept which, when the ancient world collapsed and science fell into a depression lasting a thousand years, was kept as a heritage and at the end of the Middle Ages was transmitted to the rising West-European culture.

There, in the sixteenth century, driven by a strong social development, astronomy gave rise to a new concept of the world. It disclosed that what seemed the most certain knowledge of the foundation of our life—the immobility of the earth—was merely an appearance. It showed moreover, that our earth was only one of several similar planets, all revolving about the sun. Beyond was endless space with the stars as other suns. It was a revolution, opening new ways of thinking. With hard effort and much strife mankind had to reorientate itself in its world. In those centuries of revolution the contest over astronomical truth was an important element in the spiritual struggle accompanying the great social upheavals.

Astronomy, like the study of nature in general, now entered a new era. The next century brought the discovery of the fundamental law controlling all motions in the universe. Philosophical thinking was for the first time confronted with an exact and strict law of nature. The old mystical, astrological connection between the heavenly bodies and man was replaced by the all-pervading mechanical action of gravitation.

Then, at last, in the modern age of science, the concept of the universe widened to ever larger dimensions, expressible only in numbers, against which to speak of the smallness of the earth is a meaningless phrase. Again—or still—astronomy is the science of the totality of the universe, though now merely in a spatial sense. Whereas in the ancient world the idea of the unity of the heavenly and the human worlds exalted the hearts of the students of nature, now men are stirred by the proud consciousness of the power of the human mind, which from our small dwelling place is able to reach up to the remotest stellar systems.

In early times, when physical theory was but abstract speculation, astronomy was already an ordered system of knowledge giving practical orientation in time and space. In later centuries, astronomical research was directed more and more towards theoretical knowledge of the structure of the universe, far beyond any practical application, to satisfy the craving for truth, i.e. for intellectual beauty. Then the mutual relation of the sciences became the opposite of what it had been before. Physics, chemistry, and biology shot up with increasing rapidity.

Through technical applications they revolutionized society and changed the aspect of the earth. In this revolution astronomy stood aside. The stars cannot contribute to our techniques, our material life, or our economic organization. So their study became more and more an idealistic pursuit tending toward a physical knowledge of the universe. While the other sciences won brilliant triumphs in a transformation of the human world, the study of astronomy became a work of culture, an adventure of the mind. Its history thus remains what it has always been, an essential part in the history of human culture.

Whoever penetrates into the past participates in the development of the human race as his own experience. It is the aim of this work to unfold in this past the development of our astronomical world concept as a manifestation of humanity's growth.

PART ONE

ASTRONOMY IN
THE ANCIENT WORLD

CHAPTER 1

LIFE AND THE STARS

THE origin of astronomy goes back to prehistoric times from which no records have survived. At the dawn of history we find indications in the oldest written records that astronomical phenomena had already captured man's attention, just as we find today among the primitive peoples a certain knowledge of stars and of heavenly phenomena.

What caused primitive man to raise his eyes from the earth toward the sky above him? Was it the beauty of the starry heavens, of the countless radiating points in a wonderful variety of brightness, colour and pattern, that caught his eye? Did the stately regularity of their motion across the vault, with irregularities superimposed, provoke his curiosity as to the cause? In later times these may have been driving and inspiring forces, but primitive man had so hard a struggle simply to make his life secure that there was no room for luxury incentives. To maintain himself he had to fight for his existence incessantly against the hostile powers of nature. The struggle for life occupied his thoughts and feelings entirely, and in this struggle he had to acquire such knowledge of the natural phenomena as influenced his life and determined his work; the better he knew them, the more secure his life became. It was in this way, therefore, that astronomical phenomena entered his life as part of his environment and as an element in his activities, capturing his attention. Science originated not from an abstract urge for truth and knowledge but as a part of living, as a spontaneous practice born of social needs.

The astronomical phenomenon of alternating day and night regulates the life of man and beast. Primitive tribes often read the time of day from the height of the sun with great precision; they need it to regulate their day's work. Missionaries relate how on clear days the aborigines turned up at morning assemblies at the exact hour, whereas on cloudy days they might be hours wrong. European peasants were also able to do this until church clocks and pocket watches made it superfluous. When meanwhile a more precise method was needed, the length of a shadow was paced off.

The other main phenomenon determining human activity is the alternation of the seasons. At high latitudes it is the alternation between a dead winter and an abundant summer; nearer the equator, between a dry and a rainy season. Primitive hunters and fishermen were dependent on the life cycles and migrations of the animals. Peasants and herdsmen regulated their work, their sowing and harvesting and their wanderings with the flocks, according to the seasons of the year. They were guided by their own experience of the changing aspects of nature.

Attention to the heavenly phenomena themselves became necessary when labour developed more complicated forms and new needs were felt. When the nomads or fishermen became travelling merchants they needed orientation, using for this purpose the heavenly bodies—in the day time the sun, at night the stars. Thus eye and mind were directed to the stars. The stars had names in the earliest literature of the Arabs. In the Pacific the Polynesians and Micronesians, who were experienced sailors, used the stars to determine the hour of the night; their rising and setting points served as a celestial compass, and they steered their vessels by them at night. In their schools young people were instructed in the art of astronomy by means of globes. Through contact with the Europeans, which ruined them physically and morally, this autochthonous science was lost.

The need to measure time intervals was a second incentive that led to a close observation of celestial phenomena. Time-reckoning was, apart from navigation, the oldest astronomical practice, out of which science later developed. The periods of the sun and moon are the natural units of time-reckoning, but sometimes, as curious products of a higher state of knowledge, other periods appeared, such as a Venus-period with the Mexicans and a Jupiter-period with the Indians. The sun imposes its yearly period by the seasons, but the period of the lunar phases is more striking and, because of its shortness, more practical. Hence the calendar was generally dominated by the moon, except where the climatic and agricultural seasons imposed themselves irresistibly.

Nomadic peoples regulate their calendar entirely by the synodic period of 29½ days in which the phases of the moon return. Every time the new moon—called a 'crescent' because of its growth—appears as a slender arc in the western evening sky, a new month of 29 or 30 days begins; hence the hours of the day begin at the evening hour. Thus the moon became one of the most important objects of man's natural environment. The Jewish Midrash says: 'The moon has been created for the counting of the days.' Ancient books from different peoples tell that the moon serves for measuring the time. This was the basis of the moon-cult, the worship of the moon as a living being, which by its waxing and waning regulated the time. Its first appearance and still more its fullness, when it dispelled the darkness of night, were celebrated with ceremonies and offerings.

Not only did worship result, but also closer observation, as is shown in a curious division of the zodiac into 27 or 28 'moon-stations'. They are small groups of stars *ca.* 13° distant from one another, so that the moon, in its course around the celestial sphere, every successive night occupies the following one. They were known to the Arabs under the name *menazil* or *manzil*; to the Indians as *nakshatra* (i.e. 'star'); and to the Chinese as *hsiu* ('night-inn'). Since the moon-stations in these three cases are mostly identical, it has been supposed that these peoples borrowed them from one another. Opinions have differed as to which of the three was the original inventor or whether each had received them from the Mesopotamian centre of culture. But an independent origin in each case does not seem impossible, since many of these moon-stations constitute natural groups of stars, more natural often than the twelve larger constellations of the zodiac. The head of the Ram, the hind-quarters of the Ram, the Pleiades, the Hyades with Aldebaran, the horns of the Bull, the feet of the Twins, the Twins Castor and Pollux themselves, Cancer, the head of the Lion with Regulus, the hind part of the Lion with Denebola, are all conspicuous groups. However, there is also an identity of less natural and less obvious groups, which would suggest mutual borrowing. It is well known that many cultural interchanges took place between China and India, and tablets with old Sumerian texts have been excavated in Sindh.

So the lunar period is the oldest calendar unit. But even with pure moon-reckoning, the year—the powerful period of nature—appears in the fact that there are twelve months, i.e. twelve different consecutive names of the months, names indicating a seasonal aspect: the month of rains, of young animals, of sowing or harvesting. Usually a tendency develops towards a closer co-ordination of lunar and solar reckoning.

Agricultural peoples, through the nature of their work, are strongly tied to the solar year. Nature itself imposes it on peoples living in high latitudes. The Eskimos of Labrador have no period names for the dark winter time, when outdoor work is at a standstill, but they have fourteen names for the remainder of the year. Many peoples leave nameless the months when agriculture is suspended; thus the Romans are reported to have had originally only ten months (the names from September to December mean the seventh to tenth months) and to have added January and February later. Our modern calendar, without any relation to the lunar phases, still embodies in its twelve months the tradition of ancient moon-reckoning.

In their calendars most agricultural peoples use both the month and the year, especially in southern countries, where the seasons of the year are less extreme. Thus the Polynesians and the African Negroes have their regular ceremonies at full moon, but their harvest festivities depend on the season. They know how many moons have to pass between harvesting and sowing and what are the right months for gathering wild fruits and for hunting.

Here, however, a difficulty arises because the solar-year dates of new and full moon move up and down, so that the phases of the moon cannot indicate a determinate seasonal date.

Then the nightly stars, already known through movement and orientation, offer a better solution. Attentive observation reveals that the position of the stars at the same hour of the night regularly changes with the season. Gradually the same position comes earlier in the night; the most westerly stars disappear in the evening twilight, and at dawn new stars emerge on the eastern horizon, appearing ever earlier in each following month. This morning appearance and evening disappearance, the so-called 'heliacal rising and setting', determined by the yearly course of the sun through the ecliptic, repeats itself every year at the same date. The same happens with the moment at which a star rises in the evening twilight (the end of the observable risings) or sets just before dawn, the so-called 'acronychian' rising or setting. The aborigines of Australia know that spring begins when the Pleiades rise in the evening. In the Torres Straits, to find the time for sowing, a close watch is kept at dawn for the appearance of a bright star called Kek, probably Canopus or Achernar. On the island of Java the ten *mangsas* (months) are determined by the position of Orion's belt; when it is invisible work in the fields ceases, and its morning rise indicates the beginning of the agricultural year.

Stellar phenomena are not the only means of fixing the dates of the solar year; the sun itself can also be used. The Kindji-Dayaks, living in Borneo at $2\frac{1}{2}°$ northern latitude, make use of the length of the shadow of a vertical stick at noon; the first month begins with a shadow of zero length, the second and third months with a length equal to $\frac{1}{2}$ and $\frac{2}{3}$ times the upper arm. At the Mahakam River the festivities for the beginning of sowing are determined by the sun setting at a point on the horizon marked by two big stones. The Zuni Indian priests fix the longest and the shortest day, celebrated by many ceremonies, by careful observation of the northernmost or southernmost points of sunrise, and it is the same with the Eskimos in their country, where, because of the high latitude, the method is more accurate than elsewhere.

The necessity to divide and regulate time has in various ways led different primitive peoples and tribes to observe the celestial bodies, and hence to a beginning of astronomical knowledge. We may be sure, confirmed by historical tradition, that such knowledge had also developed in prehistoric times. From these origins, science, at the rise of civilization, emerged first among the peoples of most ancient culture, in the Orient.

CHAPTER 2

AGRICULTURE AND THE CALENDAR

FROM the multitude of tribes living in a condition of barbarism, large states arose between 4000 and 1000 BC in the fertile plains of China, India, Mesopotamia and Egypt. They represented a higher stage of culture, of which written documents have been transmitted to posterity. The formerly independent peasant communities and townships, with their own chieftains and urban kings, their local gods and forms of worship, merged into larger political units. The extraordinary fertility of the alluvial silt, which yielded a surplus product, provided a living for a separate class of rulers and officials. This organization first arose from the need for the centralized regulation of the water. The large rivers irrigating these plains—the Nile, the Euphrates, the Hwang Ho—filled their beds with silt, overflowed in certain months and flooded the fields, devastating or fertilizing them, or at times excavating a new bed. The water had to be directed continuously and checked by dykes, by a deepening of the rivers, or by the digging of canals. Such control could not be left to the individual districts with their often conflicting interests. Centralized regulation was necessary, and only a strong central authority could guarantee that local interests would not prevail over general interests. Only then were fertility and prosperity assured. But when the country fell asunder into petty warring principalities and the dykes and canals were neglected, the soil dried up or was flooded, and the people starved: the wrath of the gods then lay on the land.

A strong central power was needed, secondly, for the defence of the fertile plains against the war-like inhabitants of the adjacent mountains or deserts. These people could find only a meagre living on their own land, so they made it their business to plunder and to force tribute from their prosperous neighbours. A division of tasks was necessary; a caste of warriors developed who, with their chief as king, became a ruling class controlling the surplus product of the farmers. Or the roving neighbours turned from marauders into conquerors and settled as a ruling aristocracy in the midst of the farmers, protecting them from other aggressors. In either case the result was a strong centralized state power.

This story, repeated from time to time, represents in brief outline the history of these lands. Over and over again barbarian peoples invaded and subjected them, sometimes remaining as a thin upper layer, like the Manchus in China, the Mongols in India, the Hyksos in Egypt, and sometimes as an entire people mixing with or replacing the former inhabitants, like the Aryans in India and the Semites in Mesopotamia. Though much culture was destroyed in the conquest, the invaders afterwards adopted and assimilated the existing higher state of civilization, and often imparted fresh vigour to it. After some generations, having lost their savage strength, the conquerors themselves became an object of attack by new aggressors.

In such empires the prince was the legislator, the chief justice and the head of a government of officials, who as leaders of the civil departments constituted a second ruling class alongside the military. Usually it consisted of the priests, who had been the local intellectual leaders, and now were organized into an official hierarchy. The priesthood held the spiritual leadership of state and society in its hands. It held the theoretical and general knowledge needed in the process of production; this was the source of its prestige and its social power. Where agriculture was the main occupation, knowledge of the calendar and of the seasons was their domain. Religion too—in those times one with state and society—was centralized; the local deities of the chief townships were conjoined into a pantheon under an upper god, whereas the local gods of smaller or conquered sites were degraded to a lower order of spirits or were incorporated into other gods. Thus in ancient Babylonia the goddesses Innina, Nisaba and Nana were united with the later Ishtar. The fact that Borsippa became a suburb of the rapidly growing city of Babylon found its theological expression in its city god Nabo becoming the son of Marduk. Whereas the deities of the oldest centres of culture—Eridu, Uruk and Nippur—always remained the most highly venerated gods Ea, Anu and Enlil (mostly called Bel, i.e. 'Lord'), after 2000 BC the political hegemony of Babylon made its local god Marduk the senior god of the pantheon. When in later centuries Assyria acquired the overlordship of the Mesopotamian world, Ashur took this place. When combined into a hierarchy, the gods lost some of their local character, and their character as personifications of natural powers became more prominent. The deities Sin of Ur and Shamash of Sippara in later times were always venerated as the moon god and the sun god.

The rise of a group which, as ruling class, no longer needed to secure its life by heavy toil, led to new conditions of existence. The social structure became more complicated, spiritual leadership demanded higher qualifications and advanced wider claims. Trade and commerce engendered new material and spiritual needs, and with the wealth and luxury of the monarchs and lords came a zest for art and science. Thus, for the first time on earth, along with the new social structures, higher forms of culture arose, exceeding those of the most developed prehistoric barbarism. The era of civilization began.

This involved a higher development of astronomy. It proceeded directly from the demands of time-reckoning, and more especially from the problem of adapting the moon calendar to the solar year.

One period of the moon is, on the average, 29.53059 days; a solar year is 365.24220 days, i.e. 11 days more than 12 lunar periods, which amount to 354.3671 days. After three years the moon calendar is 33 days behind the sun's progress. In order to remain in accordance with the sun, every third year, sometimes oftener, an extra month had to be added, so that the year had 13 instead of 12 months. The calendar problem consists in finding a larger period which is a common multiple of the month and the year; then after this period sun and moon will return to the same mutual position. An exact common multiple, of course, does not exist, but more or less satisfactory approximations may be found. With our precise modern knowledge of the solar and the lunar period, we are able to deduce such approximate calendar periods theoretically. We do so by converting their ratio into a continuous fraction and writing down its consecutive approximations. Thus we find $\frac{8}{99}$ and $\frac{19}{235}$, indicating first that 8 years are nearly equal to 99 months (viz. 2921.94 and 2923.53 days), so that, of these 8 years, 3 must have a 13th month and 5 have 12 months. Yet this approximation is not very good; after only 24 years the moon-date will already be 5 days late relative to the solar season. Far more precise is a period of 19 years which contains 235 months (6939.60 and 6939.69 days); a 13th month must be intercalated here seven times. Of course, in earliest times peoples did not have this exact knowledge of periods; the finding of a good calendar period for them was a difficult practical problem, only to be solved by a laborious process of adapting solar and lunar reckonings. So this problem acted as a driving force toward more careful observation of the celestial phenomena.

The question may be posed as to why such precision was necessary, since it was far beyond the needs of agriculture, which owing to the weather fluctuations, is rather irregular in its activities. We should bear in mind, however, that in those times agricultural activities were accompanied by religious ceremonies and festivities. The agricultural festivities, like all great and important social happenings, were at the same time religious ceremonies. The gods, as representatives of the dominant natural and social forces, took their part in the life of man; what was necessary or adequate socially became a commandment of the gods, strictly fixed in the rites. What by nature took place at a determinate season, e.g. a harvest-home, as a religious celebration was fixed at a certain date, e.g. an aspect of the moon. The service of the gods did not allow of any carelessness; it demanded an exact observance of the ritual. The calendar was essentially the chronological order of the ritual. This made the calendar an object of continual care on the part of the acting officials, especially the priests, but at the same time, because of their monopoly of knowledge of the favourable times, a source of their social power.

Some striking examples have come down to us of how such religious-agricultural practices led to the establishment of a luni-solar calendar. The Jewish calendar, because it had its origin in the desert, was based upon the moon; but when the Jews arrived in the land of Canaan agriculture became the chief occupation, and the calendar had to adapt itself to the sun. In spring the *massôth* was a harvest festival; the first sheaves of barley were offered and unleavened bread was made from the first grain. This celebration coalesced with the nomadic *passah*-festival, the offering of new-born lambs to Jehovah at full moon in the spring time. Thus it was located at the full moon in the first month, Nisan. How this was brought about we read in Ginzel's great textbook of chronology. 'Towards the end of the last month the priests inspected the state of the crops in the fields to see if the barley could be expected to ripen in the next two weeks. If they saw that this would be so, the *Massôth-Passah* was fixed at the month beginning with the next new moon; but if ripening could not be expected, the festival month was postponed one lunar period. This determined the other festival days.'[1]

This empirical method could serve the Israelites so long as they lived together in Palestine. When later they were dispersed, it could no longer be used; but by that time astronomy had progressed so far that they could adopt from their neighbours the knowledge of calendar periods. The period of 19 years, with a fixed alternation of 12 years of 12 months and 7 years of 13 months, then became the basis of the Jewish calendar.

Another instance is given by developments in Arabia before Mohammed. In the holy months blood vengeance was postponed and caravans could travel without danger, because this was necessary for economic life. The endless strife between the tribes was quite natural, and blood vengeance, as a primitive juridical form of solidarity between tribal members, was necessary, but its unlimited sway would have made the marketing and the provision of food impossible. Masudi says: 'The month Safar takes its name from the market in Yemen . . . here the Arabs bought their grain, and whoever omitted to do so would die of hunger.'[2] The holy month was the month of markets; from all sides the caravans travelled to the great market towns, especially to Mecca, famous for its incomparable cool well, called *Zemzem*, the source of its importance as a pilgrimage centre in Arabia. Here the tribes met and talked, and the gathering place became the political and religious centre and capital. So it was essential that the holy month should always fall at the same time of the year, when the products were harvested and available. A later Moslem scholar, Albiruni, writes: 'In pagan times the Arabs dealt with their months as the Moslems do now, and their pilgrimage moved through all the four seasons of the year. Then, however, they resolved to fix their pilgrimage at a time when their wares, hides and fruits, were ready for marketing; so they tried to make it immovable, to have it always in the most abundant season. So they learned from the Jews the system of intercalation, 200 years before the Hegira. And they applied it in the same way as did the Jews, by adding the difference between their year and the solar year, when it had increased to a month, to the months of the year. Then the *Kalammas* (the Sheiks of a certain tribe, in charge of this task) at the end of the pilgrimage ceremonies came forward, spoke to the people, and intercalated a month by giving to the next month the name of the present one. People expressed their approbation of the *Kalammas'* decision by their applause. This procedure they called *Nasi*, i.e. shifting, because every second or third year the beginning of the year was shifted. . . . They could judge it after the risings and settings of the *menazil*. Thus it remained up to the flight of the Prophet from Mecca to Medina.'[3]

In the 9th year after the Hegira Mohammed forbade the shiftings, perhaps to break the spiritual power of the *Kalammas* by taking away their function, and, in addition, to separate himself more sharply from the Jews. So now the Mohammedan calendar is based upon lunar years of 354 days, each consisting of 12 lunar months, which in 33 years run through all the seasons. Here we have a calendar which, divorced from social practice by formal prescript, has turned into a petrified religious tradition of a primitive Bedouin way of life.

This may offer an example of how calendar and astronomy developed in different ways among different peoples, in consequence of different economic conditions and political history.

CHAPTER 3

OLD BABYLONIAN SKY-LORE

IN the earliest times from which data have reached us, the land of Shinar, comprising the plains between the Euphrates and the Tigris, was inhabited by two distinct peoples: the northern part, Akkad, by a Semitic race, and the southern part by the Sumerians, two peoples with widely differing languages, appearances and customs. The Sumerians were the original inhabitants or the first immigrants, whereas the Semites later immigrated in increasing numbers from the western deserts. The Sumerian language shows no relationship to Semitic or Indo-European languages, but rather to Turanian; anyhow the racial character of this people of most ancient culture is unknown. In the centuries following 3000 BC the Sumerian cities in the south (Eridu on the sea, Ur at the desert border, Uruk, Lagash, Nippur, Larsa) showed the highest culture; one of them usually had the hegemony. The Semites in the northern towns (Agade, Sippar, Borsippa, Babylon) adopted this culture, and after further immigration became increasingly the dominant race. When about 2500 BC Sargon of Agade, and afterwards his son Naram-Sin, reigned over the whole of Mesopotamia, the military element consisted of Semites, while the scribes and civil officials were Sumerians. In the following centuries the overlordship returned to the south, where the rulers of Lagash and of Ur called themselves 'Kings of Sumer and Akkad'. After 2000 BC the Semites, through the immigration of Amoritic tribes from the west, gained definite preponderance. Then, under a dynasty of which Hammurabi is best known, the city of Babylon became the capital of a large empire and a centre of commerce and culture.

The Sumerians were the inventors of the cuneiform script, in which each sound consisting of a vowel and one or two consonants was represented by a special character. These characters were produced by impressions, at one side broad, at the other narrow, made by a stylus—a metal wedge—in a soft clay tablet which was then hardened by baking in a fire. This rendering of syllables by characters was entirely suited to an agglutinating language such as the Sumerian, where the separate words and roots were simply linked together. The Semites adopted this cuneiform script, though it was rather cumbersome for their language, with its inflection of the roots. When Babylon became the cultural centre for the entire Near East, its language and cuneiform script were used for international intercourse as far as Egypt and Asia Minor.

The deciphering of the cuneiform script and the languages written therein was one of the great achievements of the nineteenth century; it brought to light an entirely lost period of history and culture. Previous knowledge, mostly from Greek sources, mutilated and anecdotal, did not go back beyond about 700 B.C. When Henry Layard in 1846, influenced by the first excavations of the French consul Botta, went digging at the site of ancient Nineveh, he disinterred magnificent sculptures, bas-reliefs of hunting scenes, winged bulls and dragons. He was struck by the multitude of sherds with cuneiform inscriptions lying around, and, suspecting their value, he sent cases of these sherds, together with the works of art, to the British Museum. Many years later their importance was realized when George Smith deciphered some of the texts containing a narrative of the Flood, and special expeditions were sent out to collect as many of them as possible. They proved to be remnants of King Ashurbanipal's 'library' and consisted of new inscriptions as well as copies of texts from earlier centuries, dictionaries and lexicographic materials. They gave a strong impulse to the study of ancient Mesopotamian culture (since called 'Assyriology'). At first, the Assyriologists were embarrassed by the intermixture of two entirely different languages; usually every character was used as a sound, borrowed as a syllable from the Sumerian for the Semitic language; but often it was used as a so-called 'ideogram', a sign for a thing, a concept, after its meaning in Sumerian. In this way two languages were gradually unravelled at the same time; thus these literary fragments, completed by texts excavated from the ruins of other ancient towns, brought out ever more clearly the history and culture of ancient times.

An unknown and ancient history of astronomy also came to light thereby. In this rediscovered ancient world the heavenly bodies had assumed a greater importance than in any other country or era. The reconstruction of this old knowledge from the small and sparse fragments was certainly difficult. Only through vague suppositions, hazardous conjectures, false interpretations and untenable theories could the truth be gradually approximated. For many years Hugo Winckler's theory—so-called 'Pan-Babylonism'—had a great reputation; it proclaimed that in the earliest times, between 3000 and 2000 BC, a highly developed astronomical science already existed there, based on a thorough knowledge of the celestial periods and of the shifting of the aspect of the constellations through the precession of the equinoxes. It set forth how this primeval world concept, bearing the character of an 'astral-mythology', by asserting close correlation between earthly and celestial phenomena, was the origin of all later Oriental and Greek systems of thought and determined legends, lore and customs even until modern times and in distant parts of Europe. This alluring but highly fantastic theory proved untenable when the texts were later subjected to careful study. Our present real knowledge of Babylonian astronomical science is based chiefly on the work of three Jesuit scholars, the Assyriologist J. N. Strassmaier and the astronomers J. Epping and F. X. Kugler.

The era of the first Babylonian kingdom was a culmination of economic and political power and of cultural life. Commerce and industry prospered; Babylon, a most important commercial city, was not only the metropolis of a large Mesopotamian empire but a culture centre radiating over the whole of Near-Asia. Here the results of the preceding centuries of Sumerian culture were consummated, and the system of theology assumed the form valid for later times. The structure of civil life is well known to us by the famous 'code of Hammurabi', engraved in a stone which was excavated at Susa in the nineteenth century. Among the many clay tablets dug up we find numerous civil contracts on the buying and selling of land, money-lending, rent and service, which were deposited and preserved in the temples as if they were notarial offices. Through them we have made certain of the calendar, and of the sequence of the names and dates of the kings, which serves as a framework of history.

Among the Sumerian names for the twelve months in use under King Dungi of Ur—they were different in different places—we find the name for the 4th month composed of the characters for seed and hand, that for the 11th month of the characters for corn and cutting, and that for the 12th month of the characters for corn and house: thus the season is clearly indicated. So we can understand why the 11th month was duplicated there for intercalation. After the rise of Babylon the Semitic names came into use: Nisannu, Airu, Simannu, Duzu, Abu, Ululu, Tishritu, Arach-samma, Kislimu, Tebitu, Sabatu, Adaru. In the astronomical texts they are indicated by single characters, the first syllables of their former Sumerian names.

In the old Babylonian kingdom the intercalated month was a second Adaru, at the end of the year. The lack of regularity in these intercalations, as exhibited by the civil contract dates, shows that they took place empirically, according to the ripeness of the crops, or when it otherwise appeared necessary; even two consecutive years occurred with 13 months, when the calendar was too far out. A few times, when apparently a deviation had to be corrected rapidly, the 6th month was duplicated. There exists a document of such a case, reading: 'Thus

Hammurabi speaks: "Since the year is not good, the next month must be noted as a second Ululu. Instead of delivering the tithes to Babylon on the 25th of Tishritu, have them delivered on the 25th of Ululu II." '⁴ The delivery of food for the court could not, of course, be postponed a month.

Observation of some celestial phenomena was necessary for the calendar. To fix the exact first day of the month, the new moon had to be caught at its first appearance. This was not so very difficult; in that marvellously radiant climate (some winter months and occasional sand-storms excepted) with its clear visibility across the wide plains from horizon to horizon, the priest-astronomers, from their terraced towers, could easily detect the first slender lunar arc in the evening sky. They had to give attention to the full moon, too, for ceremonial purposes, and to observe, for the sake of prediction, the last appearance of the morning crescent toward the end of the month. Can one imagine that in so doing they would not have noticed the stars silently tracing their courses, progressing onward every successive month? Or that they were not struck by the brilliant planets appearing occasionally among them, and most of all by the peerless evening star? There are few texts dealing with them; even with those clear skies the mind of the early Babylonians was not so entirely occupied with the stars as has often been supposed. But it is highly probable—and this is supported by some texts—that this regular observation of the moon must gradually have led to an increasing practical concern with the stars.

The first phenomenon manifesting itself in the observation of the young moon is the regular progression and shifting of the constellations in the course of a year. The stars visible in the western evening sky are characteristic of the season; hence they can verify the calendar and the intercalations. This also applies to the first appearance of the stars in the eastern morning sky. Intercalation by means of the regular heliacal rising and setting of stars must therefore have gradually superseded the irregular empirical method of observing the crops. A positive indication may be found in a list (an Assyrian copy of an earlier original from an unknown century) of 36 names of stars or constellations, three for each month, of which every first one is clearly related to the heliacal rising in that month. For the month Nisannu we find *Dilgan* (i.e. the Ram and the Whale); for Airu the first name is *Mulmul* (the Pleiades; *mul* means 'star'); for Simannu it is *Sibziannu* (Orion); and so on. Then there is another text, much damaged, which records: 'Star Dilgan appears in the month Nisannu; when the star stays away the month must . . . star Mulmul appears in the month Airu . . .'⁵ The appearance or the delay in the appearance of the star clearly indicates directions for certain actions.

A great Creation epic from the time of the first Babylonian empire tells how Marduk, the local god of Babylon, won supremacy over the gods by beating down the monster Tiamat (Chaos), and how out of the parts of its body he built heaven and earth.

He made the stations for the great gods,
The stars, their images, the constellations he fixed;
He ordained the year and into sections he divided it.
For the twelve months he fixed three stars.

The Moon god he caused to shine forth, the night he entrusted to him.
He appointed him, a being of the night, to determine the days.[6]

So we see that many stars and constellations were already known and named. It does not appear that the twelve constellations of the zodiac were the first known or occupied a special place. Some evidence could perhaps be found in the fact that the Gilgamesh Epic, with many characteristics of a solar myth, is divided into twelve songs, each corresponding to a zodiacal sign. But this division may be that of a later version. The planets do not seem to play a role; the calendar deals only with fixed stars.

This knowledge of the stars persisted through the succeeding centuries of political decay, when Babylon was dominated by eastern conquerors, the Cassites, and the western countries were a battlefield of Egyptian and Hittite expansion. From this time data are known from field boundary stones (*kudurrus*), which were made secure from being moved by placing them under the protection of the gods whose pictures were engraved on them. We find there, besides figures representing the sun, moon, and perhaps Venus, others supposed to represent constellations: a bull, an ear of corn, a dog, a serpent, a scorpion, and a fish-tailed goat, as the constellation Capricorn was later drawn. Excavations in Bogazköy in Asia Minor, once the site of the Hittite capital, produced bricks with inscriptions invoking Babylonian deities; in this enumeration many names of stars and constellations appear, e.g. of the Pleiades, Aldebaran, Orion, Sirius, Fomalhaut, the Eagle, the Fishes, the Scorpion. The character of the Scorpion—*gir-tab*—consists of two characters meaning sting and pincer, with the sting represented by the pair of bright stars λ and υ Scorpii, just as it was afterwards drawn by the Greeks, who borrowed this constellation from the Babylonians. A much discussed text from Nippur—which states that the distance from the Scorpion's sting and from its head to the star Arcturus, given in many sexagesimals, are 9 and 7 times a certain amount—seemed to imply that at that time celestial distances were measured with some accuracy. When from other tables it was completed by analogous ratios 9 to 11, to 14, to 17, to 19 for other stars, it appeared to be a simple mathematical text, in which stars had been inserted as ready-made examples. The supposition of some Assyriologists that linear distances in space were meant is certainly not in accord with what we know about the Babylonian world concept.

There exists another, a truly astronomical document, which we can say with certainty derives from the first Babylonian dynasty. It is a text from Ashurbanipal's library and now preserved in the British Museum, with data concerning the planet Venus, and is a copy of earlier texts. In 1911 Kugler succeeded in deciphering its contents. One part consists of many groups of lines describing the phenomena of the planet (here called *Nin-dar-anna*, 'mistress of the heavens') as astrological omens, followed by corresponding predictions. They cannot be observations or astronomical computations because the same time-intervals always recur and the imaginary dates of beginning, regularly alternating between morning and evening phenomena, are always of one month and one day. For example the fifth group reads:

> In the month Abu on the sixth day Nin-dar-anna appears in the east; rains will be in the heavens, there will be devastations. Until the tenth day of Nisannu she stands in the east; at the eleventh day she disappears. Three months she stays away from heavens; on the eleventh day of Duzu Nin-dar-anna flares up in the west. Hostility will be in the land; the crops will prosper.
> In the month Ululu on the seventh day Nin-dar-anna appears in the west . . .[7]

The intervals between eastern appearance and disappearance are always 8 months and 5 days; then come 3 months until the western appearance; again 8 months and 5 days until disappearance, then 7 days until appearance in the east. The total is 19 months and 17 days, the correct Venus period. These intervals and dates of appearance and disappearance have been deduced from observational data contained in the remaining part of the text, but among the dates there are many erroneous figures wrongly copied, or incorrect phenomena. It is a curious fact that intervals from erroneously copied values have been used to deduce the mean values applied in the schematic predictions. A fault in copying put the long period of invisibility (after the eastern disappearance, which is normally about 2 months) as 5 months and 16 days and the average became 3 months. This shows that the man who later handled the materials to insert the omens had so little astronomical knowledge—or paid so little attention—that he included the grossly deviating value in the average. Since we know the meaning of the numbers, some of the errors in these observational data—which comprise 21 years—can be corrected.

Among the misplaced phenomena of Venus is found, in the eighth year, a line reading 'year of the golden throne'. The same designation is found in the texts of civil contracts of the eighth year of King Ammizaduga, the penultimate king of the dynasty, who reigned for 21 years. So the obvious conclusion was that the 21 years of the Venus text corresponded exactly to this king's 21 years' reign. This was confirmed by the fact that the years of 13 months in the Venus text corresponded to the civil texts. Since the Venus phenomena for that time can be accurately computed from modern data, this afforded a means of ascertaining the precise dates for that king's reign, and the entire chronology of about 2000 BC was put on a firm basis. The method followed by Kugler was to use the statement that in the 6th year, on the 26th day of the month Arachsamma, Venus had disappeared in the west and, on the 3rd day of the next month, Kislimu had reappeared in the east. Hence its conjunction with the sun nearly coincided with the conjunction of the moon with the sun, in a season roughly corresponding to December or January. He found that these conditions were best fulfilled by January 23rd in 1971 BC. From this it would follow that the first Babylonian dynasty reigned from 2225 to 1926 BC and that Hammurabi reigned from 2123 to 2081 BC. The mean date of Nisannu 1st thus corresponded to April 26th of our calendar.

Such precise identification of such ancient dates was generally recognized as an admirable example of astronomical-chronological research, but it is interesting that the first result was entirely erroneous. There are several dates in accordance with the data. The conjunctions of Venus with the sun return every 8 years, but 2.4 days earlier; those of the moon with the sun return after 8 years, but 1.6 days late. Hence two dates 8 years apart may sufficiently conform to the phenomena, after which the conjunctions move increasingly apart, the Venus conjunctions coming earlier, the moon conjunctions later. After 7 periods of 8 years, the Venus conjunctions come 17 days earlier, and the moon conjunctions 11 days later, so that a coincidence of the conjunctions again occurs, but now $29\frac{1}{2}$ days earlier. Hence we find a series of dates (or pairs of dates 8 years apart) following one another at intervals of 56 or 64 years, and the astronomer depends entirely on the historian to indicate the right century. When Kugler made his investigations, historians agreed to fix the first Babylonian dynasty at about 2000 BC; other investigators who tried to correct Kugler's work did not dare to put the events more than 120 years later. But there existed an old chronicle of kings by Berossus, a Babylonian priest teaching in Greece, which placed the dynasty four centuries later. This had always been rejected by historians, but recently they reversed their opinion. So now it is deemed most probable that the date of coinciding conjunctions

34

was December 25, 1641 BC, that Hammurabi reigned from 1792 to 1750 BC, and that the entire dynasty ruled from 1894 to 1595 BC—or perhaps 64 years later, for there are still uncertainties in the historical documents.

However this may be, it appears that the planet Venus was already being observed with special attention during the first centuries of Babylonian power, perhaps even earlier. Had other planets been observed with the same care, references to such observations would certainly have been preserved in Assyrian copies, but Venus alone is mentioned. It is quite clear that the priests, watching for the moon's crescent, were struck by this most brilliant of stars and that it must have appeared to them as an exceptional luminary. In later texts sun, moon and Venus are often named together as a triad of related deities, distinguished from the other four planets. Can it be that the Babylonians knew about the crescent of Venus in her great brilliancy before and after lower conjunction? There is a text, the meaning of which is a point of controversy and which is read by some Assyriologists as follows: 'When Ishtar at her right-hand horn approaches a star, there will be abundance in the country. When Ishtar at her left-hand horn approaches a star, it will be bad for the country.'[8] It does not seem impossible that in the clear atmosphere of these lands the horns of the Venus crescent were perceived; modern observers, too, have mentioned such instances. An American missionary, D. T. Stoddaert, in a letter to John Herschel from Oroomisha in Persia in 1852, wrote that at twilight Jupiter's satellites and the elongated shape of Saturn could be seen with the naked eye and that through a dark glass the half-moon shape of Venus immediately struck the eye.[9] Thus it can be still better understood that in these ancient times the Babylonian priest-astronomers devoted special attention to Ishtar as a sister-star of the moon. Not only did they watch its appearances and disappearances with religious zest, but they were struck by its regularities and tried, though in a primitive manner, to find the periods and use them for prediction. But this may have been the work of later centuries of political depression, e.g. between 1500 and 1000 BC.

CHAPTER 4

ASSYRIAN ASTROLOGY

EARLY in the first millennium BC a new Semitic power arose in the north on the upper course of the Tigris; Ashur was the name of the capital of the supreme god. In continuous struggles against surrounding states, Babylon among them, Ashur expanded and became the ruling power in the Mesopotamian plains. About 800 BC Assyria became the most powerful state of Near-Asia. Under the kings Tiglath-phileser (745–727 BC), Shalmaneser (726–722), Sargon (722–705), Sennacherib (705–682), Asarheddon (682–668), and Ashurbanipal (668–626 BC), Syria, Palestine, Phoenicia, and at times even Egypt, were conquered in a series of great and often cruel wars, and the boundaries were extended towards Asia Minor, Armenia and Media. According to the American Assyriologist Olmstead, the use of iron for weapons was the chief material factor in the Assyrian conquests. The new capital Nineveh was the political centre of a large military empire and as such was adorned with magnificent buildings. But Babylon, as a great and rich commercial centre whose wealthy citizens largely governed themselves, retained its rank as a venerable seat of ancient culture. The Assyrian kings recognized its importance by going themselves to Babylon 'to seize the hands of Marduk', i.e. solemnly to take over the government, or to nominate a relative as dependent king.

The Assyrians, as rough and warlike conquerors often do, adopted the culture of the conquered and carried on its forms and traditions. The Assyrian pantheon was identical with that of Babylon, except that now Ashur was the new god of gods. The same priesthood thought, worked and wrote in the old sanctified forms in the service of the new rulers, fulfilling the same social functions as before. The plastic arts flourished anew, not only because the workers were a new people with fresh energy and the new rulers were rich through conquest and plunder, but also because, instead of the bricks of the lowlands, stone from the surrounding mountains served as building material, excellent stuff for the beautiful reliefs which today adorn the western museums. This development reached its peak when, with the increase of wealth and culture, the first hardy warriors were succeeded by princes who loved and protected the arts and sciences. It was then that Ashurbanipal installed a library in his palace and ordered the old texts from all the ancient sites and temples of Babylonia to be assembled and copied. The thousands of clay tablets were deposited in orderly rows, with their titles at the sides, supplemented and explained by catalogues, dictionaries and commentaries, continuously augmented by new archive items, the reports to the king and his correspondence with officials. The more than 13,000 fragments dug up from this site alone and preserved in the British Museum, and the many thousands from other sites, give us a good picture of the customs and ideas, business life and culture, religion and economy of that society, and of its astronomy also.

The calendar of former times is still found in the numerous texts: the lunar month, beginning on the evening of the first appearance of the crescent, and the 12 months, now and then completed with a thirteenth. Though it is not expressly mentioned, it is almost certain that the intercalation was regulated by means of stellar phenomena, generally by their morning rising. The list of 3 × 12 month-stars mentioned above, in a copy in Ashurbanipal's library, may be an indication. Another method is indicated in a text published by George Smith: 'When at the first day of Nisannu the moon and star Mulmul [the Pleiades] stand together the year is common; when at the third day of Nisannu the moon and star Mulmul stand together the year is full.'[10] The last part of the sentence means that the Pleiades are visible long after sunset, and as this is so early in spring, it is necessary to add a thirteenth month. The stars and constellations are grouped into three divisions, the northern, the middle and the southern constellations, known respectively as the domains of Enlil, Anu and Ea.

The calendar, however, was no longer the principal motive for observing the stars. In Assyrian times it was astrology, the idea that the course of the stars has a significance for events on earth, that most strongly determined the thoughts and practice of man. The heavenly phenomena were now studied with far deeper interest for omens concerning the fate of men, but especially kings and empires.

This faith in omens existed in primitive man as the natural consequence of his belief that he was surrounded by invisible spiritual powers which influenced his life and work. It was a vital question for him to win their favour and aid, to appease or avert their hostility and to discover their intentions. Exorcism, offerings, incantations, charms and magic connected with his work occupied his daily life. Most of these spirits had their abode in the heavens. Thus in early times the conception of a close connection between the stars and man's destiny had already arisen in the minds of the Mesopotamian priests. Heaven was not so far off; the earthly rulers and their people were near to the gods

37

above. King Gudea of Lagash (about 2500 BC), in an inscription on a stone cylinder, describes the building of a temple and tells how in a dream the goddess Nisaba, the daughter of Ea, appeared to him: 'She held the shining stylus in her hand; she carried a tablet with favourable celestial signs and was thinking.' And further on: 'She announced the favourable star for building the temple.'[11] An omen is found after each of the planetary phenomena in the Nin-dar-anna text from the first Babylonian empire and there can be no doubt that the astronomical data were obtained in order to supply these predictions.

Such beliefs were a living force among the priests of the temples in Babylon and other cities and spread to all the regions influenced by Babylonian culture, including Assyria. In the courts of powerful monarchs, intent on expanding their empires by conquest, the need to foresee the future was especially great and astrology found a fertile field. The court astrologers had to find omens for every large enterprise; and from all important temple sites the king received regular reports of what was happening in the sky and its interpretation. These reports were preserved in the archives of what we call Ashurbanipal's library, and copies of all the old data were assembled there for the sake of their interpretation. Of course, not all the omens were derived from the stars. Omens were to be found everywhere: in the livers of the sacrificial victims, a variable and thus prolific object; in the flight of birds; in miscarriages, earthquakes, clouds, rainbows and haloes. There were manuals which recorded every phenomenon and what action or experience of the legendary king Sargon of Agade was in any way related to it.

Astronomical phenomena, of course, occupied a large place among them. Celestial bodies most suitable for presages were those that presented great diversity and irregularity in their phenomena. The most important, therefore, were not the sun and the fixed stars but the moon and the planets, the fixed stars forming the constant background for the phenomena. The sun-god Shamash, the all-seeing guardian of justice, day after day performs his constant course across the sky; at most he could be duller or redder than usual or invisible because of clouds or an eclipse. The moon-god Sin, on the contrary, shows an abundance of different ever-recurring phenomena. The same holds true for the planets which wander among the stars in the most unexpected ways, sometimes standing still or reversing their courses, sometimes combining among themselves or with the bright stars into ever-changing configurations. They seemed like living beings spontaneously roaming through the starry landscape, and they increasingly became the chief object of the Babylonian priests' attention. They were the stars of the great gods who ruled the world and manifested themselves in these brilliant luminaries.

The determinative character preceding the names of the planets, usually the character for star, was now often the character for god. Venus (called Dil-bat) was the star of Ishtar; Jupiter (Umanpa-udda, later mostly Sagmegar) was the star of Marduk; Saturn was the star of Ninib, a solar deity too, and Mars was the star of Nergal the god of pestilence. Red Mars was deemed an evil star, Jupiter a lucky star. But luck or evil for whom? That depended on time and place.

Different months, different compass directions and different constellations were attributed to each of the four countries: Akkad (Babylonia), Elam (the eastern mountains), Amurru (the western desert, afterwards Syria), and Subartu (the north). 'We are Subartu,' said an Assyrian astrologer. The appearance of the planets in different constellations, the duration of their stay or their rapid course, their meeting with one another or with the moon, offered an absolutely endless variety of phenomena, leaving the astrologers a wide area for tradition and ingenuity, for combining power and fantasy as well as for prudent flattery in their personal deductions.

As an instance, an assortment of texts is given here, taken from the collection published by R. C. Thompson for the use of students of Assyriology.*

When the star of Marduk appears at the beginning of the year, in that year corn will be prosperous. Mercury has appeared in Nisannu. When a planet [Mercury] approaches star Li [Aldebaran] the king of Elam will die. . . . Mercury appeared in Taurus; it had come down [?] as far as Shugi [the Pleiades, Perseus].†

Venus disappears in the West. When Venus grows dim and disappears in Abu there will be slaughter in Elam. When Venus appears in Abu from the first to the thirtieth day, there will be rain, and the crops of the land will prosper. In the middle of the month Venus appeared in Leo in the East. From Nirgal-itir.‡

When Venus fixes its position, the days of the king will be long, there will be justice in the land. When Venus is in the way of Ea . . .§

Mars is visible in Duzu; it is dim. . . . When Mars culminates indistinctly [?] and becomes brilliant, the king of Elam will die. When god Nergal in its disappearing grows smaller, like the stars of heaven is very indistinct, he will have mercy on Akkad. . . . When Mars is dim, it is lucky; when bright, unlucky. When Mars follows Jupiter that year will be lucky. From Bullutu.¶

---

* *The Reports of the Magicians and Astrologers of Nineveh and Babylon*, 1900.
† Thompson, 184.   ‡ Thompson, 208.   § Thompson, 206.   ¶ Thompson, 232.

Mars has entered the precincts of Allul [Cancer]. This is not counted as an omen. It did not stay, it did not wait, it did not rest; speedily it went forth. From Bil-na-sir.*

When Jupiter passes to the west there will be a dwelling in security, kindly peace will descend on the land. It appeared in front of Allul.

When Jupiter assumes a brilliance in the way of Bel and (becomes?) Nibiru, Akkad will overflow with plenty, the king of Akkad will grow

Transcription

mul*Dil-bat ina* ilu*Šamši ṣît ir-ti-bi*

Ana mul*Dil-bat mušḫa irši(ši) la damikti*

*ša ûmi*pl *ša la ú-šal-li-mu-ma*

    ir-bu-u

Ana mul*Dil-bat ina* arḫu*Nisanni*

*ultu ûmi I*kam *adi ûmi XXX*kam

    ina ilu*Šamši ṣît it-bal*

ú-ru-ba-a-ti

      *ina mâti ibaššu*pl

Fig. 1. Text of Tablet K725 (B.M.). See plate 1

powerful. . . . When a great star like fire rises from the East and disappears in the west, the troops of the enemy in battle will be slain. At the beginning of the reign Jupiter was seen the right position; may the lord of gods make thee happy and lengthen thy days. From Asharidu, the son of Damka.†

* Thompson, 236.   † Thompson, 187.

The fiery star mentioned here was probably a meteor. Just as in the next text:

> After one kas-bu [two hours] of the night had passed, a great star shone from north to south. Its omens are propitious for the king's desire. The king of Akkad will accomplish his mission. From Asharidu the greater, the king's servant.*

In a letter of Mar-ishtar to King Asarheddon (668 BC) we read:

> . . . in the first month . . . on the 29th Jupiter was taken away . . . now he has stayed away one month and five days in the heaven [i.e. was invisible]: on the sixth of the third month Jupiter became visible in the region of Orion; five days over his time he had been passing. This fits to the omen: when Jupiter appears in the third month the land will be devastated and the corn will be dear . . . when Jupiter enters Orion the gods will devour the land.[12]

In these and analogous texts—where the meaning of some expressions is still uncertain—we find astronomical observation far more varied and detailed than was formerly required for calendar purposes or orientation. Astrology connected the life of man so closely with the heavens that the stars and their wanderings began to occupy an important place in his thoughts and activities; it was his destiny that the luminous gods wove in those wonderful orbits. Now that his eyes had been opened and his interest awakened, man grew more and more conversant with the course of the planets, as we saw in the reports: Jupiter goes to the west, Mars stood in Scorpio, turns and goes forth with diminished brilliance; Venus stays at its place, and Saturn has appeared in Leo. Though there is no theory of periodicities (year and date are missing from all these statements) a notion of familiar and expected regularity must have arisen in the minds of the observers, expressed in the statement that the planet stood at the right place or that it was past its time.

Next to the planets, the moon was the principal celestial body for the astrologer. Its chronological significance had now to give way to its astrological interest. The shortcomings of chronology even became an element of prediction for the astrologers; that the moon was not on time was mostly a bad omen. Because of the invisibility of the moon through clouds or perhaps because of negligence in times of trouble, it is conceivable that the new crescent could appear on the 28th or 29th of a month; in the same way, the full moon, which normally should appear on the 14th, could occur on the 13th, the 15th, or the 16th.

* Thompson, 201.

When the Moon appears on the first day there will be silence, the land will be satisfied. . . .*

When the Moon appears on the thirtieth of Nisannu, Subartu will devour Achlamu; a foreign tongue will gain the ascendancy in Amurru. We are Subartu. When the Moon appears on the 30th day there will be cold in the land. The Moon appeared without the Sun on the 14th of Tebitu; the Moon completes the day in Sabatu. . . .†

When the Moon and Sun are seen with one another on the thirteenth day, there will not be silence; there will be unsuccessful traffic in the land, the enemy will seize on the land. From Apla.‡

When the Moon reaches the Sun and with it fades out of sight . . . there will be truth in the land and the son will speak the truth with the father. On the 14th the god was seen with the god. . . . When the Moon and Sun are seen with one another on the 14th, there will be silence, the land will be satisfied; the gods intend Akkad for happiness. . . .§

When the Moon does not wait for the Sun and disappears, there will be a raging of lions and wolves. . . . On the 15th it was seen with the Sun; afterwards in Tishritu the Moon will complete the day. . . . From Balasi.¶

On the first day I sent to the king thus: On the 14th the Moon will be seen with the Sun. . . . On the 14th the Moon was seen with the Sun.‖

When the Moon and Sun are seen with one another on the 16th day king to king will send hostility. The king will be besieged in his palace for the space of a month. The feet of the enemy will be against the land; the enemy will march triumphantly in his land. When the Moon on the 14th or 15th of Duzu is not seen with the Sun the king will be besieged in his palace. When it is seen on the 16th day, it is lucky for Subartu, evil for Akkad and Amurru. From Akellanu.**

In order to understand such observations and conclusions, we have to consider what phenomena are seen at about full moon. If the month begins normally, i.e. if at evening when the crescent is visible for the first time (normally 1.4 days after conjunction), the first day begins, full moon will fall in the night of the 14th, since it comes 14.7 days after conjunction (on the average; through the irregularities of its motion it may be $\frac{3}{4}$ day less or more). On the night of the 13th the moon is not yet full; at sunset the moon is already visible in the east, and she sets before sunrise; she does not wait for the sun. On the night of the 14th she is also already visible at sunset, but she has not yet set when the sun rises, and has become dim and faint in the west. The Assyrian astrologers

* Thompson, 1, 2, 4, etc.   † Thompson, 62.   ‡ Thompson, 120.   § Thompson, 124
¶ Thompson, 140.   ‖ Thompson, 154.   ** Thompson, 166.

expressed this by saying: 'the Moon was seen with the Sun'; or 'the god [Sin] appeared with the god [Shamash]'; or, 'the moon reaches the sun'. Such normality was connected with all kinds of favourable omens. But if 'the Moon did not wait for the Sun' but set before sunrise, full moon had to come later, and not until the next night, the 15th, could the moon be seen in the morning with the sun.

In these observations astrology had merged with chronology. The priest-astronomers noted that all was in order in the sky; hence there would be peace on earth. Or they noticed a disorder in the calendar, which, besides the bad omens, indicated that something had to be corrected. From the day of first appearance, full moon was to be expected on the 14th; but through irregularities this might fail. If full moon came on the 13th, they inferred that the month might have only 29 days. When the moon was seen with the sun after the 14th, the month had to have 30 days; then she 'completed the day'. When they were seen together on the 14th and the preceding month had 30 days, the next one would have only 29.

> The Moon completed the day in Adaru; on the 14th the Moon will be seen with the Sun; the Moon will draw back the day in Nisannu. . . .*

Thus the astrological motive, by demanding greater attention in observing the moon, provided for better foundations in chronology. Order and regularity were perceived and used for predicting the future.

Many more phenomena could, of course, be observed on the moon: its colour and brightness, the shape of its horns, the earth-shine ('the moon carries an "agu", a tiara or royal cap'), a nebulous corona or a ring (halo). Such a ring was often regarded as a fence enclosing a sheep-meadow with the moon as shepherd in the centre; the Babylonian character for planet, *lubat*, means 'stray sheep'. Or it was seen as a river; or it meant a siege, with the planets or stars within the ring indicating who was besieged. When the ring was not closed, it meant, of course, escape.

> When a halo surrounds the Moon and Jupiter stands within it, the king will be besieged. The halo was interrupted; it does not point to evil. . . . From Nabu-shuma-ishkun.†

> When a halo surrounds the Moon and Sudun stands within it, a king will die and his land be diminished; the king of Elam will die. Sudun is Mars, Mars is the star of Amurru; it is evil for Amurru and Elam. Saturn is the star of Akkad, it is lucky for the king, my lord. From Irasshi-ilu, the king's servant.‡

* Thompson, 53.   † Thompson, 95.   ‡ Thompson, 107.

43

Eclipses also were regarded as highly important omens. At all times their unexpected occurrence impressed people deeply, the more so because the regular phenomena were given such great significance. They afforded a multitude of prognostications. For a solar eclipse, the month was noted and the place in the sky, the aspect and direction of the horns when the sun 'assumed the figure of the moon'. Besides astronomical eclipses, dust storms may also have darkened the sun when eclipses are reported on dates of the month other than the 27th or 28th.

Omens from lunar eclipses were far more abundant. When the moon was eclipsed it was a sign for all countries; when it was partially obscured, each of its four sides related to a separate country, and the significance was different as to month, day and hour (night-watch). Hence the instruction given in the great collection of ancient omens called *Enuma Anu Enlil*:

> When the Moon is eclipsed you shall observe exactly month, day, night-watch, wind, course, and position of the stars in whose realm the eclipse takes place. The omens relative to its month, its day, its night-watch, its wind, its course, and its star you shall indicate.[13]

Thus we find in the reports on a certain eclipse in the month Simannu at the end of the night:

> An eclipse in the morning-watch means disease. . . . The morning-watch is Elam, the 14th day is Elam, Simannu is Amurru, the second side is Akkad. . . . When an eclipse happens in the morning-watch and it completes the watch, a north wind blowing, the sick in Akkad will recover. When an eclipse begins on the first side and stands on the second side, there will be a slaughter of Elam; Guti will not approach Akkad. . . . When an eclipse happens and stands on the second side, the gods will have mercy on the land. When the moon is dark in Simannu, after a year Ramanu [the storm-god] will inundate. When the Moon is eclipsed in Simannu, there will be flood and the produce of the waters of the land will be abundant. . . .*

A rich enumeration of omens; every detail had significance. With such detail as this we may speak of a fairly accurate observation of the eclipse. No wonder, therefore, that the first observed eclipses used many centuries afterwards by Ptolemy for the derivation of the moon's period, proceeded from this time: March 19, 721 BC, March 8, 720 BC, and September 11, 720 BC. All that was wanted here was an exact knowledge of the number of years and days elapsed since these dates. It

* Thompson, 271.

44

is possible that the lack of this item prevented the use of earlier eclipses and that for this reason Ptolemy began his list of kings and their consecutive years of government with the Babylonian king Nabonassar, 747 BC.

The Babylonian watch-keepers themselves will surely have perceived regularity in such carefully observed phenomena. Their reports show that they expected eclipses and announced the events:

> On the 14th an eclipse will take place; it is evil for Elam and Amurru, lucky for the king, my lord; let the king, my lord, rest happy. It will be seen without Venus. To the king, my lord, I say: there will be an eclipse. From Irasshi-ilu, the king's servant.*

> To the king of countries, my lord thy servant Bil-usur. May Bel, Nebo, and Shamash be gracious to the king, my lord. An eclipse has happened but it was not visible in the capital. As that eclipse approached, at the capital where the king dwells, behold, the clouds were everywhere, and whether the eclipse took place or did not take place we don't know. Let the lord of kings send to Ashur, to all cities, to Babylon, Nippur, Uruk and Borsippa; whatever has been seen in those cities the king will hear for certain. . . . The great gods who dwell in the city of the king, my lord, overcast the sky and did not permit to see the eclipse. So let the king know that this eclipse is not directed against the king, my lord, nor his land. Let the king rejoice. . . .†

Is it not a brilliant idea that, while the eclipse took place, the tutelary gods of Ashur, knowing that no evil threatened the country and its king, should draw a curtain of cloud over the city, so that the king might not be frightened? The greater part of these reports, according to the Leiden Assyriologist, De Liagre Böhl, falls between the years 675 and 665 BC, a critical time for the Assyrian empire, when the king, ill and full of apprehension, returned from a more or less unsuccessful Egyptian campaign. Here we have a fairly accurate date for which the state of astronomical knowledge can be given. Knowledge it was indeed, for in these reports predictions regarding eclipses were made and verified with great clarity.

> To the king, my lord, I sent: 'An eclipse will take place.' Now it has not passed, it has taken place. In the happening of this eclipse it portends peace for the king, my lord. . . .‡

There is no indication in these texts of the bases on which the priests made their predictions. There is, however, such a simple regularity— first pointed out by Schiaparelli—in the occurrence of lunar eclipses

---

* Thompson, 273.   † Thompson, 274.   ‡ Thompson, 274 F.

that it must soon have caught their attention. They could see that a second eclipse never followed at an interval of less than six months, and that it was often followed by a third after another six months. Sometimes even four or five followed one another, always at intervals of six months. Often there was a gap in the continuous series; this happened when the eclipse occurred in daytime, so that the full moon was below the horizon. Once the observers understood that this was the cause of the gap, they could fill it by adding an invisible eclipse at the right moment; thus a regular and continuous series of five or six visible or invisible eclipses was formed. Then the series finished; but after a year or two a new series began, at a time one month earlier than what would have been a continuation of the former series. So if they observed an eclipse not preceded by another 6, 12, or 18 months earlier, they could infer that a new series had started, and could thus foretell new eclipses 6 or 12 months later.

From modern knowledge it is easy to see the basis for this regularity. After six lunar periods (177.18 days) the sun, and hence also the position of the full moon, has progressed on the average 174.64° in longitude. Because of the variable velocity of sun and moon the real progress can be some few degrees more or less. Simultaneously the nodes of the lunar orbit have receded 9.38°; hence the opposite node is situated at a longitude 170.62° longer. Relative to the node, the full moon has then progressed 4.023°. Had it been near a node, then six months later it would again be near a node. The full moon can pass through the earth's shadow when its distance from the node is not more than 10° to 12°; for a total eclipse the distance should be not larger than 5° or 6°, otherwise the eclipse is only partial. We can make these conditions clear by computing the distances for a number of consecutive full moons at intervals of 6 months, starting from an arbitrary first value and increasing it each time by 4.02° or 4.03°:

$$-14.50° \mid -10.48° \mid -6.45° \mid -2.43° \mid +1.59° \mid +5.62° \mid +9.65° \mid +13.67°$$

for which the aspects will be:

| no eclipse | partial? | partial | total | total | total? | partial | no eclipse |

The accompanying figure (2), in which the large circles represent the shadow and the small circles represent the moon, shows these aspects.

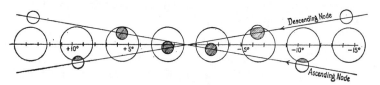

Fig. 2

46

In Mesopotamia, where the climate is favourable except in a few winter months, the priest astronomers could almost always see what was happening and must have perceived this regularity, although difficulties arose when part of the eclipses remained invisible, so that the series was broken. In the century between 750 and 650 BC, however, a series of five consecutively visible eclipses occurred four times, and a series of four eclipses four times also. Moreover, by their observations of the phenomena about full moon, they could detect the cause of a gap: when the moon in the morning did not 'wait' for the sun, the moment of exact opposition would be later, in daytime, and the eclipse would not be visible.

> The eclipse passes; it does not take place. If the king should ask: 'What omens hast thou seen?'—the gods have not been seen with one another. . . . From Munnabitu.*

Thus, already in Assyrian times astronomical phenomena were not only predicted and checked, but their unexpected absences were explained by natural causes.

* Thompson, 274 A.

CHAPTER 5

NEW-BABYLONIAN SCIENCE

CONDITIONS from which a science of astronomy could arise were present for the first time during the centuries of Assyrian supremacy. Astrology, which strongly associated man's life with the stars, directed his close attention towards the celestial luminaries and their motions. Detailed knowledge of the heavenly phenomena thus obtained could not have been the result of observations for mere time-reckoning and orientation, but there was still no trace of scientific purpose. So the early forerunners should not be compared with modern scientists who are animated by the spirit of research and explanation. Throughout antiquity man's attitude to natural phenomena was marked not by causality but by finality. The connection between the phenomena was seen not as cause and effect but as sign and meaning.

Hence no regularity or law was sought. Nevertheless, regularities imposed themselves. These caused no surprise; for the regularities in the sun's course and the moon's aspects were known of old. Regularities were now gradually perceived in the motion of the planets and the occurrence of eclipses, at first as an instinctive expectation, or an unconscious knowledge such as accompanies the experience of life in general, and then increasingly as conscious knowledge accompanied by greater skill in attempting prediction. Thus the beginnings of science arose as a systematization of experience, chiefly in the form of periods, after which the phenomena recurred in the same succession.

External conditions favoured this development. The energy of action manifest in the Assyrian conquests also asserted itself in other realms. The town of Babylon, wealthy by its trade, felt sufficiently powerful to revolt time and again against its Assyrian masters, eventually with success. When Assyrian power was weakened by heavy wars against barbaric tribes from Europe, the Cimmarions, the empire collapsed under the combined assault of the Babylonians and the Medes, and Nineveh (606 BC) became a heap of ruins. Babylon again became the undisputed capital of a new empire which under Nebuchadnezzar (604–561 BC) extended its power widely over Near Asia. All ancient Babylonian traditions were restored, the town was embellished with numerous temples and the priesthood felt itself the spiritual power of a large empire. This was still so when the Persian King Cyrus, probably called in by the priests of Marduk against rulers favouring other deities, conquered the town in 539 BC. He and his successor Cambyses acted as the elects of Marduk, as Kings of Babylon, continuing the traditions of the ancient capital. A revolt of the Persian nobles under Darius was needed to overthrow the Babylonized rulers, after a difficult war to reduce the proud city to the rank of the other Persian capitals. After Alexander's conquest of the Persian Empire, Babylon was once more the metropolis, and with its old cultural splendour strongly influenced Greek science. Throughout these centuries, from Ashurbanipal to Alexander, while the dynasties themselves changed, Babylon remained the great commercial city, capital of a flourishing empire and centre of Near Eastern culture.

Under such conditions it is not to be wondered that a strong development took place in astronomical activities, which led to the first rise of science. When those little warring countries had been absorbed into the peace of the large Persian Empire, the old omens of luck or evil had become senseless. The Persian monarchs, worshippers of Auramazda, had no use for omens from Marduk and Ishtar. This did not imply that the priestly observers could cease their work; but their attitude towards celestial phenomena changed. They were no longer ignorant earthlings, anxiously reading in the sky the messages inscribed there by the gods. They knew, partly at least, and with increasing sureness, what would be found there; they began to predict it. They knew the ways of the gods and could prove their knowledge; and this higher standing inspired them to greater zeal and persistence in their work of observation, regarding it as a ritual duty, a service to the gods. Thus it grew more detailed and precise, more complete and conscious. The distances of the planets from bright stars were noted numerically, and were perhaps measured, although nothing about instruments is mentioned in the texts.

We do not know along what lines this development took place; only some results have been preserved and not the researches. These results are scanty and fragmentary, and the discovery and deciphering of new texts may change many of our present views; yet it seems possible to deduce a probable picture of the development of astronomical knowledge in those centuries.

Let us first take the fixed stars. From the fifth century BC lists have been preserved which give a systematic survey of the relative positions of the constellations. They can be identified almost completely; a number of them date back to earlier times, as we have seen. Most of them have the names we use today, evidently borrowed by the Greeks from the Babylonians: the Bull, the Twins, the King (Regulus), the Ear of Corn, the Scorpion, the Archer (Sagittarius), the Fish-goat (Capricorn), the Eagle, the Lion, the Hydra, the Fish of Ea (southern fish), the Wolf, the Raven. But some are different: they had the Panther, the Goat and the Bowl, where we have the Swan, the Lyre and Auriga; and their Arrow-Star is our Sirius. The extent of the constellations is of course often different; in their Panther the northern part of our Cepheus was included; and the stars of our Great Dog are divided between the Arrow and the Bow. Some names still await identification.

Then there are lists of about thirty stars connected with days of the year, the days of their heliacal rising. Because of the intercalations, the dates in the Babylonian calendar relative to the stars may jump by 10 or 20 days; hence the dates of the lists, mostly rounded to multiples of five, must show a certain average or normal value. Indeed, we find (replacing the month-names by I to XII); Auriga I 20 (computed for Capella I 16), Pleiades II 1 (computed II 6), Aldebaran II 20 (computed II 18), Orion III 10 (Betelgeuse computed III 13), Twins III 10 (Castor computed III 13), Sirius IV 15 (computed IV 10), Regulus V 5 (computed IV 26), and so on. There is generally a fair measure of agreement; a large deviation, as for the last-named star, may be due to an error in copying.

Another kind of list connects the morning rise of one star with the culmination of another one, as follows: 'When you put the pole of your observation in the month Airu on the first day in the morning, before sunrise, west at your right hand and east at your left hand, and your eye southwards, then the "Breast of the Panther" [ε Cygni] stands in the midst of the sky before you and the Pleiades will appear.'[14] Something analogous is indicated in a fragmentary Assyrian text of a lunar eclipse, where a pole is mentioned and 'the star Kumaru [γ Cygni] right above you'. For every successive month there is such a sentence. To the Babylonians an astrological connection between such stars was probably implied. Another relation, a kind of antagonism, exists between pairs of stars: when one rises as the other sets. We read, for example, that the Pleiades rise, the Scorpion sets; Aldebaran rises, Arcturus sets; Orion rises, Sagittarius sets.[15] Thus we understand the importance in Babylonian observations of the horizon as a great circle dividing the celestial sphere into two halves. From their terraced towers the observers could see it in its entire circumference, and the clear sky must have permitted them to see the stars in full lustre only a few degrees above it.

We find no measurements in these lists; but there exists a list of stars all around the sky, with numbers added, which evidently stand for the consecutive distances.[16] These are given in three columns of proportional numbers, clearly denoting the same quantity expressed in different units. First in the sexagesimal units of weight (*biltu, mana, shikulu,* corresponding to the Greek and the biblical *talent, mine, shekel*); next in units of time—*beru* (2 hours) divided into 30 *ush* of 4 minutes. From Greek sources we know that the Babylonians used water clocks, in which the passage of time was measured by the outflow of water. Hence it seems safe to consider the numbers as differences in time of transit through the meridian; and they are indeed more or less concordant with the differences in right ascension between these stars. They may have been used to read in the sky the progress of the hours of the night. For this kind of measurement no graduated circles were necessary.

The connection, in these lists, between stars and dates of heliacal risings indicates that the calendar had now acquired a certain regularity. The empirical regularity in the return of these risings at certain dates must have led to the use of a fixed rule for the intercalations, especially the 8-year and the 19-year rule. With the 8-year period, the 13-month years come at intervals of 3, 3 and 2 years; they all had a second Adaru, except that after the second 3-year interval, when the calendar dates had reached farthest into the seasons, the sixth month, Ululu, was repeated, probably so that the autumn festivals fell at an appropriate time. With the 19-year period the intervals are 3, 3, 2, 3, 3, 3 and 2; and here, for the same reason, after the third of the three consecutive 3-year intervals a second Ululu was inserted. We remember that in olden times a second Ululu was an emergency measure, to use when the calendar dates had progressed too far into the year. It was possible—though not for all times—to reconstruct from civil documents and astronomical tables which years had an additional month. It was found that in the time of Cambyses and Darius an 8-year cycle was in use; this is shown by the fact that the year numbers divided by eight always leave the same group of remainders. Below is a list of intercalary years, the first line giving the year of the king's reign, the next giving the equivalent in our calendar, and the third giving the remainder: an asterisk indicates that the year had a second Ululu.

King	Cyrus					Cambyses			
Year	2*	3	4	7	9*	3*	5	8	3*
BC	537*	536	535	532	530*	527*	525	522	519*
Remainder	1	0	7	4	2	7	5	2	7

King	Darius						
Year	5	8	11*	13	16	19	22
BC	517	514	511*	509	506	503	500
Remainder	5	2	7	5	2	7	4

At first we see no regularity; then from 530 BC there is a regular 8-year period; but after three such periods it appears that the calculations did not come right, and in 503 BC intercalation had to wait until the month Adaru. A century later we again have a series of consecutive data, and here the 19-year period is strikingly revealed:

King	Artaxerxes										
Year	18	20	24	26	29	32	34*	37	40	43	45
BC	387	385	381	379	376	373	370*	368	365	362	360
Remainder	7	5	1	18	15	12	9	7	4	1	18

King	Ochus					Darius				Alexander			
Year	2	5	8*	10	13	16	18	21	1	4*	1	4	7
BC	357	354	351*	349	346	343	341	338	335	332*	330	327	324
Remainder	15	12	9	7	4	1	18	15	12	9	7	4	1

King	Philippus		Antigonus		Seleucus						
Year	2	5	2	5*	1	4	7	9	12	15	18*
BC	322	319	316	313*	311	308	305	303	300	297	294*
Remainder	18	15	12	9	7	4	1	18	15	12	9

As the remainders show after division by 19, a 19-year period was in use after 380 BC. The use of the name of the reigning king for numbering the years was abandoned when the successors of Seleucus continued to count the years from the rise of the dynasty. These year numbers form the so-called 'Seleucid Era' (SE), the first year of which began in spring 311 BC; in the years of this era those with remainders 1, 4, 7, 9, 12, 15, 18* have a thirteenth month intercalated.

The question might here be posed as to whether perhaps in the number of days in a month, 29 or 30, some cycle could have replaced the empirical fixation by the first crescent. The texts, when consulted, show no regularity of cycles; this is due to the great irregularities in the moon's course and to the changing inclination of the ecliptic to the horizon, which in spring makes the crescent visible at a shorter distance from the sun than in autumn. In the Babylonian climate the empirical way was certainly the best and the simplest. In a later century, as will be shown, science had advanced to such an extent that these irregularities could be computed beforehand.

So many observations had been made of the moon that it naturally became an object of theoretical computation. This had begun even in Assyrian times: texts from Ashurbanipal's library have been found with regular series of numbers which give schematically the time interval between sunset and moonset for the days after the first, and similarly the interval between sunset and moonrise for the days after the fifteenth of the month. Other tables, perhaps from before 1000 BC, give schematically the increase and decrease, with the seasons, of the interval between sunset and the first crescent (or between sunset and moonrise on the day after full moon). Between a maximum of 16 *ush* and a minimum of 8 *ush* the values change uniformly, $1\frac{1}{3}$ *ush* per month.

The form that theoretical knowledge took in the centuries that followed, entirely different from ours, clearly shows its origin in ordinary observing practice for purposes of prediction. The oldest document of such scientific astronomy is the later copy of a text dated year 7 of Cambyses (523 BC). It contains data on the first and the last crescent and on the phenomena around full moon, all expressed in terms of the unit *ush* (4 minutes), computed for all the consecutive months. It contains only numbers and short characters; the data for the second month with the meaning of the ideograms are shown in the accompanying table.

Airu	30	23			
	13	8	20	*shu*	(set)
night	14	1		*lal*	(opposite)
	14	1	40	*na*	(shining)
night	15	14	30	*mi*	(night)
	27	21			
Simannu	30	18	30	*shu*	

The table reads as follows: the preceding month had 29 days, so that Airu 1 is called 30; the moon sets 23 *ush* (92 minutes) after the sun. On Airu 13 the moon *sets* $8\frac{1}{3}$ *ush* ($33\frac{1}{3}$ minutes) before sunrise. On the evening of the fourteenth the moon rises 1 *ush* (4 minutes) before sunset; hence they are seen to be *opposite*. On the fourteenth the moon is still *shining* at sunrise and sets $1\frac{2}{3}$ *ush* ($6\frac{2}{3}$ minutes) later. On the fifteenth the moon rises in the *night*, $14\frac{1}{2}$ *ush* (58 minutes) after sunset. At Airu 27 the last visible moon sickle rises 21 *ush* (84 minutes) before sunrise. Airu has 29 days; at Simannu 1 the moon sets $18\frac{1}{2}$ *ush* (74 minutes) after sunset.[17]

These are the same phenomena as those observed qualitatively in Assyrian times to describe the conjunction and the opposition of moon and sun, now given in numerical form. If we ask whence the knowledge of these numbers has come, we have an indication in later texts (from 378 BC, for the years SE 33 and 79, i.e. 273 and 232 BC) where the same kind of data are given, but here as observed quantities, dispersed in chronological lists of data about planets, eclipses, haloes, prices of victuals and political events. They may be seen as a more refined continuation of the old qualitative Assyrian observations. Besides these texts, there were in later times computed tables similar to the text from 523 BC. The deciphering and explanation of such tablets, from the years SE 188, 189, and 201, carried out by Epping in 1889, was the first step in the scientific study of Babylonian astronomy. In some way the computed tables have certainly been derived from the observed data, probably by making use of schematic tables, as mentioned above.

Knowledge of the periods in which the same phenomena repeat themselves is the first form of scientific astronomy. The periodicity of the celestial phenomena, which forces itself upon the mind with careful and regular observing, was the bridge from empirical practice to predicting theory. In periodic repetition the abstract rule evolves from concrete facts. The period was the foundation and the essence of the first science of the stars. By discovering and fixing the periods, knowledge became science. This is the short summary of the development of Babylonian astronomy from the seventh to the third century BC, not only for the moon but also for the planets.

The period in which the oppositions and conjunctions of the planets, i.e. the same positions relative to the sun, return is called in full the 'synodic period of a planet'. Since Saturn, Jupiter and Mars take nearly 30, 12, and $\frac{15}{8}$ years for one revolution along the ecliptic, their synodic periods are nearly $1\frac{1}{29}$, $1\frac{1}{11}$, and $\frac{15}{7}$ years. Venus and Mercury, which oscillate about the sun, take one year for their average revolution along the ecliptic, and their synodic periods are nearly $\frac{8}{5}$ and $\frac{1}{3}$ year. This is also the period of alternation of direct and retrograde motion along the ecliptic. It is not an exact repetition, because the velocity of progress and the length of the retrograde arc vary with the longitude. So the time of revolution along the ecliptic is a second, generally longer period. Since the revolution does not take an exact number of years, a still longer duration, a common multiple of synodic period and time of revolution, is necessary to have the motions and positions relative to the sun and to the stars return to the same values. And since there is no precise common multiple, these large periods are approximate also, and the phenomena, e.g. the oppositions in the same constellation, return some days and some degrees different. Such large periods are given in the following table:

(s.p.=synodic period; rev.=revolutions; y.=years; d.=the number of days to be added):

Saturn	57 s.p.=	2 rev.=	59 y.+2 d.	(−6 d.)
Jupiter	65 s.p.=	6 rev.=	71 y.−5 d.	(−0 d.)
Jupiter	76 s.p.=	7 rev.=	83 y.+0 d.	(−13 d. or +17 d.)
Mars	22 s.p.=	25 rev.=	47 y.−7 d.	(+2 d.)
Mars	37 s.p.=	42 rev.=	79 y.+4 d.	(+7 d.)
Venus	5 s.p.=	8 rev.=	8 y.−2 d.	(−4 d.)

Mercury 19 s.p.= 6 rev.= 6 y.+8 d.   (+14 d. or −16 d.)
Mercury 41 s.p.=13 rev.=13 y.+2 d.   (−4 d.)
Mercury 145 s.p.=46 rev.=46 y.+0.3 d. (−1 d.)

These years are common solar years; since the sun proceeds every day 1° in longitude, the excess in degrees over the number of revolutions is equal to the excess in days over the right number of years. If, however, we reckon with Babylonian lunar years of 12 or 13 lunar months, the excesses in days are those added in parenthesis.

That the Babylonians knew and used part of these periods appears from a text, severely damaged, from the Persian time, in which the following is read:[18]

> . . . Dilbat [Venus] 8 years behind thee come back . . . 4 days thou shalt subtract . . . Gudud [Mercury] 6 years behind thee come back . . . the phenomena of Zalbatanu [Mars] 47 years . . . 12 days more . . . shalt thou observe . . . the phenomena of Sag-ush [Saturn] 59 years . . . come back day for day shalt thou observe . . . the phenomena of Kaksidi [Sirius] 27 years . . . come back day for day shalt thou observe . . .

Here the planetary periods are expressly named. Kugler supposed that the 27 years named with the fixed star Sirius represents a calendar period of 8 +19 years. It does not however clearly occur in the series of known intercalary years.

How was it possible for the ancient astronomers to use these large periods for prediction? To know the phenomena for a planet, they only needed to go back the number of years of the large period and copy the phenomena of that year, corrected, if necessary, by some few days and degrees. If they had to construct an ephemeris, say for the year SE 140, they had to take for Jupiter the data of the year SE 57 (140 −83), for Venus the data SE 132 (140 −8), for Saturn the data SE 81 (140 −59), etc. Texts exhibiting the successive steps in this computation show that they actually did this. First the observations were put down in chronological order in a kind of diary, as the basis for further treatment. The oldest specimen of such a diary is contained in the other parts of the text of Cambyses' year 7 (523 BC) mentioned above. Here the constellations are given in which the planets stand (in their western, eastern, or middle parts), as well as their distances from the moon or from one another, in *ammat* (of 24 *ubani*); 1 *ammat* appears to be nearly 2½°.

> Year 7, V 22 Jupiter in W. of Virgin hel. setting; VI 22 in E. of Virgin hel. rising; X 27 in W. of Balance, station; year 8, II 25 in the midst of Virgin station; VI 4 in E. of Balance hel. setting. Year 7, III 10 Venus in head of Lion evening-setting; III 27 in Cancer morning-rising; XII 7 in the midst of Fishes morning-setting; year 8, I 13 in the Chariot [the

55

Bull's horns] evening-rising. Year 7, VI 3 Saturn in midst of Virgin hel. setting; VII 13 in E. of Virgin hel. rising; year 8, V 29 setting. Year 7, II 28 Mars in W. of Twins hel. setting; VI 13 in feet of Lion hel. rising; year 8, V 12 station; year 9, II 9 in E. of Lion hel. setting . . . Year 7, VI 24 Venus greatest elongation; VII 23 at dawn Jupiter 3 *ammat* E. of Moon; VII 29 at dawn Venus 2 *ub*. N. of Jupiter; VII 12 Saturn 1 *ammat* W. of Jupiter . . . Year 7, IV 14 1⅔ *beru* after beginning of night lunar eclipse, extended over N. half; X 14 2½ *beru* toward morning Moon eclipsed, entirely visible, extended over N. and S. part . . .[19]

Another text with such a consecutive list of observational data from the year 379 BC carries the title (partly broken off but restored in accordance with similar texts): 'Observations for the festivals from the month Tishritu until the end of Adaru of the year 26 of Arses who is called Artaxerxes.' Such titles show how this purely astronomical work is still permeated by priestly ideas about religious ceremonies; in the consciousness of the observers it is always service of the gods. Here we read:

VIII 30 Moon appears 14½ [58 minutes] sickle visible; night ¼. Mars toward W. retrograding below β Arietis 2 *ammat* 10 *ub*; . . . night 12 full moon surrounded by halo, Mars stood within; the king and the king's son . . . Moon ⅔ *ammat* E. of α Tauri . . . ; 16 Jupiter in the Scorpion hel. rising, 11½ [46 minutes] visible; . . . 22 morning Moon above Saturn 2½ *ammat*, more east; 22 Mercury in the morning in Sagittary hel. rising, 22 Mars in W. station . . .[20]

Such diaries, giving the place of the moon and the planets relative to certain stars, were certainly kept regularly by the astronomers as part of their official duties.

For the next step schedules were made for each planet containing the data for consecutive years, collected from the diaries. Such a text, probably extending from 387 to 346 BC, is partly preserved; it presents the risings and settings and the stations of the planet Jupiter and its position relative to standard stars. It gives, moreover, the length of each month in the usual way . . . Duzu 1, Abu 30, Ululu 1 so that there is no uncertainty as to the number of days. From these schedules an 'auxiliary table' for a certain year is composed. Such a table is represented by a text, bearing the title: 'the first day, the phenomena, the motions, and the eclipses that have been determined for the year 140.'[21] It gives data for Jupiter in the years 69 and 57, for Venus in 132, for Mercury in 94, for Saturn in 81, for Mars in 61 and 93, and, in reverse, eclipses in 122; all years of the Seleucid Era. These are exactly the years which, by the addition of the large periods enumerated above, produce the year 140. So, by copying them with small correction where necessary, an ephemeris for the year SE 140 is obtained.

56

Smaller and larger fragments of such computed ephemerides with borders broken off and names and numbers damaged, have been investigated and deciphered for the years SE 104, 120, 194, and for the years SE 129, 178, 301. All give heliacal risings and settings, stations and oppositions—or, for Venus and Mercury, the greatest elongations—as well as solar and lunar eclipses. However, the contents of the two groups of texts are somewhat different. The first-named three give constellations and angular distances of the planets to fixed stars: e.g., for SE 120:

> II 7 night, Mars above γ Geminorum 4 *ammat*; 23, evening, Mercury below β Geminorum 2½ *ammat*; XII 24 morning, Mars below β Capri 2½ *ammat*.[22]

In the other three texts distances to stars are missing; only the constellation is named. Thus for SE 178 we read:

> IV 30 Venus and Mars in the Twins, Mercury in Cancer, Saturn in Sagittarius.

This seems to be less precise; but then, between them, other data follow:

> IV 13 Mercury reaches the Lion, 5 Venus reaches Cancer, V 3 Mars reaches Cancer, V 15 Venus reaches the Lion, VI 9 Venus reaches the Virgin.[23]

What are the confines of the constellations reached? Computation of the planets' longitudes on those dates shows them to be 112°, 81°, 142°, 52° . . ., all values 22° larger than the successive multiples of 30°. All constellations have received an extent of 30° in longitude. This means that they are used as signs of the zodiac, artificial theoretical divisions of the ecliptic. Ephemerides of this kind, by giving exact longitudes for determinate days, represent a higher stage of theoretical knowledge of the planetary motions. These ephemerides were in use up to the last years for which we have records. They were reliable because by simple computations they could be derived directly from former observations. We may be sure that by frequent comparison with new observations the theory was gradually improved into a more precise knowledge of the corrections to be applied after the great periods.

Next to the planetary motions, the eclipses occupied a prominent place in the predictions. They also showed great progress in the theory. In Assyrian times they were in many cases announced, probably on the simple basis of their return at intervals of 6 months, in a series of 5 or 6, in consequence of the first decreasing and then increasing distances of the full moon from the node. The series ended when this distance became larger than 11° or 12°; but then a new series was soon started by the preceding full moon. A preceding full moon has a longitude smaller by 29.11°; hence relative to the node which recedes monthly by 1.56°, it is 30.67° back. When the distances to the node, as computed in our example (p. 46) for average velocities, have the increasing values +9.64°, +13.67°, +17.69°, they are, for the preceding full moons, −21.03°, −17.00°, −12.98°, −8.95°, producing the start of a new series. For this new series, beginning 8 ×6 −1 =47 months after the first began, we can write down the distances between full moon and node below those of the first series:

1st series	−14.50°	−10.48°	−6.45°	−2.43°	+1.59°	+5.62°	+9.64°	+13.67°
2nd series	−12.98°	− 8.95°	−4.93°	−0.91°	+3.11°	+7.14°	+11.16°	+15.18°
	no eclipse	partial	total	total	total	partial	partial?	no eclipse

In this new series the sequence of partial and total eclipses is somewhat different from the first, because all distances are 1.52° different, at first smaller and later larger by that amount. When the second series has ended, a third and a fourth will appear. We give here the distances for these consecutive series:

3rd series	−11.46°	− 7.44°	−3.42°	+0.61°	+4.63°	+8.65°	+12.67°	
4th series	−13.97°	− 9.95°	−5.93°	−1.90°	+2.12°	+6.14°	+10.16°	+14.19°
5th series	−12.46°	− 8.44°	−4.41°	−0.39°	+3.63°	+7.66°	+11.68°	
6th series	−14.97°	−10.94°	−6.92°	−2.90°	+1.12°	+5.15°	+ 9.17°	+13.19°

Thus the series follow one another, always including two or three total eclipses in the middle, preceded and followed by one or two partial eclipses, and separated by two or three times six months minus one without eclipses; the eclipses opening the series have intervals of 41 or 47 months. The values of the consecutive distances show that each of these series of eclipses has a different aspect in the sequence of total and partial eclipses. But in the sixth series the distances have returned to nearly the old values; hence the sequence in the sixth series must be nearly similar to that in the first. Thus after five series, nearly the same sequence and aspect return.

It may well be that the Assyrian astrologers had already perceived that, 41 or 47 months after a total eclipse, another total or nearly total eclipse may be expected. It cannot be proved because in their reports they did not disclose the basis of their predictions. The larger period formed by the sequence of five series, after which the aspects return, is more interesting. What is its length? Within the series six intervals of six months occur twice, seven such intervals occur thrice, and between

the series there are five intervals of five months; this gives in total
$33 \times 6 + 25 = 223$ months. This period has been known in later writings
under the name of *saros*; in Babylonian texts and in Greek antiquity,
however, this name does not seem to occur. It amounts to $6,585\frac{1}{3}$ days,
i.e. 18 solar years plus $11\frac{1}{3}$ days; in this period the nodes have made
nearly one retrograde revolution, and the moon has passed the ascending node 242 times.

The discovery of this long period may still seem to have presented
some difficulties, since the five series here are not so greatly different in
character. But it was facilitated by the irregularities in the eclipse
phenomena. The regularly increasing distances given above are computed under the supposition of constant mean velocities and mean
sizes of sun and moon. In reality, the angular sizes of the moon and the
shadow are sometimes larger, sometimes smaller, so that the distances
from full moon to node necessary for totality of an eclipse vary between
$4.75°$ and $5.83°$, and for a partial eclipse between $9.5°$ and $12.2°$.
Still more important is the variable velocity of the sun; the longitude of
the sun, hence also the distance from full moon to node, may be up to
$2.28°$ greater (in March) or smaller (in September) than assumed in
our example. Applying such corrections (starting with an arbitrary
first value for the solar longitude) the distances between full moon and
node in the six successive series are as given in the accompanying
tabulation.

1st series	$-12.37°$	$-12.70°$	$-4.20°$	$-4.61°$	$+3.87°$	$+3.37°$	$+11.87°$	$+11.51°$
2nd series	$-11.55°$	$-10.15°$	$-3.87°$	$-1.73°$	$+3.76°$	$+6.70$	$+11.36°$	$+15.18°$
3rd series	$-12.73°$	$-5.98°$	$-6.02°$	$+2.38°$	$+2.77°$	$+10.64°$	$+10.57°$	
4th series	$-11.74°$	$-12.14°$	$-3.81°$	$-3.94°$	$+4.04°$	$+4.34°$	$+11.87°$	$+12.69°$
5th series	$-12.16°$	$-8.51°$	$-4.55°$	$-0.04°$	$+3.05°$	$+8.42°$	$+10.69°$	
6th series	$-12.03°$	$-13.02°$	$-4.79°$	$-5.12°$	$+3.37°$	$+2.87°$	$+11.45°$	$+10.94°$

Because, in consequence of the six-month intervals, the positive and
negative corrections alternate, the course of the values has become
highly irregular, and sometimes even inverted. The aspects of the
successive series now become notably different; sometimes four total
eclipses, sometimes three or two occur; in our first (and sixth) series
total eclipses suddenly appear without being preceded by partial
eclipses. The similarity of the sixth to the first series is now much greater.
This is due to the fact that the *saros* period is only 11 days greater than a
full number of solar years, so that after this period the solar longitudes
approximately return. Moreover, the period of variation in the moon's
diameter and in the size of the shadow is 9 years, so that after
18 years they too return to near repetition. This may have favoured the
detection of the 18-year period by the Babylonian observers. That they
knew and used this period in later centuries is evidenced by the

'auxiliary table' mentioned above, in which eclipses were derived from data 18 years earlier.

It is demonstrated still more clearly by a famous text in the British Museum, investigated by Strassmaier and Epping, and called by them a 'Saros-Canon'.[24] It is a fragment, broken off on both sides; it contains, arranged in a number of columns, a list with nothing but years and months, two names of months each year, without any commentary. The years are the reigning years of successive kings, indicated by first syllables of their names: Artaxerxes II, Ochus (Umasu), Arses, Darius, Alexander, Philip, Antigonus, Seleucus, the last-named continued as years of the Seleucid Era, up to SE 35; hence they extend from 373 to 277 BC. The months in each column, indicated by Roman numbers I–II–XII, increase by 6, sometimes by 5. Where the syllable *dir* is placed, a thirteenth month is inserted; then there is an apparent jump of five months only. Since each column of 38 lines comprises 223 months, it is clear that this is a list of eclipse months, completed before and after with months without eclipse. The horizontal division lines separate the series, and the addition of 'five months' indicates the five-month intervals between the groups. Comparison with a modern table of computed eclipses shows that the months with total eclipses do in fact stand in the centre of each group. Within these groups we count 3, 8, 7, 8, 8, 4 lines; adding the upper 3 as their continuation to the lower 4, we have 8, 7, 8, 8, 7, exactly corresponding to our 5 series.

Here the question arises: if 5 series are meant, why then has the last of them been cut into two parts? Strassmaier suggested that an explanation might be found in the fact that the saros period is not entirely exact. In our computation of the regular distances of full moon to node it is seen that after 5 series, 223 months, the first value (-14.50) is not exactly reproduced in the first of the sixth series (-14.97). In every saros period the full moon moves back a little, until the first of the new series falls out when its distance grows too large, and a new value, formerly too large, is added as the last of the preceding series. Then the five-month interval is moved one place forward, and the division line is displaced down over one line. This will happen at different lines until after eight or nine saros periods (150 years) all have descended one place. To explain the three lines the list should have started 450 years earlier. Strassmaier, however, assumes that at the left-hand side of this tablet 10 columns have been broken off, so that originally it went back as far as 572 BC only.

This Saros-Canon affords a clear picture of the way in which Babylonian science developed. It shows how, out of the knowledge of the succession of the eclipses in a series, the succession of the series themselves gradually came to light. We do not know the exact date of this

	X		X	dir	XI		XI		XI	dir	XII
32	IV	4	IV	1Ar	IV	5	V	11	V	29	V
dir	X		X		X	dir	XI		XI		XI

33	II 5 m	5	III 5 m	2	III 5 m	6	III 5 m	12	IV 5 m	30	IV 5 m
	VIII	dir	IX		IX		IX	(dir)	X		X
34	II	6	II	1Da	III	1An	III	13	III	31	IV
dir	VIII		VIII	dir	IX		IX		IX	dir	X
35	I	7	II	2	II	2	III	14	III	32	III
	VII		VIII		VIII	dir	IX		IX		IX
36	I	8	II	3	II	3	II	15	III	33	III
	VII	VIdir	VII		VIII		VIII	dir	IX		IX

	XII 5 m	9	XII 5 m	4	I 5 m	4	I 5 m	16	I 5 m	34	II 5 m
37	VI		VI		VIa		VII		VII	dir	VIII
dir	XII		XII		XII	5	I	17	I	25	I
38	V	10	VI	5	VI		VIa	18	I		
	XI	dir	XII		XII		XII		VII		
39	V	11	V	1A	VI	6	VI		VIa		
	XI		XI	dir	XII		XII		XII		

40	IV 5 m	12	IV 5 m	2	IV 5 m	1Si	V 5 m	19	V 5 m		
dir	X		X		X	dir	XI		XI		
41	III	13	IV	3	IV	2	IV	20	V		
	IX	dir	X		X		X	dir	XI		
42	III	14	III	4	IV	3	IV	21	IV		
	IX		IX	dir	X		X		X		
43	III	15	III	5	III	4	IV	22	IV		
dir	IX		IX		IX	dir	X		X		

44	I 5 m	16	II 5 m	6	II 5 m	5	II 5 m	23	II 5 m		
	VII	dir	VIII		VIII		VIII	dir	IX		
45	I	17	I	7	II	6	II	24	II		
	VII		VII	dir	VIII		VIII		VIII		
	XIIa	18	I	1Phi	I	7	II	25	II		
46	VI		VII		VII	dir	VIII		VIII		
	XII		XIIa	2	I	8	I	26	II		
1U	VI	19	VI	dir	VII		VII		VIII		

	XI 5 m	20	XI 5 m		XII 5 m	9	XII 5 m	27	XIIa 5 m		
2	V		V	3	V		VI		VI		
dir	XI		XI		XI	dir	XII		XII		
3	IV	21	V	4	V	10	V	28	VI		

Fig. 3. Babylonian Saros-Canon

61

text; it must have been written after 280 BC, since the Seleucid Era is used. Thus the gradual development of this knowledge took place in Persian times, between the sixth and the third centuries BC.

The Saros-Canon is an important document of Babylonian science. It is not, as are so many texts, a list partly of observations, partly of resulting predictions; it is a formulation of theory in the form of a table extending equally over past and future. It combines the multitude of former and later lunar eclipses in one condensed table, which, potentially, can be extended indefinitely in both directions. This representation of a large realm of phenomena in an abstract picture shows to what remarkable heights astronomical science had attained.

The question may be asked as to whether such predictions were also possible for solar eclipses. Their number for a certain place on earth is only half that of the number of lunar eclipses; moreover, occultations of small parts of the solar disc in the daytime are not so easily detected as are lunar phenomena at night. Also, through the parallax, their visibility is more variable and it is more difficult to find regularities. In the reports of the astrologers solar eclipses are sometimes expected or announced, but we do not know on what basis. Solar eclipses occur half a month before or after lunar eclipses, and they appear most conspicuously (because in half a month the sun progresses nearly 15°) in the years between the series of the lunar eclipses. So it may have been perceived that the solar eclipses mostly occur halfway between the 41- or 47-month series intervals, hence 20½ or 23½ months after total lunar eclipses. But these are surmises, and better information can be had only from new texts.

CHAPTER 6

CHALDEAN TABLES

AFTER the death of Alexander and the foundation of the Seleucid Empire the decline of Babylon set in. Greek trade on the Black Sea and with Egypt had already captured an important part of Babylon's east-west trade in the reigns of the Persian kings, who had vainly tried to subjugate these competitors. In Hellenistic times the Mediterranean became the chief scene of world commerce; new Greek towns, especially flourishing Alexandria with its Red Sea connection with India, grew into wealthy centres of trade and industry. Babylon now lay far outside the new traffic routes and lost its prosperity. Its position as a capital was taken by the new Greek town of Seleucia, and before long Syria became the nucleus of the Seleucid Empire. The Parthian conquest of Mesopotamia in 181 BC cut it off completely from the Mediterranean; in the last centuries before our era it is hardly mentioned. Mesopotamia, of course, remained for long afterwards a fertile agricultural land, a source of wealth to the old country towns, a source of power for the local peoples and princes, but only of limited importance.

Science was not immediately affected. When the merchants and the officials moved to the new capital, the priests remained in their temples in Babylon. World history presents more instances in which energies, once aroused, continued to inspire the sciences and arts during later centuries, and cultural traditions, methods, and ideas remained active. Even higher stages were attained, notably here, long after the flourishing economic and political life which had given them their first impulse had decayed. Astronomy continued to develop during the last three centuries BC, and even attained its climax at this time. Excavations at some of the various sites have brought to light fragments bearing witness to the highest stage of Babylonian science.

It is an entirely new form of astronomical theory that we now meet. Whereas the ephemerides computed for every year continued to be made because they met practical needs, the new form met the desire to have some special phenomenon, as, for instance, the planet's opposition or station, expressed in a quite general way, by means of a table extensible at will over past and future. The phenomena were not qualitatively described here but were taken as mere quantity; instead of stating the constellation or the distance to some star in *ammat* and *ubani*, the position was given by the ecliptical co-ordinates, longitude and latitude. The units of longitude were the zodiacal sign, 30°, and its thirtieth part, the degree, with its sexagesimal subdivisions.

The Babylonians from the earliest times possessed an excellent mathematical implement in their sexagesimal system of numbers. The numerals go from 1 to 59, rendered by crochets (tens) and wedges (units); in the row of sexagesimals, each following number is expressed in a unit $\frac{1}{60}$ of the preceding one. By means of this principle of value according to position (the same as in our decimal system), the Babylonians could express any large number by higher powers of 60 and any small part by powers of $\frac{1}{60}$. The fragments of the astronomical tables known to us consist of rows of numbers without names, intermingled with characters for months and zodiacal signs. Their meaning had to be revealed by the numbers themselves. Sometimes this was easy; but often the most patient efforts, the utmost ingenuity, and the collaboration of different scholars were needed to arrive at a satisfactory interpretation, especially as the numbers are often scarcely readable on the sherds and are spoiled by copying errors detectable only in the very process of deciphering.

The origin of these tables is unknown. We may surmise that the Chaldean astronomers had instruments with divided circles to measure the longitude of the planets or the moon or the longitudes of stars with which to compare them. But nothing of the kind is found mentioned in the clay tablets. Texts with observational data which might be considered sources of these new theoretical tables are also lacking. They stand out suddenly in full perfection, separated by a wide chasm from the earlier expressions of theory or practice. Although we realize that they must have been based upon observation, the course pursued is as yet obscure.

The tables are of two kinds, lunar and planetary. The aim was to derive in a general context such quantities as had always been the chief objects of attention. The planetary tables deal with the outstanding phenomena in five sections: heliacal rising, first station, opposition, second station, heliacal setting; for all the successive years the day and the longitude of each of them are given, with some auxiliary columns. In the lunar tables the entries—the time and place of the first crescent and of the full moon—are derived by means of many columns of auxiliary quantities. The lunar tables show a number of very complicated and only partly understood methods of calculation; by comparison the structure of the planetary tables is simple.

The variations in velocity appearing in the alternations of larger and smaller displacements are the most striking phenomena in these purely numerical data. The oppositions of the planets follow one another on one side of the ecliptic with larger—on the opposite side, with smaller —intervals of time and longitude, as the other phenomena do also. Such periodic changes can best be presented by a sinusoid (a wave curve). The Greek mathematicians did this by making use of motion in space along an eccentric circle. The Babylonians had no such spatial picture; to them the heavenly phenomena did not take place along circular orbits in three-dimensional space but on a two-dimensional vault where the luminaries followed their mysterious course. They did not develop new geometrical world structures; they were not philosophical thinkers but priests, confined to traditional religious rites, and therefore disinclined to adopt new cosmic ideas which did not conform to the holy doctrines. The planets to them were not world bodies in space; they remained luminous deities moving along the heavens as living men move on earth. Up to the last, the tables open with the invocation: 'In the name of god Bel and goddess Beltis, my mistress, an omen.' Being a priestly science, Chaldean astronomy could not pass beyond the inherited world system; but all the same the refined knowledge and the huge store of observations demanded a theoretical summary.

Thus these astronomers faced the task of representing the irregularity of a planet's motion by purely arithmetical methods. At first they managed with a simple but crude method: in one half of the ecliptic a constant larger value, and in the other half a constant smaller value was assumed. Afterwards they improved their method by representing the variable quantity by a zig-zag line, an alternating uniform increase and decrease between two fixed limits, points at which the line abruptly reversed its direction as if reflected. We met the same procedure in the simple schematic tables of early Assyrian times. What is lacking here is the idea of continuity, of a gradual slowing-down of the *increase* to a standstill at maximum and then a gradual speeding-up of the reverse change; instead, there are abrupt jumps between straight rows of numbers.

A series of numbers may serve here, as an example, taken from a fragment of a Jupiter table (pp. 68–69) representing the successive longitudes of Jupiter's second station expressed in zodiacal signs, degrees, and minutes. The successive differences in the next column are the resulting movements in a synodic period. Looking at their differences as given in the next column, we see that these movements regularly increase or decrease by $1° 48'$. In the vicinity of the limiting maximum and minimum values a computation is necessary: from $29° 41'$ to the minimum $28° 15\frac{1}{2}'$, the decrease is only $1° 25\frac{1}{2}'$, so (to complete $1° 48'$) there remains $0° 22\frac{1}{2}'$, which, used now as increase, produces the next value 28° 15½' +0° 22½' =28° 38'. In the same way at the upper limit, 38° 2': the increase of 0° 24' and then the decrease of 1° 24' (total 1° 48') bring the next value 1° 0' below the preceding one. In the graph (fig. 4) all these values are situated on a zig-zag line.

	Longitude	Motion	Difference	Sine Curve Completed	Deviations	Motion
Cancer	21° 49'			21° 56'	− 7'	
		29° 41'				29° 19'
Lion	21° 30'		−1° 3'	21° 15'	+15'	
		28° 38'				29° 11'
Virgin	20° 8'		+1° 48'	20° 26'	−18'	
		30° 26'				30° 4'
Scales	20° 34'		+1° 48'	20° 30'	+ 4'	
		32° 14'				31° 59'
Scorpion	22° 48'		+1° 48'	22° 29'	+19'	
		34° 2°				34° 16'
Sagittarius	26° 50'		+1° 48'	26° 45'	+ 5'	
		35° 50'				36° 11'
Aquarius	2° 40'		+1° 48'	2° 56'	−16'	
		37° 38'				37° 7'
Fishes	10° 18'		−1° 0'	10° 3'	+15'	
		36° 38'				36° 45'
Ram	16° 56'		−1° 48'	16° 48'	+ 8'	
		34° 50'				35° 14'
Bull	21° 46'		−1° 48'	22° 2'	−16'	
		33° 2'				32° 59'
Twins	24° 48'		−1° 48'	25° 1'	−13'	
		31° 14'				30° 49'
Cancer	26° 2'		−1° 48'	25° 50'	+12'	
		29° 26'				29° 25'
Lion	25° 28'		−0° 33'	25° 15'	+13'	
		28° 53'				29° 14'
Virgin	24° 21'		+1° 48'	24° 29'	− 8'	
		30° 41'				30° 20'
Scales	25° 2'			24° 29'	+13'	

Of course, the representation of a smooth curve by a zig-zag line cannot be exact. The errors may be seen from the last three columns in our table, where in the fifth column the zig-zag is replaced by a wave curve with semi-amplitude 4°. The fifth column gives the longitudes computed with these values, and the next column shows the deviations of the Babylonian values from these smoothly fluctuating longitudes. The deviations nowhere exceed 20'; greater accuracy cannot be expected at this period. Theoretically, the Babylonian method was as good as any other used in antiquity. Care had to be taken, however, that the mean of maximum and minimum, which is the average of all the amounts successively added, was exactly right. Otherwise the continuous adding of intervals would in the long run lead to increasing errors.

The clay tablets with texts, from which our knowledge of these planetary tables is gained, are all damaged sherds, broken and in many

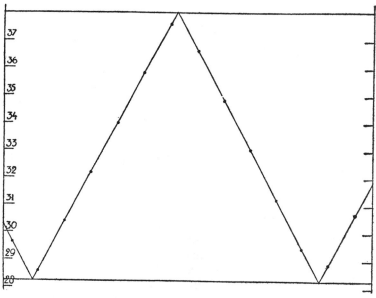

Fig. 4

places illegible, with only part of the table usable. The amplest material deals with the planet Jupiter; this is at once evident because the dates on the successive lines have intervals of one year and one month (13 lunar months and 10 or 20 days), the synodic period of Jupiter. Kugler, who was the first to decipher them, found three kinds. In the first and most primitive kind, the synodic arc (the arc covered in one synodic period) is assumed to have over part of the ecliptic, from 85° to 240° longitude (i.e. over 155°), the constant value 30°; over the other part, from 240° to 85° (i.e. over 205°), the constant value 36°. The two parts must be unequal, in order to have, for a complete revolution through the ecliptic, the right mean value 33° 8′ 45″. For an arc situated partly in one region and partly in the other, an intermediate value was computed.

Jupiter tables of the third kind were present in three fragments. The obverse side of one of these is reproduced on page 68, with the transcription in our numerals on the next page; the months are given here by the Roman numerals I to XIII.*[25] The upper line reads as follows. The numbers 3, 10 represent, in sexagesimal writing, the year 190 (SE); then follows the date Adaru 11. The next column gives the synodic arc 31° 29′ computed as a zig-zag function, as indicated in our preceding

* The hollow cuneiform characters have been copied by Strassmaier from the original sherds; they are the fundamental data of the study. The entirely black characters were computed subsequently by Kugler from his explanation.

table (page 66); then follows the longitude Cancer 21° 49', which is found by adding this arc to the longitude of the preceding line, not present here; by adding the next arc 29° 41' to this longitude, we get Cancer 21° 49' +29° 41' =Leo 21° 30'. The sign *ush* in the next column denotes the second station. Then comes the next section, with data for the heliacal setting (indicated by *shu* in the next column). It opens with a first column (43, 45, 30 as first value, in sexagesimals) serving, as did the first column in the preceding section, for computing the dates; then follow year, date, synodic arc and longitude, as in the other section.

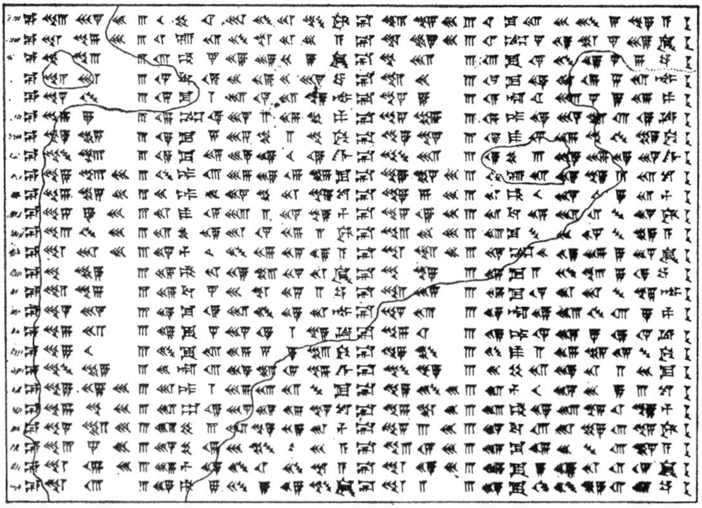

Fig. 5. Cuneiform text of Jupiter table

These tables of the third kind present a higher stage of accuracy, since the synodic arcs and the time intervals of the dates vary continuously between an upper and a lower limit; these limits are 28° 15½' and 38° 2' for the arcs. Their mean, the average of all the separate values in the long run, 33° 8' 45", is identical with the value adopted in the tables of the first kind. The tables enable us to derive more details concerning the basic suppositions applied by these astronomers. One rise and fall of the zig-zag line, after which the velocity has returned to the same value, corresponds to one revolution of the planet; together they amount to twice the distance of the limiting maximum and minimum, $2 \times 9° 46\frac{1}{2}' = 19° 33'$. One step of 1° 48' corresponds to one synodic period; hence one revolution contains 19° 33': 1° 48' $=10\frac{31}{36}$

. . 30	3 10	XII 11	31 29	21 49	♋	ush	43 45 30	3 11	IV 22	30 39	5 47	♌	shu
. . 30	3 11	XIII 22	29 41	21 30	♌	”	41 57 30	3 12	V 4	28 51	4 38	♍	”
. .	3 13	II 4	28 38	20 8	♍	”	40 32	3 13	VI 14	29 .	. 6	♎	”
42 31	. .	III 16	30 26	20 34	♎	”	42 20	3 14	VII 27	. .			
. .	. .	. 1	32 14	22 48	♏	”	44 8	3 15	VIII 11	. .			
46 7	3 16	V 17	34 2	26 50	♐	”	45 56	3 16	IX 27	34 52			
47 55	3 17	VII 5	35 50	2 40	♒	”	47 44	3 17	XI 14	36 40			
49 43	3 18	VII 25	37 38	10 18	♓	”	49 32	3 .	. .	37 36			
48 43 30	3 19	IX 13	36 38	16 56	♈	”	48 54 30	3 19	. .	35 48			
46 55 30	3 20	IX 30	34 50	21 46	♉	”	47 6 30	3 21	II 10	34 0			
45 7 30	3 21	XI 15	33 2	24 48	♊	”	45 18 30	3 22	III 25	32			
43 19 30	3 22	XII 28	31 14	26 2	♋	”	43 30 30	3 23	IV 9				
41 31 30	3 24	I 10	29 26	25 28	♌	”	41 42 30	3 24					
40 58	3 25	II 21	28 53	24 21	♍	”	40 47	3 25					
42 46	3 26	III 4	30 41	25 2	♎	”	42						
44 34	3 27	IV 18	32 29	27 31	♏	”							
46 22	3 28	VI 4	34 17	1 48	♑	”							
48 10	3 29	VI 23	36 5										
49 58	3 30	VIII 13	37 53										
48 28 30	3 31	IX 1	36 .										
46 40 30	3 32	X 18	34 .										
44 52 30	3 33	XII 3											
43 4 30	3 34	XIII 16											
41 16 30	3 36	I 27											
41 13	3 37	II .											

Fig. 6. Transcription of text in fig. 5

synodic periods. In other words: 391 synodic periods are equal to 36 revolutions, which last $391 + 36 = 427$ years. This is a more accurate approximation than the one given above (p. 54) for the planetary period; but it can be derived from the data given there: if we take six times 65 periods, each of six revolutions minus 5° 33′, increased by one period; or if we take five times 76 periods, each of seven revolutions minus 0° 57′, increased by 11 periods forming one revolution. The synodic arc found in dividing 360° by the number of periods $\frac{391}{36}$ is 33° 8′ 44.8″, for which 33° 8′ 45″ used above is a sufficient approximation.

The length of the synodic period, the interval in time between two successive lines in the table (in the first column of a section) is computed after the same principle. The values are represented by zig-zag functions proceeding by steps of 1, 48 between the limits 50, 7, 15 and 40, 20, 45 (the commas separate the successive sexagesimals), which show a difference (in days) of 9, 46, 30, the same number for the arcs (in degrees). These periods were used for computing the dates. Here the Babylonian astronomers faced the difficulty that they had no possible way of knowing the length of future months, whether 29 or 30 days, so that future *days* became entirely uncertain. Future months, however, by the 19-years calendar period, were never uncertain. So they devised this simple method: they took 30 days for every month. This means that they worked with fictitious days of $\frac{1}{30}$ month, somewhat smaller than real days. The small deviations in the time thus expressed against the real dates, at most half a day, are of no importance, since the moments of heliacal rising and setting or station, even of opposition, cannot be fixed with greater precision. It is necessary, then, that the mean excess of the Jupiter period over 12 lunar months, 44.53 days (in our decimals), be increased to 45.23 of these smaller fictitious days to add for every next line. Indeed we find that the mean value between the upper and the lower limit is $\frac{1}{2}$ (50, 7, 15 +40, 20, 45) =45, 14, the same value in sexagesimal writing.

The accuracy of the Jupiter tables of the second kind, of which a larger number of fragments is available, stands between that of the two other kinds. The synodic arc for opposite regions of the ecliptic is again taken to be 30° and 36°, but only over smaller intervals of 120° and 135° of longitude; in the intermediate regions extending over 53° and 52° of longitude, an intermediate value of 33° 45′ was adopted. Thereby the large errors remaining with the first method are reduced in the border regions. It may be mentioned that, besides tables, a text has also been found, a 'procedure text', consisting merely of directions on how to compute such tables of the second kind, clearly intended for the scribes or the apprentices of the temple.

The above statements on the accuracy of the Chaldean tables deal solely with the formal accuracy attainable through this theoretical method, not with the real accuracy attained. The real errors may be considerably larger through imperfect knowledge of the basic quantities. The errors in longitude of the second station in the tables of the third kind sometimes exceed $1\frac{1}{2}°$, not because of the inadequacy of the zig-zag function but because the amplitude was taken too large and the minimum velocity was assumed to occur at longitude about 150° instead of about 160°. For practical purposes the theoretically less perfect ephemerides, directly derived from former observations, could give better results; but the tables represent a higher stage of theoretical mastery.

Similar tables for other planets have been found, but they are less complete. For Saturn a small fragment of the third kind is known, containing oppositions from SE 155 to 167. They are based on the relation: 9 revolutions=256 synodic periods=265 years. A tablet from Uruk, dealing with Mars, is remarkable because the strongly variable velocity of this planet has been represented by more consistent application of the principle of the second kind. The ecliptic is divided into six sections, each embracing two signs; the value of the synodic arc is constant in each section and jumps at the boundaries: from 90° (in Capricorn-Aquarius), over $67\frac{1}{2}°$ (in Fishes and Ram) and 45° (in Bull and Twins) to 30° (in Cancer and Lion), then over 40° (in Virgin and Scales) and 60° (in Scorpion and Sagittarius) back to 90°. If for example, a synodic arc had to proceed from Aquarius 10°, then 20° ($\frac{2}{9}$ of the total arc) is situated in Aquarius, the remainder in the next section has a length $\frac{7}{9} \times 67\frac{1}{2}° = 52\frac{1}{2}°$, hence reaches beyond the Fishes to Ram 22° 30′, its end.

By computing the average number of synodic arcs in one circumference, we find from these data 133 : 18. Mars in one synodic period completes an entire circumference of the ecliptic plus the synodic arc; hence in its 133 periods it passes through 151 revolutions, taking 133 +151 =284 years. This relation is more accurate than the relations given in the table on pages 54–5; it is indeed the best value we can derive by means of continuous fractions from our present knowledge of the periods. The same holds for the large period of Saturn.

For Venus there are some fragments of tables indicating appearance, disappearance, and stations in the evening and morning. Because the phenomena return after eight years with only a small decrease of longitude, 2° 40′, no detailed data about the synodic arcs can be derived from them. For Mercury, in all centuries the most difficult of the planets, because of its irregularities and its difficult visibility in strong twilight, there are also some texts, but many numbers in them are nearly illegible. They give time and place for appearance and disappearance as evening and morning star in the years SE 145–53 and SE 170–85. Kugler found that similar methods were followed for the appearance of Mercury as for the other planets: for three sections of the ecliptic three different constant values were adopted for the synodic arc. They are 106 from longitude 121° to 286°; $141\frac{1}{3}°$ ($=\frac{4}{3} \times 106°$) from 286° to 60°; and $94\frac{2}{9}°$ ($=\frac{8}{9} \times 106°$) from 60° to 121°. It is a curious asymmetric variation; but the corresponding average synodic arc, $\frac{848}{2673} \times 360° = 114° \ 12' \ 31.5''$, is only $4''$ smaller than the real value; and also the corresponding synodic period, $115^d \ 21^h \ 3^m \ 51^s$, is more accurate than might be expected even after many centuries of observation. The times and places of disappearance were found by adding to the values for appearance amounts different for each zodiacal sign, varying between 44° and 12°. This, of course, is a rather poor procedure; and indeed the errors, though usually of only a few degrees, in some cases amounted to 10°. Still this success deserves respect; but Mercury presents too many difficulties to be treated satisfactorily in this way.

All these planetary tablets treated by Kugler had come, as was found afterwards, from the ruins of one temple in the town of Babylon. Tablets that were dug up later in Uruk confirmed his results and sometimes augmented them by more details.

It must be added that these were not the sole planetary tables in Chaldean astronomy. A tablet was found containing the computation of Jupiter's longitude for all the consecutive days during some months of SE 147–48, including a station as maximum.[26] The daily motions from which they were formed by summation were themselves obtained by summing a series of regularly increasing values. So the longitudes form a curve of the third degree covering the two stationary points and the intermediate retrograde motion.

In turning now to the tables of the moon, one is struck by their complicated structure, i.e. by the high state of knowledge of the irregularities of the moon's motion and the skill in treating and representing them. But hardly less is our admiration for the ingenuity and perseverance of modern investigators—chiefly the three Jesuit fathers Epping, Kugler, and Schaumberger, who from 1889 to 1935 continued each other's work—to bring to light this ancient science from no more than some few sherds with nearly illegible characters. They found two kinds of systems of computation. From one system, dating from later years, there exists, besides some small pieces, one large table, from SE 180 (130 BC), assembled by Strassmaier from separate fragments; it contains 18 columns of quantities used in the computation. From another more primitive system of computation appearing earlier

(SE 140), only a number of little fragments exist, which nowhere showed the complete computation for a single case. Yet it was possible to reconstruct the working method of this system with the help of a couple of eclipse tables and a procedure text with working directions.

The course of the computation consisted in the derivation, first, of time and place of the exact conjunctions of sun and moon (which we now call 'new moon') and, mostly on the reverse of the same tablet, of their oppositions (full moon). From the conjunctions the day and the aspect of the first appearance of the crescent were derived by the application of corrections. Both eclipses were computed by introducing the variations in latitude of the moon; solar eclipses were derived as well as lunar eclipses.

The problem was intricate because conjunctions had to be derived of two bodies, each progressing with a variable velocity. If one (here the moon) moves rapidly and the other (here the sun) moves slowly, the meeting place depends chiefly on the slow body, and the meeting time on the fast one. That the Babylonian astronomers understood the problem is apparent from their method of attacking it, namely by dividing it into separate stages of approximation. First, the variations of the solar velocity were used in computing the place of conjunction and opposition, i.e. the longitude of the invisible new moon and of the full moon; all quantities depending on this longitude could then be derived. Thereupon, the irregularities of the moon's motion were considered, in order to find the time interval between successive conjunctions or oppositions.

The first columns of the tables giving the longitude of new moon and full moon exhibit the variability of the solar velocity. In the older system it is expressed in the same crude way as in the planetary tables of the first kind; for one part of the ecliptic a greater constant speed, $30°$ per month, is adopted, and for the other half the lesser speed of $\frac{15}{16} \times 30° = 28\frac{1}{8}°$. To get the true mean speed in the long run, the two parts must be taken of unequal length; the large value holds over $194°$, from longitude $163°$ to $357°$; the smaller one over $166°$, from $357°$ to $163°$. It follows that there are $12\frac{83}{225}$ synodic months in the year. We will find later (p. 75) 29^d 12^h 44^m $3\frac{1}{3}^s$ for the length of the synodic month; so the year must be 365^d 6^h 15^m 19^s—thus 6 minutes 8 seconds too much. This is the sidereal year, the revolution of the sun relative to the stars. For the period of the seasons (tropical year) the calendar relations: $12\frac{7}{19}$ months per year, combined with the synodic month give 365^d 5^h 55^m 25^s, also too large.

How did these astronomers acquire their knowledge of the variable velocity of the sun? They may have derived it from the unequal lengths of the seasons. The Babylonians, according to Greek testimony, used a vertical pole for measuring shadow length; thus they could determine the moments of solstice and, as medium points between the solstices, the moments of vernal and autumnal equinoxes. From the Babylonian theory of the two alternative constant values we compute for spring and autumn, situated entirely within the slow and rapid parts, 94.5 and 88.6 days; and for summer and winter, which extend over the limiting points, 92.7 and 89.45 days; conversely, using the same relation, they must have been able to compute their theoretical data on the variations of velocity from the observed lengths of the seasons.

The later less primitive system is similar to that of the planetary tables of the third kind. The differences in longitude between successive new moons or full moons regularly increase and decrease and are expressed by zig-zag functions running between an upper and a lower limit. These limits are $30°$ $1'$ $59''$ and $28°$ $10'$ $39\frac{2}{3}''$; their difference (the amplitude) is $1°$ $51'$ $19\frac{1}{3}''$, their average is $29°$ $6'$ $19\frac{1}{3}''$, one step is $0°$ $18'$; these data allow us to derive the periods used. The number of synodic periods in a year is the number of steps contained in twice the amplitude, the return from maximum to the next maximum; it is $2 \times 1°$ $51'$ $19\frac{1}{3}'' \div 0°18' = 12\frac{299}{810}$. It is also the number of times that the mean arc $29°$ $6'$ $19\frac{1}{3}''$ is contained in the circumference $360°$. This does not give the same result; it is indeed a different thing. The first result is the number of returns to the same extreme of velocity, the other the return to the same longitude. Accordingly, by multiplying the mean synodic arc with the ratio $12\frac{299}{810}$ the result is not $360°$ but $29.8''$ more. Can these ancient astronomers have known that the longitude of the point of maximum velocity of the sun slowly increases? If so, they adopted too great an increase; in reality the progress is only $11.8''$ per year. The synodic arc according to modern data is $29°$ $6'$ $20.2''$; the Babylonian value is only $\frac{1}{120000}$ smaller, so that it would take three centuries to cause a difference of $1°$.

The resulting longitudes of the sun for the successive new and full moons are used to compute some quantities, dependent on this longitude. Among them is the length of daylight, the interval between sunrise and sunset. The procedure text of the older system indicates how it changes with the season, between the symmetrical extremes: 14^h 24^m in summer, 9^h 36^m in winter. This, however, is a curious statement; for it does not fit in with the well-known latitude of Babylon, $32°$ $30'$, for which these extremes should be 14^h 11^m and 9^h 49^m. A fantastic explanation was the assumption that in the distant past the town, or at least its sanctuary, was situated $2°$ farther north and tradition had retained the former values. A more natural though hardly sufficient explanation is that because of the finite solar diameter, the refraction, and the dip of the horizon for the high towers, the length of the daylight was increased at the summer solstice and that for the less well observable winter solstice the symmetrical value had been adopted.[27] Perhaps it was the simple ratio 3 : 2 of these values that persisted as the legacy from times of primitive knowledge.

Noteworthy also is the statement in this instruction: 'at 10° of the Ram the duration is 12 hours: for every next degree thou shalt add 160 seconds . . .' and so on, a line for every zodiacal sign. This implies that the equinox was situated at 10° of the Ram. In two different texts of the later system it is put at 8° and 8° 15′ of the Ram. The Chaldean astronomers did not count their longitudes from the vernal equinox, but from the beginning of the first zodiacal constellation or perhaps from a star. Because the precession changes the place of the equinoxes among the stars, different values for its longitude are given in different times and texts.

It has often been a point of surprise that the Babylonians, notwithstanding their centuries of observation, did not recognize the precession but left this discovery to the Greeks, who borrowed from them so many other data. We have to consider, however, that their minds were not bent on the discovery of new phenomena. It was their task in religious devotion and attentive perseverance so closely to follow and study the course of the heavenly phenomena that for the coming years they could calculate them in advance 'for the festivals'. Even had they perceived that the equinoxes moreover, of secondary interest to them, were displaced compared with the former records, the idea did not occur to them that this was a noteworthy special phenomenon.

Returning now to the main road, to the derivation of the time of new moon and full moon, we have next to consider the moon's variable velocity. We know that, because of the progress of the lunar perigee, the return to maximum velocity, the so-called 'anomalistic' period, takes longer than the true (sidereal) revolution. The tables of both systems have a number of columns intended for this computation. In the later more perfect system a column gives the lengths of the successive synodic periods; they are situated on a regular zig-zag line with limiting points $29^{d}\ 17^{h}\ 57^{m}\ 48\frac{1}{3}^{s}$ and $29^{d}\ 7^{h}\ 30^{m}\ 18\frac{1}{3}^{s}$, and with steps of $1^{h}\ 30^{m}$. The mean of the limiting values is the mean synodic period, $29^{d}\ 12^{h}\ 44^{m}\ 3\frac{1}{3}^{s}$; this is a very accurate value, since it depends on eclipse observations of many centuries. It is the same value mentioned by Ptolemy as having been derived by Hipparchus; it now appears to have been known in Babylon at the time when Hipparchus lived. The difference between the limiting values is $10^{h}\ 27^{m}\ 30^{s}$, and the number of steps contained in twice the amplitude is $13\frac{17}{18} = \frac{251}{18}$. One step in the table, standing for one synodic period, means that from one new moon to the next, the moon makes more than a complete return, nearly $\frac{1}{14}$ more to the same maximum of velocity. The ratio found shows that in

251 synodic periods this point of maximum has been passed an additional 18 times, hence in total 269 times. The same relation which Ptolemy ascribes to Hipparchus, that 251 synodic periods are equal to 269 anomalistic periods, here appears to be the basis of the Babylonian tables. The ensuing anomalistic period, 27^d 13^h 18^m 34.75^s, deviates only 2.7 seconds from the modern values.

The lengths of the months which must be added successively to get the next time of new moon and full moon have first to undergo one more correction, due to the variations in the sun's velocity. The interval covered monthly by the sun may deviate up to 1° from the mean, which is the distance traversed by the moon in two hours. In some further columns this correction is computed, reaching between 4 and minus 2^h 9^m 52^s. Now at last the computation is ready, and, starting from a well-observed eclipse, all further epochs of conjunction and opposition of sun and moon can be established. In this computation the days are all equal, and the hours are reckoned from midnight.

The main objective of the long computation was to find the first day of the month, i.e. the first appearance of the crescent. For this purpose five more columns follow, which have been deciphered by Schaumberger. First comes the interval from new moon to the sunset of the next day, which is usually the moment that the crescent may be expected. By means of the known daily velocity of sun and moon, their difference in longitude is found for this sunset. This quantity alone is not decisive for the visibility of the crescent that evening; what matters is how long after the sun the moon will set. This depends also on the inclination of the ecliptic to the horizon, which varies greatly with the seasons and, moreover, on the moon's latitude north or south of the ecliptic. All these corrections are given in minutes, which is amply sufficient for the purpose. The result is the time the moon remains above the horizon after sunset; it determines the visibility that evening.

Here, then, the goal is reached. The phenomenon that throughout the past centuries had been the chief and most important object of observation by Babylonian astronomers was now entirely controlled by theory as a numerically computed event. In the same way, by reckoning back from the conjunction, the visibility of the disappearing morning sickle could be derived.

It will not be necessary here to enter into the details of the older system, which was at once more primitive and more cumbersome. It is a notable circumstance that there are not only pieces of numerical tables but also a detailed procedure text on their construction, for the use of apprentices. The quantities appearing as nameless numbers in the tables here are designated by their names. It is interesting, linguistically, to see the daily movement of sun and moon given by the expression *zi sha Shamash* and *zi sha Sin*, literally 'life of the sun-god', 'life of the moon-god'. The progress of sun and moon along the sky was to the priest-astronomers the visible manifestation of the life of these light-gods, which they subjected to precise numerical operations. Numbers with six sexagesimals even occur in the tables, not to exhibit an accuracy of thousand-millionths, but because the ratio of $\frac{16}{243}$ needed here can only be expressed in the Babylonian numerical system by a very long number. They also had to face the difficult problem of how to derive the lengths of months from a column of variable velocities, a problem indeed of integration of reciprocals, and they solved it in a cumbersome way by dissecting the process into small parts, hence in a way similar to what in modern times is called 'mechanical quadrature'. But in the columns of this system there are many numbers as yet unravelled.

Computation of eclipses was the other objective of the lunar tables. Eclipses depend on the moon's latitude. Hence some few columns are interposed to serve for the computation of latitude. By their skilful structure they have presented great difficulties to modern investigation; Otto Neugebauer, with the aid of special arithmetical contrivances, was the first to succeed in deciphering and explaining those of the older systems. The Babylonians were well aware that the representation of the variations of the moon's latitude by a pointed zig-zag line was not satisfactory; if the extreme values are right, the inclination near the

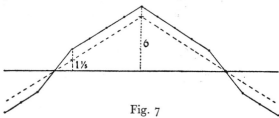

Fig. 7

nodes is $1\frac{1}{2}$ times too small, and at this point especially the correct latitude is needed for the eclipses. If, conversely, the zig-zag line at the nodes is given the correct slope, its maximum is far higher than the maximum latitude of the moon, and this impairs the computation of the crescent. The difficulty was solved by the tables giving a broken zig-zag, which has its inclination doubled at the intersection with the ecliptic. It was constructed in such a way that in a simple zig-zag line between +6 and −6 all values below $1\frac{1}{2}$ were doubled and all above it were increased by $1\frac{1}{2}$. There is no indication of the unit used. If we assume this maximum of $7\frac{1}{2}$ to correspond to 5°, the well-known greatest latitude of the moon, then one unit must be $\frac{25}{36}°$, nearly one-third of the above mentioned *ammat* which is between 2° and 2.5°.

Whereas the nodes are regressing regularly, the longitudes of the sun and, hence, of the full moon have an annual irregularity. In this system its monthly progress jumps between 30° and $28\frac{1}{8}°$; this was taken into account in such a way that in alternating half-years a larger and a smaller slope were given to the basic zig-zag line in the same ratio of 16 to 15. It is clear how difficult and complicated the computation must be when the basic line already consists of broken parts. We can, as in former cases, derive the periods used from the data of the zig-zag line. The average value of a step (per synodic month) is 2, 2, 40, 58, in the same unit in which the basic zig-zag runs between $+6$ and -6; hence the total course up and down amounts to 24. This means that one 'draconitic' period from a zig-zag top to the next top is equal to 24 units, of which one synodic period comprises 26, 2, 40, 58. This is not equal to the simple saros ratio 223 : 242; it is somewhat larger, 223.00112 : 242, and nearer to the true modern value 223.00121 : 242. It is not unexpected that a more precise knowledge is shown here, since, according to the saros-canon, it was already known in the third century BC that the saros relation was not exact. It must be added that a still more accurate value has been used in the later system. Here one course up and down is $\frac{5458}{465}$ times one step; the ratio of the draconitic and the synodic period is here 5,458 : 5,923. It is the same ratio as Ptolemy ascribes to Hipparchus, corresponding to 223.00118 : 242, hence coming still nearer to the true value. In the tables of latitude of this later system there is much of which we do not yet understand the meaning, and special eclipse tables are lacking.

For the older system, however, we can follow the computation of the eclipses to the end. In part of the fragments, following the column of latitudes, another column is inserted in which most of the lines remain blank and a number appears every sixth line only, where the latitude is smallest and an eclipse may take place. This 'eclipse-index', as we may call it, is found by subtracting the moon's latitude before the node from the value (in sexagesimals) 1, 44, 24 (which is the above unit of $\frac{25}{36}°$ which amounts to 1° 12'), and after the node adding it to the same value. Thereupon, in order to have a ten times smaller unit, the result was multiplied by 10, which makes its constant part $10 \times 1, 44, 24 = 17, 24, 0$. This constant part (of approximately 1° 12' in our units) slightly exceeds the sum total of the shadow's and the moon's semidiameter; when the moon's latitude before the node is equal to this value, the moon's disc is tangent to the shadow, a partial eclipse begins to be possible, and the 'eclipse-index' is 0. The slight excess means that the limits of possibility have been taken somewhat too wide, probably in order not to miss an eclipse because of the vagueness of the shadow border, and the small irregularities in the apparent size of shadow and moon. The eclipse-index indicates how far the moon dives into the shadow, expressed now in a unity of nearly ⅛ of the moon's diameter. Thus it starts with zero or a small negative value, increases nearer to the node until it has the value 17, 24 for latitude 0 and continues to increase past the node to 34, 48, twice the constant which indicates the limit where the moon comes free from the shadow and eclipses even of a smaller border part are no longer possible.

Such a list of all successive eclipses and nearly-eclipses exists for the years SE 138 to 160 and has been discussed by O. Neugebauer. The relevant columns are given here in the accompanying table. The first columns give the year and the month, the longitude in the ecliptic and the moon's latitude. Then follow two characters, u and *lal*, which mean 'up' and 'down' or 'positive' and 'negative' and which can best be expressed by our + and −; the first indicates whether the moon's latitude is north or south, the second whether it is near the ascending or the descending node. It is easy to see that the combination + − or − + means: before the node, and the combination + + or − − means past the node; so the 'eclipse-indices' in the last column are found according to the rule given above.[28]

ECLIPSES OF THE MOON

Year & Month			Longitude			Latitude						Eclipse-Index		
2	18	I	9	52	30 ♏	1	7	39	48	+	+	28	40	38
		VII	1	40	♉	1	17	8	12	−	−	30	15	22
	5m	XII	1	22	30 ♎	1	42	34	48	−	+	0	18	12
2	19	VI	20	36	♓	1	53	38	24	+	−	−1	32	24
		XII	20	36	♍		38	30		−	+	10	59	
2	20	VI	9	45	♓		50	9	36	+	−	9	2	24
		XII	9	32	♍		23	14	48	+	+	21	16	28
2	21	V	29	22	♊		17	7	12	−	−	20	15	12
		XI	28	28	♌	1	24	59	36	+	+	31	33	56
2	22	V	19	0	♒	1	24	24		−	−	31	28	
	5m	X	17	24	♋	1	45	47		−	+	−0	13	50
2	23	III	10	30	♑	1	25	50	36	+	−	3	5	34
		IX	6	20	♋		44	2	12	−	+	10	3	38
2	24	III	0	7	30 ♑		18	33	48	+	−	14	18	22
		IX	25	16	♊		17	42	36	+	+	10	21	6
2	25	III	19	45	♐		48	43		−	−	25	31	10
		IX	14	12	♊	1	19	27	24	+	+	30	38	34
2	26	II	9	22	30 ♐	1	55	59	48	−	−	36	43	58
	5m	VII	3	8	♉	1	50	19	12	−	+	−1	8	2
2	27	I	0	52	30 ♏		54	14	48	+	−	8	21	32
		VII	22	4	♈		49	34	24	−	+	9	8	16
2	28	I	20	30	♎		13	2		−	−	19	34	20
		VII	11	0	♈		12	10	24	+	+	19	25	44
		XIIa	10	7	30 ♎	1	20	18	48	−	−	30	47	8
2	29	VI	29	56	♓	1	13	55	12	+	+	29	43	12
	5m	XI	29	56	♌	1	43	27	48	+	−	0	9	22

This list, if we take only the first and the last columns, has some likeness to one column of the saros canon of the preceding period. It
represents a higher stage of knowledge, because, besides year and month
(merely stating that an eclipse may be expected), it gives the 'eclipse-
index' affording information about the character of the eclipse. When
this eclipse-index is between *ca.* 12 and 23, the eclipse is total, and its
duration is longer the nearer the index is to 17 or 18. For values above
23 and below 12, the eclipse is partial; if below zero or above 35, there
is no eclipse. Qualitative descriptions or arrangement in groups by
dividing lines are no longer necessary.

In the 'eclipse-indices' of the above table the variations in the sun's
motion are conspicuously shown. The irregularities found in the
computed values of our preceding chapter (p. 59) recur in the same
way in these Babylonian results; in their upward course larger values
repeatedly precede smaller ones. The irregularities have been somewhat
increased here by the primitive method of sudden jumps in the sun's
velocity. These tables show impressively how far Chaldean astronomy
had risen above the previous level of the saros canon in the numerical
and theoretical control of the eclipse phenomena.

Thus we see how in the ancient land of Shinar a theoretically highly-
developed astronomy came into being in the course of a thousand years.
It is true that our knowledge of this astronomy is as fragmentary as the
few damaged sherds from which it was derived; the meaning of many
numbers we can only guess and of their origin we know practically
nothing. But what we do know shows an admirable and unique system
of science, the only specimen of a highly-developed priestly science. Its
theory does not imply a new system of world structure, or a physical
interpretation, but merely a formal mathematical representation of the
phenomena. If the dictum is true that knowledge is science only in so far
as it contains mathematics. Chaldean astronomy certainly deserves this
name. For it managed to express the motion and phenomena of the
celestial bodies by some few numerical data which were suitable for
predicting these motions over any number of years.

The conditions which enabled the Babylonians to attain this high
performance were, first, the abundance of observational data provided
by a favourable climate and the social demands of chronology and
astrology. No less essential was the fact that they had a mathematical
apparatus, a practical system of reckoning with numbers based on the
place-value notation. As Neugebauer says: 'This gave the Babylonian
number system the same advantage over all other ancient number
systems that our modern (decimal) place-value notation holds over the
Roman numerals.'[29] It was fully developed as early as the first Baby-

Ionian dynasty, probably as the result of a flourishing commerce, and texts of that time already dealt with algebraic problems. Neugebauer points out that this algebraic character of symbolic representation of quantities was probably due to the system of writing, in which the use of Sumerian characters as ideograms played an important role.

The origin of this form of theory is unknown. Chaldean astronomers were known in other parts of the ancient world but only by their names. If these tables are predictions, then they must have been made before the first-mentioned year; which means before 188 BC and 178 BC for the older, more primitive lunar and planetary tables; before 133 BC and 157 BC for the later system with zig-zag functions. The Roman author Pliny speaks of different astronomical schools existing in different Babylonian towns, in Sippar, in Borsippa, and in Ochloe. Strabo in his description of the world mentions the Chaldean mathematicians and astronomers Kidenas, Naburiannuos, and Soudinès; a later author, Vettius Valens, mentions the first and last of these names, and Pliny names Cidenas. The border of the great moon-table of the later system contains the title: 'Computing table of Kidinnu for the years 208 to 210, [made by] Bania [son of] Nabu-balat-suikbi, and Marduk-tabik-siru, son of the priest of Bel in Sippar . . .'[30] Here the same Kidinnu, whom the Romans called Cidenas, is named as the author of the later system of lunar tables; and the third name, Soudinès, has been thought to be identical with the Chaldean Anu-shé-shu-idinna, which appears in another text. So it is often assumed that the older more primitive systems are due to an astronomer Naburianna or Naburmiannu, and that Kidinnu is the author of the more refined system of expressing the periodic irregularities of the sun and the planets by zig-zag functions. They must then have lived in the half-centuries before and after 200 BC or at some earlier time; all evidence about their dates is lacking.

The time covered by these tables extends up to 45 BC; the last tables (ephemerides) used by Kugler dealt with the year 10 BC; the fragment of a table of eclipses reaches to SE 353 (AD 42); afterwards one astronomical text from AD 75 was found. Then all testimony about Babylonian astronomy ceases. The roots of Babylonian culture had died long before; now under Parthian domination; cut off from the Mediterranean, thrown back upon the mid-Asian nomads, its spiritual life also perished. It had, however, exerted its influence upon the Western countries, where in Greece a new astronomy had arisen.

CHAPTER 7

EGYPT

GYPT consists of a narrow sinuous gorge sunk in the Libyan desert—the valley of the Nile—and of the broad delta where the distributaries flow into the Mediterranean. The sun radiates powerfully and continuously upon men and country, and at night the stars shine with unsurpassed brilliancy. No clouds come to pour their fructifying rains. Nor are they needed; the Nile provides fertility. So the sun, bounteous deity, can permeate land and people with its intense glow.

When at the beginning of summer the equatorial rains reach the sources of the White Nile, and the melting of the snow on the Abyssinian mountains fills the Blue Nile, enormous masses of water flow north, and during the following months flood the whole of Egypt. The silt that gives the country its marvellous fertility then settles down. The water is directed by canals and dykes over the entire land and is stored in lakes to provide the artificial irrigation needed for the fields in this rainless desert climate. When the water has flowed away, there comes the time for sowing and growing. The necessity for planned regulation of the water supply gave rise in early times to a central government; the first dynasty (in all there were 26) is usually put before 3000 BC, sometimes even a thousand years earlier. The need to restore the boundaries of the fields after the floods turned the Egyptians into practical land surveyors and created a knowledge of geometry (geometry literally means 'measuring the earth') at an early date. Repeatedly proofs of profound geometrical knowledge, e.g. of the golden section, have been found in the dimensions of the large Pyramids, the slope of their planes and passages, and the figures in their Chambers of the Dead.

The purpose of all early astronomical practice was the measurement of time and this was based originally on the moon. But later this moon-reckoning had entirely disappeared; the only evidence to show that in prehistoric times a lunar calendar existed is found in a few traditions and in the number of the months as 12. The yearly alternation of the phenomena in nature and in work made so strong an impression that in the earliest historic times the solar year already dominated the calendar. The year was divided into three parts of four months: the inundation from July to November; sowing and growing in the wet silt from November to March; harvesting etc. from March to July.

The year of 12 times 30, i.e. 360 days, had already in early times been found to be too short; then five days were added after the twelfth month (additional days, *epagomenes* as they were called in Greek reports) and used as festival days because they were considered to be unlucky for work. Such a year of 365 days could suffice an entire century, longer than the memory of two or three generations. The flooding of the Nile, being influenced by the weather, was irregular, sometimes varying by a month from year to year; the calendar error of ¼ day became a month only after more than a century and so for a long time could remain undetected. But in the long run the priests, who had to take care of the calendar, must have perceived that the floods arrived in later and later months. Then a distinction was made between the agricultural ceremonies of sowing and harvesting, and the calendar rites, e.g. New Year's day. The calendar continued with years of 365 days, and the Nile floods passed successively through all the twelve months.

The regularity of this progression in the agricultural years showed itself in the heavenly phenomena also. In ancient times it had been perceived that when in the capital, Memphis, the Nile began to rise, Sirius, the brightest of the fixed stars, became visible for the first time in the morning twilight just above the eastern horizon. This coincidence was too remarkable to be accidental; in it there must be some purpose of world order. Thus the divine star 'Sothis', by its reddish morning rise, became for the Egyptians the author or harbinger of the beneficent flood. The heliacal rising of the star is an entirely regular phenomenon; its period of 365¼ days is the mean agricultural year independent of the irregularities of the river's rising. Every four years it advances one day on the monthly dates; and after 1,460 ($=4 \times 365$) years, Sirius again rises on the same date. In the years 2770 and 1314 BC this fell upon New Year's day; the interval is somewhat less (1,456 years) because of Sirius' slow displacement relative to the equinoxes. In later time this period was called the 'Sothis period', which was allegorically expressed in the tale reported by Herodotus of the holy bird Phoenix returning every 1,460 years, burning itself and arising rejuvenated out of the flames. In practical chronology it was not used.

It must be remarked that Sirius could play this role only in the ancient history of Egypt. Since agriculture and flood depend on the equinoctial dates, as in our calendar, and Sirius, like all stars, is displaced relative to the equinoxes through precession and proper motion, our dates for its heliacal rising change continuously but slowly. Whereas the first rise of the water at Memphis occurs about June 25, the heliacal rising of

Sirius in 3000 BC was June 22; in 2000 and 1000 BC, June 30 and July 18. Thus in these later years it lost its annunciatory value.

The question may be raised as to why the Egyptians always held fast to this complicated calendar with two different mutually shifting years and festivals. It would have been so simple to install an agricultural Sothis year with fixed month dates by adding six instead of five days every fourth year. Sometimes, when secular rulers attempted such changes, these were frustrated by the opposition of the priests, who defended the sacred traditions. Since they were the experts in regulating calendar and ceremonies, this monopoly of subtle knowledge gave them a social power they did not want to relinquish.

The paramount significance of Sirius in Egyptian life assured it a place among the deities; Isis, the goddess of agriculture and fertility, revealed herself in the Sothis star. Other major deities of the Egyptian pantheon also found their visible representation in the stars; such as Osiris in the constellation of Orion, which had already risen high in the sky when Sirius appeared. The sun, all-powerful ruler of the sky, as the sun-god Ra, occupied the predominant place in the divine world. It is well known how King Amenhotep IV (Ikhnaton), in his struggle against the power of the priestly hierarchy, raised the sun to the rank of sole deity, under the name 'Aten'; the decline of the royal power foiled this attempt soon after his death.

It is difficult to ascertain the real state of knowledge of the starry heavens in this ancient Egypt. In the rock sepulchres and on the lids of mummy caskets lists are found of the so-called 'decans', ten-day periods or weeks, sections of the months, each under the protection of a deity represented by a star or a star group. The structure of these lists points to their use for determining the night hour. But there is considerable uncertainty about the objects represented by the names, and it is difficult to identify the pictures. There are pictures of numerous human and animal figures which probably represent constellations; but the stars themselves are missing, since only the celestial beings depicted here were of real importance to man. In the place of the Great Bear we see a bull, or only its foremost leg, the embodiment of the evil god Seth; moreover, there is a hippopotamus (the Little Bear), a crocodile, a lion, a sparrow hawk; but what stars belong to them is mostly a matter of conjecture. The most renowned and most-often reproduced picture of the sky, copied from the temple of Denderah, is from the later Hellenistic period, with a strong influence of Babylonian and Greek science.

There is in Egypt no trace of a development of astronomy to scientific stature. After the calendar had been fixed in a rigid system of years, the concern with the stars could be restricted to the heliacal risings of Sirius. Observations of the moon could no longer serve a useful purpose. The hours of daylight could be read from the sundials and at night could be determined by water clocks. Egypt can show us how little a science of the stars is fostered even by an ever brilliant sky unless that science finds a practical basis in human life and activity.

CHAPTER 8

CHINA

CHINESE authors ascribe great antiquity—earlier than 2000 BC—to astronomy in their country, but like all that is written about that time, this belief is chiefly legendary. Real history only comes into consideration after 1000 BC. The fertile loess hills and plains along the middle course of the Hwang Ho, the Yellow River, were the site of the first Chinese culture, when the south-eastern plains along the Yangtze Kiang were unreclaimed swamps and jungle. In the inhabited territories, centred in Honan or Shensi, agriculture was the chief occupation under a feudal organization of society. Under the nominal supremacy of the Chou dynasty, warring lords and princes ruled over the peasants. With the development of urban handicraft and commerce in the following centuries, the feudal order gradually dissolved and there emerged a new class of intellectuals—scribes at the courts, experts in old wisdom, officials and counsellors of the princes—mostly from the impoverished nobility. Among them were famous philosophers such as K'ung Fu-tze (Confucius, 551 to 470 BC) and Meng-tzu (Mencius, 372 to 288 BC), who transformed the old ideals of knighthood into teachings of virtuous behaviour as the basis of a Good State; virtue means good conduct according to the rites, propriety, Tao.

When the conquest and reclaiming of the oft-flooded eastern marshland demanded continual central regulation of the water by dykes and canals, feudal decentralization had to give way to a centralized empire governed by officials called 'mandarins'. The old culture was adapted to the new conditions and consolidated in fixed norms under the Han dynasty (205 BC to AD 221). A still higher stage of art and science was reached later under the T'ang dynasty (AD 618 to 907). Such epochs alternated with centuries of disruption and confusion, civil war and agrarian revolts, and with attacks and conquests by barbarian nomads. These nomads, as probably did the original inhabitants, came from the north-western Asian steppes, from which side the country is not protected by oceans and mountain ranges. These warlike Mongols infused fresh blood into the ruling class, but their relatively small numbers were soon absorbed and assimilated into the mandarin class which governed

86

the millions of farmers. Such alternations of conquest, power and prosperity with decay and confusion occurred over and over again up to modern times. The Mongolian nomads, who in western Asia acted only as destroyers of culture, built in eastern Asia a culture of their own, strong enough, through its agrarian basis, to assimilate wave after wave of conquerors. This determined the slow autonomous development and gave it a strongly conservative character.

Astronomical ideas play their important part in this culture. It is difficult to arrive at a true picture of their development because later Chinese authors try to ascribe an earlier origin to later ideas in order to give them venerability and invariable validity. A general burning of books had taken place in 213 BC at the command of the strong upstart emperor, Shih Huang Ti, probably with the purpose of breaking the influence of feudal traditions and norms. In the ensuing Han period however, the old philosophers were restored, as were the ancient books; this meant that they were largely rewritten in distorted form. In the old burned books was a note about two colleges of astronomers named Hsi and Ho which for having taken sides in a civil strife, were exterminated by the victorious party. The later edition transformed this into the anecdotal moralizing tale of two astronomers Hi and Ho who, in a merry life, neglected their duty, failed to predict a solar eclipse and were punished by decapitation. Since the day (the first day of autumn) and place (the moon-station Fang, i.e. the Scorpion's head) had been added to the tale, modern authors calculated that it related to the eclipse of October 22, 2137 BC. But it is clear that in such ancient times there could be no question of predicting a solar eclipse, and it is difficult to judge the reliability, if any, of the original tale.

The unity of time-reckoning was provided of old by the lunar period; this lunar calendar has always persisted. But since agriculture is bound to the year, intercalation was an object of incessant care. Since 350 BC the length of the year was known to be 365¼ days. From this date a rule existed that the year begins with the heliacal rise of Spica, i.e. in autumn. From later centuries a prescript was given that the full moon at the right-hand side of Spica belongs to the following year, which therefore begins in spring.

Astronomical—or rather cosmological—ideas were deeply interwoven with the Chinese world concept, which embodied only simple knowledge of the celestial sphere with its daily rotation, of the pole and the equator; the horizon played a role, but the planets and ecliptic were hardly mentioned. Heaven and earth were closely related and in complete harmony. This found expression in a state religion with its ceremonials. China is the centre of the flat earth, the 'empire of the midst', corresponding to the celestial pole, the centre of heaven; there the god Shang-ti rules as does the emperor on earth. The emperor is the 'Son of Heaven'. He maintains the harmony of heaven and earth by his actions in following precisely the ritual and the prescripts of his fore-fathers. Disorder on the part of one brings disorder on the other. This means that not only are irregularities in heaven the cause of calamities on earth but also that the evil actions of man bring about disturbances in nature and in heaven. Eclipses and comets are an indication that the emperor or the officials have sinned, governed badly or neglected the right ceremonial. 'When a wise prince occupies the throne, the moon follows the right way. When the prince is not wise and the ministers exercise power, the moon loses its way. When the high officials let their interests prevail over public interest, the moon goes astray toward north or south. When the moon is rash, it is because the prince is slow in punishing; when the moon is slow, it is because the prince is rash in punishing.' This was written in the astronomical work of Shih-shen in the fourth century BC.[31]

Chinese philosophy sought to find symmetry and relationship in life and world. The four directions, east, south, west and north are corre-lated with the four seasons, the four parts of the day, the four sections of the celestial equator. Added to the 'middle', they constitute a pentad correlated with other pentads in different realms; elements or basic matters, plants, colours, parts of the body, musical instruments, planets, flavours. Combined with another series of twelve differing kinds, it produces a period of 60 years for use in chronology. There is a kind of arid harmony of schematic order in this world doctrine with its asso-ciated rites.

The mental condition and mode of thinking of the Mongols of East Asia are entirely different from that of the Semitic and Aryan peoples of the Near East. In the countries adjacent to the Mediterranean, the rich variety of wooded mountains and fertile plains, of burning deserts and cool oases, of capricious rocky coasts and green islands, of varied landscapes and scenery, of bounteous and harsh natural forces, produced a flowering fantasy, creating a gay procession of gods and goddesses as their personifications. In the Far East it was just the contrary: the heavy weight of endless monotonous steppes and the mighty mountains of a gigantic continent discouraged fantasy and in a difficult, often cruel struggle for life produced a chilly harshness of thought and feeling. The patriarchal family system of the roaming nomads, who had no fixed sites, no villages, at most caravan towns, prevented the rise of that multitude of tribal deities which in agrarian countries formed the basis of a rich pantheon. Worship and offerings were confined to the ghosts of parents and forefathers; the smoke rose to the vault of heaven;

heaven was invoked as the highest power, with sun and moon in addition.

This Mongolian character persisted in China; religion, here a centralized state religion, did not find its expression in supernatural poetical fantasy but in practical prescripts of virtuous behaviour. In China temples are found where offerings are celebrated not to gods but to philosophers credited with holiness. The fate of man was not ruled by gods; good or bad luck was determined by virtue, by following the commands of heaven. The emperor set the example; he extolled virtue, and when he was virtuous himself, order and prosperity reigned. The imperial ceremonial providing for this was an essential part of governmental practice. The intellectuals—the cultural leaders—were not priests but mandarins, who, from their study of old literature, knew and practised virtue, i.e. the correct ritual behaviour; they were officials with a slight trace of state-priestly functioning. The often praised beauty of Chinese poetry does not consist in flowery imagination but in profound and harmonic worldly wisdom.

In the state organization the colleges of astronomers were responsible for orientation and time-reckoning, to ensure that all should be done in the right order and at the right time in keeping with prescripts. Temples and palaces were built in exact accordance with the four points of the compass. In his official acts the emperor was seated at the north side and looked towards the south; and so did the ruling officials. According to the 'book of rites', it was the duty of the chief astronomer to inform the emperor when the first month of each season was due to begin; in spring the emperor, at the head of the court, went to the eastern suburb, solemnly to inaugurate the new year. He prayed to Shang-ti for a good harvest; then on the first propitious day indicated by the astronomer, he went to the fields and with an ox-drawn plough cut three furrows and was followed in this by his ministers. The college of astronomers, in order to fulfil its task, had to keep continual watch, to attend to the celestial phenomena, to note everything they saw, and to report on the most important ones; for this purpose there was an observatory in the palace. They saw to it that their instructions were sent to all parts of the empire. They prepared the calendars, which were offered with great solemnity to the emperor, who ordered their publication. These contained the lucky and the evil days for all important actions.

Astronomical knowledge arose from the need of time-reckoning in the Chou period, the century of philosophers and literary men. The seasons were known by the positions of particular constellations: midsummer by Antares in the south at sunset, winter by the belt of Orion; the year began when the Bear's tail pointed downward. Astrology played a big role; the so-called 'astronomical' works of this time were works of astrology, collections of statements on the meaning of the phenomena. Because emperor and state were the objects of the omens, the entire terminology bore the character of the royal household; a group of stars was named a 'Palace' and the stars carried the names of court officials. In the 'Purple Palace', a group of stars of the Little Bear, near the pole of that time, the brightest (β Ursae Minoris) was called the 'Emperor'; the second in brightness (γ Ursae Minoris), the 'Prince-Royal'; a smaller star was the 'Empress'; another very faint one was called the 'Axis of Heaven'. The North Pole and its surroundings, at the top of the heavens, was the seat of the emperor (plate I).

The astrological work of Shih-shen, just mentioned, of which only some fragments are preserved by later authors, is said to have contained a list and description of at least 122 constellations with 809 stars; this catalogue, then, would be earlier than the time of Hipparchus. The constellations were other and smaller groups than ours. The stars in later maps and globes are represented by equal dots, without distinction in brightness, and the maps show differences. Thus the identification of the stars has been difficult.

Apart from the polar region, it is the 28 moon-stations, *hsiu*, already mentioned, mostly situated near the ecliptic, which occupy an important place. Afterwards they were often considered as sectors of different width, measured along the equator, all of them reaching from pole to pole. The ecliptic was not known as a circle but as a belt through which the moon and the planets travelled. A division into twelve constellations is also mentioned, partly as the monthly path of the sun and partly as the yearly course of Jupiter. That solar eclipses were connected with the moon was known to Shih-shen, but they were not explained as occultations; it was the power or essence of the moon which, when it was invisible at the end of a month, vanquished and suppressed the power of the sun. Of lunar eclipses still less was known; Shih-shen makes them occur at many different dates of the month. His notions of the course of the moon have been quoted above.

The motions of the planets were perceived only in their most general features, as variable direct retrograde velocities and stations. They were often represented rather well, but sometimes—if translation and interpretation are right—entirely wrongly. In a summary concerning Venus, the phenomena as evening star are repeated with exactly the same intervals as the morning visibility; here the Babylonian Nin-dar-anna text of a thousand years earlier was superior. At another place the succession of five Venus periods in eight years is well recorded: 'When she appears in the east, she appears in the station Ying-she [α Pegasi] and disappears in the station Kio [Spica]; she appears in station Kio and disappears in station Pi [Aldebaran]; she appears in station Pi and disappears in station Ki [γ Sagittarii]; she appears in Ki and disappears in Liu [ζ Hydrae]; she appears in Liu and disappears in Ying-she.'[32] In every cycle she proceeded 17 stations, and the next morning reappeared in the same station where she had disappeared nine months earlier. Of Mercury no motion is mentioned, only its appearance and disappearance with the moon station for every month and the astrological meaning. Concerning Mars only some quotations have survived: 'When Mars is retrograding in the station Ying-she, the ministers conspire and the soldiers revolt.'[33] In a work of a century later a more detailed description is given. The direct and retrograde motion of Jupiter and Saturn were known and used as presages; of Jupiter also the synodic period and the period of revolution, 400 days and 12 years, were known; and for Saturn they are found in a somewhat later work.

In the first centuries of Han the data became more accurate. Now the height of the sun was measured by the shadow of an 8-foot vertical bamboo pole; in the summer solstice it was 15.8 inches (each of $\frac{1}{10}$ foot); in the winter solstice 131.4 inches. These correspond to altitudes of 78.8° and 31.4° from which the obliquity of the ecliptic 23.7° is found and a polar altitude of 34.9°, indicating an observation site in Honan. The times of solstice thus determined were used for correcting the calendar. A conference of astronomers was held in 104 BC in order to improve time-reckoning; from this time calendars were published regularly. An important part in this activity was taken by Ssu-ma-Ch'ien, who published an astronomical work in which the moon's motion was treated more accurately. He was assisted by the astronomer Lo-hsia-Hung, who was the first to construct an armilla sphere consisting of rings for equator and meridian; only some centuries later it was improved by adding a ring for the ecliptic. The circles, curiously, were divided not into 360° but into 365¼°. He measured the extent of the 28 moon stations by meridian passages and a water clock. He is said to have been the first to ascertain the regularities of the planetary motions.

Ideas on world structure were said by later authors to have originated in this period. For Lo-hsia-Hung heaven was similar to an egg and it enveloped the earth, which, like a yolk, swims on the water beneath; heaven rotated around the pole, and the seasons were caused by shiftings of the earth up and down, to and fro, 'the earth moves constantly but people do not know it; they are as persons in a closed boat; when it proceeds they do not perceive it'.[34] Another doctrine, according to Yang-Hiong (53–18 BC), regarded heaven as extending like a bell glass over the vaulted earth; heaven rotated but did not come below the earth; night was caused by the sun moving so far away that its rays could not reach the earth. Later authors illustrated these ideas by computations of dimensions and distances in Chinese *li*.

In the 'Annals of Han' (*Han-shi*) there is an astronomical part[35] taken from a textbook composed in AD 25 by Liu-Hsin, known in history as an adviser in the reforms of the Emperor Wang-Mang. Compared with the work of Ssu-ma-Ch'ien of one century earlier, it shows considerable progress. The synodic month was given as $29\frac{43}{81}$ days (29^d 12^h 44^m 27^s, 23 seconds too large). The relation 19 years $=235$ synodic months $=254$ sidereal months was assumed to hold exactly so that a year was $365\frac{385}{1539}$ or 365.25016 days (11 minutes too large). Eclipses were computed according to the relation that 23 passages of the sun through the nodes were equal to 135 synodic months, so that these passages returned after $\frac{135}{23} \times 29\frac{43}{81} = 173\frac{1}{3}$ days; this eclipse period is 35 minutes too long. Extensive directions, in the form of examples, were given for computing new moon and full moon, the length of a month (29 or 30 days), the solstices, the relative position of sun and moon, the intercalation and the eclipses. The computing apparatus consisted of a system of numerals with place value to allow the writing of large numbers; and their computing method consisted in removing all fractions by multiplication of the periods, so that they had to work with entire numbers of seven and more figures. The periods of the planets are also given by the ratio of large numbers: Saturn 4320 : 4175 ($=377.93$ days), Jupiter 1728 : 1583 ($=398.71$ days), Mars 13824 : 6469 ($=780.52$ days), Venus 3456 : 2161 ($=584.13$ days), Mercury 9216 : 29041 ($=115.91$ days). The errors of these values are only -0.16, -0.18, $+0.59$, $+0.21$, and $+0.03$ days. Here also directions are given for computing the visibility, the place, the appearance, and the disappearance of the planets. Yet the practical results must have left considerable errors, for in the following centuries the need of correcting the calendar was frequently mentioned.

This considerable progress in astronomical knowledge has led to the assumption that influences from Babylonian or Greek science had been at work; similarity with Babylonian omens was sometimes noted. Even cultural interchanges with the Maya civilization in Central America have been suspected. Yet there is too much that is genuine in Chinese astronomical methods to make such borrowing probable.

The first centuries AD again brought notable progress. The inequality of the seasons and the inequalities in the moon's motions began to be known; as well as the precession, designated as $1°$ per 50 years by Yu-hsi. The inequalities in the planetary motions were treated in a later book from AD 550. In these developments Western influences certainly played a role. In the first century AD Chinese conquests in Central Asia opened up connections with Iran and led to regular overland commerce with the Roman Empire. Greek science, introduced into India by way of Bactria, could now reach China along the same routes as Buddhism. In later Chinese sources it is mentioned that in AD 164 persons

acquainted with astronomy came from Near Asia to China and that in
AD 440 Ho-Tsheng-Tien, who is known to have determined the latitudes of many places by solar altitudes, was taught astronomical science,
especially concerning eclipses, by an Indian priest.

Under the Sui and T'ang dynasties astronomical practice developed
still further. Dshang-Dsisin is mentioned as having made observations
with an astrolabe about AD 550, and the astronomer I-Hang (AD 724) as
having constructed shadow-poles, quadrants and armillas. There
were now regular connections with the Mohammedan states in Central
and Near-Asia which resulted in more accurate values of the periods,
the inclination, and the nodes of the lunar orbit. Several Chinese astronomers, authors of textbooks and observers, are mentioned; but they no
longer represent an independent autochthonous development. First
among them is Shu-ging (1231–1316), a capable mathematician; he
improved tables of the sun and the moon, secured new instruments for
the observatory and made new observations of solar altitudes and
solstices. He determined the obliquity of the ecliptic 23° 33$\frac{2}{3}$', which is
only 1$\frac{1}{2}$' too large, and determined the latitude of Peking at 40°, too
great by $\frac{1}{10}$°. This took place in the reign of Mongol Emperor Kublai
Khan (1260–94), about whom we are informed by Marco Polo. Since
at the same time the Mongols under the Emperor's cousin Hulagu ruled
in Near-Asia, there was considerable commercial and cultural exchange
between East and West. An Arabian astronomer, Dshemal-al-din, came
to China in 1250 and introduced Western instruments and observing
methods. But they did not find favour; Chinese science retained its traditional and conservative character. Notwithstanding many attempts to
improve them, the calendars remained unsatisfactory until the Jesuits
came to China in the sixteenth century and were commissioned to
establish them reliably. Thereafter there is no longer any question of
independent astronomical development in China.

This medieval tide of practical astronomy in China has provided a
great number of observational data. Conjunctions of planets with one
another and with stars were recorded, though not in exact measure.
Among the irregularities in the sky watched and carefully noted were
the comets as well as the eclipses, haloes, aurorae and meteors. The
observations of comets are interesting: the date of observation, the
constellation, and the movement were usually recorded. These Chinese
observations of comets have been important for later science because
attention was given to them nowhere else during these centuries. Thus
comets appearing in 989, 1066, 1145, and 1301 could be recognized as
earlier appearances of Halley's comet.

Highly interesting was a new star, 'Nova', which appeared in 1054
in the constellation Taurus. In the twentieth century, measurements of

the telescopic expanding 'Crab Nebula' in Taurus made it probable that it had originated from a stellar explosion eight or nine centuries before. Nowhere except from China could notice of such an event be expected, and, indeed, Chinese annals mentioned the appearance of a 'guest-star' on exactly that spot in Taurus, and described its history. It appeared in June 1054, was visible in the daytime, and disappeared two years later, in 1056.[36]

Astronomy in China, because of the remoteness of the country, has had no perceptible influence on the progress of science elsewhere. It developed independently at first and produced remarkable original methods; afterwards it became increasingly subject to the influence of Western science. Its importance is confined mainly to supplying data of observation for a time when they are missing elsewhere.

CHAPTER 9

GREEK POETS AND PHILOSOPHERS

WHEN the Greeks appeared on the scene of history, the stars held for them the same significance as they had for other primitive peoples. Celestial phenomena formed part of their surrounding world, which they needed to know for their work, for orientation in navigation as well as for time-reckoning. This is clearly seen in the great poetic works from the earliest times of tribal kings, in Homer's *Iliad* and *Odyssey* and in the *Works and Days* of Hesiod.

Homer mentions some stars by name; in the *Iliad* he speaks of the Evening Star and the Morning Star, the Pleiades, Orion, the Great Bear, and 'the star which rises in the late summer . . ., which is called among men Orion's Dog [i.e. Sirius]; bright it shines forth, yet it is a baleful sign, for it brings to suffering mortals much fiery heat'.[37] And in the other work he says of Odysseus: 'No sleep befell his eyelids while regarding the Pleiades and the late-diving Boötes, and also the Bear, which they also call the Wain, which always turns around at its own place and watches Orion and alone does not participate in bathing in the ocean. For the goddess Calypso had bidden him to keep it at his left hand when sailing over the sea.'[38] The Little Bear was evidently not yet known; at that time the celestial pole was located between the Bears, far remote from our present Pole Star.

In his description of country life, Hesiod indicates the phenomena of the stars which accompany agrarian activities in different seasons of the year. When the Pleiades rise [heliacal rising, May 10th of our calendar],* it is time to use the sickle, but the plough when they are setting [in the morning, November 12th]; 40 days they stay away from heaven [line 382]. When, 60 days after the winter solstice, Arcturus ascends from the sea and, rising in the evening, remains visible for the entire night [February 24th], the grapes must be pruned [line 566]. But when Orion and Sirius come in the middle of heaven and the rosy-fingered Eos sees Arcturus [September 14th], the grapes must be picked [line 610]. When the Pleiades, the Hyades, and Orion are setting [November 3rd, 7th, 15th], then mind the ploughing [line 615].

* Our remarks to a quotation are put in brackets.

When the Pleiades, fleeing Orion, plunge into the dark sea [morning setting, November 12th], storm may be expected [line 619]. Fifty days after the sun's turning [i.e. the summer solstice] is the right time for man to navigate [line 663]. When Orion appears [June 17], Demeter's gift has to be brought to the well-smoothed threshing floors [line 699].

Such were the beginnings from which Greek astronomy developed. What we know of it, through the scanty information reaching us from later centuries, does not deal with observations of celestial bodies but with opinions of philosophers on the world. Here we meet with a mode of thinking and a people quite different from those of Mesopotamia and Egypt, the results of completely different conditions of life.

Greece is a deeply indented, mountainous peninsula, with small areas of arable land on the coasts or along the short rivulets, isolated from one another by steep rocks and thickly wooded mountains. It is surrounded by numerous islands which form a bridge toward Asia Minor. The Hellenic tribes penetrating from the north found here a good climate and good harbours but insufficient land to cultivate for an increasing population. They became navigators, populated the islands and migrated to the opposite coasts. Overseas trade became their chief occupation and a source of wealth for such towns as Corinth on the mainland and Miletus, Ephesus and Smyrna on the Anatolian coast. By their commerce they soon dominated the Aegean and, as competitors of the Phoenicians, the eastern part of the Mediterranean. Along with the trade went adventurous people who settled on foreign shores; they founded colonies in southern Italy and on the Black Sea, as far as Marseilles and the Crimea. Some colonies later became centres of new expansion.

Here new forms of economy and politics arose. Settlers in new countries, naturally already the most energetic and independent of their peoples, are least dominated by traditions. Through changed life conditions and racial intermixing, new ideas came to them more easily than to those who remained behind. Then there were the new forms of work and ways of living; industries developed in the trading centres, and there emerged a burgher class of artisans and industrialists who manufactured products for remote markets and often travelled in person to sell them. Navigators and traders develop a state of mind different from farmers bound to their plot of land. Farmers are conservative through close contact with constantly recurring experiences and conditions; thought and feeling are bound to tradition and petrify. But for travellers the contact with various countries and peoples opens up new vistas; it makes thought mobile and free and the mind receptive to new ideas. Craft and competition make them inventive; techniques develop and physical knowledge makes them critical of traditional beliefs. Thus among the Ionians of the Anatolian coast and the Greeks of southern Italy we find the first philosophers to proclaim new theories of the world.

These Greek colonists, living between continents, between the barbarian tribes of Europe and the despotic empires of Asia, were the first 'modern' men, comparable to the burghers of later European states. They were the first to free themselves from the worship of the Olympian gods. Here for the first time a vigorous middle-class spiritual life sprang up that found expression in a flourishing poetry, and a strong middle-class independence of mind that found expression in a flourishing philosophy of nature. The strong individualism that in later history determined the character of civilized man first developed in the free citizens of Greece and first of all in the Greek colonies. Later generations could borrow this spiritual character from predecessors when their economic and social life developed similarly; thus western Europe in the Renaissance took it over from antiquity. The Greeks alone could not borrow from others; in Greece, especially in the colonies, this modern character evolved of itself autochthonously. This is the significance of ancient Greece and its culture for world history. It has sometimes been assumed that the first Greek thinkers took their ideas from Egypt or Babylon; but, considering the different social conditions, we may be sure that except for some minor details this is out of the question.

There is still another difference. Egypt and Babylonia consisted of extensive fertile plains, where the necessity for central regulation of the water by dykes and canals required a strong government. Moreover, spiritual leadership was concentrated in a powerful hierarchy of priests which practised science but bound it to strong religious traditions. In Greece no such centralization of political power was necessary, nor was it possible. The small, isolated, warring city-states remained independent; each had its local deities—collected into a hierarchy on Olympus by the poets only—and its own priests, with purely local functions and little influence. This state of affairs put its stamp upon the spiritual life in Greece, including Greek science. In the Orient science was a priest-science, chained to traditional worship. In Greece it was a lay science, freely sprouting and expanding in the versatile minds of colonists and seafarers.

In books on the history of Greek astronomy we find, first, in the seventh and sixth centuries BC the names of a number of famous philosophers, all from the towns of Asia Minor and southern Italy. But our information about their doctrines and opinions is scanty and often contradictory. This should not cause surprise when we consider how much of the writings of antiquity was lost and how little has been preserved. Certainly the works of Plato and Aristotle have been pre-

97

served; they were highly esteemed and often copied; Aristotle often gives information about the opinions of former philosphers. However, since he mentioned them chiefly to add to his own reputation, his account is seldom entirely objective. His successor, Theophrastus, wrote a book on *Physical Opinions*, and his pupil Eudemus wrote a history of astronomy; both are lost. These and some other sources were the basis for a collection of *Opinions* in the last century BC, but this was also lost; it served as the source for a work of Aëtius *On Opinions*, c. AD 100, and was extensively used by all later authors. Excerpts were published in the second century under the title *Opinions of the Philosophers*, by an unknown author who ascribed them to the famous biographer Plutarch. These and other sources served the Christian authors in the next centuries in their polemics against and their refutation of pagan philosophy. Since this Christian literature was well preserved, it could be used to restore important parts of the later collections, especially of Aëtius. Some quotations by well-informed Roman authors like Cicero may be added. But we can understand how incomplete, sometimes contradictory, often distorted, or even entirely wrong, were the ideas thus ascribed to the ancient thinkers, and how little success modern scholars could have when attempting to construct profound world conceptions from them.

These thinkers are called 'philosophers', because their speculations on the nature of things deal with the world in its entirety. They were, however, as the English classicist T. B. Farrington explains, not recluses engaged in pondering upon abstract questions; they were active men, taking part in practical life; the range of ideas, the modes of thought which they applied were those they derived from their active interest in practical affairs. When they turned their minds to wondering how things worked, they did so in the light of everyday experience, without regard to ancient myths. 'With the Milesians technology drove mythology off the field. The central conception of the Milesians was the notion that the whole universe works in the same way as the little bits of it that are under man's control.'[39] This implies that their explanations are more interesting from the physical than from the astronomical point of view.

Concerning the oldest of the Ionian philosophers, Thales of Miletus (624–547 BC), it is said in the most reliable sources that he postulated water as the first principle of all things and that the earth floats on the water like a flat disc. Later authors ascribed to him a knowledge of the sphericity of the earth and of the true cause of lunar eclipses; but this contradicts other more trustworthy information. For his compatriot Anaximander (611–546 BC), the indeterminate or the infinite was the principle of the world, and the cylindrical earth floated free in the midst of space. What he said about sun, moon, and stars, to the effect that they are hollow wheels, full of fire shining through narrow openings which are closed during eclipses, is rather obscure. According to Anaximenes, also a Milesian (585–526 BC), the flat disc of the earth is carried by the air, and the sun and moon, likewise flat discs, float in the air. The stars, he said, do not move under the earth but around the earth as a cap moves about the head; and the sun does not go beneath the earth but is lost to view behind high northern mountains and is invisible because of its great distance. Another statement, that the stars are fixed like nailheads in the crystalline vault, does not accord with this.

Still more singular, astronomically, are the opinions attributed to the poet Xenophanes of Colophon (570–478 BC): that the sun is born anew every day at its rising, and at night becomes invisible because it retreats to an infinite distance; that eclipses occur through extinction of the sun, and that the moon shines by its own light and disappears because it is extinguished each month. Similarly, Heraclitus of Ephesus (about 500 BC) is reported to have said that the sun is a burning mass, kindled at its rising and quenched at its setting; 'the sun is new every day'.[40] And these same two philosophers are considered most important and profound thinkers! Xenophanes attacked traditional religion by proclaiming broader and freer ideas, and Heraclitus, 'the obscure', is even now highly esteemed for his doctrine of eternal flux and change in all things in the world. That they expressed opinions which can be disproved by any attentive scrutiny of the sky shows that they were not astronomical observers but imaginative thinkers and as such have little importance for the history of astronomy.

Astronomy in the real sense of the word could be found in the statement that Thales predicted the solar eclipse of 584 BC. Herodotus relates that when the Medes and the Lydians fought a battle at the River Halys, the day turned into night, and that Thales of Miletus had foretold this eclipse within the period of a year. But the indefinite nature of the latter part of this statement must arouse distrust. Moreover, Thales could only have predicted such a phenomenon if he had learned either the principle of eclipse prediction or simply from Babylonian sources the expectance of this special occurrence: but at that time the Babylonians themselves, although they could predict lunar eclipses at short intervals, were not yet able to solve the far more difficult problem of solar eclipses. So this report should probably be dismissed as a fable.[41]

A more developed world picture appears with the Greeks in southern Italy. Here we first meet the prominent figure of Pythagoras of Samos (580–c. 500 BC). No writings of his are known; legends by later authors tell how in Egypt he learned the wisdom of the priests. He then founded a school or a kind of monks' order in Croton which, as part of the ruling

99

aristocracy, was later dispersed by a popular uprising. Many new scientific discoveries are attributed to him, especially in mathematics: geometrical lemmas, the golden section, the theory of numbers and the arithmetical basis of musical harmony. Aristotle never speaks of Pythagoras but of the Pythagoreans, and it is not certain to what extent their opinions were due to the master himself. Knowledge of the sphericity of the earth is generally attributed to him, as well as to the later Alcmaeon of Croton and Parmenides of Elea (probably 504–450 BC); this knowledge quite naturally arose among observant Greek navigators. It led to an understanding of different climatic zones on the earth which later authors attributed to Pythagoras and Parmenides. A sentence which has been preserved of the later Bion of Abdera states that there are regions on earth with days and nights both lasting six months; it shows that the consequences of this theory were understood. Pythagoras is said to have been the first to put forward the identity of the Evening Star and the Morning Star; he (or Oenopides of Chios) first recognized the obliquity of the ecliptic and the orbits of the planets therein. With Empedocles of Agrigentum (probably 494–434 BC) we find the doctrine of the four elements, fire, air, earth and water, moved by Love and Strife.

The cosmological ideas of Anaxagoras of Clazomenae (*c.* 500–428 BC) are more clearly defined. He settled in Athens, then the powerful political and cultural centre of Greece in its Golden Age of literature and art. A friend of Pericles, he was renowned in antiquity as the philosopher who proclaimed 'mind' as the moving principle of the universe. He was the first to state clearly that the moon shines by the light it receives from the sun and that lunar eclipses occur when the earth (or another dark body!) intercepts the sun's light. He assumed the earth's surface to be flat, the upper surface of a cylinder freely suspended in space, whereas the rotation of the celestial sphere, to which the stars are affixed, carries them below the earth. So there was a duality in his world picture: the earthly phenomena of gravity, falling bodies and fluid surfaces were arranged on a rectangular pattern, whereas the celestial phenomena showed a spherical symmetry of radii and circles around a centre.

For all these philosophers the earth rested in the centre of the universe. Another doctrine developed among the 'Pythagoreans' at the same time, ascribed especially to Philolaus of Tarentum (between 500 and 400 BC), according to which the centre of the universe is a fire, called 'Hestia' (i.e. the hearth); the spherical earth describes a daily circle around this fire, always turning its uninhabited side towards it. Thus day and night alternate; we cannot see the central fire, since (to make the number of world bodies ten) another dark body, the counter-earth, is interposed between us and the fire. The sun is a transparent globe receiving its light and heat from the central fire and from the fire outside the heavens.

Greek philosophy took a new trend with the Sophists and Socrates, and the Socratic discourses of the 'sublime philosopher' Plato (427–347 BC). The problems of nature and of world structure no longer constituted its subject matter. With the development of commerce and trade new social phenomena appeared, and through the political discussions and decisions in the citizens' assemblies, the problems of man, state and society came to the fore. The philosophy of nature gave way to the philosophy of life. Socrates spoke with some disdain of the knowledge of nature, and it occupies only a small place in his discourses. Their chief content is the culture of the individual soul through the eternal values of truth, beauty and goodness. In the works of Plato, which are the first of ancient Greek philosophy to come down to us in completeness, there are only a few parts informing us about his views on nature and the heavenly objects. Usually these are clothed in symbolical and allegorical forms, so they later became the object of an extensive controversial literature and of much modern exegesis about their meaning; this controversy still continues.

Plato's philosophy was the most direct and extreme expression of the ideas and mode of thinking of the wealthy citizens of Athens and of other Greek towns of the time. They ruled over a numerous class of slaves and artisans, and they looked down upon all manual work with utter disdain, suitable only for rough people and slaves devoid of all spiritual culture. Manual work was dishonourable for the free citizens, who should occupy themselves with intellectual work only, with mathematics, philosophy and politics. This contempt for technique was exactly opposite to the state of the mind of the early Ionian citizens and not yet present in the artists of the Golden Age, who were themselves technicians. It was typical of the philosophers throughout the centuries of decline after the catastrophe which befell Athenian power at the end of the fifth century BC. It was probably one of the main reasons why experimental natural science did not develop in antiquity.

For Plato 'ideas', spiritual entities, constituted the real world, of which our visible world is an appearance only. Reality cannot be experienced through the senses. The real world is perfect, pure and eternal and can be ascertained only by the mind as long as it does not allow itself to be dominated and distracted by the imperfect temporary phenomena. This holds also for astronomy: 'Yonder broideries in the heavens . . . are properly considered to be more beautiful and perfect than anything else that is visible; yet they are far inferior to those which are true . . . the true objects which are apprehended by reason and intelligence, not by sight . . .'

We should use the broidery in the heavens as illustrations to facilitate the study which aims at those higher objects, 'just as diagrams are used in geometry'. We shall pursue astronomy, as we do geometry, by means of problems, and we shall dispense with the starry heavens, if we propose to obtain a real knowledge of astronomy.[42] The true astronomer must be the great wise man who does not, like Hesiod and others, occupy himself with the risings and settings, but investigates the seven revolutions contained in the eighth movement.[43] Here also appearance is set against reality, though not in the way that a spiritual world is set against a material world. What in later centuries is expressed as astronomical reality against the visible phenomena—another arrangement of orbits, as seen from another point of view—is proclaimed (in Plato's doctrine) as the spiritual essence of things, evidently because it is only discoverable by the mind through abstraction from practical experience. So the practical task of the astronomer is to find out the real regular motions in perfect circles which in essence stand behind the apparently irregular wanderings of the planets. A later author mentioned that in his oral discussions Plato set the students of astronomy to solve the problem of 'what are the uniform and ordered movements by the assumption of which the apparent movements of the planets can be accounted for'.[44]

To Plato all stars were 'living beings, divine, eternal': their visible form '[the Creator] made mostly of fire, that it might be most bright and most fair to behold, and likening it to the All, he fashioned it like a sphere . . .'[45] The universe is eternal, living and perfect, a perfect globe, animated by a soul and with the movement appropriate to this form: a uniform revolution around an axis. The movements of the seven wandering stars (sun, moon and planets) are explained in an allegorical myth in the *Republic* and in a poetical description in the *Timaeus*.

In the last chapter of the *Republic*, describing what the souls saw in heaven, Plato expounds a vision of a light like a pillar, which binds the heavens together, and the Spindle of Necessity, by which all the revolutions are maintained; it shows eight whorls, fitting like boxes into one another, with their rims of different widths forming one single surface. Then follows a detailed description: the first and outermost with broadest rim of many colours [the starry sphere], which revolves with the axis; within it the other seven revolve slowly in the opposite direction. The innermost, the eighth, goes most swiftly [the moon]; then next in velocity and together come the seventh, sixth and fifth [the sun, Venus, Mercury]; then third, fourth and fifth in velocity come the fourth, third and second [Mars, Jupiter, Saturn]. The seventh is brightest, and the eighth takes its colour from the seventh, which shines upon it [sun and moon]; the second and fifth [Saturn and Mercury] are more yellow, the third [Jupiter] is most white and the sixth [Venus] somewhat less white, whereas the fourth [Mars] is pale red.

These details show most clearly that this poetical account is an allegorical description of the planets and their arrangement. Less clear is what is added on the width of the rims; after the sphere of the fixed stars, where it is broadest, the sequence in width is Venus, Mars, moon, sun, Mercury, Jupiter, Saturn. It has been supposed that the extent of the planets' deviations in latitude from the ecliptic might be indicated by this width, or that it indicated the extent of the irregularities in the planets' progress, e.g. the extent of their retrograde motion. But in both suppositions the place of the sun and moon in the series does not fit. We do not know what Plato meant by these statements. Modern scholars have tried with much ingenuity to interpret his description into a detailed bodily structure of the universe. But as T. Heath rightly remarks: 'The attempt to translate the details of the poetic imagery into a self-consistent picture of physical facts is hopeless.'[46] Not because Plato's metaphorical description of world structure lacks precision, but because it is entirely different from modern ideas of mechanical structure; strongly mystical and mathematical, it is essentially a spiritual image.

In *Timaeus* the creation of the soul and the world is described. Two circles were made, crossing one another obliquely in two opposite points and both revolving in themselves. 'The exterior motion he proclaimed to be that of sameness, the interior motion that of difference. The revolution of the circle of sameness he made to follow the side toward the right hand, and that of difference he made to follow the diagonal [i.e. an inclined line] toward the left hand; he gave the mastery to the revolution of same and similar, for he left that single and undivided. The inner circle he cleft by six divisions into seven unequal circles, divided by double and triple intervals, three of each number [which was interpreted afterward as distances 1, 2, 3, 4, 8, 9, 27]; and he bade these circles move in the opposite directions, three at the same speed, and the other four differing in speed from the three and among themselves.'[47] Here again it is quite clear that the movement of sameness is the revolution of the celestial sphere, with for all stars the same common motion toward the west, and the movement of difference represents the course of the seven moving luminaries along the ecliptic toward the east.

In order to make a moving image of abiding eternity, expressed in numbers, what we call Time, 'the sun, the moon, and the five other stars which are called planets, have been created for defining and preserving the numbers of time. . . . The moon he placed in that [orbit] nearest to the earth; in the second above the earth he placed the sun; next, the Morning Star and that which is held sacred to Hermes he placed in those orbits which move in a circle having equal speed with the sun, but have the contrary tendency to it; hence it is that the sun and the star of Hermes and the Morning Star overtake, and are in like manner overtaken by, one another.' To set forth the orbits of the others 'and the causes for which he did so . . . would lay on us a heavier task. . . . These things, perhaps, may hereafter, when we have leisure, find a fitting exposition.'[48] Thus day and night are determined by 'the period of the one and most intelligent revolution', a month by the moon, and a year by the sun. 'But the courses of the others men have not grasped, save a few out of many . . . in fact, they can scarcely be said to know that time is represented by the wanderings of these, which are incalculable in multitude and marvellously intricate.'[49]

In a posthumous work, completed by a pupil, these other planets are said to have no names, as also the third, having the same velocity as the sun [Mercury], because they were discovered by barbarians living in a better climate. But they were named after the gods: 'the morning and the evening star after Aphrodite, the one just named after Hermes. . . . Three remain, slower than these; the slowest has been named after Chronos, the next must be called the star of Zeus, and the other one, reddest of all, the star of Ares.'[50] The fact that Plato for the first time used the Babylonian names for the gods of the planets indicates that by this time oriental influences were already perceptible in Greek astronomy.

Because the movement of sameness is more rapid than all other movements and supersedes them, the daily revolution of sun, moon and planets is slower and in spirals; the bodies with the most rapid proper movement [the moon] are farther behind and appear to be the slowest. This again led to opposing appearance and reality in the heavenly motions. In the *Laws* it is said that the Greeks speak slander of the gods by saying that they go many different ways, whereas in reality they always go the same way, and also by calling the most rapid those which in reality are the slowest.

Whereas in these fragments Plato's cosmological world system in the main part presents itself without much ambiguity, there are some few sentences which give rise to the opinion—and to much discussion thereon—that Plato in his later years adhered to the idea of the rotation of the earth. First by the opposition just mentioned: only if Plato considered the daily rotation of the celestial sphere as an appearance, a delusion, could he call Saturn the slowest, the moon the most rapid; he could not do so if the celestial rotation was a reality to him. But it must be said that for Plato the separation of the planet's slower daily rotation into a common celestial rotation and an opposite proper motion of the planet along the ecliptic was a work of the mind and a mental abstraction, disclosing the deeper spiritual reality and the essence of world structure.

Secondly, because of a sentence in *Timaeus*: 'But the earth, our foster-mother, globed [rolled, wound, packed] round the axis [*polos*] stretched through the universe, he made the guardian and operator of night and day, the first and oldest of the gods that have been created within the heaven.' [51] Here, indeed the word 'operator' (*demiurgos*, 'artisan') would be more appropriate if the earth itself rotated; but the evidence of celestial rotation is too manifest in all the rest to leave any doubt, so this sentence must be understood in a broader sense. Thirdly, because of a sentence in Aristotle: 'Some say that the earth, lying at the centre, is wound and moved about the *polos* stretched through the universe, as it is written in *Timaeus*.' [52] It deals, however, not with Plato but with the Pythagoreans; and Thomas Aquinas already explained that the last words refer to the axis, for which Aristotle never, but Plato repeatedly (as in the above quotation), used the name *polos*. Nowadays there will not be much dissension as to Plato's opinion on this point.

Astronomical knowledge in Greece had gradually developed to the point that (in the fourth century BC) the separate planets and their course along the ecliptic were well distinguished. In the preceding centuries, 600–400 BC, the Greek mainland had reached a summit of economic and political flowering: at home supreme freedom of the citizens in trade and commerce, and abroad a strength that was respected in opposition to the Persian empire. It is a curious fact that, whereas literature and art reached a supremacy scarcely ever matched elsewhere, the science of astronomy rose so slowly. The first knowledge of the most simple phenomena, easily ascertainable by anyone, was ascribed to famous philosophers as their important discoveries. It shows that the mind of the Greeks in these centuries of its flowering was not directed towards practical astronomy; here, unlike Babylon, the demands of practical life brought no strong incentive for an attentive observation of the stars.

CHAPTER 10

CALENDAR AND GEOMETRY

For the Greeks, as for other peoples, travel and time-reckoning were the bases from which practical astronomy originated. Dauntless Greek mariners traversed the seas with their merchandise; yet there are few reports on the use of the stars in navigation. In the narrow Mediterranean, where coasts were always near, the stars were only required for giving the direction of sailing and to indicate the hours of the night. Later authors speak of a manual of nautical astronomy made by Thales of Miletus, the authorship of which, however, was ascribed by others to Phocus of Samos.

More frequently we find mention of how the ancient Greeks determined the time of day and regulated the calendar by means of the heavenly phenomena. The progress of the hours at night was recognized by the rising of the constellations, especially those of the zodiac. Knowledge of the twelve zodiacal signs, ascribed to Oenopides of Chios (about 430 BC), was doubtless borrowed, as the names show, from Babylon. In the course of the night five zodiacal constellations are seen to rise, beginning at nightfall with the stars opposite the sun, and ending with the stars appearing just before sunrise. So the appropriate unit of time is the double-hour, which is longer on long nights, shorter on short nights. In the daytime, when the simple estimate of solar altitude was not exact enough, the shadow of a pillar, called 'polos' or 'gnomon', was paced off. In later centuries sundials of various construction were erected; a simple specimen was a metal globule at the centre of a spherical bowl casting its shadow on the inner side, which was divided by engraved circles. Water clocks also were used.

In Greece, long-range time-reckoning was from olden times based upon the lunar period adapted to the solar year. The month began with the appearance of the crescent; hence the day began with the evening. Originally this had doubtless been established empirically; but as the climate often impaired visibility, an artificial rule in the form of a regular alternating of 'full' and 'hollow' months (of 30 days and 29 days) was established later on, in such a way that in 16 days 3 days had to be added. Solon the Athenian, in his biography by Plutarch, is said to have introduced the practice of calling the day of conjunction of sun and moon 'old-and-new', to be reckoned partly to the old, partly to the new month, and that the next day should be called 'new moon' and be the first day of the month. In Greece the calendar was an affair not of priests but of civil officials.

The twelve months of the year had different names in the different towns and regions; the Athenian names, of course, were most frequently met with in science as well as in trade. Nor did the practices of the various regions coincide; in Athens the year began at the summer solstice, in other states in different seasons. This again shows how ancient Greece was largely split up into small and isolated units. The years were named after the ruling officials, so that they too differed from city to city. The problem for later authors of how to fix the date of earlier events was facilitated by the Olympiads, the period of the great games and contests at Olympia, which occurred every four years.

The moon calendar had to be adapted to the solar year, which dominated practical life, agriculture and navigation. A later author of a manual of astronomy, Geminus, expressed this in the following words: 'To reckon the years by the sun means that the same offerings to the gods are brought always in the same seasons. . . . For they assume this to be pleasing to the gods.'[53] The time of the year undoubtedly was determined still long afterwards in Hesiod's way, by the risings and settings, also the evening risings, of conspicuous stars. The need to fix the year more exactly by astronomical phenomena may have led to the observation of the summer solstices. When the sun reaches its maximum altitude, the points of its rising and setting reach their farthest north, after which they turn back; without instruments, these positions can most easily be ascertained by comparison with earthly objects at the horizon. This probably is the meaning of a sentence preserved from the lost historical work of Theophrastus. The Athenian astronomer Phaeinos determined the solstices with respect to the Lycabettus, and Cleostratus of Tenedos determined them from Mount Ida on the nearby mainland of Asia Minor. Mount Lycabettus is a hill situated close to ancient Athens on the north-east (today at the centre of the town). The irregular profile of the mountain made it possible to find the day solstice by detecting small variations in the azimuth of the rising sun.

The lunar calendar was adapted to the sun, just as at Babylon, by intercalating a thirteenth month. This was decreed by local officials, therefore rather arbitrarily, and differed according to locality. Astronomers then tried to derive and introduce a regular intercalation by means of fixed periods. The same Cleostratus of Tenedos (about 520–500 BC) is said by later authors to have introduced (or proposed) an eight-year period (oktaëteris). This period of 99 months comprised

2,922 days; so a year was taken to be $365\frac{1}{4}$ days, and the moon's synodic period 29.55 days (0.016 too small, because 99 periods in reality are $2,923\frac{1}{2}$ days). This defect must have precluded its practical use over a greater number of years. Geminus records that this period was improved by Eudoxus of Cnidus (408–355 BC), who added three days every sixteenth year, in order to give to the month its true length in days, and then, after 10 of these periods, had to omit one thirteenth month in order to adjust to the solar year. There is no indication, however, that either the 16-year or the 160-year cycle was used in practical life.

Before the time of Eudoxus another cycle had already been devised by the Athenian astronomer Meton, pupil of Phaeinos, who is mentioned by Ptolemy as having made observations with Euctemon (about 433 BC). This is the 19-year cycle of 235 months, equalling 6,940 days. It is often called 'Meton's cycle' and is identical with the period we found in Babylon. It is uncertain whether the Greeks borrowed it from the Babylonians or found it independently; as a basis of chronology we saw how it appeared in Babylon only at a later date, after 380 BC. According to this cycle, the year is $6,940 \div 19 = 365\frac{5}{19}$ days, larger by $\frac{1}{76}$ day than the simple value $365\frac{1}{4}$; the lunar period also comes out too large, since 235 periods of the moon amount in reality to 6,939.69 days.

In such a calendar, with some years of 354 days and others of 384, the dates of solar solstice and equinox, as well as the risings and settings of the stars, make irregular leaps, now advancing 11 days, then retreating 19 days. To make these variations known to the citizens, pillars were erected with holes corresponding to dates, into which pegs were placed to indicate the phenomena. Almanacs were constructed, called 'parapegm'; because they were affixed to a pillar, and mostly containing the solar phenomena for a number of years, the risings and settings of the stars and weather predictions. Meton is reported to have published a parapegm beginning with the summer solstice observed by him in 432 BC.

Meton's 19-year cycle greatly simplified calendar activities; the knowledge of the series number of a year within a cycle at once gave the dates of the phenomena, because they returned regularly. It is not certain, however, whether it came into general use as a substitute for the local calendars. Its immediate purpose was to record the events in a fixed scale, probably also for astronomical use. It is supposed to have been in use at Athens at a later date, after 340 BC, but data on years with 13 months derived from later documents do not fit into a regular 19-year period.

In these same centuries of gradually developing knowledge of the motions of the heavenly bodies and its practical application in time-reckoning, we see a development of mathematical theory. In this field the Greek mind, through its faculty for abstract thinking, was able to establish a continually expanding structure of theorems and lemmas. These soon found their application in astronomy.

Eudoxus of Cnidus, a famous mathematician already mentioned, is known in the history of astronomy as the first to give a theoretical explanation of the planetary movements. In accordance with prevailing opinions, he supposed every planet to be affixed to a sphere revolving around the earth as centre. To explain the irregularity of planetary motion, he supposed more spheres instead of one, all homocentric; that is, all turning regularly in different ways about the same common centre. This theory offered a solution—perhaps meant as such—to the problem proposed by Plato: to represent the observed irregularity in the visual phenomena by means of perfectly regular circular motions, the only motions admissible for the divine celestial bodies.

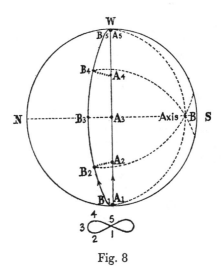

Fig. 8

In order to understand his theory, we assume a celestial globe turning about a horizontal axis directed north-south; then a body at its equator rises vertically in the east, passes through the zenith, and descends vertically in the west (case A in fig. 8). If, however, the axis is inclined, say 10°, upward to the south (case B), then the body, on rising in the east at the same time, deviates to the north, passes 10° north of the zenith, and goes obliquely down in the west. Relative to the body in A, it oscillates towards the north and back; in the further rotation, it deviates, below the horizon, the same 10° towards the south. Moreover, by rising obliquely in the east, it lags behind, at first a small amount (as shown by $B_2 - A_2$), only $\frac{1}{2}$° (proportional to the square of the inclination), but then it catches up at the highest point $(B_3 - A_3)$; again in B_4 it is the same small extent ahead. The position of B relative to A is shown in the little figure (p.109), where it is completed by a symmetrical loop for the hemisphere beneath. This picture of a figure of eight on its side was called by the Greek authors '*hippopede*' (horse fetter) because in the riding-schools fettered horses thus paced a figure of eight. So if in case B, where the rotation carries the body along B_1, B_2, B_3, etc., a backward rotation A of the sphere in which the rotation axis of B with its poles has been hinged is added, the motion east-west of the body is neutralized by the motion west-east of the outer sphere; there remains only the oscillation along the hippopede: 10° to and fro, accompanied by small deviations of $\frac{1}{2}$° up and down. This oscillation has to be superimposed upon a regular west-east rotation along the ecliptic, which takes place in the period of revolution of the planet along the ecliptic.

Here we have an irregular oscillation produced by two spheres rotating uniformly, as behoves celestial spheres; because they are inclined and rotate in opposite directions, there is left only the small hippopede. To apply this result to the motion of a planet, Eudoxus assumed four homocentric spheres: the first, the outermost, rotates in a day about the celestial poles, whereby the planet participates in east-west motion of all the stars. The second, dragged along with the first because its axis is fastened in the poles of the ecliptic, carries the planet quite regularly along the ecliptic in a west-east direction in the period of revolution of the planet. The motion along the hippopede, which is performed in one synodic period, is superimposed to transform the regular mean movement into the alternation of a long direct and a short retrograde path. For this purpose the third sphere has its axis fixed in two opposite points of the ecliptic, and it carries the fourth sphere, contrarily revolving about an inclined axis. Thus all the phenomena are represented: the daily rotation, the opposite revolution along the ecliptic, and the alternation of direct and retrograde motion in a synodic period.

This explanation of the irregular planetary motions by perfectly regular rotations of four connected spheres certainly was an admirable performance of geometrical sagacity. For the sun and the moon three spheres were sufficient, reproducing at the same time the inclined orbit of the moon. It is only through a description by Simplicius, the well-known commentator of Aristotle in the fifth century AD, that we are acquainted with some details of this theory. Simplicius gave the periods used by Eudoxus but was silent about the inclinations of the two hippo-pede-forming spheres. The Italian astronomer Schiaparelli, who in a modern study elucidated this theory, showed that by assuming this inclination to be 6° and 13° for Saturn and Jupiter, the right values for the retrograde arc, 6° and 8°, were found, whereas the deviations to the north and south were 9' and 44' only, hardly noticeable at that time. For Mars, however, it was not possible to arrive at a good representation; with the true synodic period of $2\frac{1}{7}$ years, no inclination however large could produce a retrograde motion; the oscillation was too slow. Simplicius mentioned a rotation period of 8 months and 20 days for the two spheres, which, with an inclination of 34°, gives a regression of 16°. The deviations in latitude of the hippopede going up to 5°, of course do not harmonize with the real variations of latitude of the planet. More-over, what is worse, there should be three retrograde movements in a synodic period, one in opposition to the sun and two at moment where they would be less perceptible, to both sides of the conjunction. Here the theory, geometrically so ingenious, fell short in representing the pheno-mena. The same holds for Venus. It is evident that Eudoxus had not at his disposal sufficiently numerous and accurate observations with which to compare his theory and detect its inadequacy. That knowledge in Greece at that time of the apparent motion of the planets was rather primitive is also clear from the periods given by Simplicius as Eudoxus' values: 30 years, 12 years and 2 years as the time of revolution of Saturn, Jupiter and Mars respectively, and 13 months as the synodic period for both Jupiter and Saturn. The empirical knowledge was confined mainly to the existence of retrograde motions, without quantitative details; and the theory gave an explanation only of this qualitative character. The great Greek scientists were not observers, not astronomers, but keen thinkers and mathematicians. Eudoxus' theory of the homocentric spheres is memorable not as a lasting acquisition of astronomy but as a monument of mathematical ingenuity.

Yet there was progress in practical astronomy also. As the sun had to be carefully observed for calendar purposes, its irregularities came to light; observation of the equinoxes as well as the solstices disclosed the inequality of the four seasons. In the preceding century Euctemon had given their length: 93 days for the spring (counted from equinox to summer solstice), 90, 90 and 92 days, which deviate considerably from the true values, 94.1, 92.2, 88.6 and 90.4 days. Far more accurate values were given a century later by the able astronomer Callippus of Cyzicus (between 370 and 300 BC), a pupil of Polemarchus, who himself was a pupil of Eudoxus. His lengths for the seasons were 94, 92, 89 and 90 days—hence correct to round figures.

That the sun takes a different number of days to complete the quadrants of its circle means that its velocity is alternately greater and smaller. In the theory of Eudoxus this can be represented by adding a small hippopede of 4°, which alternately accelerates and retards the sun's course. Simplicius indeed records that Callippus had added, both for the sun and the moon, two spheres to the three assumed by Eudoxus. Moreover he mentions, without further details, that Callippus added a fifth sphere for Mars and for Venus, probably to correct the defects of Eudoxus' theory for these planets. Schiaparelli has shown in his study that, with three spheres combined, a more complicated curve than a hippopede can be obtained, which affords a velocity sufficient, for Mars with a synodic period of $2\frac{1}{7}$ years, to produce a single retrogradation at opposition. It can even be done in different ways; how Callippus, an accomplished mathematician, did it we do not know.

An improvement of Meton's 19-year calendar period is also ascribed to Callippus, about 334 BC. By subtracting one day from four periods, he obtained a 76-year period of 940 months $=27,759$ days, hence with a year of exactly $365\frac{1}{4}$ days and a synodic month of $29^d\ 12^h\ 44^m\ 25.5^s$ (22 seconds too large). His first cycle began in 330 BC; but it was only used by scholars and perhaps in parapegms but not in official calendars.

CHAPTER 11

SYSTEMS OF WORLD STRUCTURE

ARISTOTLE (384–322 BC), from the Greek town of Stagira on the Macedonian coast, was a pupil of Plato, and in his first works he closely adhered to his teacher's ideas. Later, however, he developed his own world concepts, proceeding in less spiritual directions, perhaps due to his descent from a country of a more rustic middle class. Whereas for Plato ideas are the real world and visible phenomena only a deceptive appearance of them, for Aristotle the world of phenomena is the real world, in which the ideas manifest themselves as their innermost essence. The visible world of phenomena originates by the working of the 'ideas' upon unformed matter, giving it form and definiteness. The world of phenomena shows incessant change and motion, genesis and decay; but the essence of things is invariable and eternal. The aim of science and philosophy, one and the same in Greece, is the recognition through thought of this invariable essence, using the mind's power of abstraction. Abstract conceptions are derived from the concrete facts of the phenomena observed. Hence, more than his predecessors, Aristotle emphasizes the importance of careful observation and description of every detail of the phenomena; his later works constitute an encyclopaedia of all the knowledge of his time, of natural phenomena, of the animal world, as well as of the human world, politics, ethics and art. On the other hand it was necessary to give an exposition of the special phenomena as the product of the ideas and, by deriving the facts of nature from the essence of things, to demonstrate their essential truth. This is done by careful reasoning in logical deduction. Thus knowledge is presented in such a form that the phenomena described are the result deduced by argument from the presupposed essence of the world, its inner nature and general principles. By what to modern ears is often an artificial argumentation, Aristotle's science, presented as the logical outcome of general principles, acquires a dogmatic character. At the same time, by its well considered coherence, his world picture acquires a high degree of intrinsic logic and beautiful harmony.

Aristotle presents the structure of the universe as having perfectly radial spherical symmetry. The world is arranged according to spheres and spherical shells around a centre, in which the simple movements along straight lines are along radii from or to the centre. 'All change of place which we call motion is straight or in a circle or a mixture of them. These are the only simple ones . . . [Motion] in a circle is around the centre, straight is the motion up and down. Up means away from the centre, down is toward the centre. It follows that all simple movement is away from the centre, towards the centre, or around the centre.'[54]

The contrast in Anaxagoras' world picture between the radial-spherical structure of the surrounding heavens and the flat-rectangular structure of the earthly phenomena is entirely removed here; the reduction of Anaxagoras' rectangular world to a small fragment cut out of the big earthly globe means a considerable widening of the world picture.

With Plato and other predecessors, Aristotle assumed that the four elements earth, water, air and fire, which in this succession are arranged in layers one above the other, constitute the earthly part of the world. Because each of them, when removed from its natural place, tries to reoccupy it, the natural movement of the heavy elements is downward, towards the centre, and of the lighter ones, air and fire, upwards, where they come to rest. The element fire must not be considered as flame and glow. The celestial luminaries are not made of fire, as has often been supposed, but of the fifth element; they emit light and heat, because the air beneath them is ignited by the friction caused by their movement.* In the high layers, where air and fire are mixed, the vapours from below assemble, and when they kindle are seen by us as meteors and comets.

Besides the natural movements, there are the enforced movements transmitted by direct contact. There is no void in the world; each body is surrounded by air or water, and its motion is always relative to this medium. This principle involves a curious explanation of the motion of missiles after contact with the moving hand has ceased. 'The prime mover conveys to the air, as an intermediary, a power of conveying motion, and when the prime mover's action has ceased, the intermediary ceases its motion but is still able to move the missile and the air. Thus the moving power is transferred to the contiguous parts of the air that convey it to the missile, until at last the moving impulse is exhausted.'[55] This theory of the conservation of motion by some kind of propulsion by means of continuously propagated elastic force was often refuted by later writers.

---

* This opinion of Aristotle has often been felt to be inconsistent, because air does not extend into the celestial realms; it would seem to imply that all the celestial spheres carrying the luminaries form, closely packed, a relatively thin shell around the world of the four elements.

The motion of the celestial bodies is not straight and finite, but circular, invariable and eternal. So they must themselves be eternal, unalterable, divine. They are different from the four earthly elements; they consist of the fifth element (*quinta essentia*), the ether, more perfect than the other four. Changes in that world have never been observed. The realm of the ether reaches down to the moon; beneath the moon is a mixed region where the 'sublunary world' of the four elements begins.

The earth is at rest in the centre of the universe. Because all heavy parts tend towards the centre, this element has assembled here and constitutes the solid body of the earth. Lengthy arguments are used to show that any movement of the earth would be contrary to its nature. Because all the heavy parts push one another in trying to come as close as possible to the centre, the limiting exterior surface must be a sphere around the centre, so that the centre of the earth is the centre of the universe. Moreover the spherical shape of the earth is shown by visible phenomena; at lunar eclipses the boundary of the shadow is always seen to have a curvature corresponding to a globular earth. The earth cannot be very large, for a short journey north or south changes the horizon; in Egypt and Cyprus stars are seen that are not visible in Greece. The mathematicians who tried to determine its circumference gave it as 400,000 stadia.

The universe is finite and spherical. It must be finite because, were it infinite, either simple or composite, one of the elements must be infinite, and then there would be no room for the others. Furthermore, its motions must be circular and an infinite circle cannot exist; moreover, an infinite thing cannot have a centre, and the universe does have a centre. That the universe is spherical is proved, first by the universe being perfect and the sphere being the sole perfect figure and also by a sphere in its revolution always occupying the same space; outside there is no more space, there is 'neither void nor place'. Moreover, it is made probable by the spherical shape of the other bodies. That the moon is spherical we see by the boundary line between its dark and its illuminated parts. The stars also must be spherical; because they cannot move by themselves, they cannot have organs for locomotion, and since 'nature makes nothing in a haphazard fashion', the spherical shape is best fitted for them. These arguments may serve as an instance of how Aristotle derived the phenomena from general principles and ideas.

As regards the arrangement of the celestial bodies, Aristotle agreed with his friend Callippus, but on a different basis. For Eudoxus and Callippus the rotating homocentric spheres were geometrical figures without a physical character; for Aristotle, they were material objects, real crystalline shells surrounding and carrying one another along. The moving force comes from the exterior celestial sphere, which, as prime mover, in its daily rotation moves all the others in such a way that every outer sphere transfers its motion to the next lunar sphere. This means that when the planet Saturn by its four spheres has acquired its special revolution in 29 years with retrogradations, the motion of the three inner spheres must be neutralized by three contrarily moving spheres, so that only the daily rotation remains and is transferred to the next inner sphere, which is the most exterior one of Jupiter. This holds for every next body, the moon excepted because it is the last one. In addition to the 33 spheres of Callippus (4 for Saturn and Jupiter, 5 for Mars, Venus, Mercury, sun and moon), the system of Aristotle demanded 22 counter-rotating spheres, in total 55 solid crystalline spheres, to bring about the motions of the celestial bodies.

His later commentator, Simplicius, indicates that Aristotle himself was not entirely satisfied with this theory. For in earlier times it was already known, e.g. to Polemarchus, that the planet Mars in opposition during its retrograding, and also Venus when it retrograded, were much brighter than at other times, whereas the theory of the concentric spheres keeps them always at the same distance from the earth. But we have to remember that there was no other theory to explain the irregularities of the planets. Later on, other theories were devised. The general world picture of Aristotle, however, was maintained throughout the centuries that followed until the seventeenth. It shows how well this first tight and harmonious structure of the universe corresponded to the simple scientific experience of those times.

This does not mean that Aristotle's theory of world structure remained without rival. In the fifth century BC Greece had experienced an epoch of economic rise and political power that resulted in the fourth century in considerable wealth and in vivid intellectual and scientific pursuits. In this century of profound philosophic thinking we see the rise of bold and new ideas of world structure.

A contemporary of Aristotle, Heraclides Ponticus (388–315 BC, from the town Heraclea on the Pontus), is also sometimes reported to have been a pupil of Plato. Though he was famous in antiquity, none of his many writings on different subjects has been preserved. So we have to consult the reports of later authors for his philosophical and cosmological ideas. We find that Simplicius, in explaining one of Aristotle's arguments, says: 'because there have been some like Heraclides of Pontus and Aristarchus, who suppose that the phenomena can be saved when the heaven and the stars are at rest, while the earth moves about the poles of the equator from the west, completing one revolution each day, approximately . . .'[56] And at another place: 'But Heraclides of Pontus, by supposing that the earth is at the centre and moves in a circle and the heaven is at rest, thought to save the pheno-mena.'[57] This 'to save' the phenomena is a much-used term meaning to represent, hence to explain the phenomena. Another well-known later author, Proclus, in a commentary on Plato's *Timaeus*, writes: 'How can we, hearing that the earth is wound round, reasonably make it turn round as well and give this as Plato's view? Let Heraclides of Pontus, who had not heard Plato, hold that opinion and move the earth in a circle; but Plato made it unmoved.'[58] And Aëtius says: 'Heraclides of Pontus and Ecphantus the Pythagorean make the earth move, surely not in the way of progressing but in the way of turning, in the manner of a wheel, from west to east about its own centre.'[59]

Reading these quotations, we cannot doubt that Heraclides, in order to explain the daily rotation of the celestial sphere with its stars, made the earth rotate from west to east about an axis through the celestial poles. That in some of them this is expressed as moving 'in a circle' has given rise to the interpretation by some later authors—Schiaparelli in the nineteenth century and some years ago the mathematician Van der Waerden—that Heraclides had asserted an orbital motion of the earth. The quotations, however, clearly deal with a substitute for the daily rotation of the celestial sphere. A circular orbit to set the celestial sphere at rest was contained in the theory of Philolaus concerning the daily circuit around the central fire; the name also of Hicetas of Syracuse, of whom little is known, is mentioned by Cicero in this connection.

So it can be said with certainty that the doctrine of the rotation of the earth found some adherents among Greek thinkers. Greek navigators must often have observed how in their smoothly sailing ships the move-ment itself could not be felt, only its irregularities. Yet it was a big step to proceed from this experience to the idea of a motion of the earth. For this means that the most direct fact of observation, the rotation of the celestial sphere, as well as the still stronger conviction of the fixed steadiness of the earth beneath our feet, was recognized as a deceptive appearance only. That the ancient Greeks could rise to such a height of insight in overcoming prejudice testifies to that profound thinking and independence of mind that gives them an exceptional place in the history of science. Yet this brilliant idea could not maintain itself in the later centuries against the logic of Aristotle's world picture.

Another progress in astronomical theory ascribed to Heraclides relates to the motion of Venus and Mercury. There had always been a difference of opinion as to their place in the sequence of the planets; some authors placed them above the sun, so that the sun came next to the moon; others put them below the sun, between sun and moon. This question was solved by having both planets perform circles with the sun as their centre, a likely conclusion from their apparent oscillating to both sides of the sun. Aristotle did not touch on these phenomena, which indeed do not fit into his world system. Later Roman authors speak of this explanation and mention Heraclides in connection with it. Though there is much confusion in their statement of his ideas, most historians agree that it was Heraclides who first expounded the idea that Venus and Mercury describe circles about the sun.

This idea then, was a stepping stone towards a still more radical innovation regarding the world structure. It is connected with the name of Aristarchus of Samos (c. 310–230 BC), afterwards called 'the mathematician' to distinguish him from the equally famous Alexandrian philologist, Aristarchus of Samothrace. There is not the slightest doubt that he put the sun at the centre of the universe and made the earth describe a circle about the sun. We do not know this from his own writings but through others. Only one treatise of his has been preserved, but it contains nothing about world structure. It may, however, serve to indicate how he arrived at his theory and it is important in elucidating the scientific methods of the time. It is entitled *On the Size and Distances of the Sun and Moon*, and was preserved in later centuries because it was part of the much-copied *Small Composition* of half-a-dozen writings of different authors, where the problems of the celestial sphere and the phenomena of motions, risings and settings were treated in a simpler way than in Ptolemy's *Great Composition*.

Both the content and the form of Aristarchus' treatise give a peculiar aspect of the character of Greek science. Following the example of Euclid's great work on geometry, it consists of a series of 18 geometrical propositions preceded by six 'hypotheses'. Among the latter the most essential are: (2) 'that the moon receives its light from the sun'; (4) 'that when the moon appears to us halved, its distance from the sun is less than a quadrant by one-thirtieth of a quadrant' [i.e. 87°]; (5) 'that the breadth of the [earth] shadow is that of two moons'; (6) 'that the moon subtends one-fifteenth part of a sign of the zodiac'. And he continues: 'Then it is proved: (1) that the distance of the sun from the earth is more than 18 times and less than 20 times the distance of the moon (this follows from the hypothesis about the halved moon); (2), that the diameter of the sun has the same ratio to the diameter of the moon; and (3), that the diameter of the sun has to the diameter of the earth a ratio greater than 19 to 3, but less than 43 to 6; this follows from the ratio of the distances, the hypothesis about the shadow, and the hypothesis that the moon subtends one-fifteenth of a zodiacal sign.'[60] After this introduction, dozens of pages follow with propositions and their demonstrations. This Greek mathematician was able to take his stand outside the earth, where he could see it and all these celestial bodies as comparable globes; he intersected them and their enveloping shadow cone with planes, drew the intersections as circles and triangles, and applied Euclid's rigid method of demonstration to them. Thus the essential problem was treated and solved as a geometrical lemma. (See Appendix A.)

Here, for the first time in the history of astronomy, we have the pure and direct determination of the distances of celestial bodies from observational data. In a certain sense it might be called the opening-up of scientific astronomy; the conclusion that the sun is about 19 times more remote than the moon remained a regular constituent of astronomical science for 2,000 years. But the form in which this result was presented was very different from that of modern science. First, because all effort was applied to derive, by a host of geometrical propositions, a result which nowadays a schoolboy can achieve in a few minutes: the ratio of the distances of moon and sun is the sine of $3°$, i.e. $1 : 19.1$. On the other hand, the data of observation, the empirical basis of all the argument, are stated in a few lines, without any explanation of detail, as presupposed hypotheses. The procedure is entirely opposite to that of modern science, in which all attention and the largest amount of space is spent on the method, the procurement, the discussion and the communication of the empirical data.

Secondly, in the geometrical treatment one is struck by the careful precision in the indefinite form of the result: greater than, smaller than. Greek mathematicians did not work with approximate values of irrational numbers, the basis of all modern computing, though they had discovered their existence. To them, numbers were counted quantities of wholes or parts, taken exactly; a ratio expressed by sine $3° = 0.0253$ . . . was foreign to them; they worked with precise numbers: such as 18, 20, $\frac{2}{3}$, etc. In quantitative geometry they could derive only which numbers were larger or smaller than the required quantity. In this they often showed great ingenuity; but it made their geometrical determinations rather clumsy.

Important for later developments was Aristarchus' method, in his fifteenth proposition, of deriving the size of the sun relative to the earth. Its gist can be rendered briefly thus: the shadow seen at a lunar eclipse is the section of the shadow cone (tangent to sun and earth) at the moon's distance; this section, according to the fifth hypothesis, is twice the moon's diameter, hence $\frac{2}{19}$ of the solar diameter. The distance of the shadow's edge from the full moon must then be $\frac{2}{19}$ of its distance from the sun; hence $\frac{17}{19}$ remains for the distance from the full moon to the sun. One-twentieth of the latter is the distance from moon to earth, the remaining $\frac{19}{20}$ the distance from sun to earth. Expressed in the total shadow length, they are $\frac{1}{20} \times \frac{17}{19}$ and $\frac{19}{20} \times \frac{17}{19} = \frac{17}{20}$. The distance from the earth to the shadow's edge in the same unit is now $\frac{1}{20} \times \frac{17}{19} + \frac{2}{19} = \frac{3}{20}$; this is the ratio of the distances from the earth and from the sun to the shadow's edge, hence also the ratio of their diameters. Then the ratio of the diameters of earth and moon is 57 : 20.

In the treatise of Aristarchus, who did not know of the number 19, this has been worked out with careful precision in 'larger than' and 'smaller than'; the ratio, as already mentioned, is found between $\frac{19}{3}$ and $\frac{43}{6}$. It must be noted that no data are used other than 87° and the size of the shadow relative to the moon; contrary to what Aristarchus says in the Introduction, the apparent diameter of the moon is not used; it would be necessary only if the diameters were to be compared with the distances.

Nothing was said on how the basic data had been found. These Greek scholars were mathematicians rather than astronomers; the celestial bodies just happened to be the objects of their geometrical propositions. Hence the astronomical quantities were treated somewhat superficially, their precise value did not matter; ingenuity was exhibited in the solution of the geometrical problem. We do not know what observations procured the data. That the moon be halved at 3° distance from exact quadrature is a heavy overestimate; in reality, the distance amounts to no more than 10′; but it is possible that the large value depends on some rough estimate. That this quantity was noted as a means of measuring distances testified to great ingenuity and profound thinking. The matter is different with his sixth 'hypothesis'; an apparent diameter of 2° for the sun and moon so strongly contradicts their real value of $\frac{1}{2}$°, which can be ascertained by the most simple observation, that numerous explanations have been proposed: errors in copying old manuscripts (but the statement in plain words is: the fifteenth, and not the sixtieth part of a sign); Babylonian traditions (but there the real value was already known); even deliberate insertion of a wrong value to fix attention entirely on the geometrical method. We do not know; we have to accept the fact. His younger contemporary, the great mathematician Archimedes of Syracuse (287–212 BC), says that Aristarchus was the first to discover that the diameter of the sun (hence of the moon also) was $\frac{1}{720}$ of the zodiac, i.e. $\frac{1}{2}$°. This discovery, therefore, must have been made later on.

Diameters are numerical values only; the real and impressive bulk of a body is its volume. Thus Aristarchus says: 'The sun stands to the earth in a ratio larger than 6,859 : 27 but smaller than 79,507 : 216,' meaning the ratio of their volumes. If the diameter of the sun is between $\frac{19}{3}$ and $\frac{43}{6}$ times the diameter of the earth, its volume must be between 254 and 368 times the volume of the earth.

Probably it was this enormous size of the sun that forced on Aristarchus the idea that it was not proper for it to circulate around the much smaller earth, and that on the contrary, it should itself reside in the centre. We do not know his reasons; but the fact that he put the sun at the centre is stated with complete certainty by many later authors, among others by Archimedes in his *Sand-Reckoner* (216 BC). In this treatise Archimedes computes the number of sand grains required to fill the universe, in order to show that, where the Greek system of numerals was inadequate to render such immense numbers (up to 10^{63}), methods could be devised to express them in a rational way. After having spoken of the size of the earth and the planetary spheres, Archimedes continues:

'But Aristarchus brought out a book consisting of certain hypotheses, wherein it appears, as a consequence of the assumptions made, that the universe is many times greater than the 'universe' just mentioned. His hypotheses are that the fixed stars and the sun remain unmoved, that the earth revolves about the sun in the circumference of a circle, the sun lying in the middle of the orbit, and that the sphere of the fixed stars, situated about the same centre as the sun, is so great that the circle in which he supposes the earth to revolve bears such a proportion to the distance of the fixed stars as the centre of the sphere bears to its surface. . . . We must take Aristarchus to mean this: since we conceive the earth to be, as it were, the centre of the universe, the ratio which the earth bears to what we describe as the 'universe' is equal to the ratio which the sphere containing the circle in which he supposes the earth to revolve bears to the sphere of the fixed stars.'[61]

A heliocentric world structure is stated here in unmistakable words but without any detail in reasons or consequences. Neither do later authors give these details when they speak of his theory. It found no general adherence; Seleucus the Babylonian, from the town of Seleucia, who lived a century later, is the only one mentioned as having advocated the same theory. It did not, so far as we know, present itself as an explanation of the irregularities in the planetary motions; it remained a bold, ingenious, but isolated idea. The heliocentric theory could not yet enforce itself as an unavoidable necessity; astronomy first had to find new ways of practical progress.

CHAPTER 12

HELLENISTIC ASTRONOMY

ALEXANDER VON HUMBOLDT, in the second volume of his famous work *Kosmos*, dealing with the history of science, gave a vivid picture of the widening of the intellectual horizon of the Greek world through the conquests of Alexander of Macedonia. Not only was the powerful Persian Empire, until then almost inaccessible, opened to Greek commerce, exploration and culture, but the remoter regions of Inner Asia, Bactria, Turkestan and a part of India became known. New parts of the world, with their different aspects and phenomena, their mountains and deserts, their animals and plants, their different peoples with different customs, and the tidal Indian Ocean—all came within reach. Even more important still was the social change, the merging of the Greek and the Oriental world in a new economic and political development. In the rear of the conquering army came merchants and artisans from Greece, emigrating to the newly-opened regions of Near-Asia and populating both the old and the new towns. The riches accumulated in the treasuries of the Persian kings and their satraps were now dispersed as spoils, distributed among the soldiers, or spent by the new kings for town-planning, temple-building, and the construction of highways and harbours, thus stimulating trade and commerce. Easier traffic along old and new routes connected Greek commerce with India, Arabia and Africa, all a fresh source of wealth and culture. Essential in this development were the capacity and energy of the Greek citizens as officials and businessmen, providing for an exploitation of the fertile plains of the Nile, the Orontes, the Euphrates, that was more efficient than under the former rulers.

So in all these countries great and wealthy towns arose, such as Alexandria, Antioch, Smyrna, Ephesus, Nicaea and a host of smaller ones, which became centres of trade and new industry. There, beside the merchants and artisans and the well-to-do veterans, lived the land-owners and the officials who ruled the countryside. As in the Greek homeland, these towns had almost complete self-rule, hence great civil liberty, though here under the overlordship of kings who disturbed the internal peace of their states by frequent wars. Besides these monarchies a free commercial republic existed on the Island of Rhodes comparable with Venice and Genoa at a later epoch. For more than a century this republic dominated the eastern Mediterranean by its navigation and influenced all the adjacent kingdoms through its financial power. All this was the world of 'Hellenism'. Economic prosperity awakened strong spiritual activity, and these countries and towns became the seat of the new Hellenistic culture. Compared with this flourishing world, Greece proper retired into the background; it became poor and empty. Its cities declined to the level of provincial townships, with the exception of Thessalonica and Corinth, which still participated in the sea traffic, and of Athens, which remained a centre of culture, art, and philosophy, owing to the tradition of Plato and Aristotle and to its Academy and its schools.

This flowering of the Hellenistic world lasted for two centuries; then came the conquest and pillage by the Roman armies, exploitation by Rome, and absorption into the Roman Empire. The inhabitants, partly abducted as slaves, became the tutors of the Western world and Greek science became an element in the harsher and more external culture of the Roman rulers.

The strength of Hellenistic culture was due to a great extent to the merging of Greek and Oriental elements. Especially in astronomy do we see how Babylonian abundance of observed facts and Greek independence of thought combined with theoretical power of abstraction. Acquaintance with Babylonian methods, and possibly with their instruments, stimulated the Greek scholars to become themselves observers of the stars. The Babylonian results for the periods and irregularities, which had remained simply as numerical data, became in the hands of the Greeks the basis of geometrical constructions and led to conceptions of spatial world structure. Conversely, Babylonians came to be influenced by Greek theory; Seleucus, the Chaldean mentioned above, was also the author of a theory of the tides.

The centre of world commerce, Alexandria, capital of Egypt, the richest of the Greek-ruled empires, was also the centre of science at that time. There the Macedonian kings, the Ptolemies, founded a temple of the Muses, the 'Museum', famous for its library, for which manuscripts were collected from everywhere; and they called upon the most famous scholars to act as leaders and to constitute a kind of Academy of Sciences. Next to philology flourished medicine; mathematics and astronomy were also cultivated. Yet the extent and regularity of the observations at Alexandria could certainly not compare with the work of the Babylonian priests. Moreover, nothing is known with certainty about the instruments used. Euclid, in his astronomical work *On Phenomena*, only mentioned a diopter, not yet attached to a graduated circle but serving solely to fix two opposite points of the horizon. However, Ptolemy not only mentions the astronomers Aristyllus and Timocharis from the earliest Alexandrian times, between 296 and 272 BC, but also gives distances to the equator (i.e. declinations) of a number of stars, as well as differences of longitude measured by them in degrees and subdivisions. So they had instruments with graduated circles.

Another instrument they used was an equatorial ring, placed before the temples in Alexandria, in Rhodes, and perhaps in other towns, for calendar purposes. It consisted of a cylindrical belt, with its upper and lower borders exactly in the direction of the equatorial plane; the shadow of the southern half upon the inner side of the northern half left a narrow line of light at the upper or at the lower side of the equator. Thus the exact moment of the equinoxes could be fixed.

The geographer Eratosthenes of Cyrene, a contemporary of Archimedes, was one of the first directors of the Alexandrian library. Besides his geographical description of the entire known world, he was noted for having determined the size of the earth. A later writer, Cleomedes, gave an extensive description of the method used. In the town of Syene, in the south of Egypt, the bottom of a deep vertical pit was illuminated by the sun on the longest day, so that the sun then stood exactly in the zenith. In Alexandria, situated farther north, the shadow cast on a hollow sundial on that day was $\frac{1}{50}$ of the total circle. Thus the distance between these towns must be $\frac{1}{50}$ of the circumference of the earth. Since this distance was 5,000 stadia, as measured by the time the king's messengers took to run it, the earth's circumference must be 250,000 stadia. In modern times there has been much discussion on the length of the stadia used; if we take 157 metres as the most probable value, Eratosthenes' result comes very near the true figure. Later in the same book, Cleomedes mentions the learned Stoic Posidonius (135–51 BC) as having applied the same principle to the bright star Canopus, which at Rhodes just grazed the southern horizon but in Alexandria reached an altitude of $7\frac{1}{2}°$; from the distance over the sea, estimated at 5,000 stadia, a circumference of 240,000 stadia was found.

A determination of the obliquity of the ecliptic is also attributed to Eratosthenes. Eudoxus had given it as an arc of a regular quindecagon, i.e. 24°. Eratosthenes found the difference between the solar altitudes at the summer and winter solstices to be $\frac{11}{83}$ of the circumference; this corresponds to an obliquity of 23° 51', very near to the true value. The form of this statement has sometimes given rise to the supposition that graduation of circles was not yet in use. This, however, is entirely refuted by the measurements of the Alexandrian astronomers mentioned by Ptolemy.

Hipparchus of Nicaea, who lived and worked some time between

162 and 126 BC, mostly on Rhodes, is considered to be the greatest among the astronomers of ancient Greece. Of his writings only one has been preserved; but his work, consisting of practical measurements as well as theoretical innovation, was transmitted and reported by later authors, especially by Ptolemy. Three treatises by him were mentioned by Ptolemy, *On the Length of the Year*, *On Intercalation of Months and Days*, and *On the Change of the Solstices and Equinoxes*.[62] In the second treatise he improves the 76-year calendar period of Callippus by taking it four times and subtracting one day. Then the moon period is $111{,}035 : 3{,}760 = 29.53058 = 29^d\ 12^h\ 44^m\ 2.5^s$, which is comparable to the best Chaldean results and only one second too short. The length of a year is thus $\frac{1}{300}$ of a day shorter than Callippus' exact $365\frac{1}{4}$ days, i.e. $365^d\ 55^m\ 16^s$. Hipparchus makes it clear that we must distinguish between two definitions of the year; and he explains that the essential length of the year does not depend on the return to the same stars but on the return to the same equinoxes and solstices, which determine the seasons. Ptolemy quotes him as saying: 'I have written a treatise on the length of the year in which I show what is the solar year: the time it takes the sun to return from one solstice or from one equinox to the same point. It amounts to $365\frac{1}{4}$ days diminished by nearly $\frac{1}{300}$ of a day.'[63]

He mentions a summer solstice observed in 280 BC by Aristarchus, compares it with one observed by himself in 135 BC, and finds the interval half a day shorter than $145 \times 365\frac{1}{4}$ days, corresponding to the difference of $\frac{1}{300}$ of a day. As the moment of a solstice can hardly be determined with an accuracy of half a day he himself emphasizes the uncertainty of this 'small' difference; he surely must have made use of other data too, and Ptolemy himself mentions six autumnal and three vernal equinoxes, observed by Hipparchus in the years between 162 and 128 BC, probably with the equatorial ring mentioned above. Ptolemy says that errors 'up to one-fourth of a day' may also occur here, and he speaks of an insufficient stability of the instrument, since it had happened that twice on one day the illumination changed between the upper and lower rims of the equatorial ring. This, however, had another cause; it was the effect of refraction. When the sun in spring has not yet reached the equinox, refraction just after its rising can lift it from the southern to the northern side of the equator; then the sun sinks back by the rapid decrease of the refraction, and at a later hour actually passes the equator. It should be noted that the error in the 145-year interval must have been still larger, for Hipparchus' length of the tropical year given above is $\frac{1}{200}$ of a day too large (its real value was $365^d\ 5^h\ 48^m\ 56^s$), as also was the corresponding Babylonian value.

It seems probable that Hipparchus was struck by the deviation of his seasonal year from the Babylonian year, which was greater than 365¼ days and that he came to explain it by showing that they were two different things, viz., return to the same equinoctial points and return to the same stars. Thus he was led to his most important discovery, that of the progress of the stars relative to the equinoxes in the direction of the signs by a rotation of the celestial sphere about the poles of the ecliptic. This phenomenon, usually called the *precession*, may also be described as a retrogression of the equinoctial points. It appears in the tables of the Chaldeans in such a way that at different times different longitudes have been adopted as zero points; hence the Chaldeans are sometimes claimed to have been the true discoverers, from whom Hipparchus borrowed his knowledge. There can be no doubt, however, that it was Hipparchus who recognized it as a continuous regular progress; he derived its amount from a comparison of earlier Alexandrian observations with his own. Ptolemy tells the story in this way: 'In his treatise *On the Change of the Solstices and the Equinoxes*, Hipparchus, by exact comparison of observed lunar eclipses of his own time with others which had been observed by Timocharis in earlier times, arrives at the result that in his own time Spica preceded the autumn equinox by 6° and in Timocharis' time by 8°.'[64] In the middle of an eclipse the moon stands exactly opposite the sun, and the longitude of the sun can be derived from its declination, i.e. by measuring its altitude at midday; so by measuring the distance of Spica from the eclipsed moon, its longitude, i.e. its distance from the equinox, can be derived. 'And also for the other stars which he compared he shows that they have proceeded by the same amount in the direction of the zodiacal signs.'

From this change of 2° in an interval of 169 years a yearly variation of 45″ is derived. The curious thing is that this value does not occur with Hipparchus (i.e. it is not in Ptolemy's book) but that the latter quotes from his treatise *On the Length of the Year* thus: 'When by this reason the solstices and equinoxes in one year are regressing at least $\frac{1}{100}$ degree, they must have regressed at least 3° in 300 years.'[65] Then this value of 1° in 100 years, i.e. of 36″ per year (14″ too small), without the words 'at least', is used by Ptolemy farther on as the value derived by Hipparchus. That the displacement of the stars took place parallel to the ecliptic was shown by a comparison of the declinations of 18 stars measured formerly by Aristyllus and Timocharis and later by Hipparchus, as communicated by Ptolemy: at one side of the celestial sphere they had increased, by moving toward the north; at the other side they had decreased, by moving toward the south, with a maximum amount of 1°. Hipparchus concluded that the stars moved regularly about the poles of the ecliptic, or rather, according to the title of his writing, that the equinoctial points, with the equator attached to them, moved regularly back along the ecliptic.

The inequality of the four seasons, in which the sun completes the four quadrants of the ecliptic, was already well known to Callippus, whose values were given earlier. Ptolemy ascribes more accurate values to Hipparchus: 94½ days for the spring, 92½ days for the summer, so that 178¼ days remain for the half-year between autumn and spring equinox. These are very nearly equal to the values involved in the Chaldean tables: 94.50, 92.73 and 178.03 (cf. p. 74). Whether he borrowed them from Babylon or derived them from his own observations is uncertain; in any case he made observations of equinoxes and solstices, of which Ptolemy has communicated a small number. His great merit, however, consists in his theoretical explanation of this inequality, by means of an eccentric circle which the sun describes about the earth.

According to their essential nature and the need for harmony, it is assumed that the circular orbits of the heavenly bodies are performed quite uniformly. It is because the earth has its place outside the centre that we see the sun's velocity unequal, regularly increasing and decreasing between a largest value in the perigee and a smallest in the apogee. How far the earth stands outside the centre of the circle—its eccentricity—and in what direction, Hipparchus could easily derive from the length of the seasons by means of simple relations between lines and arcs in a circle (as shown in fig. 9), which form a first beginning of trigonometry. A first table of chords was ascribed to Hipparchus.

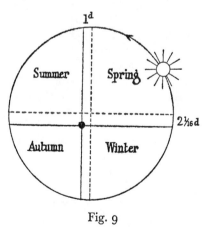

Fig. 9

His result was an eccentricity of $\frac{1}{24}$ of the radius, with the apogee in the direction of longitude 65½°. In this explanation of the irregularity of the phenomena by means of the spatial structure of the orbits, we see the Greek mind, with its faculty for abstraction and its command of geometry, in its full strength.

The knowledge of the moon's motion, that had previously received little attention in Greece, increased considerably as a result of Hipparchus' work. Several eclipses between 146 and 135 BC were observed by him. Ptolemy says that, by comparing them with earlier Chaldean eclipses, Hipparchus derived more accurate periods than those which 'still earlier astronomers' had at their disposal. It was supposed that these words referred to Babylonian astronomers; but since we know that exactly the same values were used in contemporary Chaldean tables and that they must therefore have been known still earlier, it is assumed that Hipparchus acquired his knowledge of the periods from Babylon. It is quite credible, though we have no exact details, that in these centuries there was a certain intellectual intercourse between Babylon and the Hellenistic centres of learning. The eclipses show that the return of the moon to the same node (called 'return of latitude') and the return to greatest velocity (called 'return of anomaly') take place in periods different from the return to the same star. Instead of the saros period of the Babylonian astronomers ($6,585\frac{1}{3}$ days $=223$ synodic periods $=239$ returns of anomaly $=242$ returns of latitude $=241$ revolutions in longitude $+10\frac{2}{3}°$), Hipparchus introduced a much longer interval of time: 126,007 days $+1$ hour $=4,267$ synodic periods $=4,573$ returns of anomaly $=4,612$ revolutions minus $7\frac{1}{2}° =$ nearly 345 revolutions of the sun; moreover 5,458 synodic periods are 5,923 returns of latitude. They afford a synodic period of 29^d 12^h 44^m 3.3^s (only 0.4^s too large) and a sidereal revolution of 27^d 7^h 43^m 13.1^s (only 1.7^s too large), both very accurate. It appears, from the detailed description which Ptolemy gives of his procedure, that Hipparchus did not simply copy Babylonian values but checked and corrected them by a careful discussion. The variable velocity of the moon could easily be explained, as with the sun, by an eccentric circle. In the case of the moon the direction of apogee is not constant, for in 4,612 revolutions of the moon it is passed 4,573 times, i.e. 39 times less; hence it proceeds in the same direction as the moon itself, 118 times more slowly, and completes a revolution in nine years.

Proclus in the fifth century AD described an apparatus through which Hipparchus tried to measure the diameters of the sun and the moon, and even their variations. It consisted of a long lath, provided at one end with a vertical plate with an opening to look through, at the other end a movable plate with two openings at such a distance that when the sun was low the upper and lower edges of the disc were just covered by them. Of its use and of the results nothing is known.

Eclipses were used by Hipparchus for yet other purposes. During a solar eclipse (probably in the year 129 BC),[66] which had been total at the Hellespont, only four-fifths of the sun was obscured at Alexandria. Since the distance between these places, expressed in the earth's radius, could be computed, Hipparchus was able to derive the parallax of the moon, hence its distance from the earth; he found it variable between 62 and 74 radii of the earth.

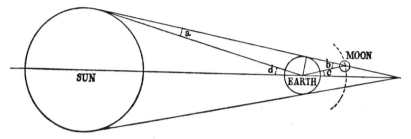

Fig. 10. Shadow Cone of the Earth

Another ingenious method of determining the moon's parallax was ascribed to him: from the measurement of the size of the earth's shadow where it is traversed by the moon. From fig. 10, representing the plane section of the shadow cone and the spherical bodies, we can see at once a simple relation of the angles designated by letters. In the triangle moon-earth-sun, angle a + angle b = angle c + angle d, or, calling them by their names: parallax of the sun plus parallax of the moon are equal to the sun's radius plus the semidiameter of the shadow, as seen from the earth. Since the parallax of the sun is very small, the moon's parallax is found, with a slight error only, by adding the apparent semidiameters of the sun and the shadow.

Hipparchus also is assumed to have made the first catalogue of fixed stars, with their place in the sky expressed by longitude and latitude with respect to the ecliptic. There are reasons to suppose that it formed the main part of the catalogue included by Ptolemy in his own work three centuries later and that it contained nearly 850 stars, to which Ptolemy added another 170. The instrument used to determine the positions is not mentioned; it was probably of the kind later called an 'armillar sphere' or 'armilla'. The Roman author Pliny (c. AD 70) gives as the reason why he undertook this work that Hipparchus 'discovered a new star, and another one that originated at that time', and for this reason counted them and determined their positions. This short sentence does not allow us to decide whether during Hipparchus' life a 'nova' really did appear, or whether he saw the appearance and disappearance of a variable star, such as Mira in the Whale. It might also be that his work resulted simply from an increasing interest in the stars and constellations at that time. This interest, as a social phenomenon, manifested itself in his critical commentary on the poetical description of the starry heavens by Aratus and in the fact that this work is the only one among Hipparchus' writings that has been preserved. The exact star catalogue may thus be considered its scientific counterpart.

This interest goes back to another Babylonian influence upon the astronomy of the Western world, that of astrology. In the centuries before Alexander, Greek astronomy in general had been free from astrology, though the belief that the weather was influenced by the heavenly bodies was often expressed (for example by Eudoxus). But in the following centuries astrology, together with all oriental science, penetrated the Hellenistic world rapidly. We have already mentioned that in the third century BC the Babylonian priest Berossus came to the island of Cos as teacher and historian of Babylonian culture. In Alexandria, about 160 BC, a treatise of astrology appeared, with the names of Nachepso and Petosiris as the alleged authors, supposedly two Egyptian priests of earlier times; all later books on astrology have borrowed freely from this early source. The philosophic school of the Stoics contributed to the spreading of astrological belief; it fitted well into their doctrine of the unity of all mankind with the universe. Thus Posidonius, already mentioned as author, scientist and philosopher of renown— whose works, however, are all lost with the exception of some fragments preserved by others—taught a universal sympathy, i.e. consonance and fellow-feeling, between the earthly and the heavenly worlds. His, too, was the well-known exposition of the difference between astronomical and physical science: for the astronomer every explanation is valid that saves (i.e. represents) the phenomena, whereas the physicist must deduce the truth, explaining them from the first causes and working forces. So, if it does not ascertain the physical truth, what is the use of astronomy, i.e. of the astronomical computation of the course of the stars? It has to serve a higher purpose—the prediction of human destiny.

Astrology, as an ever-spreading mode of thinking, permeated first the Hellenistic and then the Roman world, where it found fertile soil. Especially was this so when Rome, after conquering all the kingdoms of ancient culture, and swollen with the immense riches captured from all their treasuries, was afflicted by a century of cruel civil wars that imperilled the life and prosperity of every citizen. 'The collapse of the great Hellenistic monarchies under the pressure of Rome and the ruin of the flourishing Greek cities under the Roman domination, the exhausting wars, the repeated cruel and bloody social revolutions, the general misery of the times and the growing oppression of both rich and poor in the Hellenistic East'—this was how the historian M. I. Rostovtzeff[67] described the conditions which brought about the sudden bankruptcy of science and learning in the first century BC. Astrology now became an integral part of the world conception and of the culture of antiquity, not only among the masses in the form of crude superstition and a belief in soothsaying, but also as a theoretical doctrine of scientists and philosophers.

Astrology, in the Greek and Roman world, however, had to acquire a character far different from what it had been in old Babylon. There the gods had inscribed the signs of their intentions towards the human world in the irregular courses of the bright luminaries across the sky; but it was only the fate of the larger world, of the states, the monarchs, the peoples at large, to which the great gods—the rulers of the brilliant planets—paid attention: for his individual interests man had to go to his local deities. In the Greek and Roman world, whose citizens had a strong feeling of individuality, astrology had to acquire a more individual character; it had to interest itself in the personal lot of everybody. Moreover, now that the planets had become world bodies describing their orbits in space, calculable by theory, their character had changed; their course was not a sign but a cause of the happenings on earth. The life of every man, like meteorological and political phenomena, was subject to the stars. So the horoscope—which deduces the life-course of a person from the position of the stars at the moment of his birth—became the chief purpose and content of astrological practice. The positions of the planets relative to one another and to the stars, their risings and settings, the position of the constellation relative to the horizon, were the most important data. The oftener the supposition of simple relations was put to shame by the events, the more complicated and arbitrary became the rules and directions. As a result of the mutual cultural influences we now see personal horoscopes appearing sometimes in the Chaldean cuneiform inscriptions.

Chiefly because of this universal spread of the astrological concept of life, astronomy in antiquity stood in the centre of public interest. It was at the time the only knowledge deserving the name of science; and it was the most practical science, more than just a basis for the calendar. Through the intimate connection of the stars and human life, astronomy became man's most important form of knowledge. Allusions to celestial phenomena are common and numerous in the works of Roman poets and prose writers, which cannot be well understood without knowledge of astronomy. Moreover, there was an extensive popular astronomical literature, most of which had since been lost. Eudoxus is said to be the first to have given a detailed description of the stars and the constellations. It formed the basis of a great poem, *On the Phenomena* by Aratus, who lived (about 270 BC) at the court of the Macedonian King Antigonus Gonatus, himself a pupil of the Stoics. This poem was famous throughout antiquity; it was read in the schools and was the source of all the mythological tales concerning the heroes and animals represented in the constellations, which could be found in our books on astronomy up to the nineteenth century. From the great mathematician Euclid a work has been preserved that gives the mathematical theory to Aratus' poetry: a lucid exposition of the circles on the celestial sphere, its rotation, the ensuing phenomena of rising and setting—in short, all that afterwards was called 'sphaerica', the theory of the (celestial) sphere. The universal esteem for Aratus' poem explains why Hipparchus found it worth while to give a detailed criticism, with corrections, of the poem as well as its source, Eudoxus.

Many other popular astronomical works are known from these and later centuries: a Latin poem of Manilius (about AD 10) on the stars, highly praised as a literary work; and still later a poetical description of the heavens by Hyginus. Then there is a thorough work by Geminus (about 70 BC) on the whole of astronomy, that provides us with many valuable details on its history; and a work of Cleomedes (contemporary of the Emperor Augustus) entitled *Circle Theory of the Celestial Phenomena*. The number and spread of such books indicate how strongly astronomy as a living science was rooted in the society of the time.

Besides such books, there existed, for the use of astrological predictions, almanacs with the positions of the planets computed ahead, fragments of which have been found in Egyptian papyri. Further, celestial globes are mentioned as an aid to picturing the heavens and for instruction; for example, an Archimedes globe, brought to Rome by the consul Marcellus, after the conquest of Syracuse. On such globes the stars themselves were often omitted and only the figures of the constellations were depicted, naturally, since the astrologically important effects, as a rule, came from the constellations and not from the separate stars. The Farnese Atlas may serve as an example; in it is pictured a statue of Atlas, the giant, bearing the celestial globe on his neck; an engraving made from it in the eighteenth century by Martin Foulkes, and published in Bentley's edition of Manilius' poem, has been reproduced here on a reduced scale (see plate 2). Attention may be drawn to the four horns on the head of the dog, probably representing the fiery beams emanating from the Dog-Star Sirius. Mechanical models of the world system, a kind of orrery, seem also to have been constructed; but what they represented is an already more highly-developed world structure.

CHAPTER 13

THE EPICYCLE THEORY

THE epicycle theory offered the first satisfactory explanation for the irregular course of the planets. The planet was assumed to describe a circle (epicycle), the centre of which described a larger circle about the earth, which occupied the centre of the universe. This theory was a natural sequel to Heraclides' proposition that Venus and Mercury describe circles about the sun. Whilst they are seen to oscillate from side to side of the sun, the sun carries them along in its yearly course; seen from the earth, they must alternately go a long way in the same direction with the sun, but more rapidly, and a short way back in the opposite direction. Their apparent motion relative to the stars then has the same irregular character as that shown by the other planets. Since it is composed of two regular circular motions, it is plausible to conceive the motion of the other planets (Mars, Jupiter and Saturn) as a combination of two circles also; a larger circle (deferent, leading circle) about the earth as centre, along which the centre of the smaller circle (the epicycle) moves. This centre is a void point here, whereas for Venus and Mercury it was occupied by the sun.

Two points of revolution are thus needed to express the planet's motion. We have to bear in mind here that in Greek science the epicycle was held to be attached to the radius of the large circle; in revolving, its lowest point (nearest to the earth) always remained the lowest point. The planet's time of revolution along the epicycle is always reckoned from the lowest or the highest point until the same point is reached again; this is the synodic period of the planet. Its passage through the lowest point is the middle of the retrograde motion. For Venus and Mercury it is the inferior conjunction with the sun; for the other planets it is the opposition to the sun.

The epicycle theory offered a far simpler and more accurate representation of the variable course of the planets than did the rotating spheres of Eudoxus and Aristotle. Moreover it explained their variable brightness as a result of their varying distances from the earth. These distances could be computed easily from the sizes of the circles. The relative size of the epicycle and the deferent for Venus and Mercury

133

follows from their greatest elongation to the right-hand or the left-hand side of the sun; from 46° for Venus and 22° for Mercury, we find the ratio of the radii 0.72 and 0.37. For the other planets this elongation must be taken relative to the invisible, regularly progressing epicycle centre. For Mars 42° is found; then the epicycle radius is 0.67 times the radius of the great circle, and the distances of Mars at the highest and lowest points have a ratio of 1.67 to 0.33. This explains its great variations in brilliancy.

It has often been asked why the Greek astronomers, having become acquainted with the heliocentric world system of Aristarchus, went back to the more primitive geocentric system of the epicycles. We surely cannot find the reason in what Paul Tannery said, that ancient Greece in these centuries lacked the genius for complete renovation;[68] in this respect the Greeks have certainly shown their superiority. The reason must be, first, that the epicycle theory was the most natural course of development for Greek science. The epicycle theory *was* the renovation wanted. The heliocentric world structure devised by Aristarchus was a fantastic stroke of genius, not a necessary consequence of facts. What bore it out was the bodily size of the sun; though it was argued on the other hand that in the human body the heart, the seat of life, was also outside the bodily centre of mass.

A new world structure had to be a theory of the planetary motions. The motions of the planets at the time were known only in rough outline; they had first to be observed accurately and represented in the most natural way, which was effected in fact by the epicycle theory; it constituted the direct geometrical representation of the visible phenomena.

Secondly, social influences probably also played a role, especially the general belief in astrology, which was the practical application of science. Astrology did not need theories on the physical nature of the celestial bodies, but only practical tables for computing their apparent motions. It was not merely indifferent but flatly hostile to physical structures which might disturb the primitive belief that the stars in their courses pronounced the fate of human beings. So it is easy to understand the saying of Cleanthes, the leader of the Stoics, that 'it was the duty of the Greeks to indict Aristarchus of Samos on the charge of impiety for putting the Hearth of the Universe in motion' (i.e. the earth).[69] The epicycle theory, representing the appearances, had to take preference over Aristarchus' world structure.

The epicycle theory must have originated, perhaps gradually, in the third century BC. The first certain report on it is connected with the name of the great mathematician Apollonius of Perga (about 230 BC), the founder of the theory of conic sections. Ptolemy hands down one of his geometrical propositions which derives the stations of the planets. The Chaldean astronomers had carefully noted these stations and in their tables had computed them as important basic elements of knowledge of the course of the planets. Greek theory had to show that it was equally or better able to solve the same problem, to foretell the time and place of the stations. Apollonius did so by reducing the problem to geometrical levels, drawing a line from the earth which intersects the epicycle in such a way that the sections had a definite ratio. Because of its importance, the demonstration is given in Appendix B.

In the next century Hipparchus occupied himself with the epicycle theory and gave it its classical form. He demonstrated that the motion in an epicycle, described in the same period but in opposite direction with the concentric circle which is the orbit of its centre, is identical with the motion along an eccentric circle. This can at once be seen in fig. 11, where the points 1, 2, 3 and 4 occupy a circle shifted upward. Both models, therefore, can be used to represent the variable velocity along the ecliptic which the sun shows and which the planets also show besides their oscillations. So it was natural to render these oscillations by epicycles and to choose the eccentric circle for the variable velocity along the ecliptic.

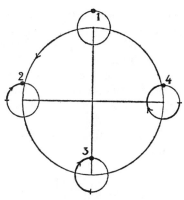

Fig. 11

According to Ptolemy, Hipparchus had indeed recognized that the retrogradations of the planets were different at opposite sides of the ecliptic; hence the motions were more irregular than had been previously assumed. From a somewhat obscure statement by Pliny, two centuries later—that the farthest apogee of Saturn is situated in the Scorpion, of Jupiter in the Virgin, of Mars in the Lion—it appears that in an earlier time, possibly in Hipparchus' own time, there was also some qualitative knowledge of these irregularities. But, Ptolemy says, he did not have at his disposal a sufficient number of observations from predecessors to work out a numerical theory completely. For that reason he restricted himself to assembling new data: 'a man who in the whole field of mathematics had reached such a profoundness and love of truth'[70] could not content himself with giving a theory in general terms only. He had to determine the numerical values in the orbital motions from the phenomena and to show that they could be adequately represented by uniformly described circles. This, however, was not possible from the available data.

This was the work of Ptolemy himself, who thus brought the epicycle theory to completion. He says that he first corrected the planetary periods of Hipparchus by means of his own observations, and found them to be as follows:

Saturn	57 s.p. = 59 y. + $1^3/_4$ d. = 2 r.	+1° 43′
Jupiter	65 s.p. = 71 y. + $^{49}/_{10}$ d. = 6 r.	−4° 50′
Mars	37 s.p. = 79 y. + $3^{13}/_{60}$ d. = 42 r.	+3° 10′
Venus	5 s.p. = 8 y. − $2^3/_{10}$ d. = 8 r.	−2° 15′
Mercury	145 s.p. = 46 y. + $1^1/_{30}$ d. = 46 r.	+1°

(s.p. = synodic periods; y. = years; d. = days; r. = revolutions)

These are the same multiples as used by the Babylonians. For the first three named planets a synodic period, i.e. a revolution in the epicycle, has passed when the sun overtakes the planet; neglecting the small remainders, these periods are $\frac{59}{57}$, $\frac{71}{65}$ and $\frac{79}{37}$ years, and the periods of revolution along the deferent are $\frac{59}{2}$, $\frac{71}{6}$ and $\frac{79}{42}$ years. For the two others the period of revolution is exactly one year. As to exactitude, these periods, with the small remainders added, are entirely comparable with the Chaldean values; the angular remainders may be inaccurate up to half a degree.

Ptolemy, in establishing the motions of the first three planets, had to split the problem up into two parts. First, the motion of the epicycle centre along the deferent (which he always calls the 'excentre'), and then the motion of the planet along the epicycle. For the first purpose he had to eliminate the oscillations due to the epicycle and to observe the planet when it was seen exactly in the same direction as the epicycle centre, i.e. when it stood in front of this point. How could he know that? The basic principle of the epicycle theory is that the radius of the epicycle, which connects its centre with the planet, turns uniformly in the same stretch of time in which the sun describes its circle and thereby has always the same direction in space as has the radius in the solar orbit. Hence the planet will stand exactly before the centre (as fig. 12 shows) when its longitude is 180° different from the sun's longitude as seen from the centre of its orbit. In other words, the planet stands in opposition not to the real sun but to the 'mean sun', which performs its yearly course exactly uniformly; where the real sun is seen from the earth does not matter. Then the observed longitude of the planet is the desired longitude of the epicycle centre.

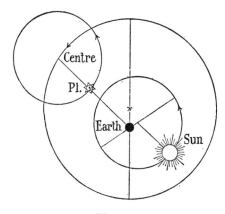

Fig. 12

For practical computation, a number of observations at consecutive days about opposition are necessary. Ptolemy's observations of the planet consisted partly in conjunctions with, or near approaches to, bright stars or the moon, partly in direct measurements with an instrument with graduated circles, which he called 'astrolabon' and which corresponds to the later armillas. From the observed longitudes the exact moment of opposition to the 'mean sun' between them can be derived, as well as its longitude.

If we know for a point (the epicycle centre) moving on the eccentric circle for three distant moments the direction as seen from the earth, the position of the earth within this circle can be determined. It can either be constructed in a geometrical drawing or computed numerically. Since the excentre is described uniformly, the three positions on the circle are known from the intervals of time. Then the problem is identical with the so-called 'Snell's problem' in geodetics: to derive the position of a station by measuring there the directions to three surrounding known stations; this problem can be solved in a direct way.

Here Ptolemy met with a difficulty. He says: 'Now we found, however, with continued exact comparison of the course given by observation and the results from combinations of these hypotheses, that the progress of the motion cannot be quite so simple. . . . The epicycles cannot have their centres proceed along such eccentric circles that, seen from the centre [of these circles], they describe equal angles in equal time. . . . But the latter bisect the distance between the point from which the motion appears to be uniform, and the centre of the ecliptic.'[71] Expressed in another way: the epicycle centre does describe an eccentric circle, i.e. a circle with its centre outside the earth, but in such a way that its motion is apparently uniform when viewed, not from this centre, but from another point (the *punctum aequans*, 'equalizing point', 'equant') situated as far to the other side of the centre as the earth is on

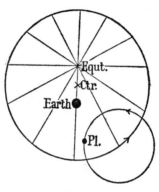

Fig. 13

this side. This means, as figure 13 shows, that in reality the excentre is not described uniformly by the epicycle centre; nearest to the equalizing point (in the apogee) it goes more slowly, at the opposite side more rapidly. Whereas the basic principle of Greek cosmology— the circular and uniform motion of all celestial bodies—is paid lip service, actually, on the pretext that the motion appears to be uniform as seen from another point, it is violated.

But in this way the phenomena offered by the planets could be represented in the most perfect manner. Ptolemy does not indicate by what argument or by what observation he arrived at this theory. He only says: 'we found that . . .' It is, however, easy to see what phenomena must have led to it. The distance of the earth from the centre of the great circle—which is its real eccentricity—determines the variable apparent size of the epicycle, visible in a variation of the oscillations of the planet to both sides of the epicycle centre. The distance of the earth from the point at which the angular motion is seen to be uniform determines with what variations of velocity the epicycle centre seems to move along the ecliptic; the precise situation of the circle and its centre in this respect is of secondary importance only. The observations must have shown that in the latter case an eccentricity is found twice as large as in the first case; so the distance of the equalizing point from the earth is twice the distance from the centre to the earth. Or, in his own words elsewhere: 'The eccentricity deduced from the greatest deviation in the anomaly relative to the ecliptic was found to be double the amount of the eccentricity derived from the retrograde motion in the cases of the largest and smallest distance of the epicycle.'[72] That the ratio of these eccentricities should be exactly two was a simple supposition which proved, however, to be a lucky hit. For the sun, itself proceeding along its eccentric circle, such a distinction could not be made; what eccentricity had been found for it by Hipparchus corresponded to the large or double planetary eccentricity.

Ptolemy's first task was now to derive the eccentricity and the direction of the apogee of the deferent for Mars, Jupiter and Saturn from three oppositions. For Mars he observed the following values:

(1) in the fifteenth year of Hadrian at 26/27 of the (Egyptian) month Tybi (i.e. December 15/16 of AD 130) at one hour in the night, the longitude was 81°;

(2) in the nineteenth year of Hadrian at 6/7 of the month Pharmuthi (i.e. at February 21 of AD 135) at nine hours in the evening it was 148° 50';

(3) in the second year of Antonine at 12/13 of the month Epiphy (i.e. May 27 of AD 139) at 10 hours in the evening it was 242° 34'.

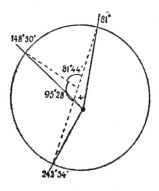

Fig. 14

From the intervals in time, after subtraction of an entire number of revolutions, the angles between the directions as seen from the equant are computed: 81° 44' and 95° 28'; between the directions as seen from the earth these angles are 67° 50' and 93° 44'. These data are represented in figure 14. What is required is the position of the earth within the circle. The problem cannot be solved directly; Ptolemy solved it by successive approximations. First he supposed a circle about the equant as centre. Then the problem, as stated above, could be solved directly by the mathematical means at his disposal: the propositions of Euclid and the table of chords for every arc, computed by himself, the prototype of the later sine tables. The result is the large eccentricity, half of which is the amount the circle must be displaced. He now computes how much the directions as seen from the earth are changed by this displacement, and with the corrected values the computation is repeated in a second approximation. It affords very small corrections for a third approximation, which entirely satisfies the original data. In these computations, which are communicated *in extenso*, Ptolemy applied the method of convergent approximation which played such an important role in later mathematics.

The results obtained in this way for the total eccentricity and the longitude of the apogee are: for Mars $\frac{72}{360}$=0.200 and 115° 30'; for Jupiter $\frac{33}{360}$=0.092 and 161°; for Saturn $\frac{41}{360}$=0.114 and 233°. If we compare them with what, according to modern knowledge, were the true values at the time—for Mars 0.186 and 121°; for Jupiter 0.096 and 164°; for Saturn 0.112 and 239°—it appears that his representation of the planetary orbits was highly satisfactory.

The second problem, the size of the epicycle relative to the excentre, had to be solved by observation of the planet outside its opposition, when it stands far to the side on its epicycle. For this derivation Ptolemy, curiously, gives an observation of Mars only three days after opposition (fig. 15): 'Since it is our further aim to fix numerically the relative size of the epicycle, we chose an observation made nearly three days after the third opposition, in Antonine year two at 15–16 Epiphy (i.e. May 30–31, AD 139), three hours before midnight.'[73] With the astrolabon directed by means of Spica, the longitude of Mars was found to be 241° 36'. Mars at the same time was found to stand 1° 36' east of the moon (which, according to the tables stood at 239° 20', corrected 40' for parallax giving 240° 0'), hence 'also in this way a position of Mars was found at 241° 36', agreeing with the other result'. The exact concordance of the minutes of arc clearly indicates that they are fitted on purpose; since the time of observation is given in full hours only and the moon moves 33' per hour, any value within this range might be chosen so as to give the same minutes as the direct measurement.

With these data it is again simply a computation by geometry, consisting of a calculation of sides in a triangle with known angles—a triangle, surely, with very small angles, 2° 43' at the earth and 1° 8' at the epicycle centre, so that errors of measurement not greater than $\frac{1}{4}$° can strongly vitiate the result. Yet the result, 0.658 for the radius of the epicycle, is almost exactly equal to the true value, 0.656. So it may be taken for granted that his value does not rest solely on this observation but has been derived from further observations at greater distances from the opposition. The derivation in his book, then, must be considered rather as an example to show the method used. Computed

Fig. 15

with modern data, the longitude of Mars at the moment of his observation was 242° 16', hence 0° 40' larger. It is well known that, owing to an error of about 1° in his vernal point, all Ptolemy's longitudes are too small; for Spica the error in his catalogue was 1° 19', so that in the measurement of Mars an error of 39' was made. For Jupiter and Saturn his derivation was based on observations far from the opposition; the result ($\frac{23}{120}$=0.192 and $\frac{13}{120}$=0.108) compares well with the values from modern data of 0.192 and 0.105.

With these three planets the epicycle theory appears in its full power. If the ancient astronomers had had extensive series of observations at their disposal, they would have seen how well a computation with this theory was able to represent them. A comparison with modern theory may convince us that this simple structure of circles in space excellently represents the details in the planetary motions which formerly looked so capricious and intricate.

With Venus and Mercury things were not so simple. Here not the epicycle but the deferent is performed in pace with the sun. The epicycle theory does not say, however, as Heraclides did, that the sun is the centre of their movements but demands solely that the epicycle centre revolves in the same period of time as the sun and that its radius remains always parallel to the radius of the sun in its circle. Whereas the greatest elongations of these planets east and west of the sun are different and variable in amount, this must be represented theoretically by uniformly described concentric epicycles.

141

For Venus the differences are small; the great elongation from the sun is always between 45° 53′ and 46° 43′. Here a circle with its centre near to the sun fits excellently as epicycle. In order to derive the figure of the deferent, Ptolemy made use exclusively of a number of greatest elongations to both sides of the sun. He first derived its apogee at 55° of longitude. In a direction 90° different—at a longitude of 325°—the greatest eastern and western elongation were found to be 48° 20′ and 43° 35′. Their difference, 4° 45′, shows that the epicycle centre is 2° 22½′ ahead of its mean place, which is identical with the place of the mean sun; hence the eccentricity is $\frac{1}{24}$. Apparently, Ptolemy did not perceive—at least he does not mention it anywhere—that this is the same eccentricity and nearly the same apogee as he had formerly derived for the sun, so that the bodily sun itself occupies the centre of the Venus epicycle. If it is a matter for wonder that the observations of Venus here give the same erroneous eccentricity as those of the sun (the real value was $\frac{1}{30}$), one has to remember that Ptolemy's observations were elongations, i.e. distances from the real sun, from which the longitudes of Venus were computed by means of reductions taken from solar tables based on the erroneous eccentricity of $\frac{1}{24}$; so that he got as 'result' what he had put in under another name.

To derive the size of the epicycle, Ptolemy took a greatest elongation at the apogee and one at the perigee of the deferent; he found 44° 48′ and 47° 20′, from which the two distances 0.705 and 0.735 can be computed. Their mean, 0.720, is the radius of the epicycle; half their difference expressed as a fraction, $\frac{1}{48}$, of the radius is the eccentricity of the observing point, the earth. It is half the amount of the total eccentricity $\frac{1}{24}$ just found; hence the Venus deferent has an equant. It is a remarkable fact that in its character of Venus deferent the solar circle exhibits an equant, that could not be derived from itself. Since Ptolemy was not aware of their identity, this fact had to be rediscovered afterward by Kepler. There are, however, many points of doubt in the data used.

With Mercury the differences and difficulties were far greater; the greatest elongations vary between 17° and 28°. So Ptolemy did not succeed in establishing a satisfactory theory. To a considerable degree this was due to the difficulty of observing the planet; it is visible only in strong twilight near the horizon, at a great distance from other bright stars, which hampers exact observation. So the time and, because of its rapid motion, the position of its greatest elongation were difficult to establish. Another source of trouble lay in the theory adopted; circular orbits were not likely to express the asymmetric oscillations of the planet. Thus Ptolemy came to assume an oval orbit for the epicycle centre, with the earth on the major axis outside the centre. Such an oval can be produced by a combination of circles, namely, by making the centre of the deferent describe a small circle in the opposite direction twice during one revolution. Some chief features of the motion of Mercury could indeed be represented in this way, but in a less accurate and more complicated way than with the other planets. It shows the many-sided possibilities of the epicycle theory; but Mercury certainly was too difficult an object for satisfactory treatment.

Finally, Ptolemy included in the epicycle theory the deviations of the planets to the south or the north of the ecliptic, expressed in their latitudes. It was done by introducing, for Mars, Jupiter and Saturn, small inclinations of the deferent to the ecliptic and of the epicycle to the deferent. He had no reason—as we have—to suppose the epicycle to be parallel to the ecliptic; so he had to determine two inclinations for each planet. He used some few crude data only: from a latitude of Jupiter of 1° in conjunction and 2° in opposition he found the inclinations to be $1\frac{1}{2}$° and $2\frac{1}{2}$°. By observation, he knew that the planet's deviation in latitude on the nearest point of the epicycle, at one side of the ecliptic was to the north, at the opposite side to the south. According to Greek theory, this deviation should remain the same in going around the ecliptic, because the epicycle was assumed to be fixed to the radius of the deferent. Ptolemy therefore had to correct it by a special contrivance: the nearest point of the epicycle revolves along a small vertical circle in one period of revolution, so that it is lifted up and down, oscillating between the extreme northern and southern deviation. To the already complicated structure of the orbits of Venus and Mercury he added oscillations up and down, two of the epicycle relative to the deferent and one of the deferent relative to the ecliptic, all directed by small vertical circles. And for those who should think these mechanisms too complicated for the celestial bodies he added some philosophical consolations: 'Let nobody, looking at the imperfection of our human contrivances, regard the hypotheses here proposed as too artificial. We must not compare human beings with things divine. . . . What more dissimilar than creatures who may be disturbed by any trifle and beings that will never be disturbed, not even by themselves? . . . The simplicity itself of the celestial processes should not be judged according to what is held simple among men. . . . For if we should look at it from this human point of view, nothing of all that happens in the celestial realms would appear to us to be simple at all, not even the very immutability of the first [i.e. daily] rotation of heavens, because for us human beings this very unchangeableness, eternal as it is, is not only difficult, but entirely impossible. But in our judgment we have to proceed from the immutability of the beings revolving in heaven itself and of their motions: for from this point of view they would all appear to be simple, and even simple in a higher degree than what is regarded on earth as such, because no trouble and no pains can be imagined with regard to their wanderings.'[74] It must be added that afterwards, in tables for practical use, he considerably simplified the structure of the orbits.

Chaldean astronomy did not bother about the latitude of the planets. The Chaldean tables dealt with longitudes, in progress and retrogradation; the ecliptic as central circle of all planetary orbits seems to have remained unnoticed. Latitude occurs only for the moon, because here it is necessary for the eclipses, and for the exact computation of the crescent. In this point the superiority of Greek over Babylonian astronomy is manifest; it saw the celestial luminaries as bodies having definite orbits in space. The epicycle theory, in the definite form given by Ptolemy, stands out as the most mature product of ancient astronomy.

CHAPTER 14

THE CLOSE OF ANTIQUITY

A T the commencement of our era, all civilized peoples of antiquity living around the Mediterranean, the ancient world sea, were assembled in the Roman Empire. A lively sea traffic connected them into one economic unit; from the opposite shore and from the conquered East riches and foodstuffs flowed to privileged Italy, especially to the ruling capital, Rome. The peoples who lived beyond the frontiers, barbarian tribes in Europe and Africa and the Asiatic empires in the Orient, had to be repelled by strong armies in continual frontier warfare.

Inside the enormous realm peace reigned under the emperors of the first two centuries, interrupted only once by a contest between a couple of generals over the emperorship. The growth of agriculture, trade and commerce, extending ever more evenly over the provinces, led to the spread of intellectual culture; originating from the eastern lands that remained Greek in language and Hellenistic in character, it now reached the rustic Western conquerors. Here it remained an imitation; in the field of natural sciences and art the Romans produced little that was original.

This holds for astronomy too. The contributions of the Romans and their subjects are soon enumerated. Cleomedes—his name is Greek—has already been mentioned for his manual of astronomy; the simultaneous visibility of the sun and the eclipsed moon opposite above the horizon was explained by him in the right way: refraction of the rays of light near the horizon. Ptolemy used observations of occultations of stars by the moon made originally by Menelaus at Rome in AD 92 and by Agrippa in Bithynia in AD 98; the first name, Menelaus, is also Greek. From Plutarch, the author of the famous *Lives*, in the second century AD, we have a dialogue *On the Face in the Moon*, wherein the moon is described as an earthlike body, with mountains and depths casting shadows, a view far more modern than that of Aristotle. Not as an extension but as an application of science, Julius Caesar's calendar reform must be noted here. To do away with the confusion caused by arbitrary changes, all relation to the moon and all intercalation of months was abolished. The historian Suetonius described the situation in these words: 'Then, turning his attention to the reorganization of the state, he reformed the calendar, which the negligence of the pontiffs had long since so disordered, through their privilege of adding months or days at pleasure, that the harvest festivals did not come in summer, nor those of the vintage in the autumn [consequently, objects for offerings were lacking]. And he adjusted the year to the sun's course by making it 365 days, abolishing the intercalary month, and adding one day every fourth year.'[75] What constituted the character strength of the Romans, their sense of social-political organization, created a mode of time-reckoning destined to dominate the entire future civilized world.

The nature of the Roman Empire, however, gave a new and special character to the scientific work of these times. In former centuries the rise of trade and commerce and the rivalry of the small states and towns in Hellas and the Orient had awakened creative initiative in the growth of new ideas. Now that civilized mankind had become an all-embracing unity, the character of scientific work also tended towards an all-embracing unity. The study of nature became the assembling of all knowledge into one body of science. Instead of the fresh ingenuity of original thinkers, came the all-encompassing learning of the compilers. Instead of the ebullition of new ideas, came the organization of all that past centuries had wrought—often, of course, elucidated by genuine ideas—into encyclopaedic works. Scientifically the centuries of the Roman emperors were the centuries of the great collective works: of Strabo, and later of Ptolemy, on geography; of Pliny on natural history; of Galen on medicine. They constituted the completion of ancient science.

It was the same with astronomy. As a compendium of Greek astronomy there appeared Claudius Ptolemy's *Thirteen Books of the Mathematical Composition* (*Matematikè Suntaxis*). He lived at the time of the emperors Hadrian and Antonine, a contemporary of Plutarch. His years of birth and death are unknown; observations by him are known from the years AD 127 to 151. He lived in Alexandria and belonged entirely to the Greek cultural world. His work was far more than a compilation of former knowledge. Ptolemy was no compiler but a scientific investigator himself; with Hipparchus he was the greatest astronomer of antiquity. He improved and extended the theories of his predecessors; he added to science through his own observations and explanations. We have already seen how he brought the epicycle theory to completion by giving it a precise form and numerical values.

Ptolemy's work is a manual of the entire astronomy of the time. It is true that he deals with the celestial sphere only—the stars, the sun, the moon and the planets—and does not speak of comets. Those, he considered, did not belong to astronomy; although the philosopher Seneca (about AD 70) in a much-quoted sentence spoke of them as celestial bodies, whose far-stretched orbits, on which they were mostly invisible, would certainly be discovered in later times, Ptolemy sided with Aristotle, who considered them to be earthly phenomena in the higher realms of the air.

Also in harmony with Aristotle is the basic structure of the universe which Ptolemy expounded in his first chapters. The heavenly vault is a sphere, and as a sphere it rotates about its axis; the earth, too, is a sphere and occupies the centre of the celestial sphere; as to its size, the earth is as a point relative to this sphere, and it has no motion to change its place. The arguments are the same as Aristotle's; moreover—without even mentioning Aristarchus or Heraclides—he argued against those who had expressed an opposite opinion:

'Some philosophers think that nothing prevents them from assuming that the heaven is resting and that the earth in nearly a day rotates from west to east. . . . As to the phenomena of the stars, nothing would prevent this by its greater simplicity from being true; but they did not perceive how very ridiculous this would be with regard to the phenomena around us and in the air. In that case, contrary to their nature, the finest and lightest element [the ether?] would not move at all or not differently from those of opposite nature—whereas the bodies consisting of atmospheric particles show a tendency to more rapid motion than the earthly matter. Moreover, the coarse-grained heaviest bodies would have a proper strong and rapid uniform motion—whereas, as everyone knows, heavy earthly things hardly can be put in motion. If we should concede this, they certainly would have to admit that a rotation of the earth would be more violent than all motions that take place on her, and in a short time would have so rapid a reaction that everything not fixed to her would appear to have only one single motion, contrary to hers. And we would never see a cloud, or anything that flies or is thrown, move towards the east, because the earth would outdo them in the motion towards the east, so that all the others, outdistanced, would appear to move towards the west.

'If they should say that the air is carried away with equal velocity, then also the earthly bodies in it would be seen to lag behind. Or if carried away by the air as if firmly attached to it, they would never be seen to move forward or backward, and all that flies or is thrown would have to stay at its place without being able to leave it; as if the motion of the earth would deprive them of any ability to move, be it slow or quick.'[76] These are the arguments which Ptolemy advanced for his geocentric world system.

In a geometrical introduction a table of chords for angles increasing by half a degree is first computed and presented; the proposition used, on quadrangles in a circle, still appears in our modern textbooks as 'Ptolemy's theorem'. Since the Greek numerical system did not know of decimal fractions, the chords are given in sexagesimals, with the diameter taken as 120 units, a token of Babylonian influence (so for an arc of 90° the chord is given as 84, 51, 10, i.e. reduced to our notation, 0.707107). Throughout the work fractions in the length of lines are given in sixtieths.

Then propositions on plane and spherical triangles are derived which are needed farther on. Connected with them, quantities on the rotating celestial sphere are computed: viz. the time and duration of the rising of the different zodiacal signs, as well as the inclination of the ecliptic to the horizon, the meridian, and other vertical circles—all necessary in astronomical and astrological computations. Because they depend on the latitude of the place of observation, they are given for standard latitudes specified by the maximum length of daylight (from sunrise to sunset): from 12 hours at the equator, over $12\frac{1}{4}$, $12\frac{1}{2}$, $12\frac{3}{4}$, 13 hours, etc. (corresponding to latitudes of 4° 15', 8° 25', 12° 30', 16° 27', etc.), increasing up to 23 and 24 hours (at 66° and 66° 8' 40" of latitude).

Then the sun's motion is dealt with. The relevant quantities are the obliquity of the ecliptic, the length of the year, and the eccentricity of the sun's circular orbit. Ptolemy describes two instruments used to determine the obliquity; one is a graduated circle on a pedestal, within which a smaller circle with notches and index can be turned, so that, by means of the shadow, the meridian altitude of the sun can be read; a picture of this instrument was given later on by Proclus. The other is a graduated quadrant, serving the same purpose. He found the difference between the meridian altitudes of the summer and winter solstices always to be between 47° 40' and 47° 45', and he states that it is nearly the same value as that found by Eratosthenes and used by Hipparchus: $\frac{11}{83}$ of the circumference. Half of it, 23° 51$\frac{1}{3}$', is the value adopted for the obliquity of the ecliptic.

As to the orbit of the sun, he first mentions that Hipparchus stated the intervals between spring equinox, summer solstice and autumn equinox to be $94\frac{1}{2}$ and $92\frac{1}{2}$ days, and had derived from them an eccentricity of $\frac{1}{24}$. 'We, too, came to the result that these values are nearly the same today. . . . For we found the same intervals from exactly observed equinoxes and an equally exactly computed summer solstice in the 463rd year after Alexander's death' (i.e. AD 139–140).[77] These are AD 139, September 26, one hour after sunrise; 140, March 22, one hour after noon ($178\frac{1}{4}$ days later); and 140, June 25, two hours in the morning. 'The last interval is $94\frac{1}{2}$ days; for the interval from this solstice to the next autumn equinox there remain $92\frac{1}{2}$ days.' So he derives an eccentricity of $\frac{1}{24}$ and an apogee at 65° 30', both identical with Hipparchus' values. In reality, at his time the intervals were, according to modern data, 93.9 and 92.6 days.

The same equinoxes are used to derive the length of a year by comparing them with Hipparchus' values for 147 and 146 BC, 285 years earlier. He finds the interval to be $70\frac{3}{10}$ days more than 285×365 days; since $285 \times \frac{1}{4}$ days is $71\frac{1}{4}$ days, he concludes 'that in 300 years the return of the sun to the vernal equinox takes place nearly a day earlier than would correspond to a year of $365\frac{1}{4}$ days'.[78] And again he states his complete accord with Hipparchus' length of the year, 365 days 5 hours 55.2 minutes. This length, however, was seven minutes too great; hence his interval surely was one day too long. The moments of his equinoxes, when computed from modern data, are in fact found to have taken place one day earlier than he reports as his observational result.

The length of the year or, rather, the difference between the return to the same star (the sidereal year), and the return to the equinoxes (the tropical year) is directly connected with the precession. Whereas other authors do not mention this discovery by Hipparchus, Ptolemy understands its importance and confirms its amount. 'In comparing the distance of the stars to the solstices and equinoxes with those observed and noted by Hipparchus, we also found that a corresponding progression in the direction of the signs had taken place.'[79] Comparing an observation of Regulus made by himself (AD 139) with one by Hipparchus, he finds that it had proceeded 2° 40' in the intervening 265 years, hence 1° per 100 years. Then he compares a number of observations of occultations or conjunctions of different stars (Pleiades, Spica, β Scorpii) with the moon, made by Timocharis in Alexandria, with analogous observations made by Menelaus at Rome and Agrippa in Bithynia, and again he derives an increase in longitude of 1° in 100 years. Yet we know that the amount given is far too small and that its real value is 1° in 72 years; the real displacement since Hipparchus was 1° greater and since Timocharis it was $1\frac{1}{2}$° greater.

Because of these contradictions, modern astronomers have often severely criticized Ptolemy, claiming that not only was he so possessed by blind faith in his great predecessor that, without criticism, he adopted his values, but that for this purpose he even fabricated or fashioned and doctored his own observational results, i.e., he falsified them to make them agree. In Delambre's great work, *Histoire de l'astronomie ancienne* (1817), we read: 'Did Ptolemy himself make observations? Are not those which he says he made but computations from his tables and examples for the purpose of understanding his theories better?' Farther on he adds: 'As to the main question, we cannot see how to decide it. It seems hard to deny absolutely that Ptolemy made observations himself. . . . If, as he says, he had in his possession observations in greater number, we may reproach him that he did not communicate them and that nowhere does he tell what might be the possible error of his solar, lunar, and planetary tables. An astronomer who today acted in this way would certainly inspire no confidence at all. But he was alone; he had no judges and no rival. For a long time he has been admired on his own word.'⁸⁰ It is clear, however, that it is not right to judge Ptolemy's work by the usages and standards of modern science. The scientific outlook in antiquity was different from ours; there was no regular experimental research with acknowledged standards of judgment; observational results were not considered documents. Ptolemy's work was essentially theoretical; his aim was to develop and expound a geometrical picture of the world. Observation and theory, as we saw with Aristarchus, were at that time differently related to each other. Observation was simply an extension of experience, finding out where the celestial body happened to be. Theory was the new wonderful view of the world and the deeper insight into its structure; it was philosophy inquiring into the essence of things. The data used were instances or specimens, chiefly, but not necessarily, taken from observation with all its uncertainties, not intended as important new knowledge but often simple verifications, easily accepted, of respected earlier knowledge.

In the case of the precession, moreover, the sources of real error must not be overlooked. In Ptolemy's observation of Regulus, the setting sun was first compared with the moon, and the low moon was later compared with the stars; the effect of refraction must have been that too small a longitude of the star was found relative to the sun. If the sun's longitude was taken from the tables, their errors due to errors in the equinoxes were transferred to the result. In deriving the longitude of stars from occultations by the moon, the tables and parallaxes of the moon had to be used, which contained a considerable number of errors. With the time roughly noted, the velocity of the lunar motion, $1°$ per 2 hours, allows large adaptations of the moon's assumed position to the expected value. Many of Ptolemy's surprising data might be due to such causes.

Another set of data is worth mentioning here. In order to show that the precession is really a motion parallel to the ecliptic, so that on one side of the celestial sphere the stars move to the north, on the opposite side to the south, Ptolemy states the declinations (the distance in latitude from the equator) for a number of stars, as measured by Timocharis and Aristyllus, by Hipparchus and by himself. If from the extents of displacement in declination we now compute the value of the precession in longitude along the ecliptic, we find $46''$ per year, $1°$ in 78 years, not so very different from the true value. The measurements, not vitiated by the errors and complications introduced by solar and lunar tables,

appear to be good and reliable, with a mean error of not more than 8'.
Ptolemy, however, did not possess the trigonometrical formulas to
make such a computation.

After the sun, Ptolemy deals with the moon. Here he does not content
himself with confirming Hipparchus' results; he goes new ways of his
own. First he checks and corrects its motions. As to the mean daily
motion in longitude, expressed in sexagesimals (separated as usual by
commas) 13°, 10, 34, 58, 33, 30, 30, and the motion relative to the sun,
12°, 11, 26, 41, 20, 17, 59, corresponding to a synodic period of
29^d 12^h 44^m $3\frac{1}{2}^s$ and to a tropical year of 365^d 5^h 55^m he finds that no
correction is needed. To find the return to the apogee (the 'anomalistic
period'), he made use of three Babylonian lunar eclipses from 721 and
720 BC and compared them with three observed by himself in AD 133,
134 and 136.

To find the period of return to the same node, he compared an
eclipse from 491 BC with another in AD 125, selected in such a way that
all the other determining quantities were the same in both cases. Thus
he found a correction of 0° 17' in 854 years ¦for Hipparchus' return to
apogee, which was based on the ratio 269 : 251, and a correction of
0° 9' in 615 years for Hipparchus' return to the node. The values in
degrees and its sexagesimals per day became in the first case 13; 3, 53,
56, 17, 51, 59, in the second case 13; 13, 45, 39, 48, 56, 37. They correspond to a backward motion of the node in a period of 6,796.26 days
and a forward motion of the apogee in a period of 3,231.62 days. In all
these daily motions the two last sexagesimals are not warranted.

The basic orbit of the moon, in this way, is a circle inclined to the
ecliptic, with the points of intersection—the nodes—uniformly regressing in a period of 6,796 days, i.e. a good 18 years. To represent the
variations in velocity between apogee and perigee, an eccentric circle
is not used (for reasons given below), but an epicycle. By taking the
motion along the epicycle a little bit slower (6' 41" a day) than the
motion of the epicycle, a regular progression of the apogee in longitude
is obtained. The radius of the epicycle, corresponding to the eccentricity in an eccentric orbit, was found from the eclipses to be $\frac{1}{11.49}$ =
0.087, producing a maximum deviation of 5° 1' from uniform motion.

All these results are based upon lunar eclipses; the time of mid-
eclipse determines the position of the moon far more accurately than
could any direct measurement. Ptolemy, however, did not content
himself with them; he wished to know the moon's position at other parts
of its course. So he constructed an instrument which he called 'astrolabon', which has no connection with what in later times was called an
'astrolabe' but is identical with the later *sphaera armillaris* or, in short,

armilla. He gives an extensive description of the instrument, which, because of its importance for astronomy, is reproduced in figure 16. Two solidly connected rings represent the ecliptic and, perpendicular to it, the colure, i.e. the circle through the summer and winter points of the ecliptic and the poles of equator and ecliptic. An inner circle can turn around two pins at the poles of the ecliptic; its position, a longitude, is read on the graduated circle. Another graduated circle sliding along it within, and provided with sights, enables the observer to read the latitude of the star towards which they are pointed. This system of rings must be placed in such a way that its circles coincide with the circles at the celestial sphere. For this purpose it can revolve about two pins which are fixed in the colure ring at the poles and are attached to a fixed ring representing the meridian.

First of all, therefore, the meridian and the poles are set in the right position by placing the pedestal correctly. After the colure ring has been set at the longitude of the sun, the system of rings is revolved until the sights are pointed at the sun. Thus the position of the circles is the same as at the sphere, and the longitude and latitude of any star can be read by pointing the sights at the star. Instead of the sun, a well-known star

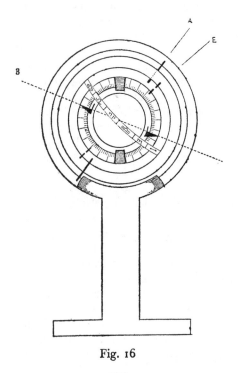

Fig. 16

can also be used to put the circles into the right position. In later manuscripts a picture of this instrument was given, with all its circles—the
ecliptic of course excepted—placed in one vertical plane. To distinguish
their functions, one must look at the connecting pins.

This instrument, afterwards also used to determine the positions of
the fixed stars, served Ptolemy to measure repeatedly in the daytime
the longitude of the moon relative to the sun. Then it was found that the
measurements did not agree with theory. At first and last quarter, the
maximum deviations from the uniform course were not 5° 1′, as at full
moon, but 7° 40′.

It was for the purpose of representing and explaining this 'second
anomaly' of the moon (in modern times called 'evection'), which
depends on the position relative to the sun, that an epicycle for the first
anomaly had been introduced. The mechanism devised by Ptolemy
makes its distance from the earth alternately smaller and larger, at the
quarter moons one and a half times larger than at full moon and new
moon. Actually it comes down to the epicycle's centre describing in a
monthly period not a circle but an oval; both the maximum and the
minimum distance occur twice a month, and the great axis turns slowly,
always being directed to the sun. Formally, to keep in line with the rule
of Greek astronomy that all motions must be circular, Ptolemy assumes
the deferent circle itself to revolve in the opposite direction, so that the
epicycle's centre in its monthly revolution twice meets with both the
apogee and the perigee. The eccentricity of the lunar deferent determines the ratio (the ratio of 7° 40′ and 5° 1′) of the distance in perigee
and apogee at 1 −0.21 to 1 +0.21; the values themselves depend on the
size of the epicycle, 0.106 times the radius of the deferent (see fig. 17).

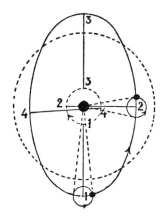

Fig. 17

153

Thus in an ingenious way the two irregularities in the moon's motion are represented by a system of circular motions, but with the result that the distance of the moon itself from the earth may change between 1.21 +0.11 and 0.79 −0.11, so that its apparent size also would change in the same ratio, i.e. as 33 to 17. Now everyone who by chance looks at the moon can state, without needing a measuring instrument, that this is not true; that the moon is not sometimes (at first and last quarter) double the size of the full moon. So the theory certainly cannot represent the real motion through space. It is said that this does not matter if only the apparent, observed motions are well rendered; the space structure is only a formal means to represent the visible course. This, however, is not true in the case of the moon; the parallaxes, which affect the visible position of the moon, would also be variable in the same unacceptable ratio.

Ptolemy then describes an instrument for measuring the parallax of the moon (fig. 18). It is an instrument for measuring in the meridian the distance of an object from zenith: an inclined lath, its upper end hinged at a vertical pole so that it can describe a vertical plane, is directed toward the moon by two sights. Its inclination is not read on a circular arc but on a graduated rod supporting its lower end, hinged at a lower point of the pole, and so representing the chord of the arc between the moon and the zenith. Ptolemy first used it to determine the meridian altitude of the moon at its maximum distance north of the ecliptic, which gave the inclination of the moon's orbit to the ecliptic as 5° 0'. Then, to find the parallax of the moon, he measured its zenith distance when, at its highest point, the parallax is negligibly small, and again at its lowest point, when the parallax is great; because the orbit

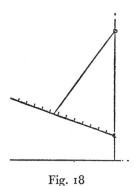

Fig. 18

is assumed to be symmetrical north and south of the equator, the parallax can be deduced. 'From a number of parallax observations in such positions made by us, we will communicate one to show the course of the computation and to derive the further consequence.'[81] This is an observation made on October 1 of AD 135, which afforded a parallax of $1°\ 7'$ at zenith distance $49°\ 48'$, i.e. a distance of $39\frac{3}{4}$ radii of the earth. The cusps of the orbital oval then are at a distance of 59 radii and its flat sides at $38\frac{43}{60}$; these values are the foundations on which his tables of the parallax are based. A modern reader will ask with surprise whether all his other observations for parallax, not communicated, could agree, and why he chose just this one. But the ancient philosopher rather will be exalted to see how such sublime and intricate things as the distance and motion of the moon can be treated by such simple and strict calculation. The extreme values of parallax appearing in his tables are now $53'\ 34''$ and $63'\ 51''$ (highest and lowest points of the epicycle) at the cusps (full and new moon) and $79'$ and $104'$ at the flat sides (in the quarter). The range of these numbers, of course, far exceeds the real extremes of the moon's parallax. In this theoretical structure Ptolemy became a victim of his principle of explaining all the peculiarities in the moon's course by a system of uniformly described circles.

The derivation of the dimensions of the celestial bodies and of the distance of the sun was the task that remained. By using a diopter instrument like that of Hipparchus he found that the diameter of the sun appeared to be constant and equal to the full moon at its greatest distance. This means that annular solar eclipses cannot occur. For the determination of diameters, he thought this method not sufficiently accurate; so he now proceeded to derive them by theory, by means of the known elements of the orbits. He took two lunar eclipses observed at Babylon, so chosen that the moon was at its greatest distance from the earth. One was the eclipse of April 22, 621 BC; one-fourth of the moon's diameter was eclipsed; computation showed the moon to be at a distance of $9°\ 20'$ from the node, so that its centre was $48'\ 30''$ north of the ecliptic. At the other eclipse, July 16, 523 BC, half its diameter was eclipsed; and, with a distance of $7°\ 48'$ from the node, the latitude of its centre was $40'\ 40''$. This, then, is the radius of the shadow, and the difference from the preceding value, $7'\ 50''$, is half the radius of the lunar disc. 'Hence the semidiameter of the shadow is only slightly $(4'')$ less than $2\frac{3}{5}$ times the semidiameter of the moon, $15'\ 40''$. Since from a number of similar observations we got almost concordant numerical results, we have used them in our theoretical researches on eclipses as well as in deriving the distance of the sun.'[82]

This derivation of the distance of the sun by Ptolemy, which appears as an intricate manipulation of triangular and circular sections of the spherical bodies and their shadow, comes down to the same relation as that derived by Hipparchus mentioned earlier (p. 129), but it is now used in another way: the sum total of the radii of the shadow and the sun, 40′ 40″ and 15′ 40″, diminished by the lunar parallax at the greatest distance of the full moon, 53′ 34″, leaves 2′ 46″ as the remainder for the solar parallax, so the sun's distance is 1,210 times the radius of the earth. It implies that the diameter of the sun is 5½ times greater, its volume 170 times greater, than that of the earth. For the moon they are 3⅘ times and 39 times smaller.

For a modern reader this entire derivation is illusory. The rough statements that one-fourth or one-half of the lunar diameter was obscured—rough inevitably because the border of the shadow is ill-defined—may be several minutes in error, and then the solar parallax may vanish completely. For Ptolemy, however, the matter must have been quite different. The character and purpose of the derivation are theoretical, an exposition of the geometrical connections of the phenomena and quantities. As such, it may be said that this derivation also exhibits the high point of view of Greek astronomy. That the astronomer is able, in theory, to derive the distance of the sun from observation of lunar eclipses shows how far man has proceeded in broad insight into world structure.

The sixth book of Ptolemy's work is devoted to the computation of the solar and lunar eclipses on the basis of this theory of the sun and the moon. In this fundamentally different treatment of eclipses as compared with the Chaldean method, we see reflected the different character of both astronomies. In Babylon the continuous series of eclipses was constructed as a totality, completed by an equally continuous, though irregular, row of eclipse indices. In Ptolemy's treatment every eclipse is computed individually, on the basis of tables from which the longitude differences, the anomalistic arcs, and the distances to the nodes can be derived for any moment; a special table indicates the distance to the node corresponding to eclipses of 1, 2, etc., digits' magnitude. The great errors in his lunar parallaxes mentioned above do no harm here, because the eclipse phenomena occur in the vicinity of full moon and new moon.

After sun and moon, he dealt with the stars. Ptolemy's work contains the first published catalogue of fixed stars, consisting of 1,022 stars, for which the co-ordinates relative to the ecliptic, longitude and latitude, are given, as well as the brightness, here indicated by the word 'magnitude'. The stars are arranged in constellations (part of the stars, called 'formless', stand outside) and are described by the parts or limbs they occupy.

The degrees of brightness are given as first, second, third, sixth magnitude, while for some stars the words 'greater' or 'smaller' are added for greater precision; these designations have been maintained throughout subsequent centuries. The longitudes generally are about 1° too small; this has often been explained by Ptolemy having borrowed his catalogue from Hipparchus and having corrected the longitudes for precession by a correction of 1° too small, 2° 40' instead of 3° 40'. As stated above, however, it is quite possible that his precession rests on observations made by himself. It has been conjectured that a list of a good 800 stars observed by Hipparchus were taken over by him without new measurements and that he added 170 mostly smaller stars. This is presumed on the grounds that the latitudes are mostly given in sixths of a degree, but for some stars in fourths of a degree; these sixths and fourths must therefore have corresponded to the graduation of their instruments. So far, this cannot be decided. The accidental errors are, of course, greater than these units of circle reading; comparison with modern data shows that the longitudes have a mean error of 35', the latitudes of 22'.

For six stars Ptolemy adds a remark on their colour; he calls them *hypokirros*, i.e. yellowish. They are Aldebaran, Betelgeuse, Arcturus, Antares, Pollux and Sirius; the first five of these are the first magnitude stars now called 'red' or 'reddish'. In this 'red' the real colouring is exaggerated, just as Mars may appear to us fiery red, though as seen in a telescope it is only yellowish. So Ptolemy's description is more correct, and he must have seen them as we do, with the exception, however, of Sirius, which we know as bluish-white. That Ptolemy described it as having a reddish colour has always caused surprise in later centuries, and modern authors have often concluded that Sirius must have changed its colour since antiquity. That it was not simply a copying error may be inferred from the fact that in Roman literature *rubra canicula* is often spoken of, the red Dog Star, the fiery, burning star that brings the heat of summer. In a cuneiform text the star Kak-si-di is mentioned, that rises in the late autumn evenings and 'shines like copper'.[83] Such a catastrophic change however, from a red to a blue star is entirely excluded by modern astrophysics. Moreover, we find in the astronomical poem of Manilius a line on Sirius reading: 'Since it is standing far away, it throws cold rays from its azure-blue face.' The most probable explanation may be that Sirius, already visible when just rising, is coloured by the long distance which its rays must travel through the atmosphere; this holds good especially for Egypt, where it was observed only at its heliacal rising as a reddened star close to the horizon. Moreover, the colour attributed to the stars and planets by Roman authors often indicates their astrological, rather than their physical, character; thus was Saturn referred to as 'black'.

After the star catalogue Ptolemy gives a detailed description of the

Milky Way, which in all later centuries up to the nineteenth has not been repeated or improved; only in the second half of that century was it surpassed by more careful researches. And then an entire chapter is devoted to the construction of a celestial globe to depict the catalogued stars. 'For the background we choose a darker hue as corresponds not to the sky at daytime but more to the dark at night.'[84] He describes in detail how, by means of graduated rings, the stars are inserted as points according to their longitude and latitude. 'Finally for the yellow and otherwise coloured stars we put up their special colour in such measure as corresponds to the magnitudes of the stars.' The figures of the constellations are indicated by faintly visible lines only and not by striking colours; the Milky Way also is depicted with its bright parts and its gaps. So must Ptolemy's celestial globe have surpassed all later ones in precise representation of the sky. He used it to study and read, for any place of observation, the phenomena and positions of the stars relative to the circles, and especially to the horizon, i.e. their risings and settings.

The remaining and most important part of Ptolemy's work, the last five books, is occupied by the planets, in an exposition of the epicycle theory of their orbits, as explained above. Tables are given for all the movements in deferent and epicycle, allowing rapid computation of their longitude; in the last book tables are also given for deriving the latitudes. Finally, Ptolemy treated the computation of special positions and conspicuous phenomena such as the stations, the heliacal risings and settings, and the greatest elongations of Mercury and Venus, which stood at the origin of astronomy and played a major role in Babylon. The elaborate theory representing all irregularities is able to derive them in full detail; here Greek theory had surpassed the Chaldean rows of numbers. Tables are given of the elongations in the stations for consecutive values of the longitude, also tables of the greatest elongations of Venus and Mercury and of the heliacal risings and settings for the first point of every zodiacal sign.

Thus in this great manual of ancient astronomy, the *Mathematikè Suntaxis*, the world of heavenly bodies stands before us as a universe geometrically portrayed. It is a picture of eternal continuous motion in circular orbits, obeying determinate laws, a picture full of simple harmony, a 'cosmos', i.e. an ornament. It is expounded in a straight progression of exact demonstrations, without disturbing irregularities. Data are so selected or fashioned with admissible limits that the demonstrations tally exactly, and no incidental deviations—needing explanation, which might leave a certain feeling of doubt to disturb the harmony of the construction—distract the reader's attention. Moreover, the book, the work of the author, had to conform to a standard of perfection in form and workmanship. Mental work in those days stood not so far from handicraft; just as the craftsman carefully removed from his product any rough irregularity which might disturb the eye in enjoying the pure harmony of form and line, so the piece of mental work had to captivate the eye and mind by the pure presentation of the universe in its mathematical image.

Geometry occupied a paramount place in Greek culture as the only abstract exact science of the visible world. Because of its rigidly logical structure, proceeding from axiom and proposition to proposition, it stood out as a miracle of the human mind, a monument of abstract truths, outside the material world and, notwithstanding its visibility to the eye, entirely spiritual. The small particle of utility in its origin in Egyptian geodesy hardly carried weight against Euclid's great theoretical structure. Certainly the *sphaerica*, the theory of the sphere and its circles, found a wide practical application in astronomy in the description of the celestial sphere, and the risings and settings of the stars, and so was used and taught during all later centuries. But spherics was only a small part of geometry. The entire science of lines and angles, of triangles, circles and other figures, with their relations and properties, was a purely theoretical doctrine, studied and cultivated for its intrinsic beauty.

Now, however, came the science of the planetary motions, the work of Ptolemy, as a practical embodiment of the theory. What otherwise would have been imagined truths, existing in fantasy only, here became reality in the structure of the universe. Here it acquired form and specific value and size. In the world of planets the circles moved, distances stretched and shrank, angles widened and dwindled, and triangles changed their form, in endless stately progression. If we call Greek astronomy the oldest, indeed the only real, natural science of antiquity, we must add that it was geometry materialized; the only field, truly, where geometry could materialize. Whereas outside this world of astronomy the practice for students of geometry would have been restricted to an idle working with imaginary self-constructed figures. In astronomy they found a realm, the only one but the grandest, where their figures were real things, where they had definite form and dimension, where they lived their own life, where they had meaning and content: the orbits of the heavenly luminaries. Thus the *Mathematical Composition* was a pageant of geometry, a celebration of the profoundest creation of human mind in a representation of the universe. Can we wonder that Ptolemy, in a four-line motto preceding his work, says that 'in studying the convoluted orbits of the stars my feet do not touch the earth, and, seated at the table of Zeus himself, I am nurtured with celestial ambrosia'?

His task, however, was not finished. Rising from the table of Zeus, he had to enter the council room of the gods, to hear how they make known their ordinances to the mortals. It was not disinterested eagerness for knowledge of the heavenly motions that incited him, as well as his predecessors and contemporaries. The knowledge of these motions was a means toward the higher purpose of practical knowledge of the future happenings on earth, of human life and destiny. Hence the thirteen books are followed by four more books, which were mostly edited separately and acquired the name *Tetrabiblos*. In modern times attempts have often been made, by doubting the authenticity of his authorship, to absolve the great astronomer from the blame of having believed in that astrological superstition. But in his purely astronomical work also we can find a statement of this kind: at the end of the eighth book, speaking of heliacal risings and settings, he says that their influences upon the weather are not invariable but depend also on the oppositions to the sun and the position of the moon.

In the four books, however, Ptolemy gives a general theory of the influence of the celestial bodies upon earthly happenings and on man. There he says: 'That a certain power emanates and spreads from the eternal world of the ether upon all that surrounds the earth and that is totally subject to change; that the first sublunar elements, fire and air, are encompassed and changed by the motions in the ether; and that they then encompass and change all else, too, earth and water and the plants and animals therein—this is entirely clear to everybody and needs little comment. For the sun, with its surroundings, in some ways always imposes its order upon everything on earth, not only by the changes accompanying the seasons of the year, the procreation of the animals, the fruit bearing of the plants, the flowing of the waters, and the changes of the bodies, but also by its daily revolutions. . . . The moon, as the body nearest to the earth, bestows her effluence upon the earth; for most things, animate and inanimate, are sympathetic to her and change together with her, the rivers increase and decrease their streams with her light, the seas turn their own tides with her rising and setting, and the plants and the animals in whole or in some part become full or diminish together with her. . . . Also the passages of the fixed stars and the planets give abundant presages of hot, windy and snowy conditions of the surroundings, by which also all that lives on earth is conditioned. Then, too, their positions [aspects] relative to one another, by the meeting and mingling of their dispensations, bring about many and complicated changes. For though the sun's power prevails in the general ordering of quality, the other [bodies] can add to or retract from it in details; the moon does it more obviously and continuously, as in her conjunctions, at quarter or full moon; the stars do it at greater intervals

1. The Assyrian Tablet, K725, British Museum (see page 40 for the transcription)

Chinese Star Map (pp. 88–90)

2. The Constellations on the Farnese Atlas (the blank spaces, other than the South Pole, are the damaged Ursa areas and the places where the hands hold the globe) (p. 132)

and more obscurely, as in their appearances, occultations and approaches.' Of course, other causes are also at work on man: descent, place of birth, nationality, rearing; moreover, lack of ability and knowledge often makes predictions fail. But 'it would not be fitting to dismiss all prognostication because it can sometimes be mistaken, for we do not discredit the art of the pilot for its many errors. . . . Nor should we gropingly in human fashion demand everything of the art but rather join in the appreciation of its beauty.'[85]

Then detailed expositions follow, first on the characters and working powers of each of the planets, then on the fixed stars and zodiacal signs which are concordant or discordant in their effects with different planets, and, further, on what countries are affiliated with these different planets and stand under their influence. In subsequent books the method and basic predictions are explained, first for the weather and then, by means of the horoscopes, for men, not only as to quality and events of body and soul but also as to their material fortunes and happiness. Since the fundamental data—the positions of the celestial bodies at every moment—are provided by astronomical computations laid down in the former books, science also in its practical aims is now completed.

Thus the work of Ptolemy stands before us, a great monument to the science of antiquity. During his life and work, under the emperors Hadrian and Antonine, undisturbed peace and prosperity reigned in the Roman Empire. It was as if, at long last, after all the previous strife and chaos of wars of conquest and civil war, civilized mankind, now united, had entered on an era of sunny peace, of harmonious culture and of tranquil security. But this society was rotten to the core. The sunset glow of antiquity shed its last radiance upon a worn-out world. Before long, towards the end of the second century, the storms broke loose that within the space of a single century were to undermine the power of the world empire and, after another century, were to lay waste the foundations of the ancient world and its culture.

The totality of the causes of this downfall cannot be discussed here; there is, moreover, much diversity of opinion on the connection between its various aspects and phenomena. First, there was the fact that its labour system was based on slavery, which like a cancer corrupted the entire life-system of antiquity and which, when the wars of conquest ended and the supply of slaves was checked, resulted in a primitive serfdom. Another important cause was the disappearance of gold and silver coin—by the exhaustion of the Spanish mines and by the importation of luxuries from India—with ensuing commercial paralysis and a stoppage of production for sale, with a retreat to primitive agriculture for private use. It brought about a great decline in the financial power of the state. Then came a terrible pestilence, brought back in AD 188 by the armies from Asia, with a toll of 2,000 dead per day for Rome alone. It swept over all countries, was rampant for many years and shattered the military power of the Empire. Now the barbarian tribes pressed in and, after devastating raids, settled in the depopulated areas. Mention has also been made of the extermination of the ruling classes, the nobility and the propertied urban citizens, by the peasant armies and their soldier-emperors. Finally, after a century of continuous decline there followed the establishment of a despotic government of officials, until, another century later, the Western Empire was entirely conquered by armed Teutonic peoples.

In this chaos of devastation and utter ruin of ancient society the culture of antiquity also collapsed. In the masses, devoid of hope and future, a new life-conception arose which turned away from the actual world and, while striving to make it tolerable through mutual succour and philanthropy, found comfort and refuge in the belief of a better life hereafter. Christianity increasingly superseded the old religions, fought successfully against the pagan systems of philosophy, and, with the withering of the old state power, took its place as the social and spiritual organization of society.

In this new world concept, concealed from the wise but revealed to the simple-hearted, there was no longer room for the most highly developed science of antiquity. Its cosmologic world scheme was superseded by the original biblical teaching of the flat earth, which, indeed, fitted in with the return to the primitive modes of production—agriculture for home and for tribute and tithes. Whereas the Christian authors of the second and third centuries were well versed in their polemics against the pagan philosophers, the ideas of the Church Fathers in the next centuries became more primitive. Lactantius ridiculed the doctrine of a spherical earth, and Cosmas Indicopleustes, the traveller to India, described a square, flat earth in the form of the poetic effusion in the Book of Ecclesiastes.

So progress in astronomy after Ptolemy was out of the question. Works on astronomy in these centuries are summaries and comments, giving explanations of the classical works, including Ptolemy. They are of value to us because they often give details from works themselves lost. Such commentators are not investigators but scholars; they were praised among their contemporaries not for what they did but for what they knew. Among them was Proclus, surnamed Diadochus (the successor), who worked in the fifth century in Athens as the last of the pagan philosophers; he wrote a useful commentary on the more elementary parts of Ptolemy. Shortly afterwards there was Simplicius, famous

THE CLOSE OF ANTIQUITY

commentator on Aristotle, who, when expelled by the Byzantine Church in AD 529, found a temporary refuge in Persia. He mentions that his teacher Ammonius of Alexandria had made observations of Arcturus by means of an astrolabe. In these commentaries the most simple principles form the main contents; the scientific level reached by Ptolemy was no more. The astronomical part of the works of Isodorus, Archbishop of Seville in the seventh century, highly praised for his learning, wherein it was said that the stars receive their light from the sun and that the moon revolves in 8 years, Mercury in 20, the sun in 19 and Mars or the Evening Star in 15, shows to how low a level astronomical knowledge in Western Europe had fallen.

But in the next century a new power—Islam—appeared in the Near East, and an offshoot of ancient astronomy sprang up in those parts.

CHAPTER 15

ARABIAN ASTRONOMY

In the middle of the seventh century AD the Arabs broke out of their deserts and conquered the surrounding civilized countries, first Egypt and Syria, which then were parts of the Byzantine Empire, and then Mesopotamia, the prosperous nucleus of the New-Persian Empire. This was a repetition of analogous former expansions, such as the Semitic conquest of Babylonia after 2000 BC, and the later Aramaic invasion of Syria. We do not know exactly the cause of such migrations; often climatic events such as severe droughts were the underlying factor. In the most highly developed commercial and trading centres, in the Hejaz, influenced by the Jewish and Christian religions, Mohammed had founded and propagated his doctrine that was to bind the ever-divided and interwarring tribes into one powerful unity—the doctrine of Islam, the brotherhood of all the faithful, before which all the old ties of tribe and family had to give way.

The Arabs, a people of strong and hardened men, camel-tending nomads, proud warriors and robbers, imaginative possessors of a flowery language and of rich poetry, now became the masters of Near-Asia. With their fresh primeval force and the intensive new ideology of Islam, the Arabs gave a powerful impulse to the economic and cultural development of this part of the world, bringing it to a rich flowering. Through their conquests they created a world empire extending from the Atlantic coast in Spain and Morocco to India and the mid-Asian steppes. Trade and commerce unfolded here, bringing far distant regions within one large economic unit. As a religious community the entire realm remained a cultural unit even when politically split up into a number of independent sultanates. In all these countries, mostly in the fertile plains of Mesopotamia, Syria, Egypt and Andalusia, artistic handicrafts developed in a number of new flourishing towns. Commerce, carrying the products of China and India to the West, of Byzantium and Europe to the East, brought about an interchange of culture and science. Powerful rulers, first the caliphs at Baghdad, later the sultans of smaller countries like Egypt, and still later the Turk and Mongol conquerors from Central Asia, became promoters and protectors of the arts and science.

It was the concern for the security of their own lives and future that aroused in sagacious princes a direct interest in science. Medicine to maintain health and life, astronomy to ascertain the future and their destiny, held a first place for them. Mohammedan scientists—we call them Arabians although by birth they were Syrians, Persians, Jews and, later, natives of other countries—were most famous as physicians and astronomers, and in this connection they cultivated mathematics, chemistry and philosophy. Mostly these sciences were combined; physicians, as in later centuries in Europe, had to know astrology in order to find the propitious time for different treatments. Moreover, in the Mohammedan world there was a direct need for astronomical knowledge; for the pure moon-calendar observation of the moon was necessary. A precise timekeeping by means of water-clocks and sun-dials had to indicate the times prescribed for prayer; because when praying the face had to be directed towards Mecca, astronomical experts had to indicate that direction in the mosques throughout the world.

In the first flourish of the Baghdad caliphate in the eighth century, ancient science made its entry. Primary knowledge was borrowed from the Nestorian Christians, who had found in Persia a refuge from the persecutions of the Byzantine Church and had founded schools there. What was borrowed from India was more important. After the conquests of Alexander and of the later Macedonian rulers of Bactria, Greek science had grown an offshoot in India. Under the Gupta dynasty in Hindustan (about AD 400–650) there arose a literature of mathematical and astronomical writings, called 'Siddhantas', proceeding from different authors, amongst whom Brahmagupta is the best known. In these works one meets the Greek world picture: the spherical earth and the epicyclic orbits of the planets, less detailed in comparison to Ptolemy and without the equant. Sometimes even a rotation of the earth is mentioned.

From India this influence now turned back to the west. It is mentioned that in 773 there appeared, before the caliph Al Mansur, a man from India who was acquainted with the stars and could calculate eclipses. Whereupon the caliph ordered the translation of the Indian books. The first astronomical tables were published in the next century by Muḥammad ibn-Mûsâ al-Khwârizmî. That these tables had been translated from Indian originals was evident since all data were given for the meridian of Udshain, the capital of one of the states in central India and seat of an observatory. Theory was lacking in these works; they consisted only of numerical tables with instructions for use, intended for calendar and astrological purposes. They introduced a valuable innovation into western arithmetics: the Indian system of numerals (with place-value and zero for open place) which afterwards in Europe were called 'Arabic numerals'. Interest once awakened, Syrian translations of many Greek authors which persisted through the Nestorians, were translated into Arabic. Caliph Harun al-Rashid ordered Greek manuscripts to be assembled; more eager still was his son and successor Al-Ma'mûn (reigning from 813 to 833), who in a peace treaty with the Byzantine emperor stipulated the handing-over of numerous Greek manuscripts. Among them was Ptolemy's work, which since late antiquity had been known as 'the very great (*megistè*) composition', and, now, by combination with the Arabic article, got the name *Almagest*, by which it was known in Europe many centuries later. An Arabic translation from 827 is in the Leiden University library.

Thus were laid the foundations for astronomical studies. A work on the elements of astronomy, consisting of an intelligible summary of Ptolemy, was written by Al-Farghânî (in medieval European books, 'Alfraganus'). More widespread was the contemporary work of Jafar Abû Ma'shar (in Europe, 'Albumazar') on astrology, dealing with the positions of the signs and planets and the meaning thereof. During the entire Middle Ages it was considered a standard work and it was one of the first books to be printed, in Augsburg in 1486. Numerous other books on astrology appeared in the ninth century, the first period of the flowering of Arabian learning. Among the many authors on astronomy, mathematics and medicine, Thâbit ibn Qurra (826–901) is known chiefly by his theory on 'trepidation'. Since Ptolemy had stated the precession to be 1° per century, whereas later a more rapid movement had been found, the conclusion was reached that it was variable and consisted in an oscillation of the equinoxes. Another oscillation was indicated by the diminution of the obliquity of the ecliptic from 23° 51⅓', Ptolemy's value, to 23° 34' or 35' as measured by the Arabian astronomers. Tobit combined these oscillations into the theory that, superimposed on the uniform recession, the zero point of the ecliptic describes a small circle of 4° radius in 4,000 years. This theory was accepted throughout the Middle Ages and, because of its complicated character, has given much difficulty to astronomers. It shows that the Arabian scholars were not simply imitators of ancient science but were independent thinkers.

This is apparent in all their work. Caliph Al-Ma'mûn, in order to check the ancient statement on the size of the earth, ordered his astronomers to measure a degree of latitude in the plain of Palmyra— the Arabian world had no objections to assuming the sphericity of the earth, since the Koran was silent on this matter. About this measurement it is reported that from one and the same point observers went to the north and to the south until they found the latitude to have changed by 1°; so they found that 1° was equal to 56⅔ Arabic miles, each of

4,000 'black ells'. The Arab astronomers had instruments corresponding to Ptolemy's descriptions, such as armillas and quadrants. For simple measures of altitude they used the astrolabe, a graduated circle freely suspended from a ring kept in the hand, provided with sights on a revolving arm; circles were engraved on the back, to enable them to read the hour angle for special stars instead of computing it. These instruments were finished with great artistic skill.

Al-Battânî (in Europe, 'Albategnius'), whose name in full was Muḥammad ibn Jâbir ibn Sinân Abu-'Abdallâh al-Battânî (died 928), is considered the greatest among Arabian astronomers. At his time the caliphate was in decay and powerless; but he himself was of lordly descent and had no need of royal favour. In the town of Rakka, between 877 and 919, he made numerous observations, with results often deviating from Ptolemy. His astronomical work was afterwards repeatedly translated and published in Europe. He was the first to introduce into his tables half-chords for half-angles instead of chords, i.e. what we call 'sines'. He was also the first to give computing methods for spherical triangles, which were further developed by later Moslem mathematicians.

By comparing his observation from the year 880 with Ptolemy's, he derived $365^d\ 5^h\ 46^m\ 24^s$ for the length of a year; if Ptolemy had not given his equinox one day wrong, the result would have been $1^m\ 58^s$ larger, nearly right. From the length of the seasons ($93^d\ 14^h$ for spring; $93^d\ 0^h$ for summer) he derived a longitude of the apogee of $82°\ 17'$ and an eccentricity of 0.0346; the latter value is exactly right. His apogee was $16°\ 47'$ larger than Ptolemy's; he stated expressly that his results were different from Ptolemy's value, but he did not speak of a regular increase. This increase considerably surpassed the increase by precession ($10.4°$ in 740 years); so he was the first to discover the progress of the sun's apogee relative to the stars. He himself noted only the progress with the stars and applied the precession to other longitudes of planetary apogees also. His catalogue of stars was Ptolemy's with the longitudes increased by precession.

Of astronomy Al-Battânî said that the science of the stars comes immediately after religion as the noblest and most perfect of sciences, adorning the mind and sharpening the intellect, and that it tends to recognize God's oneness and the highest divine wisdom and power.[86] He did not speak of the utility of astronomy in recognizing the influence of the stars on earthly happenings. Since, however, he wrote elsewhere on the conjunctions and gave a commentary to the *Tetrabiblos*, it is obvious that astrology was self-evident to him.

Al-Battânî's tables, of course, were complicated, because they were based upon the full theory of Ptolemy. It is probably for this reason that the old, less perfect, but more handy tables of Al-Khwârizmî were re-edited by the Spanish astronomer Maslama ibn Ahmed (died 1008), transferred to the meridian of Cordoba, and completed by an excerpt from Al-Battânî.

From the same time (the tenth century) 'Abd al-Raḥmân ibn 'Umar called Al-Sûfî (i.e. 'the wise') must be mentioned, because in a book on the fixed stars, where he takes the longitudes from Ptolemy, increasing them by 12° 42', he gives the magnitudes of the stars after careful observations by himself. So his work is a valuable independent source of knowledge of their earlier brightness. At the same time in Egypt, protected by Sultan Al-Hakim of the Fatimid dynasty, the astronomer Ibn Junis not only published new tables (the 'Hakemite Tables') with theory and computing methods, but also communicated a large number of observations of eclipses, conjunctions and altitudes, partly taken from the records of former Arabian observers as early as 829, and partly from his own work in Cairo 977–1007. About the same time the Turkish prince Sharaf Al-Dawla ordered the construction of an observatory, in his garden at Baghdad, to be provided with many new instruments. Here a number of astronomers, among whom Abu'l Wefa is best known, made observations of equinoxes, solstices, and the obliquity of the ecliptic.

Astronomy now makes its appearance in the more distant realms of Islamic culture. Spain in the eleventh and twelfth centuries, under the Cordoba caliphate, had a high tide of culture, of arts and sciences. Among the Spanish astronomers of this time, Ibn al-Zarqâla ('Arzachel', 1029–87) was prominent; he published the *Toledo Tables* with a description of the instruments and their use, especially the astrolabe. He made many observations, from which he deduced a solar apogee at 77° 50', smaller and less accurate than Al Battani's value. At a somewhat later date lived Jâbir ibn Aflaḥ (sometimes mistaken for Gabir ibn Haijan, the alchemist) whose astronomical work was translated into Latin and often published in medieval Europe. In the thirteenth century the power of Islam in Spain came to an end. The religious zeal of the Christian Castilians—in need, moreover, of southern winter pastures for profitable sheep-breeding, to produce the fine wool for Flemish cloth—conquered the fertile Andalusian plains with the brilliant capitals Cordoba and Seville, thus putting an end to the flowering of Arabic horticulture.

The work of the Castilian King Alfonso X, surnamed 'The Wise', may be considered as a last offshoot of Arabian astronomy in Spain. He assembled around him a number of astronomers to construct new astronomical tables; these 'Alfonsine Tables' (1252) were in use for three centuries, up to the middle of the sixteenth century. The leader of this group was the Jewish scholar, Isaac ben Said; it is said that, owing to the traditions of Jewish jubilee years, the period of precession in the tables was assumed to be 49,000 years and the period of trepidation, 7,000 years.

In the same century the Mongol ruler, Hulagu il Khan, a grandson of Genghis Khan, founded an observatory at Maraga (near Tabriz in Persia), under the direction of his counsellor Nâsir Al-Dîn al-Tûsî (died 1274), an able astronomer from Khurasan. At great expense a library of 400,000 manuscripts was installed and many instruments, partly of new design; among them was a great quadrant of 10 feet radius, constructed by Al-'Urdî. After a dozen years of assiduous observation of the planets by Nasir ud-din and his assistants, they were able to construct the 'Ilkhanic Tables'.

Another two centuries later astronomy sprang up in Samarkand. Ulugh Beg, a grandson of the Mongol conqueror Tamerlane, during his father's reign founded there a richly endowed observatory, where he himself took part in the observations. One of the instruments was part of a circle, probably a quadrant, 60 feet in radius, placed vertically in the meridian between masonry, in order to determine accurate altitudes of the sun. Ulugh Beg also determined the positions of the stars; he was the only oriental astronomer known not to have copied Ptolemy after correcting the longitudes for precession; he founded them upon his own observations. His catalogue of stars observed in 1420–37, first became known in Europe a century afterwards and was not printed until 1665, when it had already been surpassed by European catalogues.

Some of the most important names in Arabian astronomy have been mentioned here. In his list of astronomers and mathematicians, H. Suter gives nearly four hundred names, all of whom were praised by contemporaries and successors as great scholars. There are many among them whose chief merit lay elsewhere, in the realm, for example, of medicine or chemistry. Ibn Sînâ ('Avicenna', 980–1037), famous as a physician and philosopher, also made observations, wrote on astronomy and edited a compendium of Ptolemy; in those days it was possible to master many sciences. But it cannot be assumed that to be a scholar always implies being an investigator. Arabian scientists certainly observed diligently; they constructed new instruments, and in astronomy they seem to have displayed more practical activity than did the Greeks. Also, the accuracy of their work often surpassed the results of antiquity. Their aim, however, was not to further the progress of science—this idea was lacking throughout—but to continue and to verify the work of their predecessors. The continual compiling of new tables did not necessarily mean—in fact seldom meant—progress in more precise values. The tables were needed for astrological purposes. Hence the astronomers had to compute and publish them ever anew; and if they were scrupulous people, they took care to check the positions of the planets by their own observations and to complete them by their independent results.

In their theoretical ideas, however, they did not go beyond antiquity. They often were not content with Ptolemy's theory; but when they deviated, it was from preference for Aristotle. Thâbit ibn Qurra is reported to have assigned to each planet a space between two eccentric spheres. Ibn al-Haitham ('Alhazen'), known by his work on the refraction of light, had 47 spheres, all turning in a different way about and within one another. Ptolemy's conception of circles and centres, existing in fancy only, was not concrete enough for them.

In the twelfth century Aristotle's philosophy was diligently studied and developed by Moslem thinkers, especially in Spain. The most famous among them, Muḥammad ibn Rushd ('Averroës'), used it as the foundation for a pantheistic philosophy which spread through Europe and was condemned as a dangerous heresy by the Church. He and his followers thought that circular motion about a centre was possible only when a solid body, like the earth, occupied that centre. The Jewish scholar Moses ben Maimon ('Maimonides'), as well as the Moroccan astronomer Al-Bitrûjî ('Alpetragius'), rejected the epicycle theory; the latter considered the motion of the sun, the moon and the planets as a lagging behind the daily rotation and so came back to the ancient ideas of Plato.

So there was a brilliant rise in Arabian astronomy, but no significant progress. After some centuries it died down. This indeed was the case with Mohammedan culture as a whole. A mighty impulse of conquest, borne by great religious enthusiasm, had built a world empire in which trade and crafts under social and economic prosperity engendered a special civilization. But an impulse towards continual progress was lacking; minds were dominated by a quiet fatalism. Then came devastation by the Mongol inroads from the steppes of Asia; for in their irresistible attacks the Mongols razed towns, exterminated the inhabitants, destroyed the irrigation works and thereby turned flourishing, thickly-populated regions into lifeless deserts. The power and the flower of Islam declined, and along with them its culture and its astronomy. Of the rich libraries little was saved; no Arabian manuscript is known to exist today of the tables of Al-Khwârizmî. The importance of Arabian astronomy lay in the fact that it preserved the science of antiquity in translations, commentaries, interpretations and new observations and handed it down to the Christian world. Thus it considerably influenced the first rise of astronomy in medieval Europe.

PART TWO

ASTRONOMY IN REVOLUTION

CHAPTER 16

DARK EUROPE

I N the ninth and tenth centuries, when under Islam commerce and trade, culture and science attained their highest flowering, Europe was a barbaric land, sunk in utter impotence. The Arabs controlled the Mediterranean; they sacked the city of Rome in AD 846, and occupied not only Spain and Sicily but also Provence, whence they raided France. From the north the Vikings came looting and conquering; from the east the Magyars invaded and devastated Western Europe. Economic life had dwindled to a miserable agricultural activity carried on by ignorant serfs, who were ruled by lords and priests hardly less ignorant than themselves. A scanty knowledge of Latin and of some Latin authors maintained itself with difficulty in some few monasteries. Latin, a living language down to the seventh century, had now become a clerical language, restricted to the Church. 'For centuries following,' said Pirenne, 'there was to be no science but within the Church.'[87]

Yet the connection of man with heaven and science was not lost altogether. In some monasteries a limited knowledge of constellations was preserved because their rise in the night indicated the hour for religious services. The rules for fixing the date of Easter formed part of the Christian doctrine; some ecclesiastics were therefore needed who could understand something of the course of the stars and who were able to make computations according to precept. As a thin rivulet of science, *computus* (i.e. computation, which in this context always means calendar-computation) ran through these centuries. At an even earlier date the English monk Bede, surnamed 'Venerabilis' (died AD 735), had computed lists of Easter dates; his writings show that he was acquainted with Pliny and Seneca, and also with the sphericity of the earth. The mysterious variations of the date of Easter were considered by the monk Notker (of the St Gall monastery, a renowned centre of studies) as part of the general domination of the heavenly bodies over earthly life.

About the year 1000 the European world began to recover. The Northmen and the Magyars were repelled, Christianized and absorbed into the community of European culture. The Italian seaports conquered their Mediterranean trade routes and became centres of flourishing commerce. Towns developed as market places, or around the castle, or where the trade routes between Italy and the North intersected. At newly-founded monasteries, where productive labour was combined with intellectual study, a more profound cultivation of the Christian doctrine and a stricter pattern of life developed. In the eleventh century, under the leadership of Hildebrand (afterward Pope Gregory VII), the Church grew into a well-organized clerical hierarchy, directed from Rome. The Papacy took its place as a spiritual power parallel to the temporal power of feudal lords, the kings and emperors. It assumed the spiritual leadership of Christian Europe and felt strong enough to take the offensive against Islam in protection of Christendom. These Crusades brought the rude and primitive world of knights and monks into contact with the refined, but already declining, Arabian culture.

European Christendom now began to raise itself spiritually with the help of Arabic science, chiefly imported from Moslem Spain. Even before this time the learned Gerbert, afterward Pope Sylvester (99–1004), is reported to have made inquiries at Barcelona about books on astrology; he is credited with the authorship of a book on the astrolabe. A century later, Athelhard of Bath personally went to Spain to study Arabic wisdom at its source, which he extolled and explained after his return. In 1126 he edited a Latin translation of the astronomical tables of Al-Khwarîzîm, in Maslama's version, thus proceeding from the first instead of from the later more highly developed forms of Arabic astronomy. He also translated the astrological work of Albumazar. Such a translation generally amounted to a difficult search for, and a forging of, more or less appropriate terms. The first tables of European astronomy bear witness to the awkwardness of the tiro; Arabic words appear untranslated in the Latin text or in the headings of the tables, e.g. 'inventio elgeib per arcum' (the finding of the sine for the arc).[88] Often we come across Latin words which are quite different from the more precise terms adopted later, thus, e.g. 'obliquatio' for what was afterward called 'declination' of the sun. Some Arabic words have remained in use as technical terms, e.g. 'azimuth', 'zenith', 'nadir'. In the same way the names of some stars, which simply designated certain objects or parts of the body in Arabic, became proper names in European astronomy, e.g. Betelgeuse ('the giant's shoulder' —bat al-dshauzâ'), and Algol (the 'monster's head'—ra's al-ghûl).

A number of other translators of Arabic science now came forward. The most famous among them was Gerhard of Cremona (1114–87), who went to Toledo to search for Ptolemy's work and there found a wealth of books not only by Arabian authors but also by the scholars of antiquity. He set himself to edit Latin translations, first of the

Almagest (1175)—a translation made shortly before in Sicily, direct from the Greek, was little known—and then also of Euclid, Galen, Aristotle, Archimedes and many others. Universities were now founded in Bologna, Paris and Oxford, where the new sciences were taught and began to pervade men's minds. But not without opposition, for at first Aristotle was banned at the University of Paris. Irresistibly, however, the new ideas spread all over the European world which had now outgrown the primitive agrarian economy and its spiritual counterpart, the primitive cosmology.

At the end of the twelfth century the consolidation of European Christendom was accomplished. The Church had by now become the leading spiritual power in the feudal society of princes and knights, of abbeys and monasteries, of peasants and urban citizens. All over Western Europe the spires of Gothic cathedrals soared, symbols of burgher freedom in ecclesiastical garb; and at the courts the poetry of chivalry flourished. The clergy constituted the intellectual class in this simple society of agriculture, craft and commerce, and performed the social functions of spiritual, administrative and scientific leadership.

Under Innocent III, the Papacy rose to the summit of its power and became a universal monarchy of Christendom, commanding and deposing kings and emperors. The Franciscan and Dominican orders were founded as the strong moral and intellectual militia of the Church. Through their preaching and propaganda, soon reinforced by persecution and inquisition, all dissenting ideas were opposed and suppressed, and even forcibly exterminated; in this way, unity of doctrine was established. It was from these circles that the students and exponents of science emerged.

At the beginning of the thirteenth century the work of translation was finished; it was followed by assimilation, critical discussion and independent creative work. The study of astronomy as a special doctrine was mainly restricted to 'sphaerica', the doctrine of the celestial sphere and of the phenomena resulting from its daily rotation. A textbook on this subject, written by Johannes de Sacrobosco (John of Holywood), who died in Paris in 1256, was still widely used three centuries later. A new astronomical world-picture arose, which formed part of the general conception of the world, framed by the scholars of the thirteenth century.

The ablest and most original thinker and teacher of this century was the Dominican friar Albertus Magnus (1193–1280), who in his writings expounded the conceptions and doctrines of Aristotle and also referred to Ptolemy as an astronomer and astrologer. Even greater fame was won by his fellow-friar and pupil, Thomas Aquinas (1225–74), a less original but more methodical mind. In his works he united Aristotle's cosmology with the doctrine of the Church into one system of thought which, under the name of Scholasticism, was to dominate the minds of men for many centuries. By their side stood the Franciscan friar Roger Bacon, who also praised Aristotle as the source of wisdom, and sharply criticized his own age and its learning. He made fantastic references to future machines, and recommended experience, *experimentatio*, as the true method for the acquisition of knowledge; on this account he has often been acclaimed as a precursor of the later principle of inductive science.

Thus the astronomical world-picture in Europe had risen to the level of Greek antiquity again. About the spherical earth in the centre of the world the planets and stars move in the celestial sphere. That is how we find it described in Dante's *Divina Commedia*, in which he placed Hell in the deepest depths, that is, at the centre of the earth. With his companion, the poet descended to the centre, where Lucifer was undergoing his punishment, and then on the opposite side ascended to the earth's surface again. There, under a milder sky—which Columbus later on believed he had found in the mild climate of the West Indies—he saw four brilliant stars, symbols of the four principal virtues, which later commentators, probably wrongly, identified with the not very bright stars of the Southern Cross. He then climbed the mountain of Purgatory, from which the planetary spheres were ascended.

It must be added, however, that the world concept of those days was not confined to the sober structure of Aristotle's cosmos. It was pervaded and dominated by astrology. The dominion of the stars over terrestrial events caused medieval man to look upon astronomy as the supreme doctrine of the world. The belief in occult forces and in magic was universal then, even among the most famous scholars. Their knowledge of the secrets of nature made them magicians; the study of nature in its early stages was intimately associated with magic, and all great scholars, Ptolemy and Galen as well as Avicenna, Gerbert and Albertus Magnus, were regarded by contemporaries and posterity as sorcerers and miracle-workers. So little was known of the laws of nature that the whole of creation appeared as a miracle, a world of wonder in which everything was possible or might be found true, and in which naïve credulity accepted all so-called 'facts'. Along with the science of antiquity its superstitions were taken over, and among them astrology appeared as the all-embracing doctrine of the world.

This also applies to the three great thirteenth-century thinkers referred to above, whose views about astrology were substantially similar: the stars rule the earthly bodies, God rules the lower creatures by means of higher beings, and all things on the earth by means of the celestial spheres. Thus the movement of the stars dominates life on earth, and the conjunctions of the planets disturb the regular order of events. But for man this is no ineluctable fate; he is not entirely subject to it. His will is free, because his soul, as a higher being, proceeds from the Supreme Being. It is only when he fails to offer resistance that he is carried along by nature, and thus is swept away by the power of the stars. Horoscopes were therefore useful as warnings, and prognostications had to be kept within general terms; hence severe censure of astrological fortune-tellers was imperative. After having thus reconciled itself with the Church, astrology could maintain itself through the following centuries, notwithstanding occasional scepticism and criticism from opponents.

CHAPTER 17

THE RENASCENCE OF SCIENCE

IN the fourteenth century the dream of the Church as a universal monarchy of European Christendom faded. The medieval world had developed into new forms. In all countries the towns had grown up as centres of crafts and commerce; the urban dwellers, increasing in prosperity and power, emerged as the class that more and more determined the aspect of society. Kings, supported by the financial power of the burgher class, established centralized state powers with laymen as civil officials. In continual struggles, the powers of the kings of France, England and Castile increased, and they superseded the power of the popes, who were often little more than dependent bishops of Rome. Secular power superseded ecclesiastical power. The clergy were no longer the spiritual leaders; intellectual leadership of society passed into the hands of the laity. Interest in science increased among the urban dwellers, who were eager for knowledge with which to promote the development of trade, and desirous of raising their status as masters of the new world. Their sons attended the universities, which were steadily growing in number, studied Roman Law, and became officials and advisers of the princes, assisting them in undermining feudal law and feudal society by means of this new judicial doctrine.

The study of science was still directed to assimilating ancient knowledge and the entire culture and science of antiquity. But a new spirit arose here and there, a spirit of independent research, of new and bold ideas, and of desire for further progress. At the University of Paris, a group of philosophers, Jean Buridan, Albert of Saxony and Nicolas Oresme, precursors of the scholars of the sixteenth century, attacked Aristotle's physics, especially his theory of motion. But when shortly afterwards France was increasingly ravaged by the Hundred Years' War with England, their influence succumbed to the power of scholasticism.

It was chiefly Italy and Germany which now formed the vanguard of the revival. In Italy the urban dwellers had acquired great wealth and power during the preceding centuries because they were nearest to the sources of oriental trade. It was here that the arts and sciences began to flourish. Germany was at the crossroads of trade between Italy and the north, and between east and west; in prosperous towns like Nuremberg, Augsburg and Cologne a strong and independent middle class ruled. Here, in the fifteenth century, great fortunes were amassed in the hands of the captains of finance, like the Fuggers and the Welsers, who became important powers in world politics. Whilst Paris lost its leading position in science, new universities were founded in Prague, Heidelberg, Vienna and Leipzig.

In the fifteenth century the sources of knowledge of antiquity began to flow more abundantly. The Byzantine Church, seeking help against Turkish aggression, came into closer contact with the Roman Church; knowledge of Greek thus spread among the Western scholars. Numerous Greek manuscripts were brought to the West, especially to Italy, where they were ardently collected and studied; they presented direct and pure texts of the ancient writers, instead of the corrupt and often unintelligible translations through Syriac and Arabic. The Western World was ripe to take possession of the entire spiritual heritage of antiquity. The spirit of humanism with a hint of paganism inspired the scholars, clerics as well as laymen, and began to supersede scholasticism. In astronomy the study of Ptolemy took a prominent place.

Two ways lay open and were followed for the study of astronomy: the collecting and studying of incorrupt manuscripts of the ancients, and the making of new observations. Observations with simple instruments, borrowed mostly from Arabic writings, had already been made in earlier centuries; the most widely used instrument was the astrolabe, many specimens of which can still be seen in our museums. More accurate instruments were used by Guillaume St Cloud, who in Paris at about 1290 measured solar altitudes, from which the latitude of his observation post and the obliquity of the ecliptic (23° 34') could be derived; in 1284 he also observed a conjunction of Jupiter and Saturn. Paolo Toscanelli (1397–1482), afterwards geographical adviser to Columbus, systematically noted the positions of comets among the stars, in 1433, 1449, 1456 and later years. In Vienna, Georg Purbach (1423–61), who took his name from his Austrian birthplace, taught at the university after having travelled in Germany and Italy. He was first in Western Europe to expound Ptolemy's epicycle theory in a book called *New Theory of the Planets*, in which he inserted it into Aristotle's world system by separating the region of each planet from its neighbour by solid spherical shells.

To him came as pupil, afterwards as assistant, Johann Müller of Königsberg, a village in Franconia, who later called himself Johannes de Monte Regio and in astronomical literature is known as Regiomontanus. He lived from 1436 to 1476. During the years 1456–61 Purbach and he made many observations of eclipses, comets and solar altitudes, in the course of which they perceived that the Alfonsine Tables were several degrees in error. Their desire to obtain better manuscripts of Ptolemy was stimulated by a diplomatic visit to Vienna by Cardinal Bessarion, who had held high rank in the Byzantine Church. Their plan to attend him on his return to Italy was frustrated by Purbach's early death; Regiomontanus alone now accompanied the cardinal to Italy, where he learned Greek, collected and copied Greek manuscripts, and lectured on astronomy. After a short visit to Hungary, whose King Matthias Corvinus had acquired manuscripts during his wars against the Turks, Regiomontanus settled in 1471 in the town of Nuremberg. Here, in this centre of Middle European trade, flourishing commerce and handicrafts offered the most favourable opportunities for the construction of instruments as well as for the printing of books.

Indeed, the newly-invented art of printing opened up new possibilities for science. The printing of books, with careful correction of the text, put an end to the annoying evil of the numerous copying errors in manuscripts. The new process, it is true, did not yet include the printing of tables and figures. Regiomontanus had therefore to found a printing office himself and to instruct the compositors, thus acting as a pioneer of the printing trade. A circular letter from him is still extant, in which he enumerated the titles of the books he intended to print and publish. This list of 22 items, all in Latin, mostly editions of ancient astronomers and mathematicians, includes Ptolemy's *Geography* and *Astronomy* (*'Magna Compositio Ptolemaei quam vulgo vocant Almagestum, nova traductione'*, i.e. in new translation); also Archimedes, Euclid, Theon, Proclus, Apollonius and others, followed by his own works, almanacs and minor writings. He began by publishing the planetary theory of his teacher Purbach and the astronomical poem by Manilius; after this, carefully computed almanacs, in Latin and in German, appeared. He won great fame with his *Ephemerides*, in which the positions of the sun, the moon and the planets had been computed for 32 years, from 1475 to 1506. Then, in 1475, the Pope summoned him to Rome to seek his advice on the urgently needed reform of the calendar. Here he died the next year. His great projects remained unfinished; the printing office was not continued, and his manuscripts were scattered. His own works were not printed until forty years later; he had not been able to accomplish the translation of Ptolemy, and it was not until 1505 that an older Latin translation was printed in Venice, while the first printed Greek edition of Ptolemy appeared in Basle as late as 1538.

It was not merely through his printing works, but even more because of his practical astronomical work, that Regiomontanus gathered around him in Nuremberg a circle of admirers and students of science, who also provided money for the printing business. Among them were the patricians Willibald Pirkheimer and Bernhard Walther, both humanists well versed in Greek. Walther became his pupil in practical astronomy, and at his house equipped a room for mounting instruments, the first real observatory, where they made observations together. After Regiomontanus's death, Walther assiduously continued to observe the celestial bodies. By the time he died in 1504 he had made 746 measurements of solar altitudes and 615 determinations of the positions of planets, moon, and stars. It was the first uninterrupted series of observations in the new rising of European science; a century later Tycho Brahe and Kepler utilized them in their work.

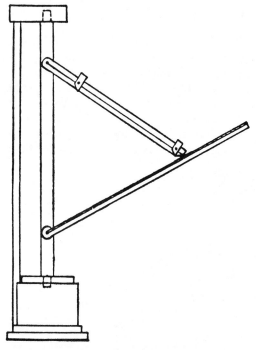

Fig. 19. Regiomontanus's three-staff

The instruments, made of wood after Regiomontanus's design, were of simple construction. First there was the so-called *Dreistab* (three-staff), also called *triquetrum*, already described by Ptolemy. It consisted of a lath about 9 feet long (with two sights to direct it towards a star) which was hinged at the top of a vertical pole; the lower end was pressed against a second lath, graduated and hinged at a lower point on the pole; the distance from this point to the lower end of the first lath indicated its inclination. This instrument was used mainly to measure midday altitudes of the sun; one inch on the divided lath corresponded to nearly half a degree. A more widely used device was the cross-staff for measuring the distance between two celestial objects (fig. 20). Along a graduated lath, which the observer took in his hand and directed at the mid-point between the two objects, a cross-lath was adapted to slide up and down, until its two ends, as seen from the lower end of the lath, coincided with the two stars. The reading of the cross-lath combined with its constant length gave the angular distance between the two stars. For several centuries the cross-staff was the most common instrument for navigators to measure the altitude of the sun or a star above the horizon. Afterwards, in 1488, Walther had an armilla made, also after Ptolemy's description (p. 152), on which, after careful adjustment, he could directly read the longitude and latitude of a planet. In an attempt to measure distances by time intervals he also made use of clocks, although at the time they were imperfect, being regulated by friction only. The care with which he made his observations is shown by the fact that he discovered the upward displacement of the sun near the horizon, which he rightly explained by atmospheric refraction. His carefulness is still more clearly shown by the accuracy he attained; his positions for the planets, measured by means of cross-

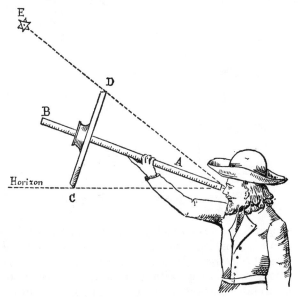

Fig. 20. The Jacob's staff

staff and armilla, had a mean error of only 5', and the errors of his solar altitudes were usually below 1'. His younger collaborator, Johannes Schoner, who continued his work, published all these observations later.

Thus we see how, in the fifteenth century, science took a new trend. In the preceding centuries the most highly praised scientists had been scholars, not investigators. They reproduced science as they found it but did not produce new science. Books and writings, not experiments and observations, were the source of their knowledge. Now, however, a new era opened up, in which observation of new phenomena became the source of continuous scientific progress.

Science in Europe had now risen to the highest level reached in antiquity and had even surpassed it. Ptolemy's planetary theory was completely understood, and the observations of Hipparchus and Ptolemy were exceeded in accuracy by those of Regiomontanus and Walther. But the main point was this: what in antiquity had been the farthest end, what in the Arabian world had just been reached, was here in Europe the starting point for a continuous and rapidly increasing progress.

Whence this difference? The question is important not only for astronomy but for the entire science of nature. The rise of science in these and subsequent centuries, which formed the basis of modern society and its culture, began with the rise of astronomy. When we look for causes and connections, it immediately strikes us that now in this Europe there was a different culture, a different society, a type of man different from that of decrepit antiquity or from the bygone grandeur of the fatalistic orient. There was now a vigorously growing society, a burgher world full of energy, a new kind of people who took up the thread of history where it had dropped from antiquity's hands. Was it a matter of racial genius? But the Greeks and Romans were a kindred race, and the Arabian Semites were surely not inferior in talents and ability. Shall we say, with Huntington, that it was the stimulating temperate and stormy climate that awakened man's energies? In this respect Renaissance Italy cannot have been much different from Roman Italy. We shall not be far wrong if we assume that among all the interacting influences, the social structure and the forms of labour and trade played an important part.

The barbarians, the Teutonic and Sarmatic tribes who destroyed the Roman Empire and whose descendants built medieval society, had known the degradation of slavery not as a regular economic system but at most incidentally. Their sturdy working power, which found scope for initiative also in serfdom, made the economic system of feudalism the starting point of medieval peasant revolt and burgher freedom. The driving force of craft and commerce had roused them to fresh energy in developing technical methods, trade routes, the organization of the shops, and the assembling and investing of capital, thus revolutionizing society. A strong individualism pervaded the work of the artisans, the enterprises of the merchants, the ideas of the thinkers. It constituted a new culture which sought support in a revival of the antique world in order to liberate itself from the bonds of medievalism. Man's highest aspiration was directed at discarding medieval barbarism by the study of ancient culture. Renaissance to them was the enrapturing revelation of the beauty, the wisdom, the joyful happiness of antiquity, which in their eyes was the acme of human civilization, and which they passionately absorbed.

But their minds were too full of tension to remain content with the antique pattern. Once they had completely assimilated it, it became the basis for further development. The same spirit of adventure and daring that incited man to discover new worlds across the oceans, also drove them to make discoveries in science. Thus they outstripped antiquity. To these men this was no break with antiquity, but its continuation. The spirit of the Renaissance, pervading at the dawn of the new century became manifest in a renewal of science.

The principal science to which this applied, and, properly speaking, the only science worthy of the name, was astronomy. Strong social needs put it into the foreground of public interest. First of all, there were the demands of chronology. The calendar was in disorder; the ancient intercalation rules had not been sufficiently accurate, and the discrepancies had become unduly large. The vernal equinox fell on March 11th instead of on March 21st, and the full moons came three days too early; thus the computation of Easter Day was all wrong. Regiomontanus had been obliged to inform the Pope that new observations would be needed to provide a reliable basis for improved rules.

Next came the needs of navigation, which on the wide oceans made more severe demands than did ancient Mediterranean traffic. The stars, sun and moon were now needed to find the position of a ship by the determination of geographical latitude and longitude. The Iberian peoples, first the Catalans and then the Portuguese and the Spaniards, as seafarers and discoverers, opened up the routes across the oceans. They made use of the astronomical knowledge handed down from Arabian science, chiefly by Jewish scholars. In the fourteenth century King Pedro III of Catalonia had astronomical tables made by the Jewish astronomer Jacub Carsono. The cross-staff was said to have been invented—or at least introduced—at the same time by Levi ben Gerson from Provence. Zacuto, who taught at Salamanca University, published a 'perpetual almanac' in Hebrew. One of his pupils, José Vizinho, court physician to King João II of Portugal, was a member of the 'Junta dos Mathematicos', the scientific body for navigation. For the use of the Portuguese navigators this Junta published a *regimento* with instructions on the manipulation of instruments, and a nautical year-book with directions for computation. Thus Arabian astronomy contributed to the discoveries of the Portuguese. Wherever they went on their voyages to the south and to India, their pilots measured solar altitudes with the astrolabe in order to deduce their latitude by means of the solar declination in their tables—they evidently did not use the cross-staff. At this time the works of the German astronomers also began to be used, especially Regiomontanus's *Ephemerides*. Columbus, on his voyages, used these as well as Zacuto's almanac; and it was the lunar eclipse of February 29, 1504, predicted therein, which he used in a perilous situation to make the aborigines pliable. Probably Martin Behaim from Nuremberg, geographer and merchant in the Azores, contributed, as a member of the Junta, to the introduction of the results of German astronomy into Portugal.

Conversely, the practice of navigation now greatly stimulated the interest in astronomy. The sailors now became acquainted with the sphericity of the earth as a personal experience. Along with the unknown regions of the terrestrial globe, the unknown parts of the starry sphere were opened up to man's view. European navigators perceived the unknown stars and constellations in the southern sky with amazement. From one of the early Portuguese voyages along the African coast, at the Gambia river in 1454, Cadamosto wrote: 'since I still see the north polar star I cannot yet perceive the southern pole star itself, but the constellation I behold in the south is the Southern Wain.'[89] This belief in the symmetry of the stellar arrangement about the two celestial poles was a curious belief of the first explorers. Probably what Cadamosto saw was a combination of the two bright Centaurus stars (α and β) with the same quadrangle of stars which Christian zeal afterward regarded as the Southern Cross, and in which Vespucci on his travels south imagined that he recognized Dante's 'four brilliant stars'. Certainly, astronomy in those days was not a theoretical doctrine but a living science.

But interest in astrology was more universal than these practical activities. The medieval doctrine, that from the stars emanated the forces that determined events on earth, was now spread by almanacs through all classes of the population. Such almanacs, as well as single calendar sheets, printed in large numbers, contained besides the celestial phenomena, the eclipses and conjunctions, also predictions of the weather, of natural catastrophes, and of favourable and unfavourable times for various activities, even bloodletting and haircutting. Regio-montanus wrote a book on medical astrology, containing similar instructions.

Attention was directed particularly to the conjunctions of the two mighty planets Jupiter and Saturn, which take place every 20 years, each succeeding occurrence at a longitude 117° back. So they occur alternately in three zodiacal signs in trigonal position, until, slowly advancing, after two centuries the next triad of signs becomes the scene of the occurrences. In 1488 a famous astrologer, Johannes Lichtenberger, wrote: 'Attention must be paid to the important constellation of the weighty planets Jupiter and Saturn, whose conjunction and coincidence threaten terrible things and announce many future calamities . . . and to this terrible conjunction the horrible house of the ill-fated Scorpion has been assigned.'[90] This refers to the conjunction of 1484, which was assumed to extend its influence over twenty years. It was followed in 1504 by a conjunction in Cancer and in 1524 by one in the Fishes. The belief that great floods would then devastate the earth brought panic to Europe; 133 writings on this theme by 53 authors are known. The times were full of terror and fright; only an isolated few among the scholars, such as the Italian humanist Pico della Mirandola, combated astrology as a superstition. In a preface to a book by Lichtenberger, printed at Wittenberg, Martin Luther wrote: 'The signs in heaven and on earth are surely not lacking; they are God's and the angels' work, warn and threaten the godless lords and countries, and have significance.'[91]

The sixteenth century was an age of great social and spiritual revolutions. What had developed gradually in preceding centuries, the growing forces of burgherdom, the rise of monarchical power, the contest within and the opposition against the Church, all resulted in a rupture with the past. The opening-up of the entire earth by the navigators liberated human minds from the old narrow world conceptions but also, owing to the influx of American gold and silver, caused a rise in prices and general impoverishment, which exploded in revolts of exasperated peasants and urban artisans. New conceptions of life expressed themselves in new religious systems, in Lutheranism, in Calvinism, and in the renovation of the Catholic Church. Their struggle was mingled in a chaotic way with the fight between the princes, the nobles and the burghers in a series of political and religious wars. The quiet certitude which the Church formerly could guarantee through its unchallenged authority had vanished; the basis of life was shattered. In these unstable and turbulent earthly conditions, the bond linking everything on earth to the stars with their fixed and calculable course offered to man his only refuge. Stronger than ever before grew the need for, and the confidence in, astrology; more than ever it occupied the thoughts of everyone. The princes employed 'mathematicians', chiefly as astrological advisers (the word *mathematica* at that time meant 'astrology'). Municipal and provincial councils, too, appointed mathematicians, not only to teach the sons of wealthy citizens and nobles in the schools but also to compile almanacs for peasants and citizens, containing prognostications on weather and on political prospects.

Astronomy in the fifteenth and sixteenth centuries stood in the centre of practical life and occupied man's attention more than any other science. Thoughts dwelt mainly on the stars; any ingenuity and scientific initiative in the human mind were directed primarily to the science of the celestial bodies. So it is not surprising that in those days people no longer confined themselves to respectfully repeating the views of the ancients, but struck out upon new tracks.

CHAPTER 18

COPERNICUS

THE new world system, as it appeared briefly in ancient Greece, had been developed by different individuals in two distinct stages: first the rotation of the earth, as an explanation of the apparent daily rotation of the celestial sphere, and second the yearly orbit of the earth. But the old theory, based on the positive premise of the immobility of the earth, had an inherent unity and was seen by everyone as a coherent whole. This probably was the reason why the new world system, in replacing the old picture, sprang up complete in its entirety. This was the work of Nicolas Copernicus (plate 3).

Copernicus (Niklas Koppernigk), born in 1473 at Thorn (now Polish: Torun), was descended from a family of German colonists, who a century before had been called into the country by the Polish king. German immigrants had already settled in these eastern regions, first under the rule of the Knights of the German Order, then under the Polish kings. They had founded a number of prosperous towns, such as Danzig, Thorn and Cracow, which became flourishing centres of trade and commerce and seats of urban culture. Such pioneers, as we remarked in a former case, are more open-minded, less prejudiced, and less bound to tradition than people staying on in their old homes.

Having become acquainted with astronomy in its close connection with astrology at the University of Cracow, Koppernigk was sent to Italy in 1496 for juristic studies and again in 1501 for medical studies. Here, while studying astronomy with Domenico Maria Novarra, a native of Ferrara, he came into close contact with the strongly pulsating life of the Renaissance. He studied Greek and became acquainted with what Greek and Latin sources reported on the dissentient opinions of ancient philosophers. These reports doubtless awakened or strengthened his ideas of a different world structure. When, later, he was dedicating his work to Pope Paul III, he mentioned the following as the sources which gave him the courage to work out his new theory: Nicetas, who according to Cicero made the earth move; Philolaus the Pythagorean, who made the earth describe a daily orbit around the central fire; and Heraclides and Ecphantus, who made it rotate about its axis.

Having returned in 1503 to Poland, where his uncle, the Bishop of Ermland, had procured him a seat in the Chapter of Frauenberg, he soon, probably about 1512, formulated his ideas in a *Commentariolus* ('short comment') which he sent to a number of friends and astronomers. Concisely and emphatically, the pillars of the new world system were erected in seven theses: (1) There is no single centre for all celestial orbs or spheres. (2) The earth's centre is not the centre of the world but only of gravity and of the lunar orbit. (3) All orbs encircle the sun, which as it were stands in the midst of all, so that the centre of the world is situated about the sun. (4) The relation of the distances of sun and earth to the height of the firmament is smaller than that of the earth's semi-diameter to the solar distance, so that its ratio to the altitude of the firmament is imperceptible. (5) What appear as motions in the firmament are not due to it but to the earth; hence the earth with its closed elements rotates in daily motion between its invariable poles, whereas the firmament is immobile, and the last heaven is permanent. (6) What appears to us a motion in relation to the sun is not due to the sun itself but to the earth, with which we are revolving just as any other planet; so the earth is carried along by several movements. (7) What shows itself in the planets as retrogradation and progress does not come from their own but from the earth's part; its single motion therefore suffices to [explain] many different phenomena.

Thus he was already sure of the basis of his new world system. But his main task occupied the years to come, in his leisure moments between official duties, participation in the administration and political direction of the diocese. That task consisted in obtaining by observation an exact numerical derivation of the orbits of the planets, as the basis for the computing in advance of future phenomena. In this way the practical needs of astronomers could be satisfied, now that the Alfonsine Tables were increasingly in error. In this way, too, his work could replace that of Ptolemy. However, when he had finished all this research, he hesitated to publish it for many years. Though influential friends among the clergy—Tiedeman Giese, Bishop of Kulm, and Cardinal Schoenberg of Capua—strongly encouraged him, he delayed publication because he foresaw the opposition due to prejudice, and his placid mind shrank from strife. Lack of fervent zeal against the new Lutheran doctrines had already made him suspect by the clerical zealots. He overcame his hesitation only when a young mathematician from Wittenberg, Georg Joachim, called 'Rheticus' (i.e. a native of the Grisons), visited him in 1539 to learn about his theory, and had given it an enthusiastic review in a *Narratio prima* ('First Communication'), about 1540.

He entrusted the manuscript, *De revolutionibus, libri VI* ('Six Books on the Revolutions'),* to Rheticus to have it printed in Nuremberg; it appeared in 1543, the year of his death. To this first publication, Osiander, a Lutheran minister in that town, who supervised the printing, had added—probably to meet the opposition of the Wittenberg theologians—an anonymous preface, entitled 'On the Hypotheses of this Work'. Also the extension of the title into *De revolutionibus orbium coelestium* ('On the Revolutions of the Celestial Orbs') could convey the idea that the earth was not necessarily included. Not until many years later did it become known that these additions were not by Copernicus himself.

In this work of Copernicus we may note three different aspects, distinguishing it from Ptolemy and from other books on astronomy written since Ptolemy. These aspects are the heliocentric world structure, the introduction of new numerical values, and a new mechanism to represent the details of the planetary motions.

Firstly he sets forth the new basis of the world system, and in the First Book provides the arguments against Ptolemy. He explains that the world is spherical and that the earth, too, has the shape of a sphere. The mobility of a sphere consists in its turning around an axis in a circle which has no beginning and no end. The celestial bodies exhibit various different motions, yet their motions must be circular and uniform, or consist of circular motions; for only thereby does what was before return in fixed periods. Since reason refuses to accept irregularities in what is arranged in the best order, we have to assume that uniform motions appear to us irregular because of a difference of poles or because the earth is not in the centre of the circles. Most authors, surely, assume the earth to be at rest in the middle of the world and regard other views as ridiculous; but on closer consideration it is seen that the question is not yet decided. For every observed change originates either from the motion of the object, or of the observer, or of both. If a motion of the earth is assumed, this must appear, although in opposite direction, in all that is outside, as if things filed past her; and this holds especially for the daily motion. Since heaven contains all, it is not conceivable that motion should not be attributed to what is contained therein rather than to what contains all. If then someone should deny that the earth occupies the centre of the world but admits to her a distance not large enough to be measured against the sphere of the fixed stars, but comparable to the orbits of the sun and the planets, he could perhaps indicate the cause for the apparent irregularities in the different motions as due to another centre than the earth's.

The ancient philosophers tried to show that the earth rests at the centre chiefly because all heavy matter tends to move to the centre of

---

* Revolutions means here circular movements.

the world and remains there at rest. According to Aristotle, earth and water move downwards, air and fire upwards, whereas the heavenly bodies revolve in circles. A rotation of the earth, Ptolemy says, would be contrary to this, and by so violent a retraction everything would be torn asunder. Objects falling down vertically would not reach their intended place because it would have been carried ahead from under them with great velocity; and we would see clouds and everything suspended in the air always moving westward.

If, however, one assumes a rotation of the earth, he certainly will say that it is a natural and not a violent motion; and what happens by nature is contrary to the effect of outer violence, and remains in perfect order. Why did Ptolemy not fear the same thing with heaven, which, according to him, has to rotate at a much more tremendous speed

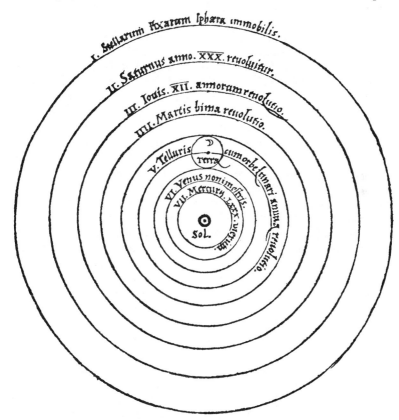

Fig. 21. The Solar System according to Copernicus

because it is larger than the earth? Since the earth is a globe enclosed between its poles, why not attribute to it the motion that is natural to a sphere rather than to assume that the entire world, of which the boundaries are unknown, is moving? A large part of the air, in which the clouds are floating, is drawn along with the earth, so that to us they appear to be at rest; whereas the remote realms of the air, in which the comets are seen, are free from this motion. The motion of falling or rising things is always a double one relative to the universe, composed of linear and circular motions. To a simple body when it is in its place belongs a simple motion, and this is the circular motion through which it appears to be at rest. When it is removed from its natural place, this is against the natural order and gives rise to an accelerated linear motion. To this must be added that the state of immobility is considered to be more noble and divine than unrest and variability; hence it belongs to the entire world rather than to the earth.

Now, since there is no objection to the motion of the earth, we have to investigate as to whether it has other motions and may be regarded as a planet. That it is not the centre of all circular motions is shown by the irregularities of the planets and their variable distances that cannot be explained by concentric circles around the earth. There are other centres besides the centre of earthly gravity; gravity is the tendency of cognate particles to combine into a sphere. It is credible that the same tendency is present in the sun, the moon and the planets, giving them their spherical shape while they describe their circular orbits. If the earth also has other motions they must be found in foreign motions presenting a yearly period. When we assume the immobility of the sun and transfer its yearly motion to the earth, the risings and settings of the stars, whereby they become evening and morning stars, follow in the same way, and the oscillations and stations of the planets appear to be motions lent by the earth to them. Then we must assume that the sun occupies the middle of the world.

The succession of the planets was always assumed from their periods of revolution: Saturn, Jupiter, Mars. The old contest about whether Mercury and Venus should be placed above or below the sun is now solved in this way, that both, as Martianus Capella wrote in antiquity, revolve about the sun. For the other three also the sun must be the centre, because in opposition, as is shown by their greater brightness, they come nearer to us, and in conjunction are fainter and more remote. Between these two groups the orb of the earth is situated, with the moon and all that is below the moon. At the outside, highest and most remote, is the sphere of the fixed stars, immobile and so large that the dimension of the earth's orbit is negligible against it. Then follow Saturn, Jupiter and Mars, finishing their orbits in 30, 12 and 2 years respec-

3. Copernicus (p. 188)

Tycho Brahe (p. 204)

Two drawings of the moon from Galileo's Day Book

4. Galileo (p. 227)

tively; then comes the earth with 1 year; then Venus with 7 months and Mercury with 80 days. In the middle of them all stands the sun. 'Who in this most beautiful temple would put this lamp at a better place than from where it can illuminate them all? Thus, indeed, the sun, sitting as on a royal throne, leads the surrounding family of stars.'[92]

These are the arguments (given mostly in his own words but much shortened) which Copernicus adduced for his new system. His arguments against Ptolemy were mostly philosophical, belonging to modes of thinking handed down from the ancient world. There remains some vagueness in the exposition of his system, as appears from the use of the Latin term *orbes* (of which the English equivalent is 'orbs'). This name, as is seen in the first thesis of the *Commentariolus*, sometimes means spheres with the sun as centre; by their rotation they carry the planets attached to them around in circular orbits. Sometimes the name means the orbits themselves, especially when numerical dimensions are derived.

The new theory was not the result of experience or observation; it did not contain new empirical facts that would compel man to relinquish the old concepts. Observations can only show motions of other bodies relative to the earth. These were the same in both world systems. What gave strength to the new system was its simplicity and harmony. With its simple basis, however, its consequences were enormous, even beyond the vision of its author. In their full extent they were only drawn by his later followers. Whereas Copernicus spoke of the celestial sphere as a real object, immovable and containing all, its original function was dissolved into nothing since the stars at rest in space no longer needed to be connected by a material sphere. The unfathomable depths of space were opened to man. At the same time—a still heavier demand on man's imagination—the most solid basis of his existence, the fixed earth beneath his feet, was drawn along in whirling motion with furious velocity, unimaginable, unacceptable, contrary to the most direct and certain experience. The doctrine of Copernicus meant a complete upheaval in man's world conception, which, as the new truth spread, was to determine modern thinking ever after.

The renovation of cosmic theory might have consisted in a simple transcription of Ptolemy's system, copying all numbers and measures but giving them new names: what was called the 'deferent' of Mars or Jupiter was now called its 'real orbit'. But Copernicus at the same time revised Ptolemy's numbers and measures—this was the second aspect of his work. He used new observations, mostly made by himself to derive the orbits for the present time with more accurate periods of revolution, and he computed new tables. Thus he produced a new manual of astronomy suited to replace Ptolemy in every respect.

Ptolemy's theory, moreover, could not satisfy him in the explanation of the variable velocity of the planets. His fundamental principle was that all motions of the celestial bodies consist of uniformly described circles. He thought it inadmissible to disregard this principle, as is actually done with the assumption of a *punctum aequans*. In criticizing this theory, he said: 'It is certain that the uniform motion of the epicycle must take place relative to the centre of its deferent and the revolution of the planet relative to the line through this and the centre of the epicycle. Here, however [in the old theory], they allow that a circular motion may take place uniformly about a foreign and not its own centre. . . . This and similar things have induced us to consider the mobility of the earth and along such other ways that the uniformity and the principles of science are preserved and the reason for the apparent inequality is rendered in a more constant way.'[93]

This then was the third aspect in which his work deviated from Ptolemy, a new and most ingenious mechanism to replace the equant. He faced the problem that in aphelion and perihelion (in 1 and 3 in the figure) the distances to the sun should be affected by the single

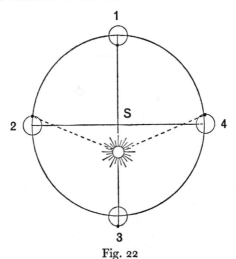

Fig. 22

eccentricity e, whereas at 90° anomaly (in 2 and 4) the direction sun – planet should be affected by the double or total eccentricity $2e$. He solved it by making the distance of the sun from the centre of the circle $1\frac{1}{2}e$ and in addition having the planet in the same period describe a small epicycle of radius $\frac{1}{2}e$. Then in aphelion and perihelion, as fig. 22 shows, the two effects of eccentricity and epicycle subtract, whereas in the sideways positions they add.

Following this principle Copernicus made a new computation of Ptolemy's data for the oppositions of Saturn, Jupiter and Mars, as well as of his own observations of three oppositions of each of them, made between 1512 and 1529. In accordance with Ptolemy, it was not the opposition to the real sun but to the mean sun that was used. In Ptolemy's system this was unavoidable, as we saw above; in the new system it was not. It means that not the sun itself but the centre of the earth's orbit is taken as centre of the world, around which the planets describe their orbits; it is this centre that occupies the point S in our figure. It seems as if the proud proclamation of the sun's kingship in the first chapter was forgotten. The computations were made in the same way as with Ptolemy, by successive approximations; first the small epicycle was neglected, then from the resulting distance taken to be $2e$ the fourth part was taken as radius of the epicycle, and the observed longitudes were corrected for its effect. For the ancient data he arrived, of course, at the same numerical results as Ptolemy, only expressed in a different way: for eccentricity and epicycle radius he found for Saturn 0.0854 and 0.0285; for Jupiter, 0.0687 and 0.0229; for Mars, 0.1500 and 0.0500; and the longitudes of the aphelion (former apogee) were also identical.

From his new observations he found for Saturn the eccentricity 0.0854, the epicycle 0.0285, in total 0.1139, exactly corresponding with Ptolemy's $6\frac{5}{6}$ sixtieths $=0.1139$; the longitude of aphelion $240° 21'$ had increased by $14°$ in the fourteen intervening centuries. For Jupiter he was led astray by an error in computation;[94] so he assumed Ptolemy's value for the eccentricity and derived therewith the longitude of aphelion $159°$, which had advanced $4\frac{1}{2}°$ since Ptolemy. For Mars he found the eccentricity 0.1460 instead of 0.1500, which would have fitted Ptolemy's 0.200, and the aphelion advanced by $10° 50'$. 'The distance of the centres of the orbits we found to be 0.0040 smaller than he did. Not that Ptolemy or ourselves have made an error, but as a clear argument that the centre of the orbit of the earth has come nearer to that of Mars, whereas the sun has remained immobile.'[95] If we consider that this difference corresponds to $14'$ only in the first opposition, it seems questionable whether he did not put too much confidence in the reality of such small differences.

For Venus and Mercury, now called the 'inferior planets', the new world system could have removed many of Ptolemy's difficulties simply by having the planet describe a smaller eccentric circle within the larger circle of the earth. But first, by his insistence upon the centre of the earth's orbit instead of the sun itself as the world centre, Copernicus made his task more difficult and his system more complicated than was necessary. He followed Ptolemy so closely that his exposition often seems to be a copy of his predecessor's in somewhat altered language. Moreover, it was difficult to transform a centric epicycle into an eccentric orbit. He made the centre of the orbit of Venus describe a small circle in half a year; and the many irregularities inserted by Ptolemy into both the epicycle and the deferent of Mercury in the new theory also made Mercury's orbit in space dependent on the motion of the earth. That the other planets in their true course should depend on the earth makes a strange impression; here something remained of the old geocentric ideas.

To derive exact results for the orbits of these planets, just as for the superior ones, new observations besides the data from antiquity were necessary. For Venus he could make use of his own observations; but for Mercury he did not succeed. 'Indeed, this way of investigating the course of this star was shown to us by the ancients, supported, however, by a clearer sky, since the Nile, it is said, does not breed such vapours as the Vistula does here. Nature denies this facility to us, who live in a harsher country, where tranquillity [we might expect here the word transparency] of the air is rarer, and moreover, owing to the more inclined position of the celestial sphere, Mercury can be seen more rarely . . . so this star has caused us much trouble and labour in investigating its wanderings. Therefore we have borrowed three positions from those observed carefully in Nuremberg.'[96] Of these observations, one was made by Walther in 1491 and two by Schoner in 1504.

In respect of the inclinations of the planetary orbits, Copernicus may seem to have been at an advantage with Ptolemy in having to determine one inclination only for each planet. Since, however, he took over the data laid down by Ptolemy in his tables, there was no other way for him than to assume it to be variable. Thus he said that the latitude 'changes most where the planets, near opposition to the sun, show to the approaching earth a larger deviation in latitude than in other positions of the earth. . . . This difference is greater than would be required simply by approaching or retreating from the earth. From this we recognize that the inclinations of their orbits are not fixed but vary by some oscillating movements related to motions of the earth's orbit.'[97] With Venus and Mercury matters were still more complicated, as they had also been for Ptolemy. In his theory of the latitudes of the planets, Copernicus adopted Ptolemy's theory almost literally, with all its inadequacies, only expressing it in a different way in order to adapt them to the heliocentric basis. All this was a consequence of his respect for his great predecessor and his unjustified trust in Ptolemy's observational data and theoretical deductions.

For the moon, however, he could, by his system of circular motions, free Ptolemy's theory from the defect that the moon in the quarters should be at times twice as near to the earth as at full moon. To explain his 'second inequality', Ptolemy had made the entire lunar epicycle approach to and recede from the earth. Copernicus achieved the same purpose by having the moon twice a month describe a small circle, whereby the distance of the moon changed by $\frac{1}{10}$ only.

There were more complications in his world structure. The precession was given by Ptolemy as 1° in 100 years and this was derived from the change of stellar longitudes since Hipparchus. In later times a more rapid displacement had always been found, and Copernicus assumed, as did the Arabian astronomers, a *trepidatio*, an alternation of a slower and a faster precession. Like Thâbit, he connected this alternation with the gradual decrease of the obliquity of the ecliptic, clearly shown by the measurements since antiquity (Ptolemy 23° 51⅓′; Arzachel, 23° 34′; his own measures in 1525, 23° 28½′), which, according to these data, should even be variable itself. He combined them all into an oscillating movement of the earth's axis of rotation, to which he had already given a movement. The ancient idea that in the case of an orbital revolution of a body some fixed connection with the centre must be assumed implied that, in such a simple case of movement, the axis is always kept inclined in the same way towards the central body. To explain the fact that the earth's axis kept its direction in space, Copernicus had to give it a yearly conical motion, such as we often see with a rapidly rotating top— he called it the third motion of the earth. The precession could now find a simple explanation if the period of the conical motion was taken somewhat smaller than the orbital period, so that each year there remained the small precessional displacement of the axis. Copernicus now superimposed on this system another oscillation in a period of 3,400 years to make up for the variations in the precession and the obliquity. The result was that the length of the tropical year now showed periodical variations. A second consequence was that Copernicus, through lack of a regularly moving equinox, had to count his longitudes of the stars from an arbitrarily chosen star (he took the first star of the Ram: γ Arietis) and to reduce Ptolemy's catalogue to this zero point.

The same over-confidence in the observations and data from antiquity brought about another complication. Since Ptolemy gave an eccentricity of the solar orbit of $\frac{1}{24}$, and later authors found $\frac{1}{30}$ and different apogees, Copernicus assumed them to be irregularly variable. He represented them by supposing that the centre of the earth's orbit in the same 3,400 years describes, relative to the sun, a small circle with a radius of 0.0048, making the eccentricity vary between 0.0414 and 0.0318 and the apogee oscillate by 7½° to both sides. 'But if someone should suppose that the centre of the yearly orbit is fixed as centre of the world and that the sun is mobile by two movements similar and equal to those which we have derived for the centre of the eccentric circle, everything would appear as before, with the same numbers and the same demonstration. So there remains some doubt as to which of them is the centre of the world, for which reason we expressed ourselves from the beginning in an ambiguous way on whether it is situated within or about the sun.'[98]

Thus, the new world structure, notwithstanding its simplicity in broad outline, was still extremely complicated in the details. This, on first impression, gives to Copernicus's book a strange and ambiguous character. In the first chapters a new world system is proclaimed and explained which subverted the foundations of astronomy, brought about a revolution in science and in world concept, and for many centuries made the name of Copernicus a war cry and a banner in the struggle for enlightenment and spiritual freedom. Then, on studying the later chapters, we feel completely transferred into the world of antiquity; on every page his treatment shows an almost timidly close adaptation to Ptolemy's example. Nowhere the breath of a new era, nowhere the proud daring of a renovator, nowhere the symptoms of a new spirit of scientific research!

In reality, however, the contrast is not so great. The first chapters also breathe the spirit of antiquity. We have already seen this in his arguments on the earth's motion; they belong entirely to ancient philosophy. Copernicus did not consider his work as a break with the ancient world concept but as its continuation, and he appealed to ancient precursors. Through the desire to lean upon venerable authorities, the struggle between adherents and opponents in the years that followed was carried on under the names of 'Pythagorean' and 'Ptolemaic' world systems respectively. It all remained within the realm of ancient science; Copernicus was wholly a child of the Renaissance.

The astronomers of the sixteenth century regarded the addition of all those complicated circular motions as a refinement of the old theory. Copernicus was highly esteemed among them as one of the foremost astronomers, the man who had improved and replaced Ptolemy. This, however, was only on account of the details and the improved numerical values; his heliocentric system was considered an ingenious theory but was not accepted as truth. His numerical values were the basis of new astronomical tables computed by the Wittenberg mathematician Erasmus Reinhold. These tables, which Reinhold, in honour of his patron the Duke of Prussia, called the 'Prutenic Tables', soon superseded the Alfonsine tables in use until that time.

CHAPTER 19

ASTRONOMICAL COMPUTING

THE growth of science consists not only in the development of ideas and theoretical explanations but also in the improvement of the practical working methods. The practical work of the astronomer is twofold—observation and computation. From the numbers read on the instruments, which are the direct results of observation, the desired values of the astronomical quantities must be derived by computation. Thus in the fifteenth and sixteenth centuries the construction of the mathematical apparatus was as important a part of scientific progress as was the construction of the technical apparatus, the instruments for observation.

Greek geometry had already taught how from given lines and angles other lines and angles could be computed. For practical use Ptolemy had computed and inserted in his great work a table of chords. For all angles or arcs increasing by $\frac{1}{2}$° from 0° to 180°, the length of the chord was given in sexagesimals of the diameter. Since the sine of an angle is half the chord belonging to the double angle, his table corresponds to a table of sines for angles increasing by $\frac{1}{4}$° from 0° to 90°. The Arabian astronomers then derived practical relations between sides and angles in a triangle which we know as 'trigonometric formulae'.

Computation means practical handling of numbers and numerals. The Greek system of numerals was highly unsuitable for this purpose. They consisted of 27 letters of the alphbet, the first 9 for the units from 1 to 9, the following for the tenths from 10 to 90, the remainder for the hundreds from 100 to 900, simply placed one after the other; for the numbers from 1,000 to 1,000,000 the same letters were primed. These numerals could be used to write down the numbers, but they gave no aid in computing, i.e. in addition, subtraction, multiplication or division of large numbers. Decimals were unknown; some simple fractions like $\frac{1}{2}$, $\frac{1}{3}$, $\frac{2}{3}$, $\frac{3}{4}$, often occurred and had their special signs. Far more practical was the Babylonian sexagesimal system, where the units in which the successive numbers from 1 to 59 were expressed were the successive powers of $\frac{1}{60}$. Therefore Ptolemy gave the chords of his table in sexagesimal parts of the radius; three numbers secured an accuracy of $1 : 60^3 = \frac{1}{216000}$.

Europe in the Middle Ages had inherited the clumsy Roman system of numerals, so little suited to practical computing that in commerce a counting-frame or abacus always had to be used. Then gradually the Indian system of numerals, with place value and zero for the open places, penetrated from the Arabic world into Europe. The Italian merchants in their commerce with the Orient came to appreciate this system, though at first it was distrusted by less skilled dealers as a kind of unfair secret art. In the beginning of the thirteenth century a manual, *Liber abaci*, appeared, written by the much-travelled Leonardo of Pisa, afterwards called 'Fibonacci', that gave instructions for computing with arabic numerals and was used for many centuries. The translators of Arabian astronomical works of course copied the tables with their arabic numerals, though at the very first, in the manuscripts of Athelhard's translation, they were changed to Roman numerals.[99] Gradually the arabic numerals spread over Europe, drawn in different shapes (as may be seen on the figure of an astrolabe of the year 1547) before they got their modern forms. In the fifteenth century German commercial clerks went to Venice to learn the new calculation and the new ('Italian') book-keeping. In the sixteenth century division of numbers was still deemed such a difficult operation that Melanchthon at the University of Wittenberg had to give special lectures on the subject. In the book of Copernicus all numbers in the text are given in Roman numerals.

With the first rise of astronomy in Western Europe, Purbach began with the computation of a more extensive and accurate table of sines for angles increasing by 10'; the sines are given in seven figures, with the radius put at 6 million. Regiomontanus expanded it by means of interpolation into a table with the angles increasing by 1', so that for angles measured (in degrees and minutes) the sine could simply be copied. He also computed a table of tangents, for every full degree, to be used for measurements with the cross-staff. In a work on plane and spherical trigonometry, not printed until 1561, long after his death, he derived many formulae for the computation of lines and angles.

Though the extension and accuracy of these tables could suffice for the astronomical practice of that time, mathematicians continued to make them more perfect, thus anticipating future application. Rheticus (1514–76) began in 1540 with the computation of sines, tangents and secants in 15 figures (in printing afterwards cut to 10) for angles increasing by 10". At his death the tables were not yet ready; his disciple Otho had to finish them, publishing them in 1596. They served as the basis for many later tables; other computers later on corrected some remaining errors.

Theory was further developed at the same time. The system of relations between the goniometric functions and of the formulae for computation of all the elements of triangles was built up into a complete trigonometry. This was done by Rheticus, and then in a more complete way it was perfected by the acute mathematician Vieta (1540–1603). So it became a handy and usable apparatus with which the astronomers could work. Most important were the complicated formulae for spherical triangles to derive angles and distances at the celestial sphere. Formerly the problem had been met by dividing oblique-angles into right-angled triangles and applying their simpler relations repeatedly. Now the more general formulae were adapted to universal use.

The first problem for which they were needed was given by the old practice of measuring the sun's altitude to find the time. In the thirteenth century Sacrobosco had provided a method, borrowed from the Arabian astronomers, of reading the desired values from curves engraved on the back of the astrolabe. But this method was not exact and could be used only because of the small demand for accuracy. Celestial globes also were used to solve the spherical triangles; but the accuracy in reading the quantities on globes was limited. Walther often fixed the exact moment of a phenomenon or an observation by measuring the sun's altitude; so he could not do without exact formulae.

There were other applications of trigonometry. Because the positions of the planets relative to the ecliptic were needed, they were expressed in ecliptical co-ordinates, longitude and latitude; so the same system was used for the stars. These co-ordinates could be read from armillas—complicated structures of many rings turning one within another—which with some difficulty had to be brought into the right position. So the results were less accurate than would have been the case with a simpler structure. In the sixteenth century the conviction grew that it was better to measure equatorial co-ordinates, right ascension and declination, with a more simply constructed instrument, which allowed greater accuracy. Longitude and latitude then could be derived from them by means of trigonometric formulae. Trigonometry in this way led to the practice of measuring such quantities as were directly, and therefore accurately, measurable and deducing the other quantities by computation.

To determine the position of a planet, the simplest way was to measure its distance from bright stars, e.g. with the cross-staff, as had been done by Regiomontanus, Walther, and probably also by Copernicus. Two such distances fix the position of the planet. The problem of how to deduce the longitude and latitude of the planet from the measured distances and the position of the stars was again a problem of spherical trigonometry; for the astronomers of that time it was usual practice. Another problem to be solved by trigonometry was to find the difference in right ascension of two stars with known declinations, by measuring their distance from each other. Direct measurement of right ascension was not feasible, whereas the declinations of the stars could be directly determined by measuring their altitude in the meridian when they were highest, and subtracting the altitude of the equator. Thus trigonometric formulae combined with tables of the sines and other goniometric functions became the most important auxiliary apparatus for sixteenth-century astronomers.

In the same sixteenth century the art of computing by means of arabic numerals found its completion in the introduction of decimal fractions. The chief impulse came from Simon Stevin, a native of the Flemish town of Bruges, who is mostly known as 'the inventor' of this system of numeration. In his book *De Thiende* ('The Tenth') written in Dutch, and appearing in 1585, he explained how all computations may easily be performed 'by means of entire numbers without fractions'; hence he considered the numerals after the decimal point as integers, counting the successive powers of $\frac{1}{10}$ as units. Now the road was open not only to all kinds of astronomical computing but also to the practical handling of approximate and irrational numbers.

Of course, the practical use of the formulae in astronomy demanded long and tedious computations. The invention of logarithms, at the beginning of the next century, brought an enormous saving of time. Its principle is well known; if we append to the numbers of a geometric series a corresponding arithmetical series (hence, beside 10, 100, 1,000, 10,000 . . . or beside 2, 4, 8, 16 . . . we place 1, 2, 3, 4, . . . which now are called their 'logarithms'), any multiplication is now simplified into an addition, any division into a subtraction ($8 \times 4 = 32$ is reduced to $3 + 2 = 5$). The problem was which numbers had to be put, as their logarithms, beside 11, 12, 13. About 1580 this idea had arisen in the mind of Joost Bürgi, assistant at the Cassel observatory, but he did not grasp its great importance; it was not published, and only much later did a first table of logarithms computed by him come to light. The credit for giving them practical introduction to science falls to the Scottish scientist, John Napier, or Neper, as he wrote his name in Latin. His tables, published in 1614, had not yet the necessary practical utility, because they were not adapted to the decimal system. That was given to them by Henry Briggs, mathematician at Oxford, who came to Napier to propose a better arrangement, especially by the introduction of 10 as the basic number. In 1618 Briggs published the first, still incomplete, table of logarithms based on the decimal system, in 8 decimals. They were completed and published in 1628 by Adriaan Vlacq, bookseller at Gouda in Holland, assisted by his friend, the skilful computer, Ezekiel de Decker. These tables, with the logarithms given to 10 decimals, were the source of many tables published afterwards. Already Neper's first tables contained also the logarithms of the goniometric functions, which constitute an integral part of all later tables. In later times computers and computing offices have made new and still more accurate computations of all the logarithms, for some small prime numbers up to 64 decimals.

Logarithmic computation as an indispensable aid found its way into all practical sciences. But more than any other science, astronomy has profited by this invention. By executing the long and tedious computations of the logarithmic tables, the first inventors and computers, so to speak, lengthened the life of all later astronomers. They made possible researches which, because of the immense computation work involved, could not have been performed without this aid.

CHAPTER 20

TYCHO BRAHE

O NE of the many persons who in the sixteenth century ardently devoted themselves to the new study of nature was the Danish nobleman Tyge (latinized into Tycho) Brahe (1546–1601). As a youth, sent to Leipzig for juridical study, his passion for astronomy was shown in his secret nightly studies and observations. Like most of his contemporaries, he was deeply convinced of the truth of the astrological doctrines, and he often computed horoscopes. But to him this doctrine meant more than making prognostications. The most prominent thinkers of these times, in the confusing chaos of social and political strife and the dark uncertainty of future and fate, sought support in their confidence in an inner connection and harmony of the entire world. The stars, they said, ruled the earth, and the course of the stars is dominated by eternal laws. Our knowledge, however, of this connection and of the course of the stars was insufficient. The belief in an indissoluble bond between the precarious happenings on earth and the regularity of the stars was the guiding principle in their researches. There was a better astrology than the practice of prognostics; but it was still in its infancy, and only by careful study could it develop from this primitive condition into reliable science. This science, being knowledge of the inner unity of the world, would give man power over his fate.

This was also Tycho Brahe's world concept. He expressed it in later years in a public lecture in Latin (then the international language of scholars) given in 1574 at the University of Copenhagen. Its title, *De disciplinis mathematicis*, though dealing entirely with astrology, shows how in his view, which conformed to the views of the time, astrology was the main practical object of mathematical discipline.

In this lecture he said: 'To deny the forces and the influence of the stars is to undervalue firstly the divine wisdom and providence and moreover to contradict evident experience. For what could be thought more unjust and foolish about God than that He should have made this large and admirable scenery of the skies and so many brilliant stars to no use or purpose—whereas no man makes even his least work without a certain aim. That we may measure our years and months and days by the sky as by a perpetual and indefatigable clock does not sufficiently explain the use and purpose of the celestial machine; for what it does for measuring the time depends solely on the course of the big luminaries, and on the daily rotation. What purpose, then, do these five other planets revolving in different orbits serve? . . . Has God made such a wonderful work of art, such an instrument, for no end or use? . . . If, therefore, the celestial bodies are placed by God in such way as they stand in their signs, they must of necessity have a meaning, especially for mankind, on behalf of whom they have chiefly been created . . .

'But no less do they [who deny the influence of the stars] violate clear evidence which for educated people of sane judgment it is not suitable to contradict. Who does not perceive that the difference in quality between the four seasons is caused by the rising and the retreating of the sun and its ordinary course through the twelve parts of the zodiac? We see, too, that, with the waxing of the moon, everything cognate to her nature—such as the brains of living beings, the marrow in the bones and the trees, the flesh in the lobsters and shells and many things more—increase also; but when she wanes, these also diminish. In the same way the flux and reflux of the wide ocean is affixed as with a chain to the moon's motion, so that immediately, with the rising of the moon, the sea also begins to rise. . . . These and many kindred phenomena are known even to uneducated people. For sailors and peasants, by numerous experiences, have noted the risings and settings of certain stars, from which they can foresee and predict yearly recurring storms. The scholars, however, who are trained in this abstract science, deduce the influences of the configurations of the other wandering stars: with one another or with the Luminaries, or with the fixed stars. They have observed that the condition of the air in the four seasons of special years is affected by them in various ways. So it has been perceived that conjunctions of Mars and Venus in apt parts of the sky raise rains, showers and sometimes thunderstorms. That important conjunctions of the big planets cause vast changes in this lower world has often been shown by experience. Thus in 1593, when a great conjunction of Jupiter and Saturn took place in the first part of the Lion, near to the nebulous stars in Cancer, which Ptolemy calls the smoky and pestilent ones, did not the pestilence which swept over the whole of Europe in the years that followed, and caused innumerable people to perish, confirm the influence of the stars by a very certain fact?'[100]

Then, after dealing with the opinions of those opposed to astrology, Tycho proceeded: 'But we, on the contrary, hold that the sky operates not only on the atmosphere but also directly upon man himself. Because man consists of the elements and is made out of earth, it is necessary that he be subjected to the same conditions as the matter of which he consists. Since, furthermore, the air which we inhale and by which we are fed no less than by food and drink, is affected in a different way by the influence of the sky, as has been shown above, it is unavoidable that we should at the same time be affected by it in different ways. I leave aside what must be easily clear to every mind, that man by some hidden cause lives and is fed by the sky itself in a still higher degree than by air, or water, or by any other low elementary thing, and acquires an incredible affinity to the related stars; so that the ancient philosophers, among them Hipparchus, according to Pliny's testimony, were not wrong when they said that our spirit is a part of heaven itself.'[101]

Here all attention was directed to the stars; but in his own mind this was part of the general doctrine of the unity of the entire world, a doctrine that stimulated him to scientific research. In observing the conjunction of Jupiter and Saturn in 1563, he perceived that the Alfonsine Tables gave its time of occurrence a month wrong, and the Prutenic Tables some days wrong. How, in such a case, was a reliable judgment on the connection with earthly events possible? In the first place, a better knowledge of the planetary motions was necessary. It could be secured only by new and better observations, for which improved instruments were needed. When he visited Augsburg in 1569, he devised for the brothers Paul and Johann B. Hainzel, members of the town council and devotees of astronomy, a vertical wooden quadrant 19 feet in radius, suspended by the centre, that could be read at a plumb line to 10″. It has been used for useful measurements, but in the long run it was too unwieldly for regular work. For himself, for use on his travels, he had a lighter instrument made, which he called a 'sextant'; it was a sector of 30° with two radial diopter arms, one fixed, the other movable, to measure the distance between two stars by looking from the centre through the sights of both diopters at the same time. To read the graduation in minutes, he devised a contrivance by means of transversals, rows of ten points going up and down at constant distances between an outer and an inner circle of the border (fig. 23). This method of division he later applied to all his instruments. Having thus travelled through Europe, he returned to Denmark and lived on his uncle's estate, occupying himself chiefly with chemical experiments.

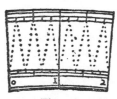

Fig. 23

Then it happened that in the evening of November 11, 1572, returning from the laboratory, his eye was caught by a brilliant star right overhead in the constellation Cassiopeia, which he had never seen there. It was a new star, a *nova*, which, as was soon ascertained, had appeared certainly after November 1st (probably between November 2nd and 6th). It was brighter than any other star; it equalled Venus in brilliancy and was even visible by day. This wonderful phenomenon stirred the entire world, first of the scholars, but also of the common people. What did it mean? This was the question posed by everyone. Beza, the friend of Calvin, supposed it to be a second star of Bethlehem, announcing a second coming of Christ on earth. Others discussed the calamities it would bring, its nature, and its place in the universe. Did it belong to the stars? Aristotle had established that in the world of stars everything was eternal and invariable. Or did it belong to the sublunar world of the variable earthly elements, and was it perhaps a singular comet condensed from fiery vapours?

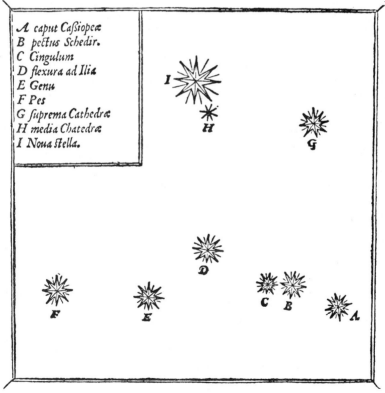

Fig. 24. The new star in Cassiopeia, from Tycho Brahe, *De Nova Stella.*

Tycho, full of the same questions, immediately began to observe the new star by measuring repeatedly its distance from the pole star and the nearby stars of Cassiopeia, both when this constellation stood near the zenith and when, twelve hours later, it stood low in the north beneath the pole. If the star had been at the same distance as the moon and so had a parallax of one degree, it would have appeared in the second case one degree lower relative to the other stars of Cassiopeia. But the distances in both cases, at high and at low altitude, were always found exactly equal, with only a few minutes of uncertainty. So it was proved that the new star had no perceptible parallax and was at a far greater distance than the moon. This demonstrated that, contrary to Aristotle's doctrine, changes did occur in the world of the stars, the realm of the ether.

Soon the star began to decline in brightness, more so the next year, while at the same time its colour changed to yellow and then to red, until at last it became colourless. It disappeared in 1574, leaving people wondering and shaken in their trust in philosophical doctrines. Tycho published a book on the new star in May 1573, after having overcome his hesitation as to whether it was proper for a nobleman to do so. He did it after he had seen the mass of nonsense written and published about the star. Here he made known his measurements and his deductions from them, and he gave *in extenso* his opinions on its nature and its astrological meaning. He deemed it probable that the star could have originated by condensation of the thin celestial matter which we see in the Milky Way, and he even designated a dark spot nearby in the Milky Way as the hole left thereby.

The appearance of the new star caused his zeal for astronomy to blaze into full strength. Things happened in the sky; who could know what great and important events the stars held in store for mankind? Tycho planned to settle in some European town and there to found an institute for the regular observation of the stars. He had Basle in mind because of its intellectual climate, so entirely different from the world of the Danish gentry.

The idea of founding an observatory for regular observation of the celestial bodies had also occurred to others at the same time. In Cassel, Landgrave Wilhelm IV of Hesse—the son of Landgrave Philip, well-known from the wars of religion and the rise of Protestantism—had during his father's reign already installed carefully constructed instruments under a movable roof in his castle, with which he himself was making observations. They consisted of simple quadrants and sextants with graduated metal rims. When in later years he became too occupied with government affairs, he took Christoffel Rothmann into his service as an observer. Later also Joost Bürgi, a talented Swiss and a skilful mechanic, was engaged, who applied himself diligently to the improvement of clocks as aids in observation. Tycho, in his travels through Germany on the lookout for a future site, visited Cassel in 1575, and this visit proved highly stimulating to both sides. There was a considerable increase in the intensity of observing the sun, the planets and the stars at Cassel in the next few years. Here the first efforts were made to introduce time as an astronomical measure for differences in right ascension; from Cassel came the first West-European catalogue of stars based on new measurements.

Another consequence of this visit was of still more lasting value. The Landgrave wrote to his colleague King Frederic of Denmark that an important and famous personality was likely to be lost to the honour and fame of the Danish kingdom. The King, himself interested in the sciences, who had already endowed Tycho for his studies, then offered him the small island of Hven, near Copenhagen in the Sound, as a fief with all revenues, in order to build an observatory there and to provide it with the best instruments. Here Tycho settled; in 1576 he began to build 'Uraniborg', a palace of astronomy, where in the following years, as his fame increased, he received princes and scholars from many countries. In that year he began the systematic series of observations which were continued there regularly for twenty years. He surrounded himself with assistants and disciples for whom he had an additional smaller observatory built, 'Stjerneborg'.

The renewal of practical astronomy at this observatory was in the first place due to the improved measuring instruments, constructed to Tycho's design. They were of varying types, each of different dimensions and make, the smaller ones for easy and rapid handling, the larger ones for utmost precision. First there were vertical quadrants for measuring altitudes, with radii of 16 inches, of 2, of $5\frac{1}{2}$ and of 7 feet, the smaller ones movable for turning to any azimuth, the larger ones fixed. In particular, a large quadrant, $6\frac{3}{4}$ feet in radius, fastened to a wall in the meridian, was much used for accurate measurements of declination; it could be read to 10″. It was made famous by a great painting often reproduced (see fig. 25), in which it is presented in full action, with Tycho himself directing, his assistants pointing, looking and noting, his dog at his feet. The sextants formed another type of instrument (also called sextants when the arcs were different from 60°): two arms, one fixed, the other movable, provided with sights, served to measure the distance between two stars. The most commonly used among them differed from his smaller travelling instrument in that two observers operated, looking from the outer arc side over the centre toward the stars. It rested with its centre of gravity on a ball-joint so that it was in equilibrium in any direction of the line connecting the stars; it had a

Fig. 25. Tycho Brahe and his great quadrant

radius of $5\frac{1}{2}$ feet. There were armillas too, but of much simpler and stabler construction, without ecliptical rings, to read equatorial co-ordinates in a more reliable way. The rings were set in the correct position by means of the right ascension of a known star; and then the right ascension of all other stars and planets could be read. The ultimate in simplification was reached in a polar axis resting above and below in masonry, adjustable by screws, to which only a declination circle of $9\frac{1}{2}$ feet in diameter, provided with sights, was attached. It could turn through all hour angles, which were then read on a somewhat larger half-equatorial circle situated on the northern side below the horizon.

All these instruments were distinguished by special precautions to secure the utmost precision. Transverse rows of points, as described above, served to read the smallest subdivisions. It was half a century afterwards that Pierre Vernier invented the method of an auxiliary arc (for instance, divided into 10 equal parts), which in the next centuries was applied to all astronomical instruments. In some countries it was wrongly called 'nonius' after the Portuguese astronomer, Pedro Nunez, whose invention was based on a different principle. For exact pointing to a star, Tycho had devised a special kind of pinnules which he applied to all his instruments. At the great quadrant, for instance, a

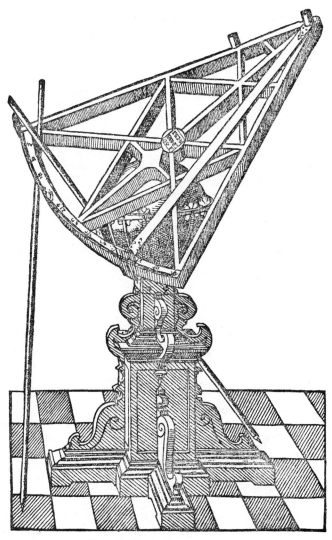

Fig. 26. Tycho's great sextant

cylindrical peg was put in the centre; before the eye was a metal plate
with two horizontal slits, one above the other at a distance exactly
equal to the diameter of the peg. If the eye, looking alternately through
the slits, saw equal parts of the star project above and below the peg,
the star was exactly pointed. Tycho gave detailed descriptions and
pictures of all these instruments in a book *Astronomiae instauratae
mechanica* ('Mechanics of the Renovated Astronomy'), published in
1598. His other book, *Astronomiae instauratae progymnasmata* ('Preliminaries for the Renovated Astronomy'), had been printed in parts
since 1588; it contained his results concerning the sun, the moon, the
fixed stars, the star of 1572, and the comets, and was published in full in
1602, after his death.

The sun was the first object. By measuring its altitudes in both winter
and summer solstices, he could derive the obliquity of the ecliptic (half
their difference) and the altitude of the equator (half their sum total). The
latter was 4' greater than what corresponded to the altitudes of the pole
star above and below the pole. He suspected the cause of the difference
to be refraction in the atmosphere, since the altitude of the sun at the
winter solstice was hardly 11°. By following the sun with his armilla on a
day in June from midday at its highest until it nearly set, he found that
it deviated ever more upwards, toward the north—as Walther had
previously perceived. At 11° altitude the sun was raised by 9'; this
explained the difference found. From a great number of measurements
he derived the refraction for different altitudes and combined them in a
correction table for refraction; at 0° altitude it was 34'; at 1° it was
26'; at 10° it was 10'; at 20° it was 4½'; at 30° it was 1' 25"; and at 40° it
was 10". Above 45° it was imperceptible; he assumed that it was caused
by vapours occupying the lower strata of the atmosphere. Since,
according to this table, the pole star, as well as the summer solstice, were
free from refraction, the obliquity of the ecliptic could be derived without
error by combining them; thus he found 23° 31½'.

Walther and Copernicus, however, had found 23° 28'. Tycho suspected that they too, as a result of refraction, had measured the winter
altitude too high and so had found the obliquity 3' or 4' too small. To
make sure, he afterwards sent one of his assistants to Frauenburg to
determine the polar altitude. It was indeed found to be 2¼' larger than
Copernicus had derived from his solar altitudes. Here we see the careful
attention given to differences of a few minutes, a mark of the higher
standard of precision in the renovated astronomy.

The curious thing is that Tycho's obliquity was still 2' too large. He
had corrected his measured altitudes not only for refraction but also
for solar parallax. According to Aristarchus and Ptolemy, this parallax,
$\frac{1}{19}$ of the lunar parallax, was 3'; and, adopting and applying this value,

Tycho got, for the solar altitude at summer solstice, a value $1\frac{1}{2}'$ too large. Apparently no doubts caused him to distrust this value—certainly it would have been difficult to make a new determination—and he simply says: 'this value appears by such subtle investigation by the Ancients to have been transmitted to us with great certainty.'[102] Consequently, many quantities determined by him are some few minutes in error. Perhaps it is also because of this circumstance that he found for the stars a smaller refraction than for the sun and gave them in a table as all $4\frac{1}{2}'$ smaller, hence imperceptible above $20°$ altitude.

To find the eccentricity of the solar orbit, he made use of the equinoxes and, instead of the not easily observable moments of solstices, of the moments that the sun, according to its declination, stood halfway between equinox and summer solstice. Because his measurements, besides being more accurate, were continued through many years, he could find the time intervals with greater precision than had former observers. He found the eccentricity to be 0.03584 and the longitude of the apogee $95\frac{1}{2}°$. He indicated why Copernicus's result was made in error, whereas from Walther's observations he found 0.03584 and $94\frac{1}{3}°$, nearly identical to his results. He assumed a regular progress of the longitude of apogee by $1°\ 15'$ per century; all the irregularities of Copernicus he simply omitted.

For the first time also the course of the moon was now followed through all the years. This provided him first with accurate values of the eccentricity in its Ptolemaic variations: $4°\ 58'$ at full and new moon, $7°\ 28'$ in the quarters, which he represented by a skilful system of epicycles. Moreover, he discovered an additional 'variation' (a name by which it has since been known): the moon in the octants is alternately $40\frac{1}{2}'$ ahead at $45°$ before full and new moon, and $40\frac{1}{2}'$ behind at $45°$ past them. He also found that the moon in spring was always behind, in autumn ahead, by $11'$; hence in winter it was slower, and in summer it was faster. It is remarkable that this deviation, called in later times 'annual inequality', was also perceived independently by Kepler, and without measurements. A moon eclipse in the spring of 1598, computed by him for a Styrian almanac, came $1\frac{1}{2}$ hours too late; when asked about the cause, he wrote that the sun had a retarding influence upon the moon, which was largest in winter, because then the sun was nearer to the earth. An explanation entirely conforming to modern views!

Further, Tycho derived from his observations that the inclination of the moon's orbit is not, as formerly assumed, $5°$ but that it oscillates between a minimum of $4°\ 58\frac{1}{2}'$ at full and new moon and a maximum of $5°\ 17\frac{1}{2}'$ at the quarters. Moreover, he found that the regression of the nodes in their 19-year period was not uniform but took place more rapidly at full and new moon, more slowly at the quarters. He could reproduce these variations by assuming that the pole of the moon's orbit, which in 19 years describes a circle of 5° 8′ radius about the pole of the ecliptic, also describes a small circle of 9½′ radius in half a year.

In deriving the position of the new star of 1572 by means of distances from the known stars of Cassiopeia, Tycho had perceived how the rough values of their co-ordinates prevented the accuracy of his measurements from showing to full advantage. He wanted accurate co-ordinates of the fixed stars; he needed them because he continually determined the positions of the moon and the planets by measuring their distances from the stars. The declinations of the stars could be found directly by measuring their meridian altitude with the great quadrant. The right ascensions had to be read from an armilla or computed with trigonometric formulae from distances measured with a sextant. Differences in right ascension were thus obtained; the right ascensions themselves had to be counted from the equinox, an invisible point of the sphere, ascertainable only by means of the sun. From its declination, measured when it was in the vicinity of the equinox, its distance to that point could be computed; if at the same time the difference in right ascension between the sun and a star had been measured, then the right ascension of the star itself was found. Since stars are not visible in the daytime, they cannot be compared directly with the sun; so the ancients used the moon as an intermediary. Tycho used the planet Venus, which moves much more slowly. In the years 1582–88, when Venus was visible by day, a large number of measurements were made of its distance from the low-standing sun and at night of its distance from the selected star. To eliminate refraction influences, an evening and a morning observation with equal altitudes of the sun were always combined. These precautions were successful; the resulting 15 values of the right ascension of the star (the brightness of the Ram, α Arietis) did not differ more than 40″, less than what could have been expected even with such careful observations. Then, by means of a large number of oft-repeated measurements, the right ascension and declination of 21 principal stars were determined; their mean error, as is found through comparison with modern data, was less than 40″. In order to ascertain whether the arc of his sextant was the exact length, he added all the consecutive differences of right ascension around the celestial sphere to see whether their sum total was 360°. The difference was only a modest number of seconds. This of course was pure chance; the sum total of a number of values, each uncertain by 20″ or 30″, can easily by mere chance be many minutes in error. Possibly Tycho had selected a number of best-fitting values, considering them to be the most reliable. The positions of the bulk of the fainter, less-often observed stars of his catalogue, numbering 788, are, of course, less accurate; their mean error, by comparison with modern values, appears to be about 1'.

Tycho's catalogue of stars was the first complete modern catalogue. It replaced Hipparchus's and Ptolemy's works, far exceeding them in accuracy. As the supreme result of the most perfect care, ingenuity and perseverance during many years, it marked the beginning of a new era of practical astronomy. For more than a century it remained the constantly-consulted source of star positions. Though the equatorial co-ordinates—right ascension and declination—had been directly measured, the catalogue, adhering to old custom, gave the ecliptical longitude and latitude computed from them. The stars were indicated, as they were with Ptolemy, by the parts or the limbs of the constellations. It was afterwards, in 1603, that Bayer, publishing under the name *Uranometria* a celestial atlas with all Tycho's stars, added Greek letters to the prominent stars; these letters, an easy designation, have since come into general use (plate 15).

Comparing his longitudes of the stars with the dates from antiquity and from the last century, Tycho derived an accurate value of the precession, 51″ per year. He assumed it to be entirely uniform; he did not speak of the 'trepidation' that had given so much trouble to Copernicus; thus this phantom disappeared from astronomy. Another phenomenon drew his attention. When he compared his stellar latitudes with those of antiquity he could ascertain that in the region of the Twins they had increased by 15′ to 20′ and diminished by the same amount in the region of the Eagle. Since the obliquity of the ecliptic was known to have decreased by the same amount, it was now clear that it was the ecliptic that had changed its position, whereas the equator there (at 90° and 270° of longitude) had kept its position relative to the stars. In this way new foundations were laid for practical astronomy with an accuracy hitherto unknown.

No less important were the observations of comets made at Uraniborg. When a comet appeared, its position was determined as often as possible, mostly by measuring the distance from prominent stars. The bright comet of 1577 was measured mostly with the cross-staff, since the new instruments were not yet ready; a modern computation of its orbit shows that Tycho's positions had a mean error of 4′. For the comet of 1585, measured with his better instruments, this error was 1′ only. One of his chief objects was to find the parallax by determining its position among the stars when it was low in the sky and comparing it with that at a great altitude. If the comet had been at an equal or lesser distance from us than the moon—as he had supposed in his youth—great differences, up to 1°, due to parallax, would have appeared. But he found no trace of parallax; the comet—according to his cautious conclusion—must have been at least six times farther away than the moon. This was sufficient to demolish entirely, in a publication on the comet of 1577, Aristotle's theory that the comets were fiery phenomena in sub-lunar regions, the upper layers of the atmosphere. Tycho has elevated the comets to the rank of celestial bodies; they belong to the realm of stars, and their orbits in space must be determined by astronomers. That he himself, by supposing them to be inclined circles, was quite mistaken cannot surprise us.

Far more extensive than this sporadic cometary work were the observations made of the planets. These had been from the beginning the chief object of Tycho's astrologically directed mind, and they constitute the chief mass of the work at Uraniborg. The treatment and discussion of these materials, in order to derive the orbits of the planets, lay ahead as his main task, to which all the preceding results were only the preliminaries.

But now the practical conditions of his life became an impediment. In 1588 his patron King Frederic died, and the government first came into the hands of a body of high noblemen, guardians of his successor, and then of the young prince himself. They were not so well disposed towards him. Tycho had been repeatedly in conflict with tenants and with others, because he tried to spend as much money as possible on his astronomical work, and the higher courts of justice had often put him in the wrong. His proud and haughty bearing had brought him many enemies among his fellow-nobles and the high officials. Some of his former prebends now being taken from him, he felt aggrieved and left Denmark in 1597. After some wanderings he found a new patron in the Emperor Rudolf II, who himself, however, was always in financial difficulties. Tycho settled in 1599 in Prague, in the emperor's residence, where with his assistant Longomontanus he continued his observations with such instruments as he had been able to carry with him from Hven. But his strength was broken; in October 1601 he died, leaving all his observations in the hands of Johannes Kepler, who had been his assistant for computations in the last years.

CHAPTER 21

THE REFORM OF THE CALENDAR

THE calendar serves for the regulation of time in social life. Like all social regulations, it had been a religious concern from early times. With the rise of Christianity in the Roman Empire, which as a universal religion was the expression of new and deeper social relations than were the old tribal and state religions, the religious festivals too acquired a new character. They lost their direct close connection with the pattern of work in its yearly cycle. As a means of regulating the Christian festivals, the calendar was now a Church affair. In the first place came the fixing of the dates of Easter, with the other connected movable festivals; for Christianity, Easter no longer meant the spring offerings of the new harvest but the yearly commemoration of the crucifixion and resurrection of Christ.

According to the Gospels the resurrection had taken place on the Sunday after the Sabbath following the Jewish Passover that fell at the full moon, the fifteenth of the first month, Nisan. Therefore, the commemoration must take place on the first Sunday after the first full moon of spring, i.e. after the vernal equinox, March 21. So astronomy was needed to fix the date, since the announcement of the date of a festival that was to be celebrated simultaneously in all the churches of the Orient and the Occident could not wait until the full moon had been observed. It must be fixed long beforehand, through knowledge of the theory. But not too difficult a theory, for the rules for computing the date had to be manageable by well-instructed priests, by means of cycles and periods. Exact astronomical computation, with inclusion of all irregularities of the moon, had to be avoided.

The large irregular changes in the date of Easter are due to the adaptation of the lunar phenomena to the Roman solar calendar. With a lunar calendar, Easter Sunday certainly would not have a fixed date, since the seven-day weeks roll on independent of sun and moon; but this date could not vary more than between one and seven days after full moon; the adaptation to the solar calendar brings the difficulties and irregularities. Of course, the 19-year cycle, the common multiple of month and year, had to be used, with easy computing directions,

217

manageable by simple minds. But if we consider that the regular alternation of 29 and 30 days goes wrong after a few years and that, by the intercalation of one day every fourth year, the regularity is also disturbed, it is clear how many difficulties, shown by much internal strife, the Church encountered in the first centuries of its existence when fixing the Easter dates.

The happy solution was the work of one Dionysius, archivist in the Pope's service, who, to distinguish himself from the great Church Father, called himself *Exiguus*, the 'little one'. In a report of about AD 520, in which at the same time he introduced the counting of the years since the birth of Christ instead of numbering them in relation to the Roman emperor eras, he fitted the Easter regulations used in the Orient into the Roman calendar. If we omit some minor complications, it comes down to this. In every year the 12 lunar months are taken alternately as of 29 and 30 days, hence in total, 354 days; but their real duration is 354.367 days. In 19 years we are then $19 \times 0.367 = 7$ days short. In this cycle a thirteenth month is intercalated seven times; if they are all taken to have 30 days—that is, on the average, half a day too much—we are $3\frac{1}{2}$ days ahead. Moreover, the Julian leap years add $4\frac{3}{4}$ days in a 19-year cycle, so that the shortness of 7 days is amply compensated. Indeed $19 \times 354 + 7 \times 30 + 4\frac{3}{4} = 6,940\frac{3}{4}$ days, whereas $19 \times 365\frac{1}{4} = 6,939\frac{3}{4}$ days. There remains one day's difference; it was smuggled away at the end of each cycle, under name of *saltus lunae*. All through the Middle Ages it remained a matter for surprise—because this human regulation was supposed to be a sacred fact of nature—that the moon made a jump every 19 years.

In the extensive medieval literature and the tiresome practice of this *computus*—then the name for Easter reckoning—many technical terms came into use which survive in the almanacs of our own day. The 'epact' of a year is the number of days by which the age of the moon at a given date is greater than at the same date in the first year of the 19-year cycle. It increases by 11 (i.e. 365–354) each succeeding year, or decreases by 19, because 30 is subtracted when it exceeds 30. In the succeeding years of a cycle it is 0, 11, 22, 3, 14, 25, 6, etc. The values 365 and 354 in reality are 365.25 and 354.37; these errors are compensated one day in the cycle, the same one day mentioned above; it lies hidden here, because the epact is 18 for the last year of the cycle, where we do not pay attention to the fact that $18 + 11$ being 29 does not return 0. The epact thus depends on the series number of the year within the 19-year cycle, called the 'golden number'; it is one more than the remainder after division of the year by 19 (the first year has the remainder 0). In the first year of the cycle the new moon falls on March 22nd; hence for every year the epact indicates the age of the moon on

March 22nd; the full moon, with age 14, falls in the first year on April 5th, and in each next succeeding year 11 days later or 19 days earlier.

To know which day of the week falls on such a date, the 'solar cycle' of $4 \times 7 = 28$ years is used, after which the series of names returns. Every next year this weekday advances by one and after a leap year by two. If we give to the days of the year the seven letters A (for January 1st), B, C to G, then the letter falling on Sunday is called the 'Dominical Letter'; it goes back by one every year, by two every leap year (after February). Knowing this letter, we find the weekday for the Easter full moon, hence the date of Easter Sunday.

Here in order to show modern progress in mathematical control of the world we may remark that Gauss in 1800 condensed this entire complicated Easter computation into a couple of simple formulae which give the result in a few minutes. Dividing the year by 19, by 4 and by 7, the remainders are called a, b, and c; put $19a + 15 =$ multiple of $30 + d$ ($d < 30$); put $2b + 4c + 6d =$ multiple of $7 + e$ ($e < 7$). Then the date of Easter is March 22nd $+ d + e$.

Conversely, the culture and mode of thinking in the Middle Ages can be measured by the fact that such a simple piece of arithmetic, by clothing it in the garment of self-made shapes, presented itself to mankind as another mysterious world with curious rules dominating the rites of the Church and the life of man. With these shapes and these rules, however, the Church succeeded in carrying on and fulfilling its task of fixing the festivals during the darkest ages of the decline of science.

When in the later medieval centuries science rose again, it was soon perceived that calendar and computus no longer agreed with reality. The Julian year of $365\frac{1}{4}$ days was 0.00780 too large; the difference amounted to one day every 128 years, so that about 1300 the vernal equinox fell on the 13th instead of the 21st of March. Moreover, the 235 months of a 19-year cycle, assumed to be $6,939\frac{3}{4}$ days, in reality were 6,939.69 days; and owing to this difference, now, after 950 years = 50 cycles, full moon came three days earlier than was computed. So the Easter computation was completely wrong, and Easter was celebrated on wrong days. Roger Bacon had already pointed out how shameful for the Christian Church it was that flesh was eaten in Lent, a cause for derision to Jews and Mohammedans. But the inner quarrels of the Church in the succeeding centuries prevented any measure being taken.

We mentioned above that in the fifteenth century Pope Sixtus IV invited Regiomontanus to Rome for this purpose; but, because of his early death, matters remained as they were. For the Lateran Council (1512-17) Paul of Middelburg, Bishop of Fossombrone, who had written a great work on calendar reform, asked the advice of Copernicus, who, however, had to answer that new data were not yet available to secure an accurate new regulation. But his own work soon supplied the deficiency; with the new data from *De Revolutionibus* and from Reinhold's tables, the reform could now be carried out. The Council of Trent (1545–63) charged the Pope to make the necessary regulations and in 1582, under Gregory XIII, it was effected.

First, the big errors that had gradually accumulated—that the equinox now came 10 days too early and the assumed moon came three days after the real moon—had to be removed by introducing sudden jumps. To keep the calendar correct in the future, new regulations had to be made. It was known that the true length of the tropical year was $\frac{1}{128}$ day less than the Julian period of $365\frac{1}{4}$ days. It was known also that the 235 months of the 19-year cycle, the basis of medieval Easter-reckoning, were $\frac{1}{17}$ day shorter than 19 years, so that after every 317 years the full moon had to be taken one day earlier.

The rather simple modification now made in the calendar was an ingenious invention of the librarian Aloysius Lilius (Giglio), offered after his death by his brother to the ecclesiastical commission installed for the reform of the calendar. It was approved and praised by the astronomical experts and at once accepted by the commission. It consisted, firstly, in the omission of three leap years every 400 years, which came down to assuming an error of $\frac{1}{133}$ instead of $\frac{1}{128}$ of a day in the Julian year; it was done by omitting the secular years that are not multiples of 400. Secondly, in keeping the 19-year cycle and having the time of full and new moon pushed back one day 8 times in 2,500 years; this should be done in the years 1800, 2100, and every 300 years to 3900, then 4300, and then again every 300 years.

The consequence of all these jumps is that at the secular years the calendar quantities undergo sudden changes: the epact decreases by 1 in 1700, 1900, 2200, 2300, 2500, and increases by 1 in 2400.

The constants in Gauss's formulae now also change considerably by the jumps at the transition; the values 15 in *d*, 6 in *e*, holding for the eighteenth century, had to be replaced by 23 and 4 during the nineteenth century and by 24 and 5 during the twentieth century.

The leading astronomer of the commission, Father Clavius, published a book to explain the principle and practice of the new calendar to the world. If the calendar reform had come before the Reformation, nothing would have stood in the way of its general acceptance. But now, in a time of great religious strife, it was established by the Pope and a College of Cardinals who had no authority in Protestant countries. Still worse, in the papal bull wherein it was proclaimed, the Pope 'ordered' the princes and republics to introduce the new calendar. Whereas it was done immediately in Spain, France and Poland, the Protestant princes and countries refused. At a later date Kepler, by very able arguments in a booklet, in the form of a discourse, tried to convince his co-religionists of the necessity for its adoption; but his purity of doctrine was suspect anyhow. When the question became urgent in 1700 in Germany, a way out was found by introducing the New Style in a somewhat different way, exhibiting and emphasizing independence from Rome. This was done by computing Easter after the real moon, not after cycles, so that Easter was sometimes a week different in Catholic and Protestant regions. In the course of the eighteenth century however the Gregorian calendar was introduced everywhere in Protestant Europe (in England in 1752).

Russia followed in the twentieth century, and the Gregorian calendar was adhered to throughout the world, though for ritual purposes local and national calendars remained in use. There is now under discussion a project for calendar reform, proceeding from the idea that in our machine age a mechanized calendar belongs to a mechanized world. In view, however, of the value attached by leading social circles to ancient traditions and customs, it seems unlikely that such a change will soon take place.

CHAPTER 22

THE STRUGGLE OVER THE
WORLD SYSTEM

However highly Copernicus was esteemed among astronomers for renovating astronomy in its fundamental numerical data, his helio-centric world system with a moving earth did not find much approval. The objections hampering its acceptance were of two kinds, the theological, arising from the authority of the Bible, and the physical, from the authority of Aristotle's doctrine, corresponding to everyday experience.

The theological difficulty held more weight for Protestants than for Catholics. Cardinals and bishops had encouraged Copernicus to publish his work; one Pope had benevolently listened to an exposition of the new theory, and another had accepted the dedication of his book. On the other hand the Protestant leaders, Luther and Melanchthon, sharply rejected it. Martin Luther, in one of his table-talks in 1539, said: 'That fool will reverse the entire Ars Astronomiae; but, according to the Scripture, Joshua bade the sun and not the earth to stand still.'[103] Melanchthon in 1550 quoted the Psalms and Ecclesiastes: the earth stands eternally, the sun rises and sets; and he added: 'Fortified by these divine testimonies, we will cling to the truth.'[104] For Protestantism the strict literal validity of the Bible was the basis of faith, whereas the Catholic Church claimed the right of interpretation. Under Paul III the general trend of the Church might seem to be hesitating and inclined to reconciliation and to the bridging-over of differences, in order to restore unity by concessions to the new idea. But conditions changed. In the second half of the century the controversy between the parties became more definite, the fight became fiercer and irreconcilable, the world became harsher. The establishment of rigorous doctrine by the Jesuit Order drove out the sunny philosophy of the Renaissance. At the Council of Trent the Church organized itself into a solid militant power, giving strict attention to any deviation from the established doctrine; thus it regained more and more of the ground previously lost.

For Tycho Brahe the theological argument was also important. An opinion contrary to the Bible, which in Protestant Denmark would certainly produce troublesome quarrels that would hamper his work of renovating practical astronomy, could not attract him. But the physical objections were decisive. First, the clumsy and heavy material earth cannot be a rapidly moving celestial body. Moreover, if the earth describes a yearly circle, this must appear in an apparent yearly circle of the stars, which, as a kind of resting epicycle, we call a yearly parallax. Copernicus had said that it would be too small to be perceptible; but Tycho was now able to measure positions and changes of position down to 2'. With a parallax of 2' the distance of the stars would surpass by a hundred times the dimensions of the planetary system. Such an enormous void between the farthest planet, Saturn, and the sphere of the stars would be entirely useless. These difficulties had induced Tycho —in 1583, he said—to devise another world system, having with Copernicus the same advantage of not needing epicycles for all the planets, and avoiding the difficulties resulting from the earth's motion. In Tycho's system sun and moon described circles about the resting earth; the sun was the centre of all the planets' orbits and carried them along with her in her yearly revolution. The celestial sphere carried all along in its daily rotation. It is clear that in this system the motions of the celestial bodies relative to the earth would agree exactly with the system of Copernicus.

In a regular correspondence between Tycho and the Cassel astronomer Rothmann, who ably defended the Copernican system, the arguments for and against each of the theories were discussed. Tell me, Tycho wrote in 1589, if the earth rotates so rapidly, how can a ball falling from a high tower hit exactly the point below? And do you think it probable that Saturn is 700 times more distant from the sphere of the fixed stars than from the sun? A star of the third magnitude—for which he assumed an apparent size of 1'—in that case would be as large as the earth's orbit. Tycho also mentioned that he made experiments on a rapidly sailing ship, by dropping objects from the top of the mast; they did not fall down precisely below but farther back.

The universities were dominated in apparently unassailable strength by Aristotle in philosophy and by Ptolemy in astronomy. Yet Aristotle's theory of motion was such an artificial doctrine that only after replacing it by more natural views directly connected with everyday appearances could the road open to scientific progress. It was this double task that made the establishment of a science of mechanics so extremely difficult. Opposition to Aristotle's theory had already appeared in antiquity and again in the Paris school of Oresme. It ascribed the continued motion of a thrown object not to a pushing of the surrounding air but to an 'impressed force', which meant that the impetus or push once acquired was preserved until exhaustion of this impetus caused the object to fall down by its own weight. In the sixteenth century, instigated by the use

223

of artillery in war and of simple mechanical devices in technical work, thinking on mechanics and opposition to Aristotle increased. Benedetti (in 1585) explained the acceleration of falling bodies by the weight continually adding new push to the existing impetus, and he contested Aristotle's opinion that the velocity of falling increases with weight. At the same time Guidobaldo dal Monte studied the weight, pushing bodies downwards along a circular track.

The adherents to the heliocentric system in the sixteenth century were insignificant in number. To be sure, immediately after its publication in 1543, Copernicus's book was diligently studied by the scholars, who used its numerical data for the computation of almanacs and tables, often praising them for their accordance with observations. But this did not involve acceptance of the new world system. In 1549 Melanchthon published his lectures on physics at Wittenberg as a university textbook; in this, by means of physical and theological arguments, Copernicus's system was refuted and dismissed as absurd. The wide influence of its seventeen impressions was enhanced through numerous books by dozens of other authors, all alike in praising Copernicus but rejecting or not even mentioning his theory. Here Catholic and Protestant universities were unanimous: Clavius, in 1570, in an often reprinted commentary to Sacrobosco, called Copernicus the renovator of astronomy but refuted his theory as absurd on the same grounds as had Melanchthon. Against this profusion of traditional instruction, Copernicus's book was not reprinted until 1566 (in Basel) and again in 1617.

Among the few adherents of the heliocentric system we find Thomas Digges, an Englishman, who in a book of 1576 speaks of an infinite world filled by mostly invisible stars. Another was Giordano Bruno, an enthusiastic apostle of the new doctrine, who in his travels all over Europe propagated the new world concept. But the boldness of his ideas—the doctrine of the plurality of worlds, all the fixed stars suns like our sun, surrounded by planets that are perhaps abodes of other men—tended to frighten rather than to attract more timorous minds; and his burning as a heretic, at Rome in 1600, was a warning. There were others who had sufficiently overcome prejudice to see the truth of the Copernican system: thus Benedetti, refuting the arguments against Copernicus, wrote: 'The celestial bodies have not been created to influence such a subordinate body as the water-covered earth with its animals and plants.'[105] William Gilbert, on finding the earth to be a large magnet, assumed it to have a daily rotation about the poles, which he supposed to be the magnetic poles; but he did not speak of the yearly motion. Simon Stevin in his *Wisconstighe Ghedachtenissen* ('Mathematical Memoirs'), published in 1605 at Leiden, sided entirely with Copernicus; he expounded the motions of the planets first with a resting earth, then with a moving earth, calling the former the 'untrue' and the latter the 'true' system. Moreover, he corrected Copernicus by omitting the third motion of the earth, the conical motion of its axis, considering its constant direction in space as a consequence of its magnetic character disclosed by Gilbert. In a letter to Tycho Brahe of April 18, 1590, Rothmann made the same remark: 'I know that in this point Copernicus is very obscure and not easy to understand. . . . This can be explained much more easily, and the triple motion of the earth is not necessary; the daily and the yearly motions suffice.'[106]

Johannes Kepler (1571–1630) became a fervent adherent after hearing, as a student at Tübingen, the exposition of the heliocentric system by Maestlin, who himself, though well-disposed towards it, was reluctant to support it entirely. Then, teaching at Graz as a provincial mathematician, in his first book *Mysterium cosmographicum* (1596), Kepler came forward as a vigorous defender with an entirely new argument. In this book he gave an explanation of the structure of the planetary system—why there are six planets at just these several distances from the sun—by connecting them with the five regular polyhedrons, called the 'Platonic figures'. If a sphere is constructed upon each of the six planetary circles, we may put between each pair of successive spheres, supposed to be exactly concentric, one of the regular solids in such a way that its edges are situated on the exterior sphere and its planes are tangent to the interior sphere. The ratio of the diameters of the outer and the inner sphere in the case of a icosihedron and a dodecahedron is 1.24, in that of a cube and an octahedron 1.73; of a tetrahedron it is 3. Among the ratios of the radii of the successive planetary circles, there are two small values, 1.4–1.5 for earth-Venus and Mars-earth; two larger values, 1.8–1.9 for Venus-Mercury and Saturn-Jupiter; and one very large, 3.4, for Jupiter-Mars. Of course there cannot be exact equality because the centres of the planetary orbits are not situated in the sun, so that the eccentricities also play a part. But the concordance is too great to be due to chance alone. Hence Kepler, guided moreover by astrological ideas, placed the five solids in his succession proceeding from the centre outwards: 8-, 20-, 12-, 4-, 6-hedron, in between the six planetary spheres. The figure in his book, representing a model of this arrangement, is reproduced in fig. 27. By disclosing this secret of world structure, Kepler elevated Copernicus's theory far above the level of a debatable opinion based upon uncertain empiricism and made it a fundamental philosophical truth.

Kepler sent his cosmological work to several astronomers, among them Galileo Galilei, who since 1592 had been teaching mathematics and astronomy at Padua. In his letter of acknowledgment of August 4, 1597, Galileo wrote: 'Many years ago I came to agree with Copernicus, and from this position the causes of many natural effects have been found by me which doubtless cannot be explained by the ordinary supposition. I wrote down many reasons and arguments, and also refutations of opposite arguments, which, however, I did not venture until now to divulge, deterred by the fate of Copernicus himself, our master, who, although having won immortal fame with some few, to countless others appears . . . as an object of derision and contumely. Truly, I would venture to publish my views if more like you existed; since this is not so, I will abstain.'[107] Present yourself with your proofs, was Kepler's answer (October 13, 1597); with combined forces we must shove the cart; with your proofs you can assist your partners who now suffer

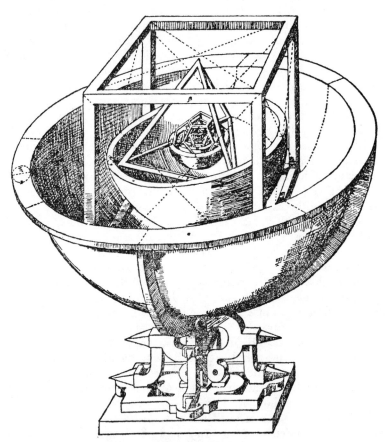

Fig. 27. Keeker's model of five regular polyhedrons between the planetary spheres

from unjust judgment. He suggested a well-directed action to create an impression of influence in order to encourage timorous minds. 'Be confident, Galilei, and proceed! If I am right, only a few of the chief mathematicians of Europe will keep aloof from us; such is the power of truth.'[108]

Galileo Galilei (1564–1642)—often called by his Christian name only—continuing in a more thorough way the first attempts of Benedetti and dal Monte, had arrived at a critical attitude toward Aristotle's theory of motion. Chiefly by clear argument, sometimes aided by experiments on swinging pendulums and on motion along an inclined plane, he gradually gained a better understanding of the laws of motion determining the phenomena of falling and thrown bodies. He could now recognize the futility of the objections against a rapid motion of the earth's surface, objections which had been brought forward by Ptolemy and repeated over and over again. He came to understand that a state of motion, just as well as a state of rest, remains when not disturbed by other influences. But his adherence to Copernicus's theory did not appear in his academic lectures; here, conforming to the imposed task, he taught the celestial sphere and the theory of Ptolemy with its arguments.

Yet, when Tycho's *Progymnasmata* appeared in 1602 with all the results on the new star of 1572, and when in 1604 another equally brilliant new star appeared in the low southern parts of Ophiuchus, he could not refrain from pointing out, in some well-attended public lectures, that they completely refuted Aristotle's doctrine of the immutability of the superlunar regions of the stars.

Then, about 1608, the first telescopes became known in Europe. We do not know the true history of the invention; apparently Zacharias Janssen, optician in the Dutch town of Middelburg, constructed one in 1604, copied from a specimen belonging to an unknown Italian—the possibility of combining lenses had already been alluded to before. Janssen, as a peddler, had shown and sold some of them at fairs in Germany and elsewhere. In 1608 Lippershey, another optician in Middelburg, originally from Wezel, offered a telescope to Prince Maurice and the States of Holland, chiefly for use in war. It was tested successfully and its author was well remunerated. The requested licence —also asked for shortly afterward by Metius at Alkmaar—was refused, because many people already knew of the telescope. Soon the report spread, and specimens were shown and sold in France. Galileo, after hearing about it, himself constructed one and offered it in 1609 to the Doge and the Signoria of Venice for use in war and navigation. Gradually improving his instruments, he directed them towards the moon and the stars. Then followed rapidly the series of wonderful discoveries that put Copernicus's doctrine in the centre of public interest. At first he communicated them only in repeatedly copied letters to his friends and colleagues. Then he published them in a little booklet, *Sidereus nuncius* ('Messenger of the Stars') which appeared in March 1610 and caused great commotion in learned circles.

On the moon he saw the border line of the illuminated part irregularly broken. In the dark part he saw, near the border line, isolated light patches, which, as the moon waxed, grew larger and merged with the illuminated part; clearly they were mountain tops (pl. 4). So the moon was not crystalline but, like the earth, an uneven broken surface with mountains and valleys; the circular walls, however, showed a different structure. The planets in the telescope looked different from the stars; they were pale discs with enlarged surfaces, whereas the stars remained strongly sparkling points, only appearing brighter than with the naked eye. Nebulous patches and the Milky Way appeared to consist of a number of small stars; in the Pleiades he counted more than 40 stars, and everywhere between the known stars smaller ones which were invisible to the naked eye were now seen. Looking at Jupiter, he perceived on January 7, 1610, three small stars, on January 13th one more; they accompanied Jupiter in its progress, but every other night in a different formation, moving to and fro. In a letter of January 30th, Galileo called them new planets revolving about a larger planet; in honour of the Grand Duke of Tuscany he named them 'Medicean stars'. That these bodies clearly described orbits about Jupiter as their centre showed that the earth cannot be the centre of all movements and, therefore, gave support to the system of Copernicus.

A great sensation was caused by these discoveries, admiration and joy on the part of friends and partisans, scholars as well as laymen, but still more doubt and opposition, increasing to hostility, among the dominant authorities of learning. The professors of philosophy, attacked in their own empire—the University of Padua was known as a stronghold of Aristotelianism—announced a sharp refutation. In a number of writings the existence of the new wandering stars was refuted on logical arguments; since Aristotle had not mentioned them, it could not be true. When Galileo in a public disputation tried to convince his Paduan colleagues by showing the stars, they would not look in his telescope, maintaining that it was appearance only and therefore illusion. As to the two leading astronomers of Italy, old Clavius dismissed the matter with a disclaiming joke; the other, Magini at Bologna, an esteemed correspondent of Tycho and Kepler, jealous of his younger colleague, wrote letters in which he expressed his doubts and declared the observations to be deceit or self-deceit. When Galileo, visiting Bologna in March, met a number of professors at Magini's and showed them Jupiter through his telescope, nearly all of them declared that they saw nothing of the new stars; this negative result was given wide publicity. It was not simply bias through unwillingness; it was also the real difficulty of seeing things through a telescope for people entirely untrained. The first telescopes were very primitive, the lenses were badly figured, the images had irregular coloured borders, and all in all they were not comparable even to a modern opera glass. In September Galileo wrote to Clavius, who had not yet succeeded in seeing them, that the instrument should be fixed, because when held in the hand its vibrations made the new stars invisible. The discoveries which opened the road to the modern world system were not fully ripened fruits, easily picked up after gently shaking the tree. They were the fruits of strenuous exertion and extreme care in observing and in asserting what was seen, of laborious and often enervating struggle, first against the inner doubts and then against the obstinate, often perfidious, defence of an anti-quated doctrine that still powerfully dominated men's minds.

Kepler at Prague gave Galileo enthusiastic support. Immediately on receiving the *Nuncius* he expressed his ideas thereon in a printed open letter. There he said that the idea of constructing a telescope for studying the celestial luminaries had never occurred to him, because he supposed that the thick blue air would prevent the details of the very remote stars from being seen; but now Galileo had shown that space was filled with a thin and harmless substance only. He pointed out—what his friend Pistorius had already maintained—that more accurate measuring with instruments would now be possible. He dealt with the consequences of the new discoveries by indulging in fantasies of possible inhabitants on the moon and of the phenomena visible on Jupiter with its four moons. He had always shrunk from Bruno's 'terrible philosophy' of an infinite world with an endless number of other solar systems; but now Galileo's discovery of innumerable small stars showed that our sun was a unique body, surpassing all the others in luminous power. Such were his thoughts on the new discoveries.

Yet, though he had full confidence in Galileo's observations, he him-self had not seen Jupiter's companions, and with his friends he was uneasy about the tidings from Bologna. Kepler was not a good observer; his sight was bad and he was awkward with instruments; so he never tried to make a telescope himself. The Dutch telescopes found in Germany did not show the Medicean stars; they magnified too little. According to Galileo, magnification of twenty or thirty times was necessary. The grinding of good lenses was a difficult art. In September for the first time Kepler succeeded in seeing the new stars through a better telescope belonging to a visitor. But finding the laws of their motion seemed to him very difficult, almost impossible. In the next months they were seen in France and also in England (by Harriot). The observers of the Jesuit College at Rome at last also succeeded in improving their telescopes, and in December Clavius sent word to Galileo—who in the meantime had settled in Florence, in the service of the Grand Duke—that he and his colleagues were convinced of the existence of the new planets. His younger colleague, Father Grienberger, in January 1611 wrote to Galileo: 'Things so hard to believe as what you assert neither can nor should be believed lightly; I know how difficult it is to dismiss opinions sustained for many centuries by the authority of so many scholars. And surely, if I had not seen, so far as the instruments allowed, these wonders with my own eyes. . . . I do not know whether I would have consented to your arguments.'[109]

In the meantime other discoveries had followed. Hidden in a letter-puzzle—to make his priority secure—Galileo ascertained in July 1610 that Saturn was treble, touched at either side by a smaller resting globe; to us this appearance testifies to the low quality of the first telescopes. In the same way he announced in December that Venus imitates the figures of the moon. Some followers of Copernicus had predicted it; others, Kepler among them, believed that the planets partly radiated their own light or were saturated by absorbed sunlight. In his letters Galileo emphasized that this result—the darkness of the planetary bodies showing the similar character of the planets and the earth—confirmed the truth of Copernicus's theory that the earth is a planet.

In order to win over the influential Fathers of the Roman College to his views by discussion and demonstration with his instruments, Galileo went to Rome in March 1611. Welcomed as an honoured guest, he gave lectures and held discussions, and in an assembly was abundantly praised for his discoveries. In a report of April 1611, to Cardinal Bellarmin of the Holy Office, Clavius and his colleagues confirmed the truth of Galileo's discoveries. But they did not confirm Galileo's theory; it seemed to them more probable that in the crystalline body of the moon there were denser and lighter parts than that its surface was unequal; others thought differently, but as yet there was no certainty. The General of the Jesuit Order had instructed members as far as possible to retain and to defend Aristotle's doctrine. The confirmation related only to the observed facts; the report did not discuss the explanation by the heliocentric system. Though, shortly before his death, Clavius wrote that now the astronomers should see by what structure of the celestial spheres these phenomena could be explained, this did not imply the acceptance of Galileo's theory. Indeed, a strict demonstration of the earth's motion could be given neither by the mountains of the moon, nor by the Jupiter satellites, nor by the crescent figure of Venus.

Galileo did not remain the only traveller on these new paths. Telescopes had come into many hands and were diligently used to study the sky; they brought a wide conviction of the reality of the new phenomena and led to fresh discoveries. This, however, implied much contention over the priority of the new discoveries and over their interpretation. Galileo naturally considered this field to be his special domain; but others, also naturally, tried to get their share of the fame.

In a calendar published in 1611, and afterward in a book published in 1614 on the 'World of Jupiter' (*Mundus Jovialis*), Simon Marius (Mayer) of Anspach tells that in 1609 he came into possession of a Dutch telescope, directed it at the stars and gradually became conscious of the importance of what he saw; that he had discovered the satellites of Jupiter independently and began to observe them somewhat later than Galileo. The latter afterward hotly inveighed against Marius, accused him of plagiarism, and imputed that he had not really seen the new stars. Independent discovery, of course, could not be proved; but from his later observations Marius had succeeded in deriving better values than Galileo for the periods of revolution and the other elements of their orbits.

The discovery of sunspots occurred at the same time. In former times it had happened that an exceptionally large spot had been perceived on a hazy sun or in a solar image projected by a small opening. But now regular detection and observation became possible, at first by directing the telescope upon the sun when much dimmed by morning or evening haze. The first public announcement came from Johann Fabricius of Emden, the son of Kepler's friend David Fabricius, in the summer of 1611. Galileo had already shown them during his visit to Rome in the spring of the same year. The researches of Chr. Scheiner, SJ, at Ingolstadt, aided by his pupil Cysat, began at the same time; they were published the next year. He had been warned by his Superior not to put trust in his observations, because nothing of it existed in Aristotle; so he had to publish his work anonymously, and in order to avoid conflict with Aristotle he explained the spots as small dark bodies circulating about the sun. Like Fabricius, Galileo declared the spots to be parts of the sun itself, proving its axial rotation. In criticisms of Scheiner's publications, made in his book, *History and Demonstrations Concerning the Solar Spots*, distributed by his friends in Rome in 1613, Galileo strongly attacked the entire doctrine of Aristotle.

The scene of the struggle had now changed; the subject of the contest was not the observations but the interpretations, not the practice but the theory. After his return from the triumphs at Rome, Galileo gradually perceived that the climate had changed. His friends, who had previously greeted his discoveries with enthusiasm, now became cautious and advised him to be content with the victories won—Campanella only, writing from his prison, strongly urged him to stand firm in the defence of Copernicus. Galileo himself considered his discoveries with the telescope only as aids to the real goal, the proof of the truth of the heliocentric theory.

The fight now concentrated upon the Copernican system, i.e. on the movement of the earth. Galileo believed he had found a direct proof of this movement in the phenomena of the tides. Kepler had supposed them to be an effect of the moon, but Galileo tried to explain them by inequalities in the velocity of the earth's surface. This velocity is a combination of the daily and the yearly revolutions; on the night side of the earth their velocities combine, and on the day side toward the sun they subtract. This, he said, makes the water of the oceans oscillate. Neglecting the obvious dependence of the tides on the moon, he thought to find in the tides a proof of the earth's motion. It was not convincing, because according to his theory high and low tide should occur only once each day.

The real fight took place on the field of theology. Ignorant monks preached in the churches against the new theory as a heretical doctrine contrary to the Bible; secret denunciations against Galileo arrived at the Holy Office. Galileo on his side dealt with the Bible texts in some letters which were widely distributed and interpreted them according to the new views—not always successfully, when he tried to make the biblical texts proofs for the Copernican system. He found supporters among the clerics themselves; a Carmelite monk, Foscarini, in a printed letter to the General of his Order, defended the Copernican system with strong arguments.

Galileo again travelled to Rome, first to clear himself from the charge of being a heretic; indeed, he was a devout Catholic, who always in his writings professed that he obeyed what the Church in its deeper divine insight should declare to be truth, though at the same time demanding that the theologians, before deciding, should acquire an exact knowledge of Copernicus's work. The justification succeeded well enough; more difficult was his second task, to convince the influential persons and the authoritative cardinals. He argued that there could be no conflict between revealed and natural truth; so when the latter had been clearly demonstrated by the facts and could not be changed by any interpretation, the former, the interpretation of the Holy Scripture, had to adapt itself. But this logic did not impress the theologians, who were firm in the ecclesiastical doctrine and for whom the results of the scientists carried no conviction. For them the issue was, that if the Church did not take a firm stand by a clear verdict an unbridled discussion by clerics and laymen on the interpretation of Bible texts would arise, and this was incompatible with its rigid discipline of doctrine. So the result was not in doubt; the Church had to take its stand against the motion of the earth. Characteristic was the sloppy and inaccurate rendering of Copernicus's doctrine in the report of the doctors of theology and the verdict of the court (February 25, 1616). As regards the first thesis, that 'the sun is centre of the world and entirely unmoved as to its place', they proclaimed that it was foolish and philosophically absurd and formally heretical. As regards the second thesis, that 'the earth is not centre of the world and not unmoved but moves relatively to itself also in daily motion', philosophically the same conclusion holds, and theologically it is at least an error of faith. This wording suffices to show the ignorance or carelessness of the judges. Galileo was summoned before Cardinal Bellarmin. He was informed that the doctrine of the earth's motion must not be taught nor held to be true—it might only be dealt with as a mathematical hypothesis—and he was admonished to abandon it. To which Galileo submitted. On the basis of this verdict, the Congregation of the Index resolved, March 5, 1616, that the books of Copernicus and all other books teaching the same ideas should be suspended and forbidden until they were amended.

Now that it was no longer permissible publicly to defend the heliocentric world system of Copernicus, Galileo concentrated his researches and criticism on Aristotle's philosophy of nature. At the same time he tried repeatedly, but ever in vain, to have the prohibition repealed by the next Pope, Urban VIII, who, when a cardinal, had shown friendly interest in his work. At last it became too much for him and he found a way to meet, as he supposed, the objections by giving to his ideas the form of a discourse in which the old doctrine imposed by the Church was formally put in the right but in such a way that the arguments of the new doctrine could be exhibited in full. His literary ability and dramatic talent made this work, which appeared in 1632 in Italian, *Dialogo sopra i due massimi systemi del mondo* ('Dialogue on the Two Principal World Systems'), an excellent popular work which completely finished Aristotle's doctrines of motion and of physics.

Because of the light it shed on the questions vehemently disputed at the same time, the book was hailed enthusiastically by his friends and all modern-minded scholars. But now the Church authorities (although the *imprimatur* had been given) were roused to repulse the attack. They did not let themselves be deceived by the outer form of submission to theological authority; the meaning behind the words was clear enough. Personal influences also made themselves felt; Jesuit scholars, stung by Galileo's severe criticism of their writings, which offended the scholarly pride of the entire Order, worked against him and succeeded in making the Pope Galileo's personal enemy. So he was summoned before the Inquisition. Formally Galileo was right, because it was permissible

233

publicly to discuss the Copernican doctrine as a hypothesis; yet he was condemned for disobedience, probably on a forged document, and was compelled in 1635 solemnly to abjure the heliocentric doctrine. A modern Catholic astronomer, Professor Joseph Plassmann at Münster in 1898, called this event 'the most fatal mistake that has ever been made by the Church authorities against science'.[110]

In the seventeenth century many writings appeared for and against the Copernican system; those against mostly adhered to the Tychonian system. But doubt and opposition among scientists were lessening. The ecclesiastical ban could not hold up the progress of science. It only made it more difficult for Catholic scientists to accept or to publish new ideas. Even in 1762 a new edition of Newton's major work by two expert Minorite Fathers was preceded by the declaration that they considered the theory expounded as a hypothesis only and that they adhered to the Church's verdict. But in the eighteenth century real opposition ceased, and the new view penetrated widely among the people. In 1744, for the first time, Galileo's *Dialogue* was allowed to be printed, though with all the 'corrections' prescribed by the Inquisition. After many attempts, the ban was finally lifted in 1822; and after 1835 Copernicus, Kepler and Galileo no longer appeared on the index of prohibited books. So the attempt of the Church to arrest the progress of science by her authority had resulted in an acknowledged defeat.

CHAPTER 23

KEPLER

B Y his first book, *Mysterium cosmographicum*, Kepler had caught the attention of Tycho Brahe, because of its independent thinking, its astronomical knowledge, and the author's clearly apparent ability and perseverance in computing. Tycho made approaches, but their personal contact became possible only after Tycho had settled in Prague, when, at the same time, through a general expulsion of the Protestants from Styria, Kepler was forced to seek a living elsewhere. Tycho saw to it that in 1601 Kepler was appointed 'Imperial Mathematician' in the service of Rudolf II, with the assignment to assist in the reduction of the observations of the planets and the construction of new planetary tables. After Tycho's death, when the work had scarcely started, after settling certain difficulties with the heirs, he inherited Tycho's observations as well as the task of finishing the work.

In Kepler, a man of another generation, the character of a new century emerged. Astrology as the doctrine of world unity dominated his mind, not as a fearful spying upon the stars to discover human destinies but as a passionate desire to penetrate into the secret of this unity. As with many of his contemporaries, the spirit of research lived in him, the intense curiosity, the desire to unravel the secrets of nature. Not in the sober sense of modern research into nature but rather in the sense of what we call mysticism. He had the intuitive feeling that the entire universe was a miracle, materially and spiritually connected with the equally miraculous creations of the human mind, geometry, the theory of numbers, music—all of which was already apparent in his first book. His astrology bore a more modern character than did that of his predecessors; he derides the belief that events might depend on the names given to the constellations in ancient times; but he did assume that conjunctions of the planets influenced earthly events, just as the sun and the moon do. He was in the first place a physical thinker; in all phenomena he looked for causal connections, sometimes surprising us by modern views, sometimes entirely in error. The previous generation had asked of any phenomenon: What does it mean? The new generation asked: What is it and what is its cause?

In a booklet on astronomical optics (*Astronomiae pars optica*) in 1604 he explained refraction in a way somewhat different from Tycho, through the transition of the light-rays from the thin ether into the air itself, so that it must be present at all altitudes, increasing from the zenith, and must be equal for all celestial luminaries. He did not succeed in finding the law of refraction by experiment; this had to wait for Snell at Leiden (1620). But with his results he was able to construct a table better than Tycho's.

He considered the moon, as also did Bruno and Gilbert, as similar to the earth, dark and having high mountains—this was before the invention of the telescope. He spoke of the penumbra at lunar eclipses and explained the ruddy light of the totally eclipsed moon through refraction of the sun's light when passing through the earth's atmosphere. He also defended Maestlin's explanation of the pale illumination of the lunar disc beside the crescent—the old moon in the arms of the young— as light reflected upon the moon by the sunlit earth. The brilliant new star mentioned above, which, after a conjunction of Jupiter and Saturn, soon accompanied by Mars, appeared in October 1604 in the 'fiery' sign of Sagittarius, led Kepler to publish in 1606, after the star had faded and disappeared, a booklet in which he discussed its physical characteristics as well as its astrological significance. Here he compared the scintillation of the stars with the sparkling of a diamond when twirled; a better explanation, that it was due to the undulating motion of the air, had already been suggested by the renowned classicist J. J. Scaliger.

All these topics were accessory occupations; they are important because they indicate the tendencies of his thinking on the celestial bodies. His chief work during these years consisted of computations from Tycho's observations of the planets. From 1601 onward he was occupied with the planet Mars, which Longomontanus had previously found too difficult an object. Kepler began by deriving from the observations a list giving accurate values of the moment, the longitude and the latitude of all the oppositions since 1580. Here he at once went his own way. Copernicus, following Ptolemy, had assumed the centre of the earth's orbit to be the centre of reference for all planetary orbits, and Tycho had adhered to it in deriving oppositions to the mean sun. Kepler in his first work had already designated the real sun itself as the natural centre of the planetary system; so he judged that opposition to the real and not to the mean sun should be derived and discussed. This was his first important modification and simplification of former methods of treatment.

The oppositions deduced from Tycho's observations, completed by those observed by himself and by his friend David Fabricius at Emden in the years 1602 and 1604, are contained in the accompanying list.[111]

OPPOSITIONS OF MARS

Date	Time	Longitude	Latitude	Longitude Computed	Difference	Latitude Computed
1580, Nov. 18	1h. 31m.	66° 28′ 35″	+1° 40′	66° 28′ 44″	−0′ 99″	+1° 45½′
1582, Dec. 28	3h. 58m.	106° 55′ 30″	+4° 6′	106° 57′ 4″	−1′ 34″	+4° 3⅓′
1585, Jan. 30	19h. 14m.	141° 36′ 10″	+4° 32⅙′	141° 37′ 46″	−1′ 36″	+4° 30½′
1587, Mar. 6	7h. 23m.	175° 43′	+3° 41′	175° 43′ 16″	−0′ 16″	+3° 37′
1589, Apl. 14	6h. 23m.	214° 24′	+1° 12¾′	214° 26′ 12″	−2′ 12″	+1° 5⅓′
1591, June 8	7h. 43m.	266° 43′	−4° 0′	266° 43′ 51″	−0′ 51″	−3° 59⅙′
1593, Aug. 25	17h. 27m.	342° 16′	−6° 2′	342° 16′ 42″	−0′ 42″	−6° 3¾′
1595, Oct. 31	0h. 39m.	47° 31′ 40″	+0° 8′	47° 31′ 54″	−0′ 14″	+0° 5⅕′
1597, Dec. 13	15h. 44m.	92° 28′	+3° 33′	92° 28′ 3″	−0′ 3″	+3° 20′
1600, Jan. 18	14h. 2m.	128° 38′	+4° 30⅝′	128° 38′ 18″	−0′ 18″	+4° 30¼′
1602, Feb. 20	14h. 13m.	162° 27′	+4° 10′	162° 25′ 13″	+1′ 47″	+4° 7⅖′
1604, Mar. 28	16h. 23m.	198° 37′ 10″	+2° 26′	198° 36′ 43″	+0′ 27″	+2° 18⅖′

The variations in angular velocity, which is great in the region of Aquarius (at longitude 330°) and small near the Lion (at longitude 150°), are conspicuous at first sight. Ptolemy had represented this variation by introducing a *punctum equans* and placing the centre of the circle midway between it and the earth (here to be replaced by the sun). Kepler wished to test whether this bisection exactly represented Tycho's results. Ptolemy, in computing the orbit, had needed three oppositions; since Kepler had to determine one additional unknown, viz., the ratio of division of the total eccentricity by the circle's centre, he had to use four data. For four chosen moments (the oppositions of 1587, 1591, 1593, 1595) he knew the directions as seen from the sun and also the directions as seen from the *punctum equans*, since the latter increased proportionally to the elapsed time. The problem of finding from these data the direction of the line of apsides (longitude of aphelion) and the two distances from the circle's centre to the sun and to the equant could not be solved by a direct method. Kepler had to solve it by trying various suppositions, in successive approximations. 'If this cumbersome mode of working displeases you,' he says to the reader, 'you may rightly pity me, who had to apply it at least seventy times with great loss of time; so you will not wonder that the fifth year is already passing since I began with Mars. . . . Acute geometers equal to Vieta may show that my method is not at the level of art. . . . May they solve the problem geometrically. For me it suffices that . . . to find the way out of this labyrinth I had, instead of the torch of geometry, an artless thread guiding me to the exit.[112]

The result of these computations was that the total eccentricity amounted to 0.18564 times the radius, that the sun was 0.11332, and the equant was 0.07232, distant from the centre, whereas the longitude of aphelion (for 1587) was 148° 48′ 55″. Modern theory shows that the two distances should be approximately $\frac{9}{16}$ and $\frac{7}{16}$ of the total eccentricity. How exactly these elements represent the data may be seen from the list

on page 237, in which the remaining differences between observation
and computation are given in the sixth column. 'So I state that the
places of opposition are rendered by this computation with the same
exactness as Tycho's sextant observations, which, through the considerable diameter of Mars and the insufficiently known refraction and
parallax, are affected by some uncertainty, surely as much as 2 minutes
(prime).'[113]

Thus Chapter 18 of his book closes. Then Chapter 19 begins with the
words: 'Whoever would think it possible? This hypothesis so well in
accordance with the oppositions, yet is wrong.'[114] Ptolemy was right in
bisecting the eccentricity. This was apparent at once when Kepler
computed the real distances by means of the observed latitudes. Computing, on the other hand, the oppositions in the case of a bisection of
the total eccentricity, he found for 1582 a longitude of $107°\ 4\frac{3}{4}'$, deviating
nearly 8' from the previous computation and 9' from observation.

'From this so small deviation of 8' the reason why Ptolemy could be
content with bisecting the eccentricity is apparent . . . Ptolemy did not
claim to reach down beyond a limit of accuracy of $\frac{1}{6}°$ or 10'. . . . It
behoves us, to whom by divine benevolence such a very careful observer
as Tycho Brahe has been given, in whose observations an error of 8' of
Ptolemy's computation could be disclosed, to recognize this boon of
God with thankful mind and use it by exerting ourselves in working out
the true form of celestial motions. . . . Thus these single eight minutes
indicate to us the road towards the renovation of the entire astronomy;
they afforded the materials for a large part of this work.'[115]

This was the enigma he had to solve: the unequal division of the
eccentricity could not be true, though it rendered all longitudes in a
perfect way; under the name of 'vicarious hypothesis' he used it later on
to compute for any moment the longitude of Mars as seen from the
sun. Moreover, the equal division did not represent the oppositions. In
this dilemma, for the first time, the trigonometric determination of
distance was employed and remained the backbone of his work. In the
triangle sun-Mars-earth the direction of each side was known: earth-sun
through the observations of the sun (rendered by Tycho's tables),
earth-Mars through the observations of Mars, and sun-Mars through
the vicarious hypothesis. From the angles known, the ratio of the sides,
i.e. of the distances could then be computed. By selecting observations
separated by exact multiples of a year, so that the distance earth-sun
was the same, the variations of the distance sun-Mars were found.
Taking observations with intervals of a full number of revolutions of
Mars, the variations of the distances earth-sun afforded the figure of
the earth's orbit.

Kepler now turned first of all to a closer examination of the latter.

It was necessary, not only because its uncertainties could spoil the exactness of his computations of Mars, but also because, with Ptolemy and Tycho, the earth had no equant, whereas as a planet it should not be different from the others. By applying the trigonometric method to observations where Mars occupied the same place, he found an eccentricity of the earth's orbit of 0.01837 (the five decimals are not an indication of precision but a consequence of Kepler's always taking the radius 100,00 instead of writing decimals). Since the fluctuations in angular velocity had provided Tycho with an eccentricity of 0.03584, twice as large, it appeared that the earth too had an equant. This enabled him to construct tables giving exact distances and longitudes of the sun. Now computing the distances of Mars to the sun accurately, he found the greatest value, in aphelion, to be 1.6678 and the smallest value, in perihelion, 1.3850 radii of the earth's orbit. Then the radius of Mars's orbit is 1.5264, and the distance of the sun from its centre is 0.1414 : 1.5264 = 0.0926, essentially the exact half of the total eccentricity 0.18564 derived from its motion.

Thus it was established from observations of Mars that for this planet and for the earth the total eccentricity between the sun and the equant was divided exactly into halves by the centre of the orbit. Kepler, however, as a physical thinker, could not reconcile himself to the idea that a void point should be able to govern the planet's motion. In aphelion the planet proceeded more slowly, in perihelion more rapidly, because farther from the *punctum equans*. In the same ratio it was in the first case farther from, in the second case nearer to, the sun. So the sun regulated the velocity of the planet; the velocity was in exactly inverse proportion to the distance. The logic of the case was now at once obvious to Kepler. According to the lever principle, a planet at greater distance is moved by the solar force with greater difficulty, hence needs more time to describe a certain arc. This physical explanation shows, said Kepler, why he was right to relate all motions to the bodily sun instead of to a void centre of the earth's orbit, and at the same time why Tycho was wrong in having the heavy sun describe an orbit about the earth. Now the importance of the sun became even greater than it had been for Copernicus; it was now not only the source of light and heat for the entire planetary system, but the source of force also. Light and force, both immaterial, expand in space, but in a different way. Since light expands over spherical surfaces, going upward and downward and to all sides, it decreases with the second power of distance. Solar force, on the contrary, expands along circles in the ecliptic, not upwards and downwards, driving the planets through the zodiac in longitude only, hence decreases with the distance itself. One can understand this solar force by assuming that the sun rotates about an axis and thus draws along the planets in the same direction, more slowly as their distance increases. As to the nature of this force, Kepler pointed out that magnetism, which is a directing force, operates as if the magnet consisted of threads or fibres. The sun, too, did not attract the planets (if it did, they would fall into the sun) but directed their course through a sideways force, as though it consisted of annular magnetic fibres. This is more than an analogy, for Gilbert had discovered that the earth,which in the same way directs the moon in its orbit, is a magnet.

From these speculations it appears that Kepler was not simply an astronomical computer; what mattered to him was an understanding of the physical nature of things. His speculations are similar to the ideas that developed later in the seventeenth century. They were the product of new inquisitive impulses, far superior to the sterile scholasticism of the preceding century that still dominated the academic chairs. Because he spoke in terms of a force proceeding from the sun, he has sometimes—wrongly as we see—been called a precursor of Newton. Rather he was a precursor of the natural philosophy of the seventeenth century; what appears in Descartes's vortex theory as a vague philosophical speculation, with Kepler has the freshness of direct conclusions imposed by the facts of experience.

It is certainly true that Kepler, in trying to explain the remaining problems by referring to the planets as if they played an active part— speaking of the 'spirit' or 'essence' of the planet that has to pay attention to the apparent size of the sun—also became vague and contradictory. But it is more important that he had developed a new method of computation. The time spent by the planet on a small arc of the orbit, inverse to the velocity, was proportional to the distance from the sun; so, to have the total time for a longer arc, all the intervening distances had to be summarized. This was a problem of integration: 'If we do not take the sum total of all of them, of which the number is infinite, we cannot indicate the time for each of them.'[116] He first solved the problem by numerical summation; but then for the sum total of the distances he substituted the area between the limiting radii; though this was not exactly the same thing, it fitted well. The area could easily be computed as a circular sector diminished by a triangle. This was formulated afterward as Kepler's second law, the *law of areas*: the radius vector describes equal areas in equal times.

Thus the elements of the orbit and the method of computation were known, and in aphelion and perihelion, as well as at 90° distance, the computed longitudes were right. But in the octants, at 45° distance from these points, the old difference of 8' again appeared, too large to be attributed to errors of Tycho's observation. To solve this riddle, he again applied his trigonometric method to derive the distance of Mars from the sun directly from observation. At longitudes 44°, 185°, 158° he found 1.4775, 1.6310, 1.66255. Computation by means of the elements found above afforded 1.48539, 1.63883, 1.66605. Observation thus showed the distances of Mars from the sun to be smaller than that which followed from the circular orbit. Sideways the planet took its course, not on, but within, the circle described through aphelion and perihelion. 'The matter is obviously this: *the planetary orbit is no circle*; to both sides it goes inward and then outward until in the perihelion the circle is reached again. Such a figure is called an oval.'[117]

Thus for the first time, and forever, the principle that for a thousand years had been accepted by all astronomers as the basis of astronomy had been destroyed, that the circle was the natural and true orbit of heavenly bodies. The exactness of Tycho's observations had shown it to be incorrect. As the first triumph of the empirical study of nature, Kepler's result stands at the entrance of modern scientific research.

His difficulties, however, had not yet come to an end. In attempting to give a physical explanation for the deviation from the circle, by proposing a propelling force of the sun variable with distance and a kind of active epicyclic motion of the planet, he was led into a year's futile computing. David Fabricius of Emden, with whom he exchanged letters and whom he praised as the best observer since Tycho's death, warned him that his computation did not tally with observation. At last he perceived that the lateral compression of the orbit, 'the maximal breadth of the deficient sickle, 0.00429 times the radius', was exactly half the square of the eccentricity ($0.0926^2 = 0.00857$). Then 'it was as if I awoke from sleep and saw a new light'.[118] Now it was clear that the orbit must be an ellipse, with the sun in one of the foci; in a nearly circular ellipse the oblateness is half the square of the eccentricity. In this result, usually called 'Kepler's first law', the *true elliptic figure* of the planetary orbits was disclosed for the first time.

The last chapters of his book are devoted to the remaining task of deriving the position of the plane of the orbit. He found the longitude of the ascending node to be 46° 46⅓' and the inclination 1° 50' 25". The latitudes observed in the different oppositions are, as is shown in the list on page 237, well represented by these elements. The remaining deviations are larger than for the longitudes, owing mostly to uncertainties in refraction and parallax. Of the accessory oscillations, which Copernicus had assumed, nothing could be perceived. Kepler pointed out that a parallax of Mars larger than some few minutes would spoil the concordance; hence the parallax of the much more remote sun must certainly be below 1'.

Most of this work had been done in the first years at Prague, under the constant pressure of financial and other worries, of delicate health, of difficulties as to the disposal of the observations, and with many other interruptions. In 1604 his results were ready, and in 1605 he could present the manuscript to Emperor Rudolf. It was not until 1609 that it appeared, the delay being due chiefly to the lack of money in the imperial treasury. The title *Astronomia nova, aitiologetos seu physica coelestis* ('New Astronomy, causally explained, or Celestial Physics'), preceding the title of the contents: 'De motu stellae Martis' ('On the Motion of the Star Mars'), indicated that he was aware that astronomy was put upon a new basis by this work.

In Tycho Brahe's and Kepler's work the new method of scientific research is embodied—the method of collecting data from experiment and observation, and from them deriving rules and laws which form the body of science. They were not the only ones; at the turn of the century, in every field of knowledge, investigators appeared: Simon Stevin and Galileo studied mechanics, the laws of equilibrium and motion, Gilbert studied magnetism, Vesalius studied anatomy, Van Helmont studied chemistry, and Clusius studied botany. In all of them—few in number—a new spirit forged its way as they replaced the old belief in authority by their own experimental research.

Kepler's work did not end with his book on Mars; it had only begun. His task, handed down from Tycho, was the construction and publication of the new planetary tables, the 'Rudolphine Tables'. By his work on Mars he had found the key, the knowledge of the rather simple laws of planetary motion. He now had to apply them to all the other planets. In Linz, the capital of Upper Austria, where he had accepted the job of provincial mathematician, he published in 1618 his *Epitome astronomiae Copernicanae* ('Survey of Copernican Astronomy'), in which for the first time the structure of the solar system was correctly expounded. In it the true orbits of Mercury and Venus were given as entirely regular ellipses, with the following numerical elements: for Mercury, eccentricity 0.210 and aphelion at 255°: for Venus, 0.00604 and 302°. Gone were all the complications which Ptolemy thought necessary and which Copernicus had taken from him. Like the other planets, their orbits were fixed, with constant inclination to the ecliptic, without additional oscillations. How Kepler derived them from the observations is nowhere explained. In a letter of May 5, 1616, to his former teacher, Maestlin, he merely says: 'In the summer of 1614 the theory of Venus followed, in the winter of 1615 that of Mercury; they are in no way peculiar as compared with Saturn, Jupiter, and Mars; I did it by means of a large orbit of the earth and a simple eccentric orbit just like that of Mars.'[119]

The *Epitome* constituted the first complete manual of astronomy constructed after the new principles. It deals, first of all, with spherical astronomy, the shape and the size of the earth and its place in the universe; then come the stars, of which he says that it is uncertain whether they are at equal or different distances, though their realm seems to be infinite; of these the sun is one, and, because of its nearness, appears particularly bright. Kepler next deals with the atmosphere, refraction, the twilight and the twinkling of the stars. Speaking of the daily rotation of the earth, he argues that nature, to attain its end, always chooses the simplest way. Then he discusses the risings and settings of the celestial bodies, the years and the days, the seasons and the climates. The second part, published in 1620–21, expounds the theory, the *physica coelestis*, explaining the motion of the planets by the same physical principles as in the Mars book. These physical explanations he considered equally as important as the numerical elements of the orbits; or rather, as more essential, because they disclose the causes. All this is given in the form of questions and answers, not as Galileo did, for the dramatic power of a discourse, but after the manner of a catechism, in which the question as a heading summarizes what is expounded in the answer. In a letter to the Estates of Upper Austria he indicated that this book was an explanation to the Rudolphine Tables with which he was occupied.

He was occupied with this work for many years, chiefly because lack of money in the imperial treasury prevented him from appointing computers; at last he found in Jacob Bartsch, afterwards his son-in-law, a devoted assistant in his computations. In 1627 the tables appeared; he had had to contribute to the cost of printing from his own small possessions. On account of their excellent basis, they superseded all former and also contemporary less perfect tables, like the one constructed by the Middelburg minister, Philip Lansbergen; and they dominated the field of practical astronomy throughout the seventeenth century. His theoretical work, on the contrary, penetrated but slowly. Notwithstanding their friendly correspondence, Galileo, most curiously, remained ignorant of the laws discovered by Kepler and in his *Dialogo* in 1632 wrote that the true figure of the planetary orbits was unknown.

Through all those years, in between the practical computations, Kepler worked at a second task imposed not by heritage or office but by his own craving for knowledge and understanding. What had inspired him in the work of his youth, what had driven him toward Tycho, persisted with him as his deepest longing: to disclose the secret of the world-structure, to penetrate into the thoughts which the Creator had followed in the creation of the world. In his work *Harmonice mundi* ('The Harmony of the World') he connected the planetary motions with all fields of abstraction and harmony; with geometrical figures, with the relations of numbers, with musical harmonies. But among all these fantastic relations we find one precious discovery, afterwards always cited as *Kepler's third law*—also mentioned in his *Epitome*: for all the planets the squares of the periods of revolution are proportional to the cubes of their mean distances from the sun. For himself, all this was the consummation of what during all his life had been the radiant goal of his efforts and pains: 'I write my book to be read, either by present-day or by future readers—what does it matter? It may wait a hundred years for its reader, since God himself has been waiting 6,000 years for one who penetrated his work.'[120]

We read this proud pronouncement with a smile of admiration, knowing that later science has accepted and preserved from the entire work on 'The Harmony' only that one page containing the third law. Must we say, then, that all other work done by Kepler in this book was a waste of time? To perform great things, man has to set himself even greater aims. The lasting fruit can grow only in a larger organic structure, first living, afterwards withering to dry straw. The strong impulses to work and struggle which man receives from his world are transformed in him into objectives and tasks largely determined by the world concept of his time. Through his lifework then there runs, as the fulfilment of the ideas absorbed in his youth, a unity of purpose which makes it a harmonious entity. But later generations—different persons with different aims in a changed world—take from it only what may serve them, discarding the framework. Thus what inspired and proved the triumph of the earlier precursor often appears to those who follow to be superfluous or a false direction. In later centuries, when scientific research took on more the character of the routine work along fixed tracks, this may be less visible. In this time of renovation, discovery and transition, Kepler's work shows better than any other the relation between general and personal elements in the growth of science.

CHAPTER 24

MECHANICS AND PHILOSOPHY

HE social storms of the sixteenth century had ceased to rage, and a
new social order had settled down. The power of urban craft had
declined, as new industrial settlements with their factories grew up
mostly in new sites free from the restraints of the guilds. Commerce
with distant continents expanded in old and new centres and became a
force in society. The need of centralized power found its political
expression in royal absolutism, suppressing both urban particularism
and the quarrels of the nobility. Foremost in these developments were
France, Holland and England. In France, at that time the most popu-
lous, the strongest and wealthiest country of Europe, royal power
attained its greatest brilliance under Louis XIV. Holland, dominated
by a class of rich merchants, was temporarily powerful through large
trading profits, its flourishing economy producing strong spiritual
impulses. England, a rising country of merchants and citizens, had first
to experience a period of civil war, in which the ambitious royal power
was repelled by parliament, before it could unleash its forces. On the
other hand, Germany, cut off as it was from the new world trade—as
also was Italy, where an earlier culture had gradually stagnated and
declined—had no part in this progress; plundered and devastated by its
neighbours, it was thrown back several centuries in development.

The science of nature now began to present itself consciously as a
means to improve the life and to increase the prosperity of man.
What formerly had been called 'science' was not directed to the material
enrichment of human life. Neither the doctrines of Aristotle and Plato
nor the superstitions of medieval times could serve this end. A true
knowledge of nature based on experience and experiment was needed.
The useful arts now entered into the view of science; what previously
had been a secret tradition of craft came to light in scientific publications, often by the craftsmen themselves, as in the art of metal mining
and working by Georg Agricola in Germany and in the art of pottery
by Bernard Palissy in France. 'In the later Middle Ages technical
inventions had been remarkably frequent and their cumulative effect
was now such as to raise visions of the possibility of a radical transforma-

tion of the conditions of human life,' thus Farrington in his study on Bacon.[121]

Experience and experiment became the basis of science, and its direct purpose was often to improve technical processes and industrial methods. The scientists of the seventeenth century applied themselves energetically to the use of tools and mechanisms; they were skilful workers, constructors and inventors. Or, to express it with more accuracy, many people full of curiosity about things around them made experiments with self-made apparatus, devised and constructed tools and made discoveries. These were mostly wealthy citizens, gentlemen or skilled artisans; sometimes professors or officials in the service of the princes were among them. They were afterwards mentioned in the history books as the scholars of that time, whereas most of the occupants of academic chairs, who were then called the 'scholars', are now forgotten.

At the beginning of the century lived Cornelis Drebbel, of the Dutch town of Alkmaar, inventor of European fame, constructor of many remarkable instruments, who navigated a submarine under the Thames and for this purpose prepared oxygen from nitre; he applied his chemical findings to a new and profitable method of dyeing cloth. Christiaan Huygens, whose great talent for mechanics had been encouraged in his youth, with his brother Constantijn was engaged for many years in grinding lenses for telescopes. Newton, as a boy a tinkerer at home, in later life with great patience ground a metal mirror for his telescope when the specialist workers could not satisfy him. Leeuwenhoek discovered his 'little animals' with small lenses made by himself. When Huygens had to outline a plan of projects for the recently installed Paris Academy, he proposed, as one of the items, the investigation of the motive power of gunpowder and of steam for use in machines. The seventeenth century was the century of mechanics, i.e. the science of forces and motions, in the same sense that the sixteenth century might be called the century of astronomy, the eighteenth century of calorics and the nineteenth century that of electricity—thus expressing which realm of knowledge at each period made the most notable progress.

This trend of thought and this tendency to construct mechanical instruments induced the minds of men to recognize such mechanisms in nature also. Nature, hitherto a world of wonder and a realm of incomprehensible and often mystical powers, now became a mechanical structure. The celestial bodies were supposed to be carried along by a moving fluid filling endless space. The ancient atomic theory of Democritus and Epicurus was revived by Pierre Gassend (mostly quoted Gassendi): all matter consists of small particles that by their mutual action produce all natural phenomena. In explaining the phenomena of light, Newton considered light rays as streams of small corpuscles ejected by luminous bodies, whereas Huygens spoke of a space-filling ether whose particles, by pushing one another, propagate the light waves.

New philosophical principles now arose in opposition to the traditional doctrine of Aristotle. Two outstanding names, Bacon and Descartes, proclaimed the new methods of science. Francis Bacon (Baco of Verulam) in his *Novum Organon* in 1620 set forth with resolute clarity that experiment and research are the sole bases of science and philosophy. He emphasized that science has to serve practical life. What in former centuries had been called science was mostly sterile; philosophy stood outside the real life of man, studying as it did books and theories instead of the reality of nature. Knowledge is power; man has to make himself master of nature, not by magic, as medieval ignorance supposed, but by experiment. The investigation of nature in England since that time has always been called 'experimental philosophy'. Bacon did not invent it or bring it to prominence. Many others, such as Gilbert, Galileo and Kepler, had already applied the same principles and in detailed scientific knowledge Bacon was their inferior. But as he expressed the new principle in the most general and precise way it has in England always been called by his name.

In his utopian tale *Nova Atlantis* Bacon described an ideal state, where a ruling community of scholars, investigators, travellers and experimenters was living and working together in laboratory and garden, performing practical experiments and technical inventions. They discussed philosophy, all of them filled with an eager desire to find the truth about nature and make it serve a happier life. In every utopia of the time, as for instance, in the *Civitas solis* of the gifted rebellious monk Tomaso Campanella, the necessity for a firm knowledge of nature as the basis of trade and labour was emphasized. It was a trend of thought characteristic of the beginning of the seventeenth century; in Thomas More's *Utopia*, a century earlier, nothing is to be found of such ideas.

René Descartes, in his works published between 1637 and 1643, placed the principle of critical doubt in opposition to the belief in traditional authority: one's own thinking alone could be trusted. The spiritual slavery of the belief in authority must give way to the spiritual liberty of free thinking. Thus his principle is seen to be opposed to Bacon's; not experiment but thinking is the source and warrant of truth. Pure reasoning is the sole basis of certainty. Thinking must be the source of all truth as the necessary result of the single principle. 'I will explain the results by their causes, and not the causes by their results.'[122]

This basic idea was also applied to the construction of a new world system, which, contrary to Aristotle's structure of philosophical abstractions, presented an intelligible mechanical picture of the world. In this theory the universe was filled with a thin fluid, consisting of fine dust, produced from the particles by their impacts; it was rotating in whorls, *vortices (tourbillons)*, around the sun, and, of course, in the same manner, far away around the stars. This rotation carried the planets along in their orbits; around the earth and Jupiter smaller vortices produced the revolutions of their moons. As to the comets, there remained for them the wide spaces outside Saturn's orbit, where they roamed, now approaching, now moving off between the stellar vortices. Not wishing to be disturbed in his philosophical studies by the Church—Galileo's trial had just taken place—Descartes artfully explained that according to his theory the earth may be said to be at rest, i.e. relative to the streaming fluid, its surroundings, just as a floating ship carried by the streams is at rest relative to the water.

The attempt by pure reasoning to drive the phenomena from primary causes, gave rise, of course, to many fantastic explanations and results refuted by later science. 'Descartes', said Bailly, the later historian of astronomy, 'dealt with nature as if it did not yet exist and had to be constructed. Bacon considered it as a vast edifice to be invaded and decomposed to discover its structure and arrive at its foundations. Also Bacon's philosophy, restricted to facts, still subsists, whereas Descartes's doctrine, too much subjected to imagination, has perished. Bacon had the greater wisdom, Descartes the greater boldness, but by this boldness he has served the human spirit well.'[123]

The contrast between the two philosophers corresponds to a contrast of wider scope, in attitude and life-system, between the two peoples. In England people in medieval times had already acquired a high degree of personal liberty and independence, certainly owing to their insular security which precluded the necessity of an armed state power. Personal initiative was not hampered by command and prescript from a ruling power. Thus the mode of action in trade as well as in research was direct practice; man was trying and experimenting, doing things in his own way. On the Continent people were more constricted, oppressed and hampered by powerful authority and, through old and new dependencies, were prevented from practical action in their own way. So the new ideas had to remain in the field of thought and, by consistent thinking, to be perfected into complete theoretical systems. Thus, philosophically, England came to be the country of empiricism, France and the Continent the country of rationalism.

Gradually Descartes began to oust Aristotle from the academic chairs which felt a need for complete philosophical systems. At the end of the seventeenth and the beginning of the eighteenth centuries, the vortices had penetrated most manuals of physics; in the Jesuit schools of the southern countries, however, Aristotle maintained his position during the eighteenth century. Astronomically Descartes's theory was not on a par with the science of the time; for Kepler's laws there was neither room nor explanation in the vortices. That these laws were also entirely unknown to Galileo, who assumed the planetary orbits to be circles, has been mentioned above. The young English clergyman, Jeremiah Horrox, who died at the age of twenty-two, was among the few who knew them; having first followed Lansbergen, he soon became an ardent admirer of Kepler. He was the first to explain the greatest inequality in the moon's course by the elliptic figure of its orbit. Kepler's Rudolphine Tables were universally used; but even at the end of the century Cassini and La Hire tried to find other explanations for the irregularities in the planetary movements.

This cannot surprise us if we consider that even the heliocentric doctrine of Copernicus was not widely accepted without difficulty. It is true that the physical objections gradually lost their importance. In 1638 Galileo's *Discourses on Two New Sciences, of Mechanics and of Motions* had been published in Holland. Blind and broken in health, in the last years of his life he had dictated the work to his pupil Viviani. In this compendium of his lifelong researches on the motion of bodies the foundations of the renovated science of mechanics have been laid down. Though Copernicus and his theory are not mentioned therein, it established a firm theoretical basis for the motion of the earth. Moreover, experiments were made by Gassend in 1640 to demonstrate the preservation of uniform motion, e.g. through balls thrown up vertically by a rapid rider, which fell back into his hands.

The theological difficulty, however, was more strongly felt, since in Catholic countries authors on astronomy had to avoid any conflict with the Church. The most learned and best known among the opponents was J. B. Riccioli, professor at the Jesuit College at Bologna, who made experiments to see whether bodies falling from a high tower arrived exactly below the starting point. In 1651 he published his *Almagestum novum*—the name indicates a modernized Ptolemy, though he seemed to prefer the Tychonic system—a large collection of astronomical facts and opinions, intended to be a refutation of Galileo's *Dialogue*: not a hard task when the other side was gagged. He enumerated and discussed 49 arguments in favour of Copernicus's theory and 77 arguments against; so Copernicus was defeated by a majority. But such artfulness in arguing could not actually impede the progress of science, though in Italy the chilling hand of clerical threat made scientific discussion well-nigh impossible and continually prevented able astronomers, such as Borelli and Montanari, from publishing and uttering their real opinions. In France, because the power of the Church was less dominant, conditions were better. In 1665 the French astronomer Auzout, in a letter to an influential prelate, agreed that the hypothesis of Copernicus was neither absurd nor false philosophically and that the Scriptures were not intended to instruct us in the principles of physics and astronomy, which are as useless for the life hereafter as they are useful for the life here; so he demanded more freedom for the scientists—of course, without avail.[124] In the next century, in the northern countries, the new truth gradually began to spread among wider circles of the population.

There now arose a certain organization in science. In the preceding centuries single individuals had stood out from among their contemporaries through special knowledge and predilection for the study of nature. In the seventeenth century a keen interest in the surrounding world developed among a numerous class of well-to-do and educated citizens; it appeared as an ardent curiosity and desire for knowledge which they felt to be salutary for society. They contacted one another, and through extensive correspondence, in which we find the germs of many new views, they discussed their opinions and discoveries. Before long they assembled in regular meetings; soon they got support and protection from the princes and acquired the official status of an 'Academy', somewhat on the lines described by Bacon in his *New Atlantis*, though with much less power and influence. A first organization had been formed in this way in England in 1645, meeting at first in secret because, being mostly royalists, they could not expect sympathy from the Puritan government. Afterwards, in 1662, through a royal charter, they became the 'Royal Society'. Lectures were given and communications were made at their meetings, new discoveries and ideas were discussed, letters from foreign scholars were read and experiments were performed. The Secretary, Oldenburg, from the German trading town Bremen, through his extensive correspondence with numerous European scholars, for a long time acted as a kind of central office for science. For many years Robert Hooke, a keen and versatile scientist, was appointed at a small salary to demonstrate at every session a new and interesting experiment. Among his papers a document was found describing the business and purpose of the Royal Society: 'To improve the knowledge of natural things, and all useful Arts, Manufactures, Mechanick practices, Engynes and Inventions by Experiments (not meddling with Divinity, Metaphysics, Moralls, Politicks, Grammar, Rhetorick or Logick).'[125] For the publication of all this work the *Philosophical Transactions* was founded in 1666, and throughout the following centuries they remained a most important scientific review.

In France, too, naturalists and scientists were already meeting in regular assemblies before Minister Colbert gave them official status in 1666 as the 'Académie des Sciences'. There was a curious difference between the two academies, typical of the different conditions in the two countries. This was clearly pointed out by Voltaire in the next century and was described in a later *History of the Royal Society* in these words: 'The members of the French Institute receive a yearly stipend; the Fellows of the Royal Society pay an annual sum for the support of their Institution and the advancement of science. It would be repugnant to the feelings of Englishmen to submit to the regulations of the [French] Institute, which require that official addresses, and the names of candidates for admission should be approved by the Government before the former are delivered or the latter elected.'[126] In France the Academicians received salaries, called *pensions*, from the king. Louis XIV felt himself to be the great European monarch, who extended his influence far beyond the frontiers of his country; he awarded 'pensions' to foreign scholars, and he tried to attract to Paris the most famous among them to enhance the splendour of his reign. Ole Römer came from Denmark and Dominico Cassini from Italy to take their seats beside the French astronomers Auzout and Picard. When Picard had come to inspect the ruins of Tycho's Uraniborg, Römer made his acquaintance and accompanied him to Paris. Cassini came to direct the building of an observatory, which was also to serve as the home of the Academy. Discussions took place at the sessions and experiments were made; when a comet appeared it was jointly observed and papers on its nature were read. The reports published in the newly-created *Journal des Savants*, often of a rather primitive character, sometimes presented important new ideas.

The example of the two prominent kingdoms was imitated in other countries. In Florence even earlier, in 1657, a dozen or so naturalists, mostly pupils or admirers of Galileo, the Grand Duke himself among them, had united into an 'Academia del Cimento', which in its assemblies initiated a systematic series of experiments on problems of physics; but when they were informed that the high Church authorities disapproved of such activities, they had to stop their work ten years later. In an impoverished and divided Germany, where petty princes tried to emulate the brilliant court at Versailles, learned societies were founded in different towns. The most important was the Academy at Berlin, due mainly to the personality of its founder, Georg Wilhelm Leibniz, a versatile scholar, an important philosopher and a mathematician of genius, who tried to realize his ideal of a republic of scientists in this entirely inadequate milieu. Also in the particularistic Netherlands every town of importance in the next century had its 'learned society'.

Thus the increasing thirst for knowledge in the rising middle class everywhere in Europe laid the foundations for this organization of science.

CHAPTER 25

THE TELESCOPE

THE progress of astronomy in the seventeenth and eighteenth
centuries was due in the first place to the new instrument, the
telescope, which was now at the disposal of the astronomers. Its discovery had been by chance, nothing more than a marvellous plaything.
Galileo had no real knowledge of how it worked. Kepler was the first,
in 1611, in a booklet *Dioptrical Researches*, to give a theory of the course
of the light rays through the lenses and of the formation of an image. He
discussed different ways of combining lenses, by placing them one behind
the other into an optical system. Among them was not only the combination of a convex object lens and a concave ocular, realized in
Galileo's telescope, but also a combination of a convex objective and a
smaller convex ocular. Since then the latter has always been called the
'Kepler' telescope, although he never tried actually to make it.

Such a telescope was impracticable in ordinary life because it reversed
the images and showed people upside down. In astronomy it came into
use during the following twenty years, whether by practical trial or
guided by Kepler's theory, we do not know. Neither do we know with
certainty who was the inventor. In 1655 Hans Zachariassen, the son of
Zacharias Janssen, made a statement to an official fact-finding commission that, together with his father, he had constructed in about 1619 a
lange buyse ('long tube'); it is supposed that this may have meant a
Kepler telescope, because here the focus is situated between the lenses,
hence the length is the sum total of the focal distances, whereas the
ocular of the Galileo telescope is situated between the objective and its
focus. Fontana said in 1646 that he had already made such a telescope
in 1608; and Christoph Scheiner mentioned in 1630 that he had been
using this type of telescope for many years to project the sun's image
upon a screen to observe the spots. In 1645 Father Schyrrle of the
Rheita monastery described how by means of Kepler's construction the
stars were visible far more sharply over a larger field of view. The
advantage of Kepler's over the Dutch construction is here stated exactly;
it is the large field of view. The light rays coming from a star to which
the telescope is not exactly pointed, passing obliquely through the tube,

fall upon an outer part of the ocular lens; if the lens is concave, it bends the rays farther outward, so that they do not enter into the pupil of the eye; if it is convex, it bends the rays inward, toward the optical axis, so that they enter fully into the eye. This greater field, with clear and sharp images, gives the Kepler telescope its superiority, so that since, say, about 1640 it has entirely ousted the older type from astronomy.

Shortly afterward the brothers Huygens occupied themselves, as did many others, with the patient and careful grinding of better lenses, in order to observe the heavenly bodies with greater enlargement. Galileo had used a 30 times enlargement; when this limit was surpassed the images became so much less distinct that nothing was gained. It was soon perceived that this was due chiefly to the strong curvature of the lens surfaces, which failed to make the light rays converge exactly. This could be remedied only by using lenses with a small surface curvature, i.e. with greater focal distance, so that the telescopes became rather long. Christiaan Huygens (1629–95), who at an early age had already shown himself to be a clever mathematician, later derived the best figure for a lens by theory: if one surface had a six times smaller curvature than the other, the 'spherical aberration', i.e. the defect in exact convergence of the rays into one point, which caused the defects in the image, was least. But from the practical point of view this gave insufficient improvement. With his brother he now constructed a 12 foot telescope, with a 57 mm. (2½ inch) aperture only ($\frac{1}{60}$ of the length); and thereafter one 23 feet long. With the first he discovered, in

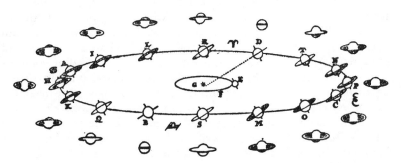

Fig. 28. Changing aspects of Saturn's ring, according to Huygens

1655, a small star near Saturn that accompanied the planet; it was a satellite describing an orbit around Saturn in 16 days 4 hours. He found even more puzzling the quaint appendices discovered and described by Galileo as two adjacent smaller companion globes. In the following years Galileo, much to his alarm, saw them diminish and disappear. Had Saturn devoured his own children?—he asked in a letter to Welser. In later years other observers saw them in different aspects. Huygens saw them in the form of two handles; he felt sure that objects near Saturn could not be at rest but must revolve; since the appendices continually kept their aspect, they must be a kind of ring around the planet. That the ring disappeared entirely in 1656 proved that it must be very thin and flat. So in 1659, in his *Systema Saturnium*, he was able to give the solution of the earlier letter-riddle: 'Saturn is surrounded by a thin flat ring not touching it anywhere, which is oblique to the ecliptic.' And he adds: 'I must say here something to meet the misgivings of those who think it strange and irrational that I give to a celestial body a figure, as none has been found until now, whereas it is held for certain and assumed to be a law of nature that only a spherical shape fits them. . . . They must consider that I have not thought out this assumption through my own invention or fancy . . . but that I see the ring clearly with my eyes.'[127]

This unravelling of Saturn's riddle was not a simple discovery with the aid of a good telescope; it was the result of thinking and deduction. For, compared with modern instruments, Huygens' telescopes were primitive and bad. A modern astronomer, examining the lens of this 12 foot telescope, described the aspect of a bright star seen through it as 'an irregular and unstable central blot from which five coloured rays extend on which parts of 5–7 diffraction rings were distinctly visible'.[128] A skilled Italian specialist, Campani, who, moreover, had better Venetian glass at his disposal, was able to make better lenses. By means of telescopes fitted out with these lenses, Cassini succeeded, in Paris, between 1671 and 1684, in discovering four smaller satellites of Saturn. In such work an improvement of the ocular was also important. First a single planoconvex lens was always used; then Huygens constructed a combination of two such lenses at a distance computed by theory. This 'Huygens ocular' has been in use up to modern times as the best fitted for telescopic view.

In order to minimize the innate defects of lenses and yet attain greater powers, telescopes were made of ever greater focal length with lenses of lesser curvature. The Huygens brothers ground lenses of a focal length of 45, 60 and 120 feet or more; Campani also made one of 130 feet for the Paris astronomers. A stiff tube of that length was difficult to construct and to handle. Hence the lens with frame was hoisted to a high pole, and the observer on the ground viewed the focal image with the ocular (plate 6). Huygens had erected such a scaffolding on his country-seat, 'Hofwijck', near The Hague, and he made observations with it. It was obviously difficult to handle such an apparatus, and we must admire the practical ability that made observations under such conditions possible. These excessive dimensions were abandoned later on, especially after Newton's researches had shown that the chief cause of the unsatisfactory images was not the figure of the lenses but the chromatic dispersion of the glass, which persisted with even the greatest focal length.

Telescopes could be used for more purposes than discovering new celestial bodies; the pioneer work of Galileo and his contemporaries had to be extended in various directions. Fontana at Naples, between 1630 and 1646, made many observations of the planets and still more of the satellites of Jupiter. Their eclipses, when immersed in Jupiter's shadow cone, were observed by Hodierna in 1652, as well as the passages of their shadows over the disc of the planet. On the basis of a number of observations, Borelli in 1665 gave a theory of their motions; and a more complete theory with tables of their movements was given by Cassini in 1668. They were based chiefly on observations of the eclipses in times of easy visibility, before and after Jupiter's oppositions. When the observations were continued at Paris, it was found that the moments of eclipse of the first satellite, the most rapid one, did not fit when Jupiter was near conjunction with the sun; they were more than 10 minutes late. The Danish astronomer Ole Römer, being in Paris in 1675, gave an explanation of this difference by means of the finite velocity of light: when Jupiter is at its greatest distance from the earth, the events are observed with greater delay. From it he derived the fact that the light takes 11 minutes to cover the distance from sun to earth.

Several observers in the seventeenth century directed their telescopes at different planets, and sometimes they thought they perceived irregular figures or spots. In his Saturn book Huygens gave a drawing of Jupiter with two equatorial streaks clearly depicted, and one of Mars with one dark band. His diary from later years contains sketches of Mars, on which some of the spots discovered later could be recognized, so that they allowed the determination of the rotation period. Cassini, with his better telescopes, in 1663 determined the rapid rotation of Jupiter (as 9 hours 56 minutes) by means of small irregularities in its equatorial bands, and in the next years found 24 hours 40 minutes for the rotation period of Mars. These discoveries were the utmost limit of what could be obtained with the rather primitive instruments of the time; interspersed were abundant announcements of pretended discoveries.

The most promising object for study with the new instrumental aids was the moon. After the first discoveries of Galileo it must have seemed an alluring task to make an exact picture or map of the moon. Here was another world, a counterpart of the earth, but easier to picture. The interiors of the far continents on our earth were inaccessible; only the coasts could be explored, so that the making of a complete map was

5. Johannes Kepler (p 225)

Christiaan Huygens (p. 254)

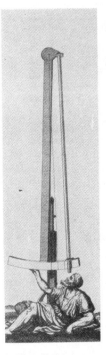

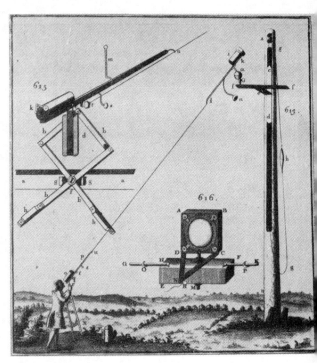

6. Picard's
   Zenithsector (p. 259)   Huygens' tubeless telescope (p. 255)

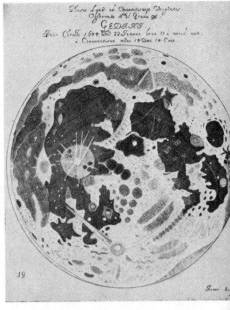

Hevelius and his Quadrant (p. 259)   Hevelius' map of the full moon (p. 257)

impossible. The lunar world, on the other hand, was entirely open to our view—of course, only the near side—so that astronomers could map it completely. The first work of this kind appeared about 1630, from the hand of the now nearly forgotten Belgian mathematician and cosmographer, M. F. van Langeren (Langrenus). More fame was won by the great atlas of the moon by Johann Hewelke (1611–87), better known as Hevelius, the Latinized name on his publications. Hewelke, patrician of Danzig, having studied at Leiden University, after his return installed an observatory on his house, where in 1641 he began to observe the moon regularly. His *Selenographia*, published in 1647, consisted of a number of self-engraved drawings and maps, on which the different features—the dark and bright regions, the mountains, the circular walls and craters—were inserted according to direct observation and were provided with names largely taken from earthly geography. Of these, only some few names of mountain ridges (Alps, Apennines, Caucasus) have been preserved in modern selenography. For all the other objects they have been discarded and replaced by another system of names devised by Riccioli in 1651 and inserted in a lunar map drawn by his pupil Grimaldi. For the mountains they used the names of famous astronomers and mathematicians; for the dark plains, regarded as seas, they took fancy character names, with geographical or meteorological meaning; hence we call some of the finest and largest ring mountains by the names Tycho, Plato, Aristarchus, and some plains are named Mare Serenitatis (Sea of Serenity) and Oceanus Procellarum (Ocean of Tempests).

The deeper thought underlying this study of the moon and the planets was conveyed in this question: If they are similar to the earth, might they not also be inhabited by living and intelligent beings? Kepler had formerly written a book entitled *Somnium*, published posthumously, the dream of a fantastic voyage to the moon; on this romantic basis a scientifically well-founded account was given, embroidered as an exercise of wit, of the celestial phenomena and living conditions on the moon. It was part of the world conception of the seventeenth-century scientists that they tried to figure out an adequate picture of the living conditions on the planets considered to be inhabited. Thus Huygens wrote a *Cosmotheoros*, published in 1698 after his death, containing reflections on the conditions and the living beings on other worlds. 'It is hardly possible,' he said, 'that an adherent of Copernicus should not at times imagine that it seems not unreasonable to admit that, like our globe, the other planets are also not devoid of vegetation and ornament, nor, perhaps of inhabitants.'[129] It is not necessary to assume that the equipment of the other planets is fundamentally different from what we know on earth. 'There appear on the surface of Jupiter certain bands darker than the rest of the disc, and they do not always preserve the same form; that is proper to clouds. . . . On Mars clouds have not yet been observed, since the planet appears so much smaller . . . since, however, it is certain that the earth and Jupiter have clouds and water, it can hardly be doubted that they are found also on the surface of the other planets.' As to the animals, 'there is no reason why their mode of sustenance and of multiplying themselves on the planets should not resemble what they are here, since all animals on this earth . . . follow the same law of nature'.[130] If there are intelligent beings, the rules of thinking and of geometry must be the same as for us.

Great popularity was won by Fontenelle's *Entretiens sur la pluralité des mondes* ('Conversations on the Plurality of Worlds') 1686. Here the planetary system of Copernicus, the inhabitants of the moon and the planets, the vortices and the comets all are treated in the light, courtly style of the reign of the Roi Soleil. 'You will not be surprised,' the author says to the Marquise, 'to hear that the moon is another earth and appears to be inhabited.' Farther on he says of the moons of Jupiter 'that they are not less worthy of being inhabited, though they have the misfortune to be subject to, and to revolve about, another more important planet.'[131] But what about the theological difficulty— which he treats at the close of the preface—that people on the moon cannot be descendants of Adam? 'The difficulty is made by those who are pleased to place men upon the moon. I do not place them there; I place only inhabitants there that are not men . . . you will see it is impossible that there are men there, according to my idea of the boundless diversity which nature must have put into its works.'

The fact that in the Kepler telescopes a real image is formed in the focal plane, which, as by a magnifying glass, is looked at through the ocular placed behind the focus, acquired a fundamental importance for astronomy. An object situated in the focal plane, be it a metal sheet, ring, or wire, is seen sharply in focus together with the image of the celestial object; by comparing them, small distances or sizes can be measured. Huygens described in 1659 how he determined the diameter of a planetary disc by exactly covering it with a metal strip in the focus. The French astronomer Auzout in 1667 described an improvement on this method by putting two parallel wires in the focal plane, one of which could be moved by means of a screw. This was the first specimen of a filar micrometer, which in later centuries developed into an ever more perfect measuring apparatus. That a young English astronomer, William Gascoigne, afterwards killed in the civil war, in 1640 had already used the same device for measuring planetary diameters was discovered much later in his manuscripts.

However important for human knowledge the discoveries with telescopes have been, the application of telescopes to measuring instruments has been far more important for science. A proposal by Morin, a notorious astrologer of Paris, in 1634, to use a Galileo telescope as a measuring instrument, of course, was impracticable and useless; only the Kepler telescope could open new ways. By bringing the image of a star as seen through the enlarging ocular exactly at the intersection of two cross-wires, the position of the telescope can be fixed far more accurately than formerly. Thus the position of a star could be determined with far greater precision. Jean Picard, the most diligent and capable of practical astronomers, in 1667 was the first to introduce this method into astronomy. His object was a new accurate determination of the dimensions of the spherical earth by measuring a meridian arc. A new method had been devised by Snellius at Leiden: a large distance on earth was derived by means of triangulation from a small, accurately measurable base-line. He had applied it to the distance between the towns of Alkmaar and Bergen-op-Zoom, separated by broad waters: and in 1617 he described it in a booklet bearing the adequate title *Eratosthenes Batavus*. Picard, following this method, measured a series of triangles with a base line in northern France; the difference in latitude between the northern and southern ends was determined by measuring at both points the meridian zenith distances of a number of stars. His instrument was a circular arc, 10 feet in radius (hence $1' = 1$ mm.), extending only some few degrees, and was provided with a telescope instead of with sights. The added figure, reproduced in plate 6, characterizing the spirit of the time, shows the observer in the guise of a philosopher of antiquity. The individual results did not deviate by more than $5''$ from the mean, testifying to the accuracy reached and the reliability of the result, 57,057 toises for one degree of the meridian. This, however, was a favourable case, owing to the relative character of the determinations. With other determinations of stellar positions, Picard found differences up to $10''$ or $15''$ in the declination, apparently dependent on the season; their origin was not detected until many years afterwards.

The new method of astronomical observation did not find general approbation. Hevelius of Danzig was working by Tycho's method. With extreme care he himself constructed accurate measuring instruments (quadrants and sextants); after the fashion of his time, he made them into show pieces of fine workmanship and installed them in his observatory (pl. 6). With his sharp vision—he could see stars of the seventh magnitude with the naked eye—he succeeded in attaining an accuracy of $1'$ and even less, thus surpassing Tycho. He measured meridian altitudes with a quadrant of 5 feet radius and distances between planets and stars with a 6 foot sextant. His observations numbered thousands, but unhappily a large part was lost through a fire that destroyed his house and his instruments. He also prepared a celestial atlas, published in 1690 after his death, in which he introduced a number of new constellations made up of small stars in the blank spaces between the ancient asterisms such as the Hounds, the Lizard, the Shield of Sobieski, the Unicorn, the Sextant, the Fox, the Lynx.

He had a sharp dispute with Hooke in England concerning the best method of observing. Hooke asserted that observation with sights only, without telescope attached, could not give sufficient accuracy. Thereupon, Edmund Halley (1656–1742), who had already used a sextant with telescopes when observing a number of southern stars at St Helena in 1676, was sent from London to Danzig in 1679 to observe with Hewelke a number of identical stars, each with his own instrument. It then appeared that the differences between their results was mostly a matter of seconds only, never reaching one minute—a proof of Hewelke's perfection in the art of observing. His precision was surely the utmost attainable without telescopes, but there was no future in it. The same precision was easily reached by the use of telescopes on the instruments and could be raised by further improvements in the instruments and methods.

From now on telescopes became a permanent part of astronomical equipment. In England they were used by Flamsteed in his first observations of 1676 in the newly-founded Greenwich Observatory. These telescopes, as may be seen in old pictures, were long narrow tubes; the objective was a small lens about one or two inches in diameter. No great brightness, but only a strong magnification was needed to point the stars with greater precision. Even with these modest objectives, the stars became brighter in relation to the background and some of them became visible in the daytime. In 1634 Morin had already described enthusiastically how with his telescope he could follow Venus for many hours after sunrise. Picard, in 1669, by observing the passage of Arcturus through the meridian in the daytime, opened up new possibilities of astronomical measurement.

CHAPTER 26

NEWTON

THE concept of attraction was not introduced for the first time by Newton. Copernicus had already spoken of the mutual attraction of the parts of the earth as the cause of its spherical shape; he assumed this faculty to be present in other celestial bodies too, causing their particles to be compressed into a sphere. Kepler, too, had spoken of gravity as a tendency of cognate bodies to approach and join one another. To him the tides were a proof that the moon exerted an attraction upon the water of the earth: 'if the earth ceased to attract the waters, all the sea water would be drawn upward and would flow to the moon'.[132] He compared gravity with magnetism: 'the earth draws along the bodies flying in the air, because they are chained to her as though by a magnetic force, just as if there existed a contact between them.'[133] This attraction had nothing to do with orbital motion; the sun, as quoted above, did not exert an attractive force upon the planets but a directive force, dragging the planets along with its rotation. Gravity and orbital motion were two different and entirely separate fields.

Nor did the seventeenth century see any connection between the vortices, which moved the planets in their circles, and gravity, working at the surface of the earth and doubtless also at the surface of the sun and the other planets. Huygens made an attempt to establish such a connection in a lecture held, in 1669, at the Paris Academy, 'On the Cause of Gravity'. Whereas Descartes had assumed that the ethereal fluid, by rotating uniformly about a certain axis through the earth, carried the moon along, Huygens made the thin fluid matter in rapid rotation move in all directions about the earth's surface. As a consequence of their centrifugal force directed outward, i.e. upward, the fine particles pressed down the larger particles of the coarse-grained visible matter, which did not participate in the rotations. This origin of gravity implied that the thin fluid matter passed freely through all heavy objects and filled the space between their particles. The velocity of this whirling motion had to be 17 times greater than the velocity of the equator, because, with a rotation of the earth 17 times more rapid than the actual one, the objects at the equator would lose their gravity.

The actual progress of science, however, went in exactly the opposite direction, not in explaining gravity by circular orbits but in explaining circular orbits by gravity. The development of the fundamental principles of mechanics had made this possible. Galileo had explained the constant velocity of a horizontal movement in the absence of friction by pointing out that such movement was part of a circular orbit about the earth's centre, which had always been considered uniform by nature. He had not been able to overcome this conception; but his researches had so perfectly cleared the way that pupils and younger scientists, like Cavalieri (1632) and Torricelli (1644), could express the 'principle of inertia' in modern form: when acting forces are absent, the motion is rectilinear with constant velocity. Then the next step was the realization that a circular orbit is not simply a natural motion—as all the preceding centuries had supposed—but a complex enforced motion. A circular motion is the result of a force directed towards the centre, continuously preventing the body from following the rectilinear motion along the tangent. This tendency to follow the tangent and move with increasing rapidity away from the centre was observed as a 'centrifugal force', a tension in the string when an object is swung around. In his work on the Jupiter satellites, Borelli in 1665 had expressed himself in this way: that the centrifugal force of the orbital motion was exactly in equilibrium with the attractive force of Jupiter. The complete theory of the centrifugal force was given by Huygens in 1673 in his work *Horologium oscillatorium*, in which he, in connection with his invention of the pendulum clock, treated a number of related mathematical and mechanical problems. He deduced that the centrifugal force is proportional to the square of the velocity and to the inverse of the radius of the circle.

So the idea became dominant that an attraction directed toward the centre of their orbit works upon the planets and the moon. It might be expected that this force decreases with increasing distance; but in what ratio? The answer to this question was given by Newton (plate 7).

Isaac Newton, a farmer's son from the hamlet of Woolsthorpe, in Lincolnshire, born in 1642, went to study in Cambridge in 1661. When the university was closed for a couple of years because of a pestilence in the town, he returned in 1665 to his native village. Here he made his first studies in what were to become the most important subjects of his later work: mathematics (the theory of fluxions), optics (the discovery that common light is composed of numerous kinds of simple light, all of different colours and refrangibility), and gravitation. The falling of bodies toward the earth caught his attention (the anecdote relates that, seeing an apple fall from the tree, he began to ponder over the cause of this falling) and raised the question as to what height gravity extended. To the moon perhaps? If so, could gravity be the force that kept the moon in its circular orbit? To settle it, he had to know in what ratio gravity decreases with distance from the earth. For this problem Kepler's third law could give a valuable indication. According to this law, a four times larger circular orbit has an eight times larger period, hence a two times smaller velocity; therefore, the centrifugal force, according to Huygens's formula, is 16 times smaller. Generally in such a planetary system the centrifugal force must be as the inverse square of the distance. Gravity compensating it must vary in the same ratio.

The moon's distance being 60 times the earth's radius, its gravity must be 3,600 times smaller than that of a stone falling on the earth's surface, or, as it was sometimes expressed, the moon falls in a minute as far as a stone falls in a second. Newton, in making the computation, assumed an arc of one degree on earth to be 60 miles, as given in a sailor's manual, the only book at hand—even today an English nautical mile is always taken to equal one minute of arc. Assuming this to be the usual 'Statute mile' of 5,280 feet, equal to 4,954 Paris feet, he computed the moon's acceleration per second to be 0.0073 feet, per minute 26.3 feet. Through Galileo's experiments, however, afterwards repeated more accurately by others, the acceleration of freely falling bodies per second was known to be 30 feet. The two values are of the same order of magnitude, but the difference, one-eighth of the amount, is too great to be acceptable. Disappointed, the story runs, he abandoned his apparently so brilliant idea. In the years that followed he occupied himself with optical and mathematical studies.

He could have used a better value, because Snellius's result, which gave, for an arc of 1°, a length of a good 69 English miles, could already be found in English books. It was confirmed by the more extensive and accurate determination of Picard in France, published in 1671, giving, for 1°, 57,065 toises or 69 English miles. Performed with this value, the new computation gave complete agreement. Thus the law of gravitational attraction, decreasing as the inverse square of the distance, was established.

Newton was not the only man to formulate this law of variation of force with distance. Part of his mathematical deductions were found in Huygens's work published in 1673. Robert Hooke, that acute and versatile but jealous scientist, asserted afterward that he had known the law for a long time—which was quite possible—and even that Newton had got the idea from him. Probably Hooke, by facing him with the problem of what the orbit of a body would be, if affected by such an attractive force, was a strong factor in drawing Newton's attention to this matter. But he himself could do nothing with the mere idea. Halley and Wren discussed the same questions, without being able to solve them. What was necessary was to demonstrate all arguments and derive all consequences of this law for the celestial orbits with exact mathematics. Newton was the only man able to do so by means of the mathematical methods he himself had constructed.

In 1684 the theory was ready in its main part; and in 1685, by solving the problem of the attraction of a solid sphere and demonstrating that it was exactly equal to the attraction by its mass if concentrated in the centre, he removed the last difficulty in the argument. Another year of severest mental exertion was needed, in which he was so entirely absorbed by his problems that dinner and sleep often were neglected and his health was badly shaken; the many anecdotes about his absent-mindedness relate to this period. Then the first part could be presented to the Royal Society in 1686. That the manuscript was not buried for a long time in its archives, was due to the unremitting care of his friend Halley, at that time assistant secretary (called 'Clerk') of the Society, who procured money for its printing, partly from his own pocket. In 1687 the work appeared under the title *Philosophiae naturalis principia mathematica* ('Mathematical Principles of Natural Philosophy').

The title of the book expresses how it could lay down new foundations for astronomy. 'Natural philosophy' was in England the name for scientific research; why mathematical principles were needed he explained in Book III, which bears the special title 'The System of the World'. There he said: 'Upon this subject I had, indeed, composed the third book in a popular method, that it might be read by many; but afterwards, considering that such as had not sufficiently entered into the principles could not easily discern the strength of the consequences, nor lay aside the prejudices to which they had been many years accustomed, therefore, to prevent the disputes which might be raised upon such accounts, I chose to reduce the substance of this book into the form of propositions (in the mathematical way), which should be read by those only who had first made themselves masters of the principles established in the preceding books.'[134] This is understandable when we consider that Newton was extremely sensitive to criticism, which, often based on shaky foundations, was set against results on which he had pondered carefully and profoundly; often he postponed publication of his results to avoid unpleasant polemics. The mathematical demonstration convinced the well-instructed and deterred the ignorant. It was at the same time that Spinoza expounded his philosophy in the mathematical form of propositions and demonstrations.

The contents of the first two Books, indeed, consist of mathematics; it is geometry applied to the motion of bodies, i.e. what we call 'theoretical mechanics'. In his preface Newton said: 'Therefore geometry is founded in mechanical practice, and is nothing but that part of universal mechanics which accurately proposes and demonstrates the art of measuring. But since the manual arts are chiefly employed in the moving of bodies, it happens that geometry is commonly referred to their magnitude and mechanics to their motion. In this sense rational mechanics will be the science of motions resulting from any forces whatsoever, and of the forces required to produce any motions, accurately proposed and demonstrated.'[135] Rational mechanics was the discipline needed to unite earthly and celestial motions into one system. Earthly motions were ruled by Galileo's laws of falling and gravity; celestial motions were ruled by Kepler's laws of planetary orbits. To connect them, Newton, as the founder of the new science, completing the work of Galileo and Huygens, began by stating its principles in the form of 'Definitions' and 'Axioms, or Laws of Motion'.

(1) Every body continues in its state of rest, or of uniform motion in a right line, unless it is compelled to change that state by forces impressed upon it. (2) The change of motion is proportional to the motive force impressed; and is made in the direction of the line in which that force is impressed. (3) To every action there is always opposed an equal reaction; or, the mutual actions of two bodies upon each other are always equal, and directed to contrary parts.[136] The concept of mass was introduced as 'the quantity of matter arising from its density and bulk conjointly'; 'the quantity of motion arises from the celerity multiplied by the quantity of matter; and the motive force arises from the accelerative force multiplied by the same quantity of matter.' Mass and weight were sharply distinguished. Hence it is that, near the surface of the earth, where the accelerative gravity, or force productive of gravity, in all bodies is the same, the motive gravity or the weight is as the body; but if we should ascend to higher regions, where the accelerative gravity is less, the weight would be equally diminished, and would always be as the product of the body, by the accelerative gravity.[137]

Because the chief aim is the treatment of the freely moving heavenly bodies, centripetal forces were introduced directly under the definitions. 'A centripetal force is that by which bodies are drawn or impelled, or any way tend, towards a point as to a centre. . . . Of this sort is gravity . . . and that force, whatever it is, by which the planets are continually drawn aside from the rectilinear motions, which otherwise they would pursue, and made to revolve in curvilinear orbits. . . . They all endeavour to recede from the centres of their orbits; and were it not for the opposition of a contrary force which restrains them to, and detains them in their orbits, which I therefore call centripetal, would fly off in right lines, with a uniform motion.' Then, after mentioning a projectile shot from a mountain horizontally with sufficient velocity, which would go round the earth in an orbit, he proceeded: '. . . the moon also, either by the force of gravity, if it is endued with gravity, or by any other force, that impels it towards the earth, may be continually drawn aside towards the earth, out of the rectilinear way which by its innate force it would pursue; and would be made to revolve in the orbit which it now describes: nor could the moon without some such force be retained in its orbit.'[138]

Then, through rigid mathematical demonstrations, Newton derived from Kepler's laws the forces determining the motion of the planets. His Proposition I (Cajori ed., p. 40) deals with the law of areas: if a revolving body is subject to a centripetal force directed to a fixed point, the areas described by radii drawn to that point will be proportional to the times in which they are described. For the demonstration, reproduced in Appendix C, Newton made use of equal finite time intervals in which the radius describes a triangle and after each of which the force gives a finite impulse to the body towards the centre. Then he proceeded: 'Now let the number of those triangles be augmented, and their breadth diminished in infinitum; and their ultimate perimeter will be a curved line: and therefore the centripetal force, by which the body is continually drawn back from the tangent of this curve, will act continually.'[139]

In these words we see that behind the geometrical form stands the spirit of his method of fluxions which pervades his geometry; it is the idea of considering quantities and motions not as definite abrupt values but as in process of originating, changing, or disappearing. Newton could be a renovator of astronomy because at the same time he was a renovator of mathematics. In his demonstrations he made use of geometrical figures of straight lines and triangles of finite size; but then he let the number of such parts be augmented and their size diminished *ad infinitum*, to fit a curved orbit and a continually working force; and he showed that the demonstrations then rigidly hold.

By means of the same figure, the reverse was demonstrated: when the succeeding areas are equal for equal time intervals, the working force is always directed to the same point. Thus Kepler's second law of the areas proportional to the time intervals proved that the planets are moved by a force directed towards, hence emanating from, the sun. For the case of a circular motion Newton showed that his method leads to the same formula for the centrifugal force as had been derived by Huygens.

Thereupon, Newton in Proposition XI derived in a general way the law of the centripetal force toward the sun from Kepler's first law that the orbit of a planet is an ellipse with the sun in a focus. By making use of the well-known geometrical properties of the ellipse, he found that the force was as the inverse square of the distance to the sun. Considering the fundamental importance of this demonstration for the history of astronomy, we have reproduced it in Appendix D (p. 500). The same rate of variation with distance—as shown above—was found by comparing two different planets (supposed, for simplicity's sake, to have circular orbits) and applying Kepler's third law. This meant that different planets at the same distance from the sun have the same acceleration and that hence the attraction exerted upon them by the sun was independent of their substance. These conclusions gave a new significance to Kepler's laws; simple empirical regularities before, they now acquired unassailable certainty as consequences of a universal law of attraction. The attempts made in the seventeenth century to find other orbits or laws of motion for the planets now lost all sense.

The mathematical propositions found their application in the third Book. From observations of the Jupiter satellites it had been derived that Kepler's laws also held for them; hence the forces that kept them in their orbits were directed to the centre of Jupiter and were inversely as the squares of the distances from that centre. The same held for the satellites of Saturn. The planets were attracted in the same way by the sun, and the moon by the earth. The acceleration of falling bodies on the surface of the earth was computed from the orbital motion of the moon to be $15\frac{1}{2}$ Paris feet, 'or, more accurately, 15 feet 1 inch $1\frac{4}{9}$ line'; whereas the same acceleration derived by Huygens from the length of a pendulum oscillating seconds amounts to 15 feet 1 inch $1\frac{7}{9}$ line. 'And therefore the force by which the moon is retained in its orbit becomes, at the very surface of the earth, equal to the force of gravity which we observe in heavy bodies there. And therefore (by Rules 1 and 2) the force by which the moon is retained in its orbit is that very same force which we commonly call gravity; for, were gravity another force different from that, then bodies descending to the earth with the joint impulse of both forces would fall with a double velocity.'[140] By Rules 1 and 2 he means the first of the 'Rules of Reasoning in Philosophy' (*Regulae philosophandi*) at the start of Book III: (1) We are to admit no more causes of natural things than such as are both true and sufficient to explain their appearances; (2) Therefore to the same natural effects we must, as far as possible, assign the same causes. These rules in modern times may look superfluous and artificial; but, in a century in which so many fantasies were offered as science, this admonition of intellectual discipline was not superfluous. And he concluded: 'The force which retains the celestial bodies in their orbits has been hitherto called centripetal force; but it being now made plain that it can be no other than a gravitating force, we shall hereafter call it gravity. For the cause of that centripetal force which retains the moon in its orbit will extend itself to all the planets.'[141]

The moons of Jupiter gravitate towards Jupiter, the planets towards the sun. There is a power of gravity tending to all the planets; Jupiter also gravitates towards its satellites, the earth towards the moon; all the planets gravitate towards one another. All bodies are mutually attracted by a force between them that moves the greater body a little, the small body much. The weights of bodies towards different planets, hence the quantities of matter in the several planets, can be computed from the distances and periodic times of bodies revolving about them; they are found, if one is put for the sun, to be $\frac{1}{1067}$ for Jupiter, $\frac{1}{3021}$ for Saturn, $\frac{1}{169282}$ for the earth. The force exerted by a celestial body is composed of the attractions of its parts, i.e. of the smallest particles of matter. This universal gravity or attraction, afterwards called 'gravitation', is a general property of all matter; all particles attract one another in accordance with Newton's law of the inverse squares of their distances. Newton demonstrated that the total attraction of a spherical body is exactly the same as though all its mass were concentrated in the centre; Kepler's laws can hold for the planets because they, as well as the sun, are spherical bodies.

The theory of gravitation was not only a more universal formula than the empirical laws from which it had been derived, for it gave in addition explanations for a number of other phenomena. Newton demonstrated that, besides elliptic orbits, parabolic and hyperbolic orbits also led to the same law of attraction, so that by this law each of these conic sections was a possible orbit, with the sun always in the focus. This result could at once be applied to the comets; their mysterious sudden appearance and disappearance were in exact accord with the infinite branches of a parabola or a hyperbola. Kepler had supposed that comets ran through space and passed the sun along straight lines. Cassini had tried, without result, to represent the observations by oblique circular orbits. Borelli, however, in 1664 suspected that the orbits were parabolas. In 1680 a great comet appeared which came close to the sun and, having rapidly made a half-turn around it, went away in the same direction whence it had come. Dörffel, a minister at Plauen in Saxony, explained its course by means of a narrow parabola with a small focal distance.

Newton gave a theoretical basis to these suspicions by stating that the orbits of the comets must be conic sections; he assumed them to be widely extended ellipses of large eccentricity, which at their tops were so nearly parabolas that parabolas could be substituted for them. He indicated a method of deriving the true orbit in space from the observed course between the stars, and he applied it to the comet of 1680. By this method Halley computed parabolic orbits of 24 comets of which two had appeared in 1337, and in 1472, and the others in the sixteenth and seventeenth centuries. In his publication of the results in 1705, he drew attention to the fact that three among them—the comets that had appeared in 1531, in 1607, and in 1682—had nearly identical orbits in space. Since both intervals were 76 or 75 years, he concluded that they were three successive appearances of the same comet, which in a good 75 years describes a strongly elongated ellipse about the sun. In a memoir of 1716 he returned to the question and pointed to comets that had been seen in the years 1456 and 1378 as possible appearances of the same body, and predicted its next return in 1758.

In his *Principia* Newton also derived the oblateness of a rotating sphere, especially of the earth. In this he had been preceded by Huygens, who, though his first computations in his diary were much earlier, about 1683 wrote a supplement to his discourse on the cause of gravity, sent it to the secretary of the Paris Academy in 1687, and himself in 1690 published both the discourse and the supplement, together with his treatise on light. In this supplement he put forward that, in consequence of the earth's rotation, a plumb line is not directed towards the centre of the earth, but is (in medium latitudes) by $\frac{1}{10}° = 6'$ inclined to the south. 'This deviation is contrary to what has always been supposed to be a very certain truth, namely, that the cord stretched by the plumb is directed straight toward the centre of the earth. . . . Therefore, looking northward, should not the level line visibly descend below the horizon? This, however, has never been perceived and surely does not take place. And the reason for this, which is another paradox, is that the earth is not a sphere at all but is flattened at the two poles, nearly as an ellipse turning about its smaller axis would produce. This is due to the daily motion of the earth and is a necessary consequence of the deviation of the plumb line mentioned above. Because bodies by their weight descend parallel to the direction of this line, the surface of a fluid must put itself perpendicular to the plumb line, since else it would stream farther downward.'[142]

In an Addendum written in 1690 Huygens computed an oblateness of $\frac{1}{578}$; this value was based on the assumption that gravity as proceeding from the vortices as its cause was constant throughout the body of the earth. Newton, however, now had a better theory; proceeding from gravity as the result of the attraction of all the separate particles, he found it to decrease regularly from the surface to the centre, where it vanishes. So he derived the ratio of the polar axis to the equatorial diameter to be 229 : 230, i.e. an oblateness of $\frac{1}{230}$. These theoretical derivations were strongly opposed by the French astronomers, who put their trust in their geodetical measurements. Careful determinations of the length of one degree of the meridian to the south of Paris had given a somewhat larger value (57,098 toises) than had been derived from the

arc of Paris–Dunkirk (56,970 toises). Cassini and his colleagues concluded that the degrees became smaller when going north and that
hence the earth must be elongated at the poles. This contradiction
between theory and practice made the French astronomers sceptical
toward the theory as a whole.

Newton dealt with other astronomical phenomena that now found
their explanation in the theory of gravitation. First he pointed out that
the attraction of the moon by the sun worked as a disturbing influence
upon the moon's orbit and was the cause of the irregularities in the
moon's course discovered by Ptolemy and by Tycho Brahe. He gave a
first theoretical computation of the regression of the lunar nodes and
found that it is strongest in the quarter-moons and zero at full and new
moon. Then he showed how the tides are caused by the different ways
in which the solid earth and the movable oceanic waters are attracted by
the moon and the sun. The precession, that regular slow increase in the
longitudes of the stars by a change in the position of the earth's axis of
rotation, could also be explained by the attraction of sun and moon
upon the flattened earth. By making a comparison with the nodes of
imagined moons revolving along the earth's equator, he could even
compute the right value 50″ per year (9.12″ by the sun, 40.88″ by the
moon).

That the planets by their mutual attraction must disturb their motion
he understood, of course, as a consequence of his theory: 'But the actions
of the planets one upon another are so very small, that they may be
neglected. . . . It is true that the action of Jupiter upon Saturn is not to
be neglected . . . the gravity of Saturn towards Jupiter will be to the
gravity of Saturn towards the sun as 1 to about 211. And hence
arises a perturbation of the orbit of Saturn in every conjunction of this
planet with Jupiter, so sensible, that astronomers are puzzled with it.'[143]
All the other mutual influences are so slight that he assumed the
aphelion and the nodes of the planets to be fixed, or at least nearly so,
and he even concluded: 'The fixed stars are immovable, seeing they
keep the same position to the aphelia and the nodes of the planets.'[144]

Newton, by his theory of universal gravitation, gave to the knowledge
of the motions of the heavenly bodies so solid a basis as never could have
been suspected. In this great scientific achievement the two formally
opposite principles of Bacon and Descartes are unified: he proceeded
from practical experience, from rules deduced from observations by
precise computation, and out of them constructed a general theoretical
principle which permitted him to derive all the separate phenomena.
And all these deductions were demonstrated by the most exact and
acute mathematics. No wonder that, after traditions had been vanquished and the difficulties of the new mode of thinking overcome, his

compatriots exalted him as an almost superhuman genius. Honours were bestowed upon him; from 1703 until 1727, the year of his death, he was President of the Royal Society. His appointment in 1696 as 'Warden of the Mint' (in 1699 promoted to 'Master of the Mint') was not simply a post of honour or a lucrative sinecure. He and his colleague, the philosopher John Locke, together with the ministers Somers and Montague, by their energetic measures in minting good silver coin, repaired Britain's deplorable monetary system, a necessary basis for the expansion of its commerce which ensued.[145]

Newton in his *Principia* did not restrict himself to an exposition of his new theory, which for us is the essential thing. For his contemporaries, criticism of the older dominant theory was equally needed. So the entire Book II is devoted to the motion of fluids and to the resistance which moving bodies experience in fluids: the first foundations for a scientific treatment of these phenomena. Here the vortices had to stand the test of science; the progress of half a century was the progress from vague philosophical talk to exact mathematical computation. The conclusion was, in Newton's words: 'Hence it is manifest that the planets are not carried round in corporeal vortices'; and in a verdict still more severe: 'so that the hypothesis of vortices is utterly irreconcilable with astronomical phenomena, and rather serves to perplex than explain the heavenly motions'.[146]

Notwithstanding this crushing criticism of the vortex theory, most Continental scientists remained sceptical towards the doctrine of gravitation. This appears most clearly in what Huygens wrote in 1690 in the above-mentioned Addendum to his Discourse on the cause of gravity. In comparing their different results on the oblateness of the earth, he said: 'I cannot agree with the Principle which he supposes in this computation and elsewhere, viz. that all the small particles, which we can imagine in two or many different bodies, attract and try to approach one another. This I cannot admit because I think I see clearly that the cause of such an attraction cannot be explained by any principle of mechanics or by the rules of motion.'[147] Of course; for in his opinion the weight of heavy bodies was caused by their being pressed down by the whirling ether outside and not through influences from inside the earth, so that the celestial bodies themselves did not act upon one another. He had nothing against Newton's centripetal force, by which the planets were heavy toward the sun, because he himself had shown that such gravity could be understood from mechanical causes. Long ago he also had imagined that the spherical figure of the sun, as well as that of the earth, could be explained by this gravity; but he had not extended its action as far as to the planets, 'because the vortices of Descartes that formerly appeared to me very probable and which occupied my mind, were opposed to it. Nor had I thought of that regular decrease of gravity, namely, as the inverse square of the distance; that is a new and very remarkable property of gravity, for which it would certainly be worth while to seek the reason.'[148]

Here it appears that what to Newton was a solution, to Huygens was a new problem. Perceiving now by Newton's demonstrations that this gravity counterbalanced the centrifugal forces of the planets and exactly produced their elliptical motion, Huygens had no doubt that Newton's hypotheses on gravity were true, as also Newton's system founded thereon. It must appear all the more probable, since it solved many difficulties that gave trouble in the vortices of Descartes; for instance, as to why the eccentricities and inclinations of the planetary orbits always remained constant and their planes passed through the sun, and their motions accelerated and retarded, as we observe, which could hardly happen if they were swimming in a vortex about the sun. And now we see also how the comets can traverse our system; it was difficult to conceive how they could have a movement opposite to the vortex which was strong enough to drag the planets along. But by Newton's theory this scruple has been removed, since nothing prevents the comets from travelling in widely extended ellipses about the sun.

'There is only this difficulty,' Huygens continued, 'that Newton . . . will have celestial space to contain only very rare matter, in order that the planets and comets meet with less impediment in their course. This rarity accepted, it seems to be impossible to explain the action either of gravity or of light, at least in the way I always used.'[149] In 1678 Huygens had already expounded, and in 1690 printed, a theory of light as a vibration, a wave motion propagating through the world ether, and in this way he had explained the phenomena of reflection and refraction. Newton had developed the entirely different theory—which was not published until 1704—that light consists in ejected particles passing through space with great velocity. Refraction of a ray obliquely falling upon a glass surface—which in Huygens's theory was due to slower propagation of the waves—in Newton's theory was easily explained by the consideration that the light corpuscles were bent toward the normal by the attraction of the denser glass matter. There was thus a profound difference in the supposed underlying world structure. For Newton, space was empty or nearly so; the light corpuscles, as well as the planets, run their course unimpeded, and gravity works through empty space from one body to another. Huygens could not agree with Newton's attraction because his theory of light required that space be filled with ether.

So in his 'Addendum' he returned to discussions of the nature, the fineness, and the tenuity of the whirling particles surrounding the earth. Newton, he said, argued to prove the extreme rarity of the ether in order that the motions due to gravity be not hampered by its resistance; but this substance, instead of hampering the motion, causes gravity. 'It would be different if we should suppose gravity to be an inherent quality of bodily matter. But I do not believe that this is what Newton accepts, because such an hypothesis would remove us far from the mathematical and mechanical principles.'[150] In the same trend of thought, Leibniz, after reading the *Principia*, wrote to Huygens (October 1690): 'I do not understand how he conceives gravity or attraction; it seems that to him it is only a certain immaterial and inexplicable virtue, whereas you explain it very plausibly through the laws of mechanics.'[151]

Here the profound basis of the controversy comes to light. Huygens admitted the exactness of Newton's computations and formulae; but they offered him no explanation. They gave no answer to the questions posed by him and his French colleagues: what is the origin of attraction? why is it that two bodies without any contact are driven toward one another? If space is filled with matter, this matter, by its contact, by pressure and attraction, transfers the motion; we see how streaming water and blowing wind drag objects with them; these are mechanical forces, easily understandable. An attraction from afar, over empty space, is entirely foreign to mechanical action.

Did Newton and his partners not see this difficulty? Certainly they did; but it did not worry them. Fundamentally, Newton, according to the general trends of thought at that time, agreed with Huygens. That he felt the same need of explanation as his contemporaries appears from a letter written in 1678 to Robert Boyle, the master of chemistry and discoverer of the law of the 'spring' of gases. Here he tried to give gravity a cause in the ether pervading all gross bodies and consisting of particles of different degrees of fineness; but his notions about things of this kind, he said, were so undigested that he was not well satisfied with them; 'you will easily discern whether in these conjectures there be any degree of probability.'[152] That he considered attraction at a distance no sufficient explanation may be seen from the letters he wrote (1692–93) to Richard Bentley, who was in correspondence with Newton on account of a series of lectures, in which he (Bentley) demonstrated the existence of God and refuted atheism by means of the law of gravitation. Newton, who was deeply occupied with theological questions and had often written on biblical subjects, in his first letter showed his agreement with this trend of thought: 'Why there is one body in our system qualified to give light and heat to all the rest, I know no reason, but because the author of the system thought it convenient. . . . To your second query I answer that the motions, which the planets now have, could not spring from any natural cause alone, but were impressed by an intelligent Agent. . . . To make this system, therefore, with all its motions, required a cause which understood, and compared together, the quantities of matter in the several bodies of the sun and planets and the gravitating powers resulting from thence . . . and to compare and adjust all these things together in so great a variety of bodies, argues that cause to be not blind and fortuitous, but very well skilled in mechanics and geometry' (letter of December 10, 1692).[153] And in his third letter of February 25th he wrote on the attraction: 'It is inconceivable that inanimate brute matter should, without the mediation of something else, which is not material, operate upon, and affect other matter without mutual contact; as it must do, if gravitation, in the sense of Epicurus, be essential and inherent to it. And this is one reason why I desired you would not ascribe innate gravity to me. That gravity should be innate, inherent and essential to matter, so that one body may act upon another at a distance through a vacuum, without the mediation of anything else, by and through which their action and force may be conveyed from one to another, is to me so great an absurdity, that I believe no man who has in philosophical matters a competent faculty of thinking can ever fall into it. Gravity must be caused by an agent acting constantly according to certain laws; but whether this agent be material or immaterial, I have left to the consideration of my readers.'[154]

In the second edition of his *Principia* (1713), in a 'General Scholium' added to the end of the third Book, to refute the criticisms that he had introduced occult qualities into natural philosophy, the same opinions were expressed in a more reticent way: 'Hitherto we have explained the phenomena of the heavens and of our sea by the power of gravity, but have not yet assigned the cause of this power. This is certain, that it must proceed from a cause that penetrates to the very centres of the sun and the planets, without suffering the least diminution of its force; that operates not according to the quantity of the surfaces of the particles upon which it acts (as mechanical causes used to do), but according to the quantity of the solid matter which they contain. . . . But hitherto I have not been able to discover the cause of those properties of gravity from phenomena, and I frame no hypotheses; for whatever is not deduced from the phenomena is to be called an hypothesis; and hypotheses, whether metaphysical or physical, whether of occult qualities or mechanical, have no place in experimental philosophy . . . and to us it is enough that gravity does really exist and act according to the laws which we have explained, and abundantly serves to account for all the motions of the celestial bodies, and of our sea.'[155]

That this was not the last word of his Natural Philosophy appears in the way he then continued: 'And now we might add something concerning a certain most subtle spirit which pervades and lies hid in all gross bodies; by the force and action of which spirit the particles of bodies attract one another at near distances' and electric bodies operate and light is emitted, 'and the members of animal bodies move at the command of the will, namely, by the vibrations of this spirit, mutually propagated along the solid filaments of the nerves. . . . But these are things that cannot be explained in few words, nor are we furnished with that sufficiency of experiments which is required to an accurate determination and demonstration of the laws by which this electric and elastic spirit operates.'[156] With these words the books of the *Principia Mathematica* close.

These sentences show that his mind was also capable of imaginative flights. But his theory remained entirely free from them. In his theory, only those relations appear which are demonstrable by exact mathematics; this is its essential characteristic. By means of the laws of gravitation, the phenomena can be derived and predicted by computation; this is the purpose of science. We meet here again with the contrast between the practical mind of the English and the theoretical mind of the Continental scientists. The latter racked their brains about the question concerning from what fundamental truths their theories followed. The former did not care and were content if they could work with the theories and derive practical results. Doubtless this was, as already pointed out, a consequence of the general mode of thinking of these peoples, rooted in their living conditions. The same personal liberty and daring energy which in the centuries that followed drove the English middle class towards commercial and industrial world power made the English scholars in their 'experimental philosophy' the pioneers of science.

Pioneers of scientific method, indeed. What in Newton's work presented itself as resignation, not asking for deeper causes but boldly applying it to further results, became the principle of modern science; a law of nature is not an explanation of the phenomena from established primary 'causes'.

CHAPTER 27

PRACTICAL ASTRONOMY

THE great extension of oceanic navigation in the seventeenth century brought the need of ever more perfect astronomical data. The harbours in the far continents, as well as the coasts and the islands, had to be explored, surveyed and mapped out by means of astronomical measurements. On the open sea also, for safe navigation, the sailors had to determine longitude and latitude. Finding the geographical latitude was no difficult problem; it could be done by measuring the greatest altitude reached by the sun or by a star in the meridian, provided, of course, that the declination of the celestial object was known. This meant that a good catalogue of stars with their declinations and good tables of the sun must be available. For the sun, the moon, and the planets there were Kepler's Rudolphine Tables, and for the northern stars Tycho's catalogue. But gradually greater demands were made, and the astronomers had to meet them. The governments now felt it their duty to foster astronomy in the interests of traffic.

As to the southern stars invisible in Europe, at first only the rough positions of a good 300 stars were available, as determined by two Dutch sailors, Pieter Dirksz. Keyzer, in 1595, at Bantam, and Frederik de Houtman in 1600, during his captivity in Achin. It was to remedy this defect that the British Government in 1676 sent young Halley out to the island of St Helena, where he succeeded in measuring accurately the positions of 350 southern stars. His results were published as a supplement to Tycho's catalogue, and Hevelius added them to his own catalogue. On this voyage and on some later voyages Halley made a number of meteorological and magnetic observations, constructed a map of the tides, and devised an explanation of the trade winds and monsoons as a consequence of the earth's rotation, which he published in 1686.

The finding of the terrestrial longitude was a far more difficult task; it remained a famous and embarrassing problem down to the twentieth century. The longitude of a point on earth is defined and found as the difference in time between that point and an adopted zero meridian. In the seventeenth and eighteenth centuries, Paris was taken as this zero point, to conform to France's economic and political rank and its nautical tables. Later on, when England became dominant on the seas, Greenwich Observatory came increasingly into use. Local time can be determined by astronomical measurements of solar or stellar altitudes. The difficulty is how to know on the high seas the standard time of the zero meridian. Galileo had proposed to use the eclipses of Jupiter's satellites, which are seen all over the earth at the same moment, and the States of Holland seriously negotiated with him in 1635. But the tables of their movement were too imperfect for an exact prediction of their phenomena. More promising was the use of the moon which every day progresses in its orbit by nearly 13.2° (i.e. $\frac{1}{27}$ of the circumference of the sky). An error of a few minutes in measuring the longitude of the moon produces an error 27 times larger in the deduced terrestrial longitude of the place of observation; but sailors were content, at the time, if they knew their position to a couple of degrees. Vespucci had already made use of the observed place of the moon to derive the difference in longitude between South America and Spain; Pigafetta, the companion of Magellan, had applied the same method, and in the sixteenth century Werner at Nuremberg and Gemma Frisius at Louvain had recommended it. The French Government of Richelieu in 1634 put up a prize for a good solution of this problem and appointed a committee for the purpose. Morin claimed the prize for his proposal to measure the distance from the moon to several stars and to build for this purpose an observatory to establish by observation the exact course of the moon. The prize, however, was denied him by the committee because the idea was not new and Morin was not able to make practical proposals for its execution. The method became practical only by the further development of astronomical practice in the following centuries.

Tycho had determined his differences in right ascension of the stars by computing them through trigonometrical formulae from distances measured in an east-west direction. So did Halley in St Helena. Earlier, Joost Bürgi at Cassel had tried to measure right ascensions directly by means of the moments the stars passed through the meridian; but this attempt was frustrated by the irregular running of the clocks. Huygens's invention of the pendulum clock solved the problem in 1656. Galileo had already perceived that a pendulum is isochronous, i.e. that (for small oscillations) the oscillation time was independent of the oscillation width. Hence it must be possible to measure durations by the number of equal oscillations of a pendulum, e.g. seconds. Many persons, first Galileo himself, and afterwards his son Vincenzo, tackled the problem of how to combine pendulum and clockwork. Huygens found a practical construction in which the pendulum regulates the turning of the cogwheels and a small impulse from the wheel at every oscillation keeps the pendulum swinging at constant width. The clock, built as an automatic counting apparatus, indicating the number of oscillations, became an accurate instrument for measuring time. This means that the clock became an astronomical instrument, an essential aid in all future astronomical measurements. The instrument of observation most fitted to be used in combination with it, constructed first in a primitive form by Römer at the Copenhagen Observatory, was the transit instrument, consisting of a telescope fixed at right angles on a horizontal axis directed exactly east-west. In revolving the axis, the line of vision of the telescope described the meridian. The moment of transit of a star, as seen through the ocular, across a vertical wire in the focus, indicates the right ascension. One second of time corresponds to as much as $15''$ in the sky; but by observing the transits over a number of parallel vertical wires, the error of the time of transit can be diminished to a small fraction of a second (plate 8).

The invention of accurate clocks provided a new means of deriving the longitude at sea; on a timepiece going with exactly the right speed (at least, since the last harbour) the time of the zero meridian could be read directly. A pendulum clock, of course, could not serve the purpose on a rolling vessel; here a portable timepiece regulated by a spring, called a 'chronometer', was needed. Hooke, together with Tompion, 'the father of English watchmaking', had first devised practical constructions, though they were not entirely successful. Then Huygens solved the problem, following the self-same principle as that of his pendulum clock, by introducing the spring balance as a regulator. If such a chronometer, carried by the vessel on a long voyage, should go one minute wrong, this would give an error of only $\frac{1}{4}°$ in longitude.

In the seventeenth century astronomy began to be a government affair. Formerly, in Tycho's time, princes had often endowed astronomical pursuits, which were personal hobbies of single individuals. In the next century, under royal absolutism, astronomy, besides being a personal scientific activity of a class of wealthy enlightened citizens, eager for knowledge, also took the form of state employment. The practical application of astronomy to the needs of navigation and geography induced the rulers to found observatories.

When Picard, in a dedication to the king, with which he introduced his *Ephemerides* in 1664, had pointed out that in France there was no instrument with which to determine latitude, the king in 1667 ordered an observatory to be built in Paris (pl. 8), later to become, at the same time, the seat of the newly-created Académie des Sciences, in which its sessions were held and experiments made. Picard was the leading expert on the practical astronomy then required, the determination of stellar positions. Yet the versatile Domenico Cassini, who had won fame by several discoveries, was called to Paris to be the first director and to enhance the glory of the prince by new discoveries; Picard became the chief observer. From 1679 onwards he published the *Connaissance des temps*, the first nautical almanac which has continued to our own day, and he felt the need to give it a firm basis of reliable observations. He proposed to have a quadrant made, 5 feet in radius, provided with a telescope, in a fixed position in the meridian to measure altitudes and times of transit. But the ostentatious edifice devoured so much money that the chief instrument was postponed and was not ready until 1683, one year after Picard's death. His successor, La Hire, thereafter used it in a regular observing practice. The observations served to correct the tables and to compute new ephemerides; but they were not collected into a new stellar catalogue, nor were they published, again through lack of money. Preserved in the archives, they could be consulted occasionally for special purposes thereafter.

In England the proposal by a French visitor at Court to have observations made for navigation purposes induced Charles II to ask for a report from the Royal Society. One of the members of the committee appointed was John Flamsteed (1646–1719), who, from his youth on, had occupied himself with astronomical observations. He it was who wrote the report in which was expounded the need of an observatory, where the positions of the celestial bodies could be determined regularly. The king then ordered, in 1675, that an observatory should be built on a hill at Greenwich, part of one of his country resorts. Flamsteed was appointed 'our astronomical observer' with a salary of £100 a year; 'Astronomer Royal' has since remained the title of the directors of Greenwich Observatory. His task was 'to apply himself with the most exact care and diligence to rectifying the tables of the motions of the heavens, and the places of the fixed stars, so as to find out the so much desired longitude of places for perfecting the art of navigation'.[157] There were no instruments. Flamsteed had to provide those himself. From his friend Jonas Moore he could borrow a sextant of 7 foot radius, and in the years 1676–88 he used it to measure distances between many stars. It was Tycho's old method which he used, with the only difference that his sextant was provided with two telescopes with cross-wires, one fixed, the other movable along the graduated arc, so that two observers were needed. Since he could not obtain from the government the money for a better instrument more suited to fundamental work, he constructed, at his own expense and with the aid of his ingenious assistant, Abraham Sharp, a mural quadrant of 7 foot radius to be used in the meridian. It was not really a quadrant but an arc 50° larger, to embrace the entire meridian from the southern horizon to the celestial north pole. The accuracy achieved with this instrument was due in large part to Sharp's skill, for he was expert at making accurate divisions on instruments.

From 1689 Flamsteed, with unremitting assiduity, notwithstanding his frail health, observed right ascensions and declinations of the stars, the sun, the moon and the planets. He was not content with accumulating observations; he reduced them to final results fit for publication. Not until 1725, after his death, however, did this *Historia coelestis Britannica* appear, a catalogue of 3,000 stars, exceeding all former catalogues in number and accuracy. Here the stars were arranged in each of the constellations according to right ascension (not according to longitude, as with Tycho); they were numbered, and these numbers afterwards were used as names of the stars. We speak of the star 61 Cygni, because it was No. 61 of the Swan in Flamsteed's catalogue. Based on this catalogue was an atlas of star maps, first published in 1729 and often reprinted afterwards.

Practical astronomy now became the regular business of specially appointed experts, often state officials, whose duty consisted in making astronomical observations. Though sensational discoveries might occur now and then, the main occupation was patient and devoted routine work. It was, however, a routine that continually renewed itself and struggled to attain greater accuracy through perpetual improvement of instruments and methods. This was the basis of the triumphal progress of astronomy in the following centuries.

The most needed and most promising work was the observation of the rapidly moving bodies—the moon and the planets. It appeared that the Rudolphine Tables did not give an exact course; small deviations showed everywhere. This could be met partly by improved numerical values for the elliptical elements. But did the planets obey Kepler's laws exactly? They did so sufficiently to confirm the truth of these laws; but there were differences. Halley in 1676 remarked that Saturn went more slowly and Jupiter more rapidly than was indicated by the tables; their periods of revolution had changed since Kepler's time. In view of all these deviations which were revealed by new accurate observations, the idea arose that the laws determined an average course only, analogous to the periodic fluctuation of temperature with the seasons, with irregular chance variations superimposed. Fortunately, Newton's law of universal gravitation appeared at the right time to establish that, by the attraction of the sun alone, Kepler's laws would be exactly followed but that, by the attraction proceeding from the other planets, deviations must occur, apparently capriciously, yet determined by natural causes.

The determining of the positions of the fixed stars, that had to serve as the basis of all study of the movements in space, brought to light new phenomena. Again it was Halley who, in 1718, comparing modern results for the latitudes of Aldebaran, Sirius and Arcturus with the data of Ptolemy, completed by those of Hipparchus and Timocharis, found that they now stood half a degree farther south than they should according to the ancient data. 'What shall we say then? It is scarcely credible that the Ancients could be deceived in so plain a matter, three Observers confirming each other. Again their stars being the most conspicuous in Heaven, are in all probability the nearest to the Earth; and if they have any particular Motion of their own, it is most likely to be perceived in them.'[158] So the so-called 'fixed stars' did not occupy a fixed position in the heavens; they had their proper motion in the celestial sphere, hence also in space. This unexpected result provided a new aspect of the world. Here was a new reason for observing the stars again and again, and more and more accurately.

CHAPTER 28

ASTRONOMERS ON THE MOVE

WITH the extension of navigation over the oceans, expeditions of astronomers now took place to solve special problems. Halley's expedition to the island of St Helena has been mentioned above. The journeys often had their origins in discussions in the newly-founded Academies. A strong initiative proceeded from the Paris Academy because its members, as *pensionnaires* of the king, could appeal to the Treasury for extra expenses.

In 1671 Academicien Richer was sent to the French colony of Cayenne to make observations of the planets and stars, with a special view to the solar parallax. But the expedition is better known for a secondary result than for its original purpose. Richer took with him a pendulum clock that had been well regulated in Paris. After his arrival in Cayenne it appeared that the clock was slow by two minutes per day, and its pendulum had to be shortened by $\frac{1}{380}$ of its length to keep pace with the earth's rotation. He soon understood that this was because gravity was diminished by the centrifugal force of the earth's rotation, which at the equator was stronger than in Paris; after his return to Paris the former length of the pendulum had to be restored. The diminution of gravity by centrifugal force could be computed exactly; it was $\frac{1}{580}$, considerably less than Richer had found. Huygens and Newton considered the difference as a proof of their theoretical result that the earth was flattened at the poles. We mentioned above that the French astronomers contested this opinion and, on the basis of geodetic measurements in France, believed the earth to be elongated at the poles.

When in the first part of the eighteenth century the theory of gravitation began to appeal to French scientists, they understood that the small difference between southern and northern France could not be decisive for the figure of the earth. The Paris Academy resolved to send out an expedition to measure a meridian arc near the equator. If the earth was flattened, the curvature of the meridian must be stronger, hence an arc of one degree must be shorter, the nearer one came to the equator. In 1735 Bouguer and La Condamine went to Peru (the northern part, now called Ecuador), and in the elongated plain of Quito, directed north-south, they measured an arc of three degrees. Their instructions were to measure, moreover, an east-west-directed arc of longitude; on a flattened earth it must be longer than the north-south degree of latitude, so that, theoretically, the problem could be solved by measurements in one region only. In this land of gigantic north-south mountain chains, the Andes, it was impossible, however, to measure an east-west arc; moreover, longitude differences could not be measured as exactly as latitude differences. Because of many difficulties and dissensions, the astronomers did not return until 1743. But the result they brought back was decisive; at Quito, Bouguer found 56,753 toises for one degree difference of latitude, distinctly smaller than the value of 57,057 found in France.

In the meantime, shortly after their departure from France, Maupertuis proposed to send out a second expedition to the far north, in order to make the evidence still more convincing. Thus in 1736 Maupertuis, accompanied by Clairaut and some other young scientists, went to Lapland. In the vicinity of Torneå, under many hardships from the excessive cold, they measured, partly alongside and partly on the frozen river, an arc of $0° 57'$. The result of these measurements, 57,438 toises for an arc of $1°$, by comparison with the result for France, also afforded a sufficient proof of the oblateness of the earth. Because Maupertuis returned directly to France in 1738, he could present himself as the man who had demonstrated the truth about the figure of the earth. The results, however, on closer consideration, showed a lack of numerical agreement; from the comparison Torneå-Paris, a flattening was derived of $\frac{1}{114}$; from Quito-Paris a far smaller flattening of $\frac{1}{279}$. Later results established that the latter value was nearly right; the observations in Peru had been made with great care and accuracy. In Lapland the difficulties had been far greater; after a severe winter, when the mercury in the thermometers was frozen and the metal instruments could hardly be touched, the observers went home as soon as possible. A remeasurement in the nineteenth century showed that the difference of latitude, found by Maupertuis, was too small.

Other expeditions in the eighteenth century dealt with the problem of the solar parallax. This is the fundamental quantity determining the distance of the earth from the sun and, hence, all distances and dimensions in the solar system. Tycho Brahe had used the traditional value of antiquity, $3'$. Kepler had derived, from Tycho's observations of Mars, that it could not be greater than $1'$. About 1630 Vendelinus made another attempt with Aristarchus's old method, this time by using a telescope to determine the exact moment that the lunar disc was bisected by the illumination boundary. He found that it took place when the moon stood at $\frac{1}{4}°$ less than $90°$ from the sun, an amount 12 times smaller than Aristarchus had given; hence the solar parallax was $\frac{1}{12}$ of Aristarchus's value, i.e. 15″. With the irregular boundary line of the illuminated lunar surface, greater accuracy could not be reached by this method.

Richer's expedition to Cayenne, mentioned above, was made especially to measure the parallax of Mars, which in the autumn of 1672 approached the earth at a distance of 0.37, so that its parallax was nearly three times that of the sun. For this purpose the declination of Mars and of the adjacent stars was measured, while Cassini at the same time made these measurements at Paris. The difference came near to the limit of accuracy then attainable. Cassini deduced that the parallax of Mars could not well be above 25″ and that the solar parallax could not exceed 10″; as a final result, $9\frac{1}{2}$″ was assumed. For the first time the solar parallax had been determined by direct measurements, though with the relatively considerable uncertainty of some few seconds, $\frac{1}{3}$ or $\frac{1}{4}$ of its amount.

The same method was once more used, later on, by the diligent Lacaille, who in 1751 went to the Cape of Good Hope and remained there for two years, observing a large number of southern stars. He also made observations of the solar and the lunar parallaxes, the former by using Mars in opposition, as well as Venus near the lower conjunction. For the lunar parallax he got an accurate value, 57′ 5″. For the solar parallax, however, the European observatories had let him down by neglecting to make corresponding observations, so that his results—10.2″ from Mars, 10.6″ from Venus—were of little value.

In the meantime however, a much more promising method had been disclosed. During his stay on St Helena, Halley had observed a transit of Mercury over the sun. With such a transit the solar disc was the background for the black disc of the planet; its positions relative to the sun, on account of its parallax, must be different for different places on earth. Halley had even attempted to derive a solar parallax from the observed moments of Mercury's ingress and egress. Of course, this was highly inaccurate, and the result, 45″, was entirely valueless. But it led him to the reflection that, if it had been Venus instead of Mercury,the conditions would have been far more favourable.

When Venus in lower conjunction is seen before the sun, its motion relative to the solar disc is so slow that it needs 7 hours to cover the solar diameter and wants 14 seconds of time for a progress of 1″. Its relative parallax, 22″, in medium conditions produces about the same value for the difference in the length of the chords described, which makes a difference of five minutes in the moments of ingress and egress. If the observer should make an error of, say, three seconds in such a moment, it would produce an error in the solar parallax of only $\frac{1}{100}$ of its amount. No wonder that Halley, when recommending this much superior method in 1691, used these words: that this will be the only kind of observation that in the next century with the highest precision will disclose the distance from the sun to the earth; what was tried in vain with different methods of parallax measurements. In this way the use of subtly graduated instruments would be entirely avoided.

In order to pass across the solar disc in its lower conjunction, Venus must be near to one of the nodes of its orbit. Since five periods of Venus are nearly eight years, there is a lower conjunction again eight years later at a longitude 2.4° different, i.e. at a latitude 8.5′ different, which, as seen from the earth, amounts to a difference of 22′ in apparent latitude. Since the solar diameter is 30′, two successive transits may be visible with an eight-year interval; thereafter the conjunction has moved in longitude too far away from the node. After more than a century a new set of two transits happens in the vicinity of the other node. Kepler

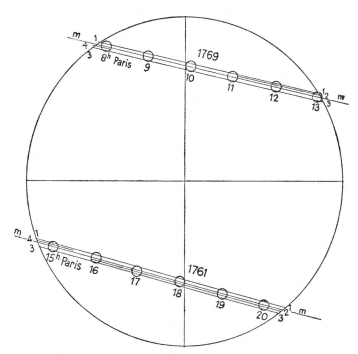

Fig. 29. Transit of Venus across the solar disc
m = track of the planet by its central point

1761: 1. Rodrigues 2. Paris   3. Tobolsk 4. Tahiti
1769: 1. Tahiti   2. Batavia   3. Vardö   4. Paris

had already alluded in vague terms to the special and remarkable character of these phenomena. Its first observation was made on December 4, 1639, by Jeremiah Horrox, who described it enthusiastically in a letter to his friend and collaborator, Crabtree. The next transits on June 6, 1761, and June 3, 1769, were those indicated by Halley, who again emphasized their importance in 1716; he pointed out the desirability of occupying as many observing stations as possible, not only far to the north and the south but also to the east and the west, where only the ingress or the egress would be visible.

When the time came near, his appeal met with a large response. A number of French and English astronomers journeyed to far-distant and little-known places. In 1761 the phenomenon could be seen in its entirety over Asia and the north polar regions; on the Australian islands the beginning only was visible; in western Europe and the Atlantic the end only. Pingré went to the Island of Rodrigues in the Indian Ocean; Chappe d'Auteroche to Tobolsk in Siberia; Maskelyne to St Helena; Mason and Dixon to the Cape of Good Hope; Father Hell from Vienna to Vardö in Norway near the North Cape; Le Gentil to India. And in still greater number astronomers set out in 1769, when the complete phenomenon was visible over the Pacific, western America, and, of course, the North Pole; the end in eastern Asia; and the beginning in eastern America and western Europe. Chappe went to California where he died from pestilence contracted in fulfilling his task. Pingré went to San Domingo, Wales to Hudson Bay, Captain Cook with some astronomers to Tahiti; several Russian observers spread over Siberia; Hell went again to Vardö to see Venus pass over the midnight sun, whereas many European and American astronomers, as well as Mohr at Batavia, observed the phenomenon at their home observatories.

The results did not come up to the high expectations. The course of the phenomenon was so slow that the first indentation of the sun's border by the black disc of Venus and the last disappearance, i.e. the moments of outer contact, could not be indicated to tens of seconds. This had been expected, but it was hoped that the second and third contact, i.e. the inner contacts of the large luminous disc and the small dark one, would be sharply observable. It was seen, however, that the small black disc remained like a drop connected with the sun's border by a black thread; and when it broke, Venus at once stood far inside the border. It was understood that this was due to some kind of irradiation or diffraction. How was one to know, then, what had been the right moment of inner contact? Observers standing beside one another differed by tens of seconds in their estimates, especially when they used different telescopes. Bewildered by the unexpected appearance, they had to be content with noting any time between the different moments.

Father Hell was even suspected, probably unjustly, of having afterwards changed and doctored his noted results.

The results for the solar parallax, deduced by different computers from different combinations of observations, consequently diverged far more than Halley had optimistically expected. Moreover the geographical longitude of a number of stations was badly known and had to be derived by the observers themselves by means of Jupiter's satellites or, in 1769, from a partial eclipse of the sun. Thus all sorts of values between 8.55″ and 8.88″ were computed and published for the solar parallax. This, however, meant enormous progress in our knowledge. Instead of by many seconds, the separate results differed by some tenths of a second only, whereas formerly the solar parallax and the distance of the sun were uncertain to $\frac{1}{3}$ or $\frac{1}{4}$ of their value. So it may be said that the transits of Venus in the eighteenth century answered their purpose completely.

Astronomical expeditions in the eighteenth century cannot be judged by modern travel conditions. They were far more difficult, harsher and more exacting, hence more adventurous also. Little was known of those distant regions and still less of their natural conditions. It has been remarked that Maupertuis going to Scandinavia would have met with less hardship if, instead of the icy cold of the Gulf of Bothnia, he had chosen the mild climate of the North Cape, situated even farther north, but washed by the Gulf Stream. Political events such as naval wars hampered free travelling. Because the vessel which carried Le Gentil to India had had to make wide detours to escape English men-of-war, the astronomer arrived in Pondicherry after the Venus transit of 1761 was over. In order to avoid another such incident, he remained in India and thereabouts until 1769, doing varied useful work; but the second transit of Venus was rendered invisible by clouds. As a sign of the times, it may be mentioned that the French Government instructed all its men-of-war to leave the ships of Captain Cook unmolested, because they were out on enterprises that were of service to all mankind.

Since the travelling astronomers were usually people well versed and interested in various departments of science they could report on many other scientific phenomena. Thus Bouguer for the first time discovered in the Cordilleras of Quito the lateral attraction of large mountains on the plumb line, a phenomenon that afterwards, when applied by Maskelyne to a Scottish mountain, made possible the first determination of the mass and the mean density of the earth. La Condamine in 1738 gave the first description of the Cinchona tree in Peru, the source of the medicine quinine already known in Europe. Many of these astronomers published diaries and books on their travels and adventures, which found a large audience, just as did the more purely geographical works of discoverers like Cook and De Bougainville. Among the rising bourgeoisie in France, England and elsewhere there was increasing interest in all aspects of science and all knowledge of nature in foreign regions. The astronomical travels, adventures on the earth, and adventures in the heavens satisfied a considerable part of this curiosity.

James Bradley (p. 289)

7. Isaac Newton (p. 262)

8. *Top left*: Römer's meridian instrument (p. 278)
Right: Bird's Quadrant (p. 292)
Left: The Paris Observatory (p. 278)

CHAPTER 29

REFINED PRACTICE

THERE was still something wrong with astronomical measurements. Accuracy surely had gradually increased, and the causes of error were more carefully investigated. Atmospheric refraction was a most annoying source of error. Cassini had derived a table of refractions from his observations; Newton had pointed out that the refraction depended on the temperature and the pressure of the air as indicated by thermometer and barometer. Several astronomers and mathematicians sought to improve theoretically and practically our knowledge of its dependence on the altitude of the star. Yet, even when corrected for refraction, the declinations of the stars, determined at Paris or at Greenwich, sometimes showed deviations of more than 10″—just as Picard had formerly perceived. Flamsteed thought that these differences, which seemed to depend on the season, were due to a parallax of the stars; but his colleagues pointed out that the variation in parallax with the season would have been different. By a yearly parallax a star is displaced towards the point of the ecliptic occupied by the sun; the actual deviations had a different character. Hooke had repeatedly measured the zenith distance of a star culminating near the zenith, so that refraction could play no part, and he found variations which he thought might be attributed to a yearly parallax.

In order to verify these results, Molyneux in 1725 had installed on his estate at Kew, near London, an instrument especially adapted to exact measuring. It consisted of a zenith sector comprising a few degrees only of a circle of large radius, 24 feet, erected in the meridian, so that the telescope attached could be pointed at a selected star (γ Draconis, the same as that used by Hooke) when it culminated near the zenith. Thus variations in its declination could be measured very accurately; the measurements were later found to have no greater error than 2″. Soon after the work was started, Molyneux stopped regular observing because he was appointed to the Admiralty; but his work was continued by his younger friend, James Bradley (1692–1762), since 1721 a professor at Oxford, who had taken part in all the preparations. When they started observations of the star in December 1725 they saw to their surprise that, although they had expected it to have reached its southernmost position, it continued to move southward until in March of the next year it had arrived 20″ farther in that direction. Then it returned, until in September it had moved 40″ northward, again to return south, reaching in December the same declination as a year earlier. It was an oscillating movement with the period of a year. But it could not be a parallax of the star, because in that case it would have been farthest south in December. To extend the observations to other stars, Bradley had another sector constructed of smaller radius (12 feet) and greater angular breadth (6° to either side of the zenith). He found that all the other stars showed the same periodical change, the extent being smaller according as they stood nearer to the ecliptic. In 1728 Bradley succeeded in finding the explanation in an apparent 'aberration' of the light rays: because the telescope is carried on by the earth in its orbital motion, whilst the telescope is traversed by the light rays with a 10,000 times greater velocity, it must be kept inclined by $\frac{1}{10000}$ towards the direction of the earth's movement. The discovery of this aberration was the first experimental proof that the earth has a yearly motion and that Copernicus was right.

This was not all. When Bradley continued his observations of γ Draconis during the following years, he perceived that there was a second oscillation: alternately for nine years its declination increased and decreased over a range of 18″. This variation, also occurring with all the other observed stars, was confirmed by information from Lemonnier in Paris. Bradley explained it in 1748 by a 'nutation' of the earth's axis, a small conical movement superimposed as a kind of small perturbation upon the large, slow, conical movement known as the precession. Its period of 18 years, which is the period of revolution of the nodes of the lunar orbit, in which its inclination to the equator oscillates between 18° and 28°, indicated the attraction of the moon upon the flattened earth as the cause of the phenomenon.

'The accuracy of modern astronomy,' said Delambre, 'owes much to these two discoveries by Bradley. This double service secures for him the highest place after Hipparchus and Kepler, and raises him above the greatest astronomers of all times and countries.'[159] Though this praise may sound too lavish, the first sentence is certainly true. So long as the position of the stars could deviate up to 30″ through these causes, it was a hopeless task to assure an accuracy down to less than 10″ by measurements. Astronomers could expect concord and results for the positions of the stars only by correcting the measurements for these two influences, aberration and nutation. Since the differences which now remained were due to unavoidable errors of observation it was worth while to reduce these errors by improving the instruments and methods of observation.

Instruments did indeed improve, chiefly because specialists with expert skill and knowledge began to construct them. Flamsteed had been obliged to make them himself with the aid of Sharp. Greater perfection was reached by George Graham, who had constructed the zenith sectors used by Molyneux and Bradley in discovering aberration and nutation. He used an ingenious method for engraving exact graduations upon his instruments, and he applied a screw for the exact reading of the smallest divisions. When in 1720 Halley succeeded Flamsteed as director of Greenwich Observatory, he found no instruments there; the instruments which Flamsteed had used were his own property and the heirs had removed them all. Halley could now order new instruments at the public expense, and all were provided by Graham. Bird, Graham's younger associate, who became his successor, afterwards made excellent new instruments in his shop not only for Greenwich but for a number of European observatories, so that the new standard of accuracy spread over the entire astronomical world. In 1767 he was given £500 by the Government for publishing in a book, for universal use, his accurate method for precisely graduating circles.

The making of precision instruments developed into a subtle handicraft by expert artisans because, chiefly for navigational needs, a market had developed with a regular demand. To determine the position of a ship on the high seas by measuring solar and stellar altitudes and moon distances, accurate instruments were needed that could be held in the hand on a rolling ship. In medieval times and afterwards people had managed with the cross-staff; but the most skilled pilot could not attain an accuracy greater than several minutes of arc. Newton had devised, about 1700, an instrument through which, by means of mirrors, the rays from two objects were brought into one telescope, each passing through half the objective. It was, however, not published and had not become known. The invention in 1730 by Thomas Godfrey, of Philadelphia,[160] seems to have remained unknown in Europe. In 1731 John Hadley published the description of a 'reflecting quadrant', later for practical reasons reduced to a reflecting sextant, in which the images of two celestial bodies (or of the horizon) were brought by reflection into exact coincidence as seen in the telescope. This sextant became indispensable to every pilot. Its regular manufacture to meet the needs of Britain's rapidly extending navigation of the seas raised instrument-making to a highly skilled and flourishing trade, and astronomy profited from that skill.

Because England in the eighteenth century took increasing precedence in commerce and navigation, the problem of longitude at sea became a matter of public importance. At Newton's suggestion (from 1701 he sat in Parliament as member for the University of Cambridge) the Government offered a high prize (£30,000) for anyone who should discover a reliable method of finding the longitude in the open sea with an accuracy of $\frac{1}{4}°$. Minor prizes were afterwards offered for partial solutions of the problem.

One of the most promising methods was the improvement of time-pieces, to keep the time of the zero meridian. Both the pendulum clock and the portable chronometer regulated by spring balances changed their speed with temperature, because of the expansion of the metal parts. Graham, mentioned above as an able instrument-maker, who had been Tompion's apprentice in clockmaking, in 1762 published a method of making the pendulum clock unsusceptible to variations of temperature by replacing the lens of the pendulum by a vessel filled with mercury. His younger colleague, Harrison, at about the same time constructed a gridiron pendulum which offered the same result by a combination of rods of different metals with different expansion coeffi-cients. The same principle was then applied by Harrison to the spring balance of the chronometers, and he succeeded in making their speed nearly insensitive to temperature. After his timepieces had been tested on ocean travels in 1761 and 1765 and their excellence had been proved, he received first £5,000 and later £10,000, on condition that he should publish a description of his construction methods. Not only in England, but also in France, in part independently, able artisans, among whom J. A. Lepaute and Ferdinand Berthoud were the most famous, con-structed ever better timepieces for nautical use.

The fact that the construction of instruments and clocks in England had now reached a level of fine precision had its consequences for technology as a whole. It was in the second half of the eighteenth cen-tury that the invention and continual improvement of spinning machines by Arkwright, Hargreaves, Crompton and others inaugurated the in-dustrial revolution in England. In the history of technology it is pointed out that many of these inventors and would-be inventors had acquired their mechanical skill as apprentices in the clockmaking trade.[161] From the clockmakers they obtained their ability to realize their ingeni-ous ideas in practical tools. Thus the needs and the practice of astronomy indirectly contributed to the rise of machine techniques in industry.

After Halley's death in 1742, Bradley was appointed to Greenwich Observatory. The observations he made with the instruments available did not satisfy him; so he ordered better ones to be made by Bird. They were a transit instrument with an 8 foot telescope, for the deter-mination of right ascensions, and a mural quadrant of 8 foot radius for the declinations (plate 8). With these instruments, from 1750 until his death in 1762, he made extensive series of observations of the stars, the sun, the moon, and the planets, which surpassed in accuracy all former work. Even more than to the precision of the readings, this accuracy was due to his carefulness in determining or eliminating the systematic errors proceeding from the instrument or from the condition of its surroundings. The small inclination of the horizontal axis of the transit instrument was regularly determined by a level; for the correct computation of the refraction, thermometer and barometer were regularly read. By such precautions he enabled later investigators to derive from the observations themselves the errors vitiating the results and to free the results from their influence. This was important because, whereas the positions of the moon were needed immediately, a careful reduction of the stars had to be postponed from lack of time.

After Bradley's death a conflict arose between his heirs and the British Admiralty as to the ownership of the 13 volumes containing the journals of observation and other manuscripts. After many years of proceedings before the courts, an agreement was reached that put all the papers into the hands of Oxford University for publication; so they were printed in full between 1798 and 1805. Then Bessel took the reduction in hand; when he published the results in 1818, he expressed in the title— *Fundamenta Astronomiae, ex observationibus viri incomparabilis James Bradley* —how Bradley's work, by its high quality, had become the basis of early nineteenth-century astronomy. In the catalogue of about 3,000 stars, in which the results are condensed, the uncertainty in both co-ordinates—right ascension and declination—is not more than some few seconds of arc.

Bradley established for Greenwich Observatory a standard of carefulness in working methods which was maintained by his successors Maskelyne and Pond. Though not equalled in his own time, his standards were nearly reached by such able and persistent observers as Tobias Mayer of Göttingen and Lacaille of Paris. In their catalogues of some hundreds of bright stars, the first improvements upon Flamsteed's, the star positions are found to have mean errors amounting at most to 4″ or 5″. Lacaille went with his instruments to the Cape of Good Hope in 1750, where he observed, besides the sun and the moon for parallax, nearly ten thousand southern stars. They were, of course, mostly telescopic stars, of the seventh and eighth magnitudes, so that now the stars of the southern sky were more completely observed and catalogued than those of the easily accessible Northern Hemisphere.

An important step forward towards even greater accuracy had been prepared at the same time by the improvement of the optical devices. To understand the development of the telescopes we have first to go back a century. In his optical studies between 1660 and 1670 Newton had discovered the different refrangibility of light of different colours.

Refraction of light through a prism is always accompanied by dispersion of the different colours in a spectrum. He perceived that this dispersion was the chief cause of the faulty and coloured images in the existing telescopes; since the different colours could not be converged into one focus, it was hopeless to try to obtain good images by improving the lenses. 'I saw that the perfection of telescopes was hitherto limited, not so much for want of glasses truly figured . . . as because that light itself is a heterogeneous mixture of differently refrangible rays' (letter from Newton to the Royal Society, February 6, 1672).[162] Thus Newton got the idea of using a concave mirror instead of a lens, because, with such a mirror, rays of all colours are strictly united in the same focus. Gregory had already devised a reflecting telescope, by which the rays reflected by the concave mirror, after a second reflection on a smaller concave mirror, reached the observer's eye through a circular hole in the centre of the larger mirror; but the optical artisans were not able to grind good mirrors for him. Newton himself set to work and, after having ground with great patience and perseverance a concave metal speculum, in 1671 he fitted it into a reflecting telescope that roused great interest when presented to the Royal Society. The image in the focus was reflected sideways by a flat mirror inclined 45° and viewed in a horizontal direction through the ocular at the side of the tube. It appeared that this small instrument, only 6 inches long, with 'something more than an inch'[163] aperture, allowed of a magnification of 40 times and showed the same details in the celestial objects as a refracting telescope of 3 or 4 feet in length.

The invention remained without practical use until about 1720, when James Short of Edinburgh succeeded in grinding concave mirrors in such a regular way that he could sell reflecting telescopes as an ordinary workshop product. Since, by the use of Gregory's construction, the observer looked straight in the direction of the object, they were easily handled. Since, moreover, because of their large opening, they gave bright images, they became the favourite instruments of amateurs and astronomers, far preferable to the larger refractors, with their pale and blurred images. It was not easy, however, to attach them to measuring instruments; so there the long narrow tubes with lenses remained in use; Bradley's 8 foot transit had an object glass of 1.6 inches only.

Newton thought that the removal of the colour dispersion was impossible; he assumed that the dispersion, though different for media of different density, was always proportional to the refraction. When, however, in the next century Euler had expressed doubts as to this proportionality, and the Swedish physicist Klingenstjerna by experiments had shown that it was incorrect, John Dollond in London, who was in correspondence with both, after much searching and experimenting, succeeded in 1757 in making a combination of glasses in which the colour dispersion was suppressed. Between two positive lenses of ordinary crown glass he placed a negative (concave) lens made of flint glass that possessed a stronger refraction and a far stronger dispersion. Thus an 'achromatic' system of lenses was realized that brought rays of different colours into one focus.

This invention was of enormous importance; it opened the way to an unlimited development of astronomical telescopes in the next century. Now that the colour dispersion was removed, the earlier work of Huygens could be continued in studying the most efficient figures of the lenses. John Dollond was a scientist and philosopher; his son Peter was a businessman. He founded a workshop, applied for a patent,* and soon his achromatic telescopes spread among astronomers and super-seded the old single-lens and also the reflecting telescopes. Their size was limited to an aperture of 3 or 4 inches, because larger flint discs could not be supplied by the glassworks. But this size, with a length of about 4 feet, was handy and usable; it gave sharp and bright images of the stars over a larger field than did the reflecting telescopes. On the Venus expeditions both kinds of telescopes—achromatic and reflecting —were used side by side.

Dollond's invention brought highly important improvements to the astronomical measuring instruments. The narrow tubes with small objective lenses that were connected with these instruments produced images of the stars too poor to be exactly pointed. With the achromatic objectives these images became fine bright points of light, allowing greater enlargement by stronger oculars, so that they could be bisected by the cross-wires with greater precision. Now the telescopes of a given length could acquire larger aperture, making faint telescopic stars easily visible and measurable. Until now, star catalogues, besides stars visible to the naked eye, contained the brightest classes only among the telescopic stars. Now it became possible to use the instruments for measuring the multitude of faint stars also.

The best instruments came from the workshop founded by Jesse Ramsden, the son-in-law of John Dollond, all of them, of course, provided with achromatic telescopes. Maskelyne in 1772 had provided Bradley's instruments at Greenwich with achromatic telescopes; since the transits of the sharp star images were observed by him at five wires and estimated in tenths of seconds (Bradley had noted them at full,

* It appeared that many years earlier, in 1733, an achromatic telescope had been constructed by Chester More Hall in Essex, but he had not made it known; the licence for making and selling them was accorded to Dollond, because, as the judge observed, 'it was not the person who locked his invention in his scrutoire that ought to profit from such invention, but he who brought it forth for the benefit of the public'. (R. Grant, *History of Physical Astronomy* [London, 1852], p. 533.)

half, or third seconds only) his right ascensions, though smaller in number, advanced considerably in accuracy. Many observatories outside England were equipped by Ramsden with English instruments. When G. Piazzi was commissioned to found an observatory at Palermo, he had a 5 foot vertical circle constructed by Ramsden, which could rotate on a 3 foot horizontal circle; both were provided with microscopes for reading. With this instrument, from 1792 to 1802, he determined with great care the positions of 6,748 stars, a number which was later increased to 7,646. Thus a higher norm of accuracy and completeness in stellar co-ordinates was established, with no greater uncertainty than a few seconds. That Bradley, by his treatment of instruments and conditions, had achieved even more did not become manifest until later through Bessel's reductions.

CHAPTER 30

REFINED THEORY

W ITH Newton's theory of universal gravitation, Descartes's vortex theory was essentially demolished. It is true that even in England the physics textbooks based entirely on Descartes remained in use; but the scholars at the universities, able mathematicians, studied Newton and taught the deductions contained in his *Principia*. To every new edition of the old textbooks further notes were added explaining Newton; so already in his lifetime his ideas gradually found general acceptance at the English universities.

This took place far more slowly on the Continent of Europe. In France, exhaustion resulting from war and the intellectual decline that set in under royal absolutism towards the end of the seventeenth century hampered scientific initiative. To begin with De Louville, an independent thinker, was the only one to come forward (about 1722) as an adherent of the new theory; for the other scholars the old doctrine was too powerful. But a change was on the way. The decline and defeat of French power by the rising English awakened a mood critical of the existing political and economic system. The increase in the social significance and power of the French middle class in the eighteenth century roused a spirit of resistance which regarded the free social conditions and political institutions of England as an example to be followed. So they were ready to accept also the spiritual basis of these conditions. One of the first spokesmen of these ideas of reform was Voltaire; in his 'Letters from London on the English' (1728–30), amongst such subjects as the Quakers, the English Church, Parliament, commerce, vaccination and literature, he also discussed English philosophy and science, especially Bacon and Newton. In his fourteenth letter on Descartes and Newton he wrote: 'A Frenchman coming to London finds matters considerably changed, in philosophy as in everything else. He left the world filled, he finds it here empty. In Paris you see the universe consisting of vortices of a subtle matter; in London nothing is seen of this. With us it is the pressure of the moon that causes the tides of the sea; with the English it is the sea that gravitates toward the moon. . . . Moreover, you may perceive that the sun, which in

France is not at all involved in the affair, here has to contribute by nearly one quarter. With your Cartesians everything takes place through pressure, which is not easily comprehensible; with Monsieur Newton it takes place through attraction, the cause of which is not better known either. In Paris you figure the earth as a melon; in London it is flattened on both sides.'[164]

Notwithstanding his lighthearted style, Voltaire made a good comparison of Newton's theory with that of Descartes. More amply still, in 1733 he informed his compatriots in a special work entitled *Elements of the Philosophy of Newton*, on the theory of light and on gravitation. By now the French world had become more susceptible to the new theory. Whereas formerly both doctrines found expression alternately in the transactions of the Paris Academy, after 1740 papers based on the vortex theory disappeared completely. You could do nothing with it, you could derive nothing from it, whereas from Newton's theory precise results could be derived merely by using mathematics. It posed a clear and great task: proceeding from the fundamental law of attraction of all particles of matter, hence of all celestial bodies upon one another, to compute their movements and to test them by observation. The theoretical development of astronomy in the eighteenth century was entirely dominated by gravitation.

Now the scientists came forward who were to continue Newton's work. Not in England, where, once liberty and self-rule had been won, self-satisfied prosperity quenched higher aspirations; all effort was directed toward practical affairs. But on the Continent, especially in France, spiritual life was roused by a strong desire for social renovation. England remained at the forefront in practical astronomy, as in all practice. But on the Continent the traditional rationalist mode of thinking, under these new impulses, developed into an outburst of theory and a profusion of mathematical treatments of the natural phenomena. A series of brilliant mathematicians sprang up; among them, as the most outstanding figures, the Bernoullis and Leonhard Euler from Basel, Clairaut and d'Alembert in France; their work was continued and completed by Lagrange and Laplace.

First the mathematical methods had to be remodelled. Newton had given all his demonstrations in a geometrical way, illustrative but requiring great ingenuity in the handling. The eighteenth-century mathematicians developed the algebraic method of analysis, in which the difficult geometrical insight was replaced by simple calculation, so that more difficult problems, otherwise intractable, could be solved. Newton himself had laid the basis by his theory of fluxions, a method of investigating changes of quantities by considering them in the limiting case of infinitely small variations. The same fundamental idea had been developed at the same time by Leibniz, but formulated in a different way, as infinitesimal calculus, by means of differential and integral formulae; in this form 'calculus' became the mathematicians' chief and most powerful tool in the centuries that ensued.

It was not only astronomical problems that occupied them. Newton had formulated the new principles of mechanics; his theory of motion established the general relations between forces, accelerations, distances and masses. The task of his successors was to apply these to all the different phenomena of moving bodies in nature. They developed new forms of the mechanical principles which were important for all kinds of motion, on earth and elsewhere. But astronomy received a large share of their exertions, first because of the difficulty of the problems posed, which was a stimulus to ingenuity, and secondly because the results of a fascinating theory in solving time-honoured problems could be verified by accurate observations. When bodies in space attract one another, the forces upon each of them—hence their accelerations—can be computed from their relative positions; by adding the consecutive accelerations, i.e. by integration, the velocities are obtained and from these, by a second integration, the changes in position. But the positions themselves resulting from this procedure are elements needed in the computation of the original accelerations. Thus the derivation of the course of these bodies was an intricate problem of analysis, designated as the solution of a system of differential equations.

For two bodies the solution was simple and was given by Newton. For three or more bodies no solution could be found. To the first eighteenth-century mathematicians, faced by the 'problem of three bodies', it proved to be unsolvable in a direct way; and thus it remained. Their disappointment can be heard in the complaint of one of the ablest among them, Alexis Claude Clairaut (1713–65), when he hit upon the problem: 'may now integrate who can . . . I have deduced the equations given here at the first moment, but I only applied few efforts to their solution, since they appeared to me little tractable. Perhaps they are more promising to others. I have given them up and have taken to using the method of approximations.'[165] And in the same way Leonhard Euler (1707–83), a mathematical genius, wrote in the Preface of his last great work on lunar theory, in 1772: 'As often as I have tried these forty years to derive the theory and motion of the moon from the principles of gravitation, there always arose so many difficulties that I am compelled to break off my work and latest researches. The problem reduces to three differential equations of the second degree, which not only cannot be integrated in any way but which also put the greatest difficulties in the way of the approximations with which we must here content ourselves; so that I do not see how, by means of theory alone, this research can be completed, nay, not even solely adapted to any useful purpose.'[166]

What these pioneers in the realm of celestial mechanics accepted as an unsatisfactory makeshift turned out to be the only way, however cumbersome and laborious, yet the most general method for solving such problems. First, the forces and accelerations are computed as they would be in the known unperturbed orbit; by integration, the deviations in position in first instance are derived. By these deviations in the position of the attracted planet, the forces and accelerations are changed by a small amount (small relative to their first values), and this gives rise to additional, still smaller, so-called 'second-order' deviations. Continuing in this way by further approximations, a final result is approached ever more closely. Since, in the first instance, the disturbing forces changed rather irregularly with the change in the relative positions of the bodies, they are split up into a number of periodical terms, depending in various ways on longitude, anomaly, nodes, and latitudes, of the disturbed as well as the disturbing body. Since in the higher orders all these terms react upon one another, their complete computation constituted from the beginning an entangled, nearly inextricable task, demanding years of work—in later times with higher standards, even an entire life of strenuous and careful work. Such work, then, was not a quiet, unimpassioned computing with ready-made formulae, as it may appear now in the textbooks. It was a forward-pressing search into the unknown country of theory, blazing a trail through the thicket now here, now there, full of adventure. Mostly the workers were driven along by actual problems, always obsessed by the question: Will it be possible to compute all the factual motions by means of Newton's theory? Is this law of attraction the precise universal law capable of explaining by itself all the phenomena observed? This question gave tension to the work of the eighteenth-century mathematicians.

The first practical problem was posed by the motion of Jupiter and Saturn. Kepler had already perceived, in 1625, that there was something wrong. Halley in his tables of 1695 had introduced a regular acceleration of Jupiter, a regular retardation of Saturn, of such an amount that after 1,000 years the planets would be displaced 0° 57' and 2° 19'. If this variation should always continue thus, with Jupiter's orbit steadily diminishing and Saturn's increasing, serious consequences for the planetary system could result. What was the cause? Could it be a result of mutual attractions, as Halley suspected, and would it be possible to compute the phenomenon by means of Newton's law? The Paris Academy of Sciences put this up as the prize problem for 1748 and again for 1752. For Euler's answer on the first occasion, though it was awarded the prize for the many important results on perturbations in general, did not solve the problem. Neither did his second memoir, in which he showed the possibility of 'secular' perturbations, always continuing in the same direction. Lagrange in 1763 published a quite new method of treating the problem of three bodies and applied it to the mutual action of Jupiter and Saturn; indeed, he found a secular term for both, but it was too small to be identified with the observed one. It was then that Pierre Simon de Laplace (1749–1827) entered the field with a careful investigation of all the smaller neglected terms of higher order in the mean motion of the two planets. He found that they cancelled out; the computed continuous acceleration and retardation turned out to be zero. Lagrange then extended this result by proving in 1776, in a general way, that the mutual attractions of the planets could not produce any secular progressive changes in the mean distances to the sun and the periods of revolution; they were subject to periodic variations only.

But what about the observed changes of Jupiter and Saturn? Some years later—in token of his embarrassment—Laplace wrote: 'After having recognized the constancy of the mean movements of the planets, I suspected that the changes observed in those of Jupiter and Saturn were due to the action of comets.'[167] Lambert opened a new track when he remarked in 1773 that the changes were different from what was assumed. A comparison of Hevelius's observations with the modern eighteenth-century results showed a retardation of Jupiter's course and an acceleration of Saturn's, just the reverse of what had been deduced from earlier observations. Hence the phenomenon was a periodical change. But a careful computation by Lagrange, including all terms containing the second power of the small eccentricities, did not reveal any term of the required amount. At last, in a memoir presented to the Academy in 1784, Laplace succeeded in solving the riddle, by the discovery that near commensurabilities of the motions produce large perturbations of long period. Five revolutions of Jupiter and two of Saturn are nearly equal, so that after 59 years (our well-known three conjunctions of ancient astrology) the two planets meet again at nearly the same place in the ecliptic. A couple of small terms of the third order, neglected because they contained the third power of the eccentricities, return after every 59 year period in the same way, and thus accumulate their effect into very perceptible changes of the longitude of the planets. This goes on until gradually the position of the conjunction shifts to other and opposite longitudes, and the effect is reversed after 450 years.

So, in reality, the perturbation is a long-period oscillation, with a period of 900 years, increasing to 49' for Saturn and to 21' for Jupiter. All the ancient and modern observations were now well represented by theory. The importance of this result was expressed by Laplace some years afterwards as follows: 'The irregularities of the two planets appeared formerly to be inexplicable by the law of universal gravitation—they now form one of its most striking proofs. Such has been the fate of this [Newton's] brilliant discovery, that each difficulty which has arisen has become for it a new subject of triumph, a circumstance which is the surest characteristic of the true system of nature.'[168] And in the next century Robert Grant wrote in his *History of Physical Astronomy*: 'By this capital discovery Laplace banished empiricism from the tables of Jupiter and Saturn, and extricated the Newtonian theory from one of its gravest perils.'[169]

Another famous object of computation was the comet whose return in 1758 had been predicted by Halley. It was now understood, as a result of Newton's theory, that the planets would disturb its motion by their attraction, so that it might return earlier or later than predicted. Between AD 1531 and 1607 the interval had been 76 years, between 1607 and 1682 somewhat less than 75 years; should the next period of revolution be 75 years, the comet could turn up in 1757. When the time came near, Clairaut set to work. It was not possible here to apply the methods used with the planets to find the perturbations for the entire orbit as a sum total of terms. It was necessary to follow the comet, computing its progress step by step as it was affected by the attraction of the planets over its entire course, the two preceding periods as well as the present one. Full of anxiety lest the comet could surprise him by coming early, he set to computing, aided by Mme Lepaute, a gifted mathematician, wife of the famous French clockmaker. They computed the perturbations by Jupiter and Saturn, struggling on day by day, hardly allowing themselves time for meals. 'The work I had entered upon was immense, and I was not able to state anything definite about the proposed object before the autumn of 1758.'[170] The comet, happily, did not surprise them, and he could reveal the reason, viz. that because of the attraction of Jupiter and Saturn, the comet would spend 618 days more on the last revolution than on the preceding one, so that its nearest approach to the sun was not to be expected before April 1759—with a margin of a month for different approximations. In a memoir to the Academy, read in November 1758, he said: 'I undertake here to show that this retardation, far from being injurious to the theory of universal attraction, is actually its necessary consequence, and that we can even go farther because I indicate at the same time its limits.'[171] The comet was first discovered at the end of 1758 by an amateur, Palitsch, living near Dresden in Saxony. It reached its perihelion in March 1759, and was visible until June. This prediction and computation of the return of Halley's comet was rightly considered a triumph of Newton's science.

The appearance of the comet was an episode. The motion of the moon, on the contrary, was the great and difficult problem which, as a touchstone of ingenuity, gave an impulse to numerous researches and new methods. The main irregularities in the moon's motion are, as Newton had described, perturbations due to the attraction of the sun. The perturbing force exerted by the sun is a rather large fraction ($\frac{1}{89}$ at new and full moon) of the attraction exerted by the earth. Since because the moon is so near small displacements are easily observed, the approximation in the computations must be extended right down to very small terms, which, by their mutual influence and dependence, engender a still more confusing multitude. The difficulty of the problem acted as a painful lesson to Clairaut at the very start of his researches in 1746, when he found for the progress of the moon's apogee only half the real amount, 20° per year instead of the observed 40°. Euler and d'Alembert found the same result. A universal formula in Newton's *Principia* had also afforded the same small amount; later on, among his unpublished manuscripts a computation was found giving the right value. Clairaut first supposed that Newton's law was not exactly true and had, for very short distances, to be completed by another small term with the inverse fourth power of the distance. But, in repeating his computation, the inclusion of a number of first neglected terms of higher order contributed so much that the first result was doubled; this also was confirmed by his colleagues. However, the true value could not be ascertained in this way with sufficient accuracy; though the motion of the lunar apogee was doubtless one of the most difficult lunar problems, in a certain way this held for all lunar perturbations. Clairaut and d'Alembert in 1754 published their lunar tables based entirely on theory and these, although they surpassed the tables of former times, did not exactly fit the moon's course. In 1745 and 1746 Euler had computed his first tables of perturbations; he gave an improved theory in 1755; and in 1772, for the third time, returned to the subject, computing further details. But agreement with observation remained unsatisfactory.

The reason was that theory could not determine the exact amount of the perturbations, though it was successful in indicating what perturbations must occur, with what periods, and how they depended on the sun, the nodes and the aphelia. For each period a large number of higher-order terms, forming an endless series, contributed to the result, so that their sum total remained uncertain. Yet the practical needs of navigation demanded tables of the moon exactly computed beforehand, and the theory of gravitation must be able to provide them.

Then it was that Tobias Mayer (1723–62) made a lucky hit by combining theory and practice. He had already acquired fame by his measurements, with primitive instruments, of the position of numerous lunar mountains, from which he derived the different librations of the moon. Then he was called to Göttingen in the kingdom of Hanover, where under the kings of England there was less narrow-mindedness than in the other petty states of Germany. He installed there an observatory and determined the positions, alluded to above, of the moon and the stars.

In 1755 he published solar and lunar tables. For the latter he took the most important perturbations from Euler's theory; but the amount of each he derived from practical data so as to render the observed positions of the moon as well as possible. By including no more than 14 terms, he secured the result that the errors in a few cases amounted to only 1½'. When the moon at sea is used as a celestial clock indicating Greenwich time, an error of 1' in the moon's position means an error of 27' in the geographical longitude, i.e. an uncertainty of, at most, 27 sea miles in the ship's position. After his death, Mayer's new tables were examined by Bradley, by order of the British Admiralty, and compared with the Greenwich observations. When they had been corrected in some points, it appeared that their errors always remained below, mostly far below, 1'. As an important aid in navigation, with instructions and methods prepared by Mayer himself, the tables were published by the Admiralty in 1770, and a grant of £3,000 was awarded to his widow by the British Government.

The problem for navigation practice was solved, but theory was still faced with a difficult and mysterious problem. When comparing the eclipses of antiquity and those of the Arabs with modern ones, Halley had in 1693 perceived that the moon's period of revolution, hence also its distance from the earth, had gradually diminished. This 'secular acceleration' of the moon was confirmed by Tobias Mayer; he first found for the amount per century 6.7″; afterwards, in the London tables he used 9″; still later Laplace derived 10″. This means that the moon had advanced 10″ after 100 years, 40″ after 200, 90″ after 300, relative to its position without this term, and that the path performed in a century ($100 \times 13\frac{1}{3} \times 360°$) increased by 20″ per century. If this diminution in the moon's orbit continued into the future, the moon would finally descend upon the earth. In 1770 the Paris Academy offered their prize for research as to whether the theory of gravitation could explain the phenomenon; Euler, in his prize treatise, could find no such explanation and wrote: 'It appears to be established, by indisputable evidence, that the secular inequality of the moon's motion cannot be produced by the forces of gravitation.'[172] In a second treatise in 1772 he supplemented this conclusion by supposing that the term probably arose from the resistance of an ethereal fluid which filled celestial space.

Such resistance could indeed explain the acceleration, but in such a way that it made a final catastrophe inevitable. After many fruitless attempts by Lagrange and Laplace, the latter at last, in 1787, succeeded in discovering the real cause. By the action of the planets upon the earth, the eccentricity of the earth's orbit was continually diminishing during some ten thousand years; because the orbit became more circular, the mean distance of the sun increased, and its perturbing effect decreased. By the attraction of the sun, the moon's orbit was enlarged; this enlargement now gradually diminished through the decrease of the sun's effect. Laplace found by a theoretical computation the same amount of the acceleration, 10″ per century, as was deduced from the eclipses. Thus again the uneasiness was removed, and the conviction that Newton's theory was capable of explaining all the movements in the solar system grew even stronger.

Laplace made a complete theoretical computation of the movement of the moon and the planets by means of mathematical developments. He collected all these researches in his great work, *Traité de mécanique céleste* ('Treatise on Celestial Mechanics'), which appeared in five volumes in 1799–1825. Here he treated all the motions in the solar system as a purely mathematical problem. He posed the problem in its most general form: each body in the world consists of small elements of mass attracting one another in accordance with Newton's law; the sum total of all these forces is the force exerted by the complete bodies, which in the case of a spheroidal figure deviates a small amount from a force proceeding from the centre. This general formulation made Laplace the spokesman, often attacked and criticized in the next century, of the mechanistic doctrine of the world. There is nothing in it of the atoms and molecules of later physics; Laplace introduced the small elements of mass only as a mathematical abstraction. Once given the initial condition of the system, i.e. the position and motion of every particle, the further course of the world was entirely determined and calculable, at least in theory. The solar system was taken here as an immense mechanism, steered and driven by the universal force of gravitation, running its calculable and predictable way into eternity. Often in these times it was spoken of as a gigantic timepiece that, once set in motion, goes on for ever—clocks then being the most perfect instruments made by human skill—in accordance with the general trend of thought that tried to understand the world as a most ingenious and perfect machine.

For the planets Laplace was able to derive the slow variations in the form and the position of the orbits, i.e. the secular perturbations of eccentricity, inclination, aphelion and node, for hundreds of thousands of years into the past and future. For the moon, he succeeded in representing its movement down to less than $\frac{1}{2}'$ by purely theoretical computation. Terms occurred in his theory of the moon—in variation of apogee and node—which were due to and depended on the flattening of the earth. Thus, conversely, he was able to derive this flattening from the values of those terms determined through observation; his result, $\frac{1}{305}$, was a welcome confirmation of the value derived from the expedition to Peru. Another term in the moon's motion, exceeding 2′, depends on the ratio of the distances of sun and moon to the earth and is called 'parallactic inequality', because it is thus connected with the solar parallax; from its empirical value Laplace derived a solar parallax of 8.6″, in close agreement with the results of the Venus expeditions. 'It is very remarkable,' he said, 'that an astronomer, without leaving his observatory, by merely comparing his observations with analysis, may be enabled to determine with accuracy the magnitude and flattening of the earth, and its distance from the sun and moon, elements the knowledge of which has been the fruit of long and troublesome voyages in both hemispheres. The agreement between the results of the two methods is one of the most striking proofs of universal gravitation.'[173]

This utterance is found in his *Exposition du système du monde* ('Exposition of the World System'), where in a popular, non-mathematical way the results of all the theoretical researches were explained. The outcome most important for the general reader was the assurance of the stability of the system. The secular perturbations of the major axes of the planetary orbits (as to the first power of the masses) are nil. Because, moreover, those of the inclinations and eccentricities remain within well-defined limits—a consequence of all planets revolving in the same direction—the inner structure will remain unchanged for ever.

However, it had not always existed in this form. The young German philosopher Immanuel Kant had already in 1755 developed the theory that all matter of the solar system had existed originally in the form of a widely extended nebulous mass. By its own inner forces, it had developed into a flat, rotating disc, which, by contracting, had produced the planets with their orbits. Newton had considered that for the existing universe to have originated in a purely mechanical way, without the intervention of a supreme intelligence, was incompatible with his world system. The progress of the eighteenth-century rationalism is now shown in that it ventured to tackle the problems of natural science. As a rational explanation of the fact that the planetary orbits are situated nearly in one plane and are run in the same direction, Kant's nebular theory was the first cosmogony with a scientific basis. Knowing nothing of Kant's work, Laplace propounded the same theory in his *Exposition*: that the solar system had originated by evolution from an entirely different original state. Now a final state of things had been reached, which could be disturbed no more. Cosmic world theory corresponded to the human world which also, after a long development from original barbarism and ignorance, had now reached—or as people then supposed: nearly reached—its final state under the rule of freedom, reason and science.

Thus at the close of the eighteenth century astronomy could look back with pride on its achievements. With an accuracy formerly unknown, it could observe the motions of the celestial bodies, and through the fundamental law of the universe which had now been discovered it could compute and predict them. It is true that this universe was the solar system only. But its limits began to be overstepped.

BOOK III

ASTRONOMY SURVEYING
THE UNIVERSE

CHAPTER 31

THE WORLD WIDENS

THROUGHOUT these centuries the fixed stars had roused interest only as a background for the motion of the planets. They were the fixed points for the determination of the changing positions of the moon and the planets. Some few details about themselves had been perceived now and then; thus a small change in position had been ascertained for some of them, so that they must have a proper motion. With some stars a periodical change in brightness had been detected, without giving great surprise, since the new stars of Tycho and Kepler had offered phenomena far more sensational. In 1596 David Fabricius perceived in the Whale a star of the third magnitude, which thereafter faded and disappeared; so he took it for another nova. But in 1638 Holwarda, of the Frisian University at Franeker, saw it again at the same place; he saw it disappear and then reappear and found that it alternately increased and decreased to invisibility in a period of eleven months. Tycho had observed the star, and Bayer had given it the Greek letter omicron; now it was named Mira Ceti, the 'miraculous one in the Whale'. Its fluctuations showed considerable irregularities; sometimes it attained the fourth magnitude only, sometimes the second, and once (in 1779) it shone as a first-magnitude star approaching Aldebaran. According to the Assyriologist Schaumberger, it probably had already been noticed in Babylon; some cuneiform inscriptions speak of the constellation Dilgan (i.e. the Whale and the Ram) in terms which meant to 'flare up' and 'extinguish'.[174] In the seventeenth and eighteenth centuries further discoveries followed; in 1672 Montanari at Bologna, who also in 1667 had perceived variations in Algol, discovered that low in the southern sky, in the constellation Hydra, a star fluctuated between the fourth magnitude and invisibility. In 1685 Kirch, in Berlin, found an analogous case in a fifth-magnitude star in the neck of the Swan, χ Cygni. Such discoveries show that there were observers who watched the stars attentively. But the interest was not yet sufficient to stimulate regular systematic observation.

Other objects belonging to this world of stars were the nebulous objects, called 'nebulae', of which the two most conspicuous had been perceived in the seventeenth century. The elongated nebula of Andromeda, easily visible to the naked eye, was first mentioned in 1612 by Simon Marius, and the nebula of Orion surrounding the fourth-magnitude star θ in the Sword, discovered in 1619 by Cysat at Ingolstadt, was in 1694 depicted by Huygens in his diary. The telescopes revealed many more and fainter ones. Because when they were first detected they were often thought to be comets and were announced as such, the French astronomer Messier, famous comet discoverer, compiled a list of over a hundred of these nebulae, which was published in 1771, to preclude false announcements of phantom comets. This was all that was known about the world of the stars. Their distance, too, was still unknown, since all attempts to measure their parallax had failed.

William Herschel (1738–1822), descendant of a German family of musicians (hence really christened Wilhelm Friedrich) from Hanover, had gone to England and at Bath had become a distinguished conductor and teacher as well as a composer. He had brought from Germany a keen interest in scientific and philosophical problems, which about 1773 turned into an increasing passion for astronomy. In that year he bought lenses for a telescope and rented a 2-foot reflecting telescope; but they did not satisfy him. He then began to make a telescope himself by casting and grinding a concave mirror. In his diary we find under September 1774: 'Attended 6, 7 or 8 scholars [i.e. music pupils] every day. At night I made astronomical observations with telescopes of my own construction.' Also on May 1, 1776: 'I observed Saturn with a new 7-foot reflector'; and on July 13, 1776: 'I viewed Saturn with a new 20-foot reflector I had erected in my garden.' By experimenting on the best metal mixture (copper with one-third tin) and grinding and polishing the surface into the right shape with the utmost care, he obtained mirrors of the most excellent quality, which produced perfectly round stellar images. On a direct comparison with the telescopes at Greenwich Observatory, they proved to be far superior. Next to this refined quality came, as a second factor, the increase in size, hence in brightness and resolving power; by using strong oculars, he could increase the enlargement from 200 and 460 times to the then unheard-of values of 2,000, 3,168 and 6,450 times. Assisted by his brother and by his sister Caroline, his devoted assistant first in his musical occupations and then in his astronomical work—in which she afterward achieved fame as a discoverer of comets—he made several 7 foot, 10 foot and 20 foot telescopes (the last-named with mirrors of 12 and 19 inches diameter) and employed them in intensive observation. His work represented considerable progress in astronomical technique, the result of skill and untiring devotion in striving to make the working apparatus as perfect as possible. Thus he had already become known in the astronomical world through his observations of the rotation of Mars and Jupiter and his micrometric measurements of the height of the lunar mountains, in the years 1777–81.

He set himself far-reaching goals—in the first place, to find the parallaxes of some stars. In a paper presented to the Royal Society, he expounded his ideas; since direct measurements of stellar positions had too large errors, he proposed to determine repeatedly the position of a bright star relative to a faint star closely adjacent to it. If the distance was some few seconds, the relative place and displacement could be estimated and expressed in diameters of the stellar discs—here is seen the importance of the regular, circular shape of these discs. Hence he had started to examine all the brighter stars attentively with his 7-foot telescope, to see whether they had faint companions nearby. Then, in 1781: 'On Tuesday the 13th of March between 10 and 11 in the evening, when I was examining the small stars in the neighbourhood of H [i.e. η] Geminorum, I perceived one that appeared visibly larger than the rest; being struck with its uncommon magnitude, I compared it to H Geminorum and the small star in the quartile between Auriga and Gemini, and, finding it so much larger than either of them, suspected it to be a comet.'[175]

With higher powers of magnification he saw the disc, increased in proportion to the power, measuring 3″ to 5″. This would not be the case with star images. During the following days it showed a slow direct motion along the ecliptic of nearly 1′ per day; it had no tail and showed a well-defined disc, which in the weeks which ensued increased in measurement, so that Herschel supposed that the object was drawing nearer to the earth. The discovery of this new and singular comet was immediately communicated to the Astronomer Royal, Maskelyne, and others, and it was also soon observed in France. When after several months an orbit was computed, it was found to be a circle 19 times larger than the earth's orbit. Hence it was a planet far outside Saturn, increasing by one the ancient time-honoured number of planets, enlarging the planetary system to double the former size. The fame of this discovery induced the King to award Herschel a salary of £200 to enable him to give up his remunerative musical trade and devote himself entirely to his passion for astronomy—though he had to add to it by grinding and selling mirrors for telescopes. It may be remarked here that among a list of 70 telescopes mentioned in 1795 as being sold to others, only one is known to have been used in valuable astronomical work: the 7-foot telescope bought by Amtmann Schroeter at Lilienthal. His gratitude to the King was shown by his giving the new planet the name of 'Georgium Sidus' (the star of George); in other countries, however, the name of Uranus came into use, and it has superseded the royal designation.

At Slough, near Windsor, where he next settled, Herschel started to make, with financial aid from the King, a still larger instrument, a 40-foot telescope with a mirror 58 inches in diameter. In 1789 it was finished, and Herschel described enthusiastically how for many hours he had observed Saturn better than before. In 1795 he gave a detailed description of the gigantic structure of heavy poles, erected on a foundation of masonry and wooden beams, in which the big tube hung and could be moved by a system of strong ropes and pulleys. It was admired and glorified as a wonder of science, and its picture was reproduced in books and magazines, and even on medals. However, it was seldom used by its author for observations. It seems that the handling of the clumsy colossus was rather laborious and that the images were not satisfactory, perhaps because the mirror was distorted by its own weight. All the important work of discovery in later years was achieved with the 20-foot telescope of 19-inch aperture (plate 9). It comprised much valuable work on bodies of the solar system, such as observations of sunspots and of the ring and belts of Saturn, as well as the discovery of the white polar caps of Mars, of two satellites of Uranus, and of two new satellites of Saturn. The main object of all Herschel's researches, however, was the world of the fixed stars.

First he completed his work on double stars and published his results in two catalogues, one of 269 objects in 1782 and another of 434 objects in 1784. For each of them the position of the faint star was given relative to the bright one; the angles of position and the larger distances were measured with the aid of a filar micrometer, and the small distances of a few seconds were estimated by comparison with the size of the stellar discs. Of course, they were not very exact, because of the coarseness of the filar micrometer. In fact, he said: 'The single threads of the silk-worm, with such lenses as I use, are so much magnified that their diameter is more than that of many of the stars'[176]; and he emphasized 'how difficult it is to have screws that shall be perfectly equal in every thread or revolution of each thread'.[177] Therefore, he constructed his 'lamp-micrometer'. Two artificial stars produced by lamps shining through pinholes, at a distance of about ten feet, were viewed with the left eye, while the right eye looked at the stars in the telescope; they could be brought into every relative position so as to appear exactly like the double star.

In the planning of his research on small companions to bright stars, Herschel proceeded from the idea that small stars looked small because they were distant. He assumed that the 'magnitude' of a star was thus a direct indication of its distance, that a fourth-magnitude star stands at a four-times greater distance than stars of the first magnitude. The faint companion could thus serve to find the parallax and distance of the bright star. However, the large number of cases in which small stars were seen in the closest vicinity to brighter stars far surpassed what was to be expected of faint stars through chance distribution. The idea that most of them must be real companions in space, with a small intrinsic brightness, must gradually have taken hold of him. The same thing had been observed by Chr. Mayer at Mannheim, who in 1777 published that he had seen planets belonging to bright stars; real planets illuminated by their sun, however, would be too faint to be visible. Many of Herschel's double stars, moreover, were so nearly equal that they could be due to chance distribution still less, and they certainly could not serve as objects for parallaxes. Surely he must soon have thought of them as real binary systems, for at the close of his first paper he added: 'it is much too soon to form any theories of small stars revolving round large ones.'[178]

Twenty years later he returned to the subject by measuring anew the relative positions of the components of a number of his double stars. In two papers, in 1803 and 1804, he described how, for about fifty of them, the position angle had changed by amounts between 5° and 51°. In a careful discussion he made sure that this change could not be caused by a motion of the sun or the proper motion of the chief star; the only admissible explanation was an orbital motion of the small star about the large one, or of both about their common centre of gravity. His discussion demonstrated—if such proof were needed—that Newton's law of mutual attraction also ruled the stars in the distant realms of space. The existence of another type of world system besides our single sun with its planets was here demonstrated: systems of two stars (some even of three or four stars) revolving about their common centre.

It was in the early eighties, too, that his attention was drawn to the problem of the sun's motion. In a paper in 1783 he said that, since we know that some stars are moving and that all stars certainly attract one another, we have to conclude that all stars are moving through space with the sun among them. A solar motion must reveal itself in an apparent opposite motion of the stars, which he called, because it depends on their distance, their 'systematical parallax'. He used the name 'apex' for the point of the sky whither the sun's motion is directed, and he pointed out that the stars situated sideways must show a maximum effect. From seven bright stars for which Maskelyne had given the yearly motion in right ascension and declination, then increased to 12 taken from Lalande, and afterwards augmented with 40 additional stars from Tobias Mayer, he deduced that the apex must be situated near the star λ Herculis. He even offered 'a few distant hints' concerning 'the amount' of the solar motion: the parallax of Sirius and Arcturus must be less than 1″; the apparent motion of Arcturus due to the translation of the solar system was no less than 2.7″ per year. 'Hence we may in a general way estimate that the solar motion can certainly not be less than that which the earth has in her annual orbit.' In 1805 and 1806 he returned to the subject, confirming his former result by making use of the accurate proper motions of 36 bright stars taken from Maskelyne. His attempt to determine the amount of solar motion could have no success, since he assumed that the distance of every star corresponded to its apparent brightness; consequently, he found a great number of bright stars (such as those which have very small proper motions) running alongside the sun in the same direction.

Perceiving with his telescopes the wealth of different objects in the heavens, he conceived the plan of collecting and cataloguing them, so as to make an inventory of the universe. Besides double, triple and multiple stars, and well-known groups like the Pleiades, he found numerous instances of what looked like nebulous spots in smaller telescopes, but which in his large telescopes appeared to consist of thousands of stars. Moreover, his telescope showed him numerous smaller nebulae; were they in reality also clusters of still smaller stars? In systematic 'sweeps' of the telescope over successive belts of the sky, he had assembled since 1783 all the curious objects he had come across. So he was able to publish in 1786 a 'Catalogue of one thousand new nebulae and clusters of stars', arranged in different groups according to their appearance and accompanied by short descriptions. In 1789 it was followed by a second catalogue of more than a thousand objects, and in 1802 a third list of 500 was added. There was no longer any confusion with comets; the nebulae had been advanced to celestial objects in their own right, as systems of suns or as worlds in the larger universe.

How large is this universe? What is its structure? Does it consist of an immense number of solar systems similar to our own? In 1750 Thomas Wright had already considered that what we see as the Milky Way encircling the sky was the projection of a gigantic system extending farthest in the plane of the luminous belt. Kant had adhered to that view. In 1761 Lambert elaborated a theory according to which the thousands of stars surrounding the sun constituted a system, and the Milky Way, because it consisted of a large number of such systems, was a system of higher order. There might be an even larger number of Milky Way systems in space, forming a still higher system. All these opinions, however reasonable they might appear, were mere speculation and fancy. Herschel was the first who attempted to determine the extent of the stellar system by systematic observations.

In the first of his two papers, 'On the Construction of the Heavens', published in 1784 and 1785, he states: 'On applying the telescope to a part of the *via lactea* I found that it completely resolved the whole whitish appearance into small stars.'[179] Hence the study of the Milky Way could restrict itself to counting the stars. Assuming the stars to be equally luminous and equally scattered in space, i.e. mainly standing at equal distances from one another, he could deduce from the number of stars counted in the field of his telescope how far the system of stars extended in that direction, i.e. the depth of the starry universe. So he called this counting 'Gaging (gauging) the Heavens' or, in short, 'star-gauges'. More than 3,000 of such counts were made, condensed in his second paper into values for a good 400 points of the sky. The result was the well-known disc—or lens-shaped star agglomeration, extending in the plane of the Milky Way over 800 times, perpendicular to this plane 150 times the mean distance of two stars, which was supposed to be the distance from Sirius or Arcturus to the sun. An often-reproduced figure of a perpendicular section of the system was added, in which the division of the Milky Way into two branches (about half its circumference) was also shown (pl. 9). 'That the Milky Way is a most extensive stratum of stars of various sizes admits no longer of the least doubt; and that our sun is actually one of the heavenly bodies belonging to it is as evident. I have now viewed and gauged this shining zone in almost every direction and find it composed of stars whose number, by the account of these gauges, constantly increases and decreases in proportion to its apparent brightness to the naked eye.'[180]

Herschel did not restrict himself in these papers to the simple determination of distances and dimensions. In his telescope he had seen so many different forms of stellar agglomerations in gradual diversity that his thoughts were naturally occupied by the question of what was their origin. He explained that, by their mutual attraction, stars that formerly were evenly spread must concentrate into regular or irregular condensations, with voids left between them. He called it the 'formation of nebulae', indicating that what is seen as a nebula in the telescope consists of a cluster of very small and distant stars. In the irregularities and cloud forms of the Milky Way this clustering tendency was clearly visible. Extensive nebulae he called 'telescopic milky-ways'. Our Milky Way is of the same kind as other nebulae. 'I shall now proceed to show that the stupendous sidereal system we inhabit . . . consisting of many millions of stars, is, in all probability, a detached Nebula.' The chapter in his paper of 1785 containing this discussion is superscribed by the thesis: 'We inhabit the planet of a star belonging to a Compound Nebula of the third form.'[181]

It is easily understood that such speculations so widely extending the realm of the universe roused enthusiastic admiration in some of his

317

contemporaries, but in many others met with sceptical doubt; no other astronomer had seen all these celestial marvels. His ideas of the evolution of stellar systems and clusters were still more strange to them. Herschel had been nurtured in his youth on the critical rationalism of the eighteenth-century Continental concept of nature that was entirely foreign to the free, but strongly conservative, English mind of those times. In the biography entitled *The Herschel Chronicle* his granddaughter, Constance Lubbock, writes: 'Herschel was the first to introduce a disturbing factor into this view of Creation by his suggestion that it was a long process, not a sudden and completed act. Perhaps one reason for the coldness with which these papers were received by the Royal Society may be the fact that thought was not yet so free in England as in France and Germany. At this time no one could hold any high position in either of the English Universities or in the teaching profession unless he took orders in the Church. The Astronomer Royal was a clergyman as well as Dr Hornsby at Oxford and the Professor of Astronomy at Edinburgh. Under these circumstances it was natural that there was a certain reluctance to show approval of Herschel's theories, which seemed to run counter to the accepted interpretation of the Biblical account of Creation.'[182]

His ideas on clusters and nebulae, however, did not remain the same when, in continuing his observations, he discovered new kinds of celestial objects. He was struck by the appearance of stars surrounded by a faint milky luminosity, uniform throughout; if it were produced by countless faint stars, they must be excessively small, or if they were normal, the central star must be an enormous body; 'we therefore either have a central body which is not a star, or have a star which is involved in a shining fluid, of a nature totally unknown to us.' Thus he wrote in a paper 'On Nebulous Stars, properly so-called' in 1791; and he continued: 'What a field of novelty is here opened to our conceptions! A shining fluid, of a brightness sufficient to reach us from the remote regions of a star of the 8th, 9th, 10th, 11th, or 12th magnitude, and of an extent so considerable as to take up 3, 4, 5, or 6 minutes in diameter! Can we compare it to the coruscations of the electrical fluid in the aurora borealis? Or to the more magnificent cone of the zodiacal light as we see it in spring or autumn?'[183] That it can exist also without a central star was shown by the no less marvellous round or elongated nebulous discs, which, because of their uniform light, he had called 'planetary nebulae'. They could now be explained, 'since the uniform and very considerable brightness of their apparent disc accords remarkably well with a much condensed, luminous fluid; whereas to suppose them to consist of clustering stars will not so completely account for the milkiness or soft tint of their light'.[184] The same holds for the great nebula of Orion, of which he wrote in 1802 that for 23 years he had seen many changes in its shape and lustre; and he added 'To attempt even a guess at what this light may be, would be presumptuous.'[185]

Now his ideas on evolution, too, were reversed. The faint shine of small nebulae not dissolved into stars, which formerly he had supposed to be far-distant agglomerations and, as such, the ultimate result of clustering forces, had been recognized as thin nebulous matter, which was now placed at the beginning of evolution as primitive matter, out of which stars are formed through condensation. We see some of these nebulous discs contracted in different degrees and with central stars of different brightness. 'When we reflect upon these circumstances, we may conceive that, perhaps in progress of time these nebulae which are already in such a state of compression, may be still farther condensed so as actually to become stars' (1811).[186] And some years later he wrote: 'We shall see that it is one and the same power uniformly exerted which first condenses nebulous matter into stars, and afterwards draws them together into clusters, and which by a continuance of its action gradually increases the compression of the stars that form the clusters' (1814).[187]

Herschel in all these researches dealt for the first time with the loftiest and most far-reaching problems of the structure and development of the universe. Other phenomena of the fixed stars at the same time drew his attention, especially the variations in brightness of some stars, which occupied two of his friends, amateurs like himself. In 1782 John Goodricke discovered the regular character of the variations of the second-magnitude star Algol (β Persei), whose variability Montanari had perceived a century earlier. Always, after a period of 2 days 21 hours, the star showed a decrease to the fourth magnitude, which Goodricke explained as a periodical obscuration by a dark body revolving about Algol in that period. Two years afterwards he discovered the regular variations of δ Cephei and β Lyrae; and Edward Pigott in 1785 found η Aquilae and a small star in the constellation Scutum to be variable. Herschel in these years also occupied himself with the magnitudes of the stars. He noticed that the sequence of brightness which he observed in a constellation often deviated strongly from the magnitudes assigned to them by Flamsteed's catalogue and also contradicted the sequence of Greek letters assigned to them by Bayer. He supposed that in many cases the stars had, perhaps gradually, changed their brightness. To investigate such changes he conceived the need for a designation of the relative brightness of the stars more precise than that given by the coarse datum of their magnitudes. So he devised a system of signs composed of points, commas and dashes to indicate their estimated greater or lesser differences in brightness. In this way he compared all the stars numbered by Flamsteed with each other; he published the comparison in 1796–99 in four catalogues, and two more catalogues, completing the constellations, were published from his manuscripts after his death. His expectation of discovering many variable stars thereby was not fulfilled; α Herculis was the only star in which he detected, in 1795, small fluctuations. His method of comparing small differences in brightness remained unnoticed until, long afterward, Argelander drew attention to it.

Through all these researches the world of the fixed stars was definitely incorporated into the realm of actual astronomy. That Herschel had been able to accomplish this was due in large part to the fact that he came into science from the outside as a self-taught man. Free from the burden of tradition, which for those educated in the profession determines the realm of their duties and the field of acknowledged activities, he could stray outside along untrodden paths. This has repeatedly happened in astronomy. Now, as the culmination of four centuries of constructive revolution, the gates to the wide spaces of the stellar world were flung open and the ways were cleared for the progress of science in the following centuries.

9. *Top left:* William Herschel (p. 312)
Right: Herschel's 20-foot telescope (p.314)
Left: Herschel's section of the Milky Way (p. 317)

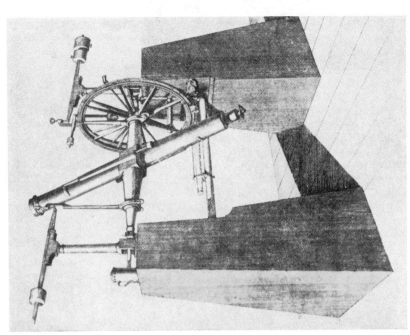

10. *Left:* Reichenbach's meridian circle (p. 324)
Right: Friedrich Wilhelm Bessel (p. 324)

CHAPTER 32

THE TECHNICAL BASIS

'WE cannot flatter ourselves that the instruments, even if still further perfected, will allow us to advance farther and to increase the accuracy of the measurements beyond one second of arc. It is quite possible that Bradley has fixed therein the limits of our knowledge.'[188] Thus in 1782 wrote the renowned historian of astronomy, Jean Sylvain Bailly, who in later years, as president of the Constituante, played a prominent part in the French Revolution. Admiration for the progress achieved—1″ is indeed a small quantity, it represents $\frac{1}{100}$ mm. on a circle of 2 metres radius—is here combined with the naïveté of the eighteenth-century citizen, for whom spiritual and scientific evolution— as also the political evolution of mankind in the near future—would be completed by the rule of reason and the knowledge of the natural world order, now well-nigh established. Who could imagine that it was only the prelude to an ever more rapid tempo of unending social development, carrying in its wake an equally unending growth of science?

The Industrial Revolution, with its profound social convulsions, began in England, in the second half of the eighteenth century. In the course of the nineteenth century it spread over the adjacent countries of Europe, over the United States of America and finally over the entire world. Technical progress was its basis; the small artisan's implements of old were superseded by more productive, ingeniously constructed machines, soon to be moved by the powerful steam engine. Out of handicraft and small business arose capitalistic big industry, controlling the economic system more and more, until finally it completely changed the aspect of the world. Man himself also changed; fierce competition on the part of the manufacturers brought about a new tension in social life and aroused a restless energy.

In the European world of absolutism and privilege, dominated by landed property and commercial capital, arose a strong middle class of industrialists and businessmen, the leaders of the new society. In fierce revolutions (as in France) or in thorough-going reforms (as in England) it built up its political power and its social dominance. Its fundamental ideals of personal initiative and unrestricted freedom of trade, thinking and action became the dominant principles of human life.

In this economic development exact science became at an increasing rate the basis of the new techniques. The old traditional working methods were replaced by the application of scientific discoveries. The rising middle class promoted the study of nature by founding universities and laboratories, because it felt that knowledge of nature was good and beneficial. Science was encouraged not only because of its technical use, to increase the productivity of labour, but also because it evoked a freer mode of thinking. It freed the mind from the bondage of tradition and became an inspiring spiritual power of knowledge and enlightenment, in which broader sections of the people gradually participated. Progress and enlightenment (*Aufklärung*) became the slogans of the new age.

That this interest in science was not due solely to its practical utility for technical progress is shown by the large share which astronomy held therein. Astronomy, next to mathematics, then the most developed and esteemed of the sciences, was the field most appropriate for the disinterested search for pure knowledge. New observatories were founded, sometimes from love of science by groups of wealthy private citizens (as the Harvard Observatory was founded in 1844 by the citizens of Boston), or in connection with universities as the first examples of what later for other sciences were called 'research institutes'. It was a token of the prominent place that pure science occupied in the mind of man in the nineteenth century.

The rise of big industry created the technical basis for the progress of astronomy. The construction of machines for factories and transportation demanded the development of new, highly perfected techniques of iron and steel, capable of producing exactly fitting parts and precisely round axles for rapidly rotating wheels. This perfected metal industry, the basis of all nineteenth century engineering, also made it possible to raise astronomical instruments from their former imperfection as products of handicraft to increasing perfection. Able technicians now came forward, and in their workshops, with all their skill and devotion, constructed carefully finished instruments out of the new materials. The renowned English firms of Ramsden, Cary and afterwards Troughton and Simms maintained the old standards; but now they found competitors in new German workshops. Influenced by the general European development, Germany in the last part of the eighteenth century experienced a new spiritual uplift in a flowering of literature, of music and of philosophy, which was followed, but not until half a century later, by an economic and political rise. The study of natural sciences participated in this rise, and in the new century German science was next in importance to those of England and France. At the outset Germany played a role in the renewing and refining of precision techniques. Workshops for the construction of instruments were established; outstanding among these were the ones founded by J. G. Repsold at Hamburg in 1802, which remained in the forefront during the entire nineteenth century, and by G. von Reichenbach in 1804 at Munich. Usually such shops began with small instruments, altazimuths for geodetic work and sextants for navigation; then gradually they ventured on larger tasks.

From the German workshops emerged a new type of instrument for the determination of stellar positions. By fixing a graduated circle perpendicularly upon the horizontal axis of a transit instrument used for meridian transit, this was transformed into a 'meridian-circle' or 'transit circle', suited for measuring right ascensions and declinations at the same time. Now the old mural quadrants could be done away with. They had always been rather clumsy instruments. Because the errors of position and graduation were difficult to determine, the constructors attempted to keep these insignificant by heavy construction and solid foundations. With the meridian circle, on the other hand, precise finish and easy determination of remaining errors were aimed at. Accuracy of the declinations had formerly been obtained by using a large radius of the circle and then making it tractable by taking a quadrant instead of a complete circle. Now the radius was taken smaller and smaller, so that the circle was less deformed by temperature influences and by flexure through gravity. The errors of complete circles turning with the axis and read by microscopes in a fixed position were smaller and could more easily be determined or eliminated. The accuracy of the readings was secured by the sharply engraved division marks viewed through a microscope and read at first by a vernier, later on mostly by a micrometer. When one revolution of the screw corresponds to 1' on the circle and the screwhead is divided into 60 parts, the tenth of a second can be directly read. The telescope attached in the middle of the axis was considerably improved by providing it with an achromatic objective of large aperture, at least four inches, so that the sharp, round images of the stars could be accurately bisected and stars down to the ninth magnitude could be seen in an illuminated field without difficulty. Through the strong magnification, a star was seen rapidly passing the wire reticle in the focal plane; by slowly moving the telescope, the observer could make the star follow the horizontal wire exactly; and listening to the ticking of the clock he estimated in tenths of a second the moments it passed the vertical wires. A different method of observing the time of transit came from America about 1844; instead of simultaneously looking at the star and listening to the clock, the observer simply registered the moments of transit across the wires on a chronograph by tapping the signal key in his hands, while the clock registered its seconds. Observation of the transits was thus made easier and more accurate; the accidental error of one transit was about 0.06 of a second of time only, and that of the combined result of many successive wires was accordingly smaller.

The refinement of the instrument would have been of no avail, in fact would have been impossible without the astronomer, who by the exigencies of his demands drove the technicians to persevere with improving the instruments, often himself directing their new designs. The pioneer in this field of precision astronomy in the first half of the nineteenth century was the Königsberg astronomer Friedrich Wilhelm Bessel (1784–1846). Like so many first-rate astronomers, he had come to astronomy from the outside. A clerk in a merchant's office in Bremen and wishing to extend his opportunities in trade by sea travel, he studied books on navigation and astronomical geography and delved ever more deeply into astronomical theory and practice. Gifted in mathematical theory and no less persistent in practical measuring and computing, he performed all his work with a thoroughness and accuracy far beyond the relatively coarse quality of the data. By his first published paper, a reduction of Harriot's observations of Halley's comet in 1607, highly praised by Olbers and Von Zach, he made his entrance into the guild of astronomers. Soon he gave up his business to become an assistant at Schroeter's private observatory at Lilienthal and in 1810 he was called to Königsberg to found there a new observatory. The masterpiece that gave him a leading place in the astronomical world was the reduction of the observations of Bradley, whose journals had recently been published in full. Because Bradley had carefully determined the errors of his instruments or noted the data from which they could be derived, this was the best material the previous century could afford. Bessel had not only to derive the instrumental errors from the observations themselves but also the astronomical constants needed for the reduction, such as aberration, nutation and refraction. So Bessel could rightly call his work, when it was published in 1818, *Fundamenta astronomiae*. The more so because the exactitude of his reduction, which went beyond the quality even of Bradley's work, put up a new and higher standard for the instruments as well as for the astronomers working with them. In Königsberg he installed in 1820 a new meridian circle made by Reichenbach, to set the example (plate 10); and after a full life of astronomical practice he added in 1841 a greater instrument by Repsold, provided with all new improvements.

Bessel proclaimed the principle that an astronomical measuring instrument never corresponds to its abstract mathematical ideal and therefore can give exact results as produced by an ideal instrument only by determining all its errors and applying corresponding corrections to the measured quantities—provided that the errors are constant because of the solid construction of the instrument. 'Every instrument,' he said in a popular lecture in 1840, 'in this way is made twice, once in the workshop of the artisan, in brass and steel, and then again by the astronomer on paper, by means of the list of necessary corrections which he derives by his investigation.'[189] The astronomer can measure with greater precision than the artisan can construct. He has to determine carefully all special quantities characterizing his instrument and all the small deviations from the ideal form and to compute the ensuing corrections. The optical axis of the telescope is not exactly perpendicular to the axis of rotation, which itself is not perfectly level and directed east-west; the steel pivots by which the axis rests in the V-shaped bearings are not exactly equal in diameter and not even exactly circular. The graduation marks do not have exactly equal distances, however small the deviations in the best instruments may be. There are margins of error everywhere of thousandths or even hundredths of millimetres. Many of these deviations depend on temperature, on weather, and on the time of day in general, or are changing gradually; so the observations have to be made in such a way that the errors may be eliminated or determined. Moreover, the observational results have to be corrected for the changes occurring in the heavens by precession, nutation and aberration. In order to facilitate their computation and to have them used by all astronomers in the same way and by the same amount, Bessel, beginning in 1830, published his *Tabulae Regiomontanae* (Königsberg Tables), which were used by every astronomer and at last became a constituent part of the astronomical almanacs.

Constructed and handled in this way, the meridian circle dominated astronomical measurement in the nineteenth century. It was the chief instrument of the numerous newly-founded observatories, of the smaller ones attached to almost every university, as well as of the large central institutes like Greenwich, Paris, Washington and Pulkovo—here supplemented by a 'vertical circle' for declinations. It was in use continually, and the catalogues of right-ascensions and declinations, in which the work of many centuries was brought together, appeared by the dozen. The standard of precision can be judged by the unit used, 0.01″ in the declinations, 0.001 seconds of time in the right ascensions. With this increase in exact work Bradley, who in Bessel's reduction had to supply all the elements, was now only needed to procure the proper motions of its 3,000 stars. A new reduction with modern data by A. Auwers, published in 1888, gave good positions—though these observations of one century earlier were of course not quite up to the new standard. Even this was no longer needed at the beginning of the twentieth century; Lewis Boss, in deriving the proper motions for a new star catalogue, had at his disposal observations of the new high quality extending over nearly a century. In all this nineteenth-century work England, which had led in the eighteenth century, was now following more slowly the progress on the Continent; not until after 1835, when G. B. Airy became Astronomer Royal, was Greenwich provided with a meridian circle of German model. This lag in quality was compensated by unbroken continuity in the observation of the stars, the sun, the moon and the planets. Greenwich could be compared with an old-established house of conservative routine, solid reputation and a fixed clientele, viz. all the world's navigation; before long, Cape Town and Washington shared in the work.

The object of this meridian work, the basis of precise modern knowledge of the universe, was the determination of the positions of the stars, chiefly in respect of accuracy and extent. Firstly, the positions of a small number of fundamental stars (36 most fundamental Maskelyne stars, or 400 almanac stars, or even 3,000 Bradley stars) were determined with the utmost precision, built up entirely independently without taking over anything from other sources. Secondly, the positions of the tens of thousands and more of telescopic stars were found by connecting them with the system of fundamental stars as their basis. The first task—to establish a reliable system—was the most difficult; the main efforts of the ablest observers were devoted to this work. The importance, on the other hand, of having proper motions of very faint stars was so great that Bessel himself, between 1821 and 1833, devoted many years of his Reichenbach circle to this work. It was called 'zone work', because the stars caught up in successively passing the meridian at the same place had nearly the same declination and formed a zone or belt. Such mass work on faint stars had already been done by Lalande in Paris between 1788 and 1803, with a more primitive instrument; other astronomers followed his example. Yet all these stars, picked up more or less by chance and irregularly distributed, did not constitute a complete collection. Completeness became possible only when, in 1871, the Astronomische Gesellschaft organized the work upon a co-operative basis. Thirteen observatories (afterwards increased to 16) were each assigned a zone of declination, 5° to 10°, or 10° to 15°, and so on, in which each had to observe, according to a common plan, all the stars in the ninth magnitude from lists prepared beforehand. It took scores of years before this programme of more than 100,000 stars was completed, because it also had to be extended down to the South Pole. For the stars of Bessel and Lalande and some others proper motions could now be derived; yet the basic thought in the work, directed more to the future than to the past, aimed at the amount of good proper motions that would be available when it would be repeated in the twentieth century.

The methods of the meridian-circle work did not, of course, remain at the level first attained. Numerous difficulties stood in the way of the realization of the ideal of precise measurement, which compelled the astronomers to investigate and unremittingly to improve their working methods. When, according to the small differences in the separate results, a high accuracy, expressed by some hundredths of a second, seemed to be reached, the disappointment was great when much larger differences, of many tenths or even entire seconds, were found between the final results of different observatories. They were of a systematic character, and it was possible, by tables of corrections gradually changing with the position in the heavens, to reduce each catalogue to some average or standard catalogue constructed from the best data. However, no guarantee could be given that such a standard catalogue did not have systematic errors of the same order of magnitude, due to errors common to all the observations.

Sources of such errors, similar for different observers and instruments, were easy to discover. The right ascensions were entirely dependent on the constant rate of the timepiece. The chief astronomical clock of an observatory does not serve to give information about the time; it is an instrument for astronomical measurement, indicating the regular rotation of the celestial sphere. Able clockmakers, with the utmost care, tried during the entire century to provide astronomers with clocks of increasing perfection. Yet fluctuations in air pressure and temperature, especially along the pendulum, as well as small irregularities, vibrations and frictions can influence its rate; and a small difference in rate between day and night can cause considerable systematic errors in the right ascensions. To diminish them, the chief clock is often enclosed in an airtight case and suspended from a strong pillar in an underground room. The timepieces themselves were improved by allowing the pendulum to swing free from the other parts as much as possible. This principle was most perfectly realized in the clock constructed by Shortt which came into use about 1924. Here a pendulum (the master-clock) swinging entirely free, had nothing to do but to keep the other pendulum (the slave clock) going at the right pace, while, in its turn, it is kept going by regular small impulses from the latter.

Another source of systematic differences in right ascension are the so-called 'personal' errors already discussed and pointed out by Bessel. Observers always wrongly estimate or register the moment the star image crosses the wire by nearly the same amount, usually late by anything up to several tenths of a second. Every observer is in error to a different extent; with increased training this personal error does not become smaller but more constant. Experiments with artificial stars showed, moreover, that the error depends largely on the brightness of the star and on its apparent velocity across the wires. Thus systematic errors in right ascension, depending on declination and brightness, were produced. In order to remedy this evil, Repsold in 1889 introduced the travelling wire; the observer by slowly turning a screw keeps the wire constantly bisecting the moving star, while the frame automatically makes the electric contacts which are registered and afterwards read. This method of observing is not absolutely free from personal error—every observer has his special way of keeping the wire to the right or the left-hand side of the true position—but these errors are a tenth or less of the errors involved in the old method, and so are the variations caused by brightness. With the personal errors reduced to some few hundredths of a second, the sources of error for the right ascensions are, to a great extent, removed.

It was more difficult with the declinations. In describing the meridian from the south through the zenith to the north, the telescope has a differently inclined position relative to gravity and to the horizon for every declination. Owing to irregularities in the metal parts complicated flexures arise, not only of the circles but also of the telescope, which can only be imperfectly derived from measurements with special contrivance in horizontal and vertical positions. Far worse are the effects of the refraction in the atmosphere, which in earlier centuries was an impediment to a good determination of the stars' positions. Able mathematicians—beginning with *citoyen* Kramp in Cologne in 1800 (Year VII of the Republic), whose work was used by Bessel—developed and improved the theory of the refraction during the entire nineteenth century and computed its variation with the altitude of the star. But large uncertainties remained, especially near the horizon, chiefly because the decrease in temperature and density in the higher atmospheric layers was not well known. What this means is shown by the fact that at an altitude of 30°, where no astronomer hesitates to make good measurements, the correction for refraction amounts to 160″, so that, to compute it correctly to 0.01″, the result must be certain to $\frac{1}{16000}$ of its amount. Here we have a main source of systematic error. Nor is this the worst; far more serious are the unknown irregularities in the refraction for which we cannot account. The air layers of different density may be inclined, or there is (as often in clear frosty weather) a rise in temperature with height. The temperature inside the observing room is usually somewhat higher than outside, in spite of attempts to equalize them, and the dividing line follows the irregular course of walls and roof. All these influences vary with the different latitudes at which the observatories are situated but nowhere are they absent. So it is not surprising that the most carefully derived catalogues show large systematic differences and that even the adopted standard system may be considerably wrong. So we can understand Kapteyn's complaint in 1922, shortly before his death: 'I know of no more depressing thing in the whole domain of astronomy than to pass from the consideration of the accidental errors of our star-places to that of their systematic errors. Whereas many of our meridian instruments are so perfect that by one single observation they determine the co-ordinates of an equatorial star with a probable error not exceeding 0.2″ or 0.3″, the best result to be obtained from a thousand observations at all of our best observatories together may have a real error of half a second of arc and more.'[190]

Kapteyn showed the probability that Lewis Boss's standard system, then considered the best, contained considerable errors because the proper motions derived for some zones of declination were directed too much to the north, and for others too much to the south. A new standard system, derived by Kopff at the Berlin Astronomical Computing Office from the best modern series of observations, in fact deviated from the Boss System at these points. To decide the question an entirely different method of observing was required, where flexure and refraction would play no role in the determination of declinations. For an observer on the earth's equator the celestial poles are situated on the horizon, and all the stars of the same right ascension with declinations from $+90°$ to $-90°$ are situated beside one another on, or a little above, the horizon. Declinations can then be measured as azimuths with a horizontal circle, free from refraction. Led by these considerations, the Leiden Observatory in 1931–33 sent two observers to Kenya in East Africa to measure declinations by means of an accurate azimuth instrument. Their results indicated the need for systematic corrections, as was expected ($+0.6″$ at $-30°$; $0″$ at $+10°$; $+0.3″$ at $+30°$ of declination), so that the Berlin system was shown to be nearer the truth. The way to a faultless system of declinations by improved repetition seems to have been opened up.

We now return to the beginning of the nineteenth century, to view the progress of technical knowledge in another domain—in optics. Dollond's introduction of the achromatic lens systems had brought about a great improvement in the measuring instruments. Because glass-making methods were quite imperfect, however, it was not until the nineteenth century that the telescopes themselves gained in perfection and power. This was due mainly to the work of a young and gifted Bavarian, Joseph Fraunhofer (1787–1826). In 1806 he entered the workshop of the Utzschneider and Reichenbach firm, where he was soon at the head of optical work. In 1817 the departments were separated, and in association with Utzschneider, a rich financier with strong scientific interests, Fraunhofer founded an Optical Institute at Munich. Two imperfections had to be overcome. In the first place, the values of the refraction indices for the different kinds of glass which were needed in computing the lenses were very imperfectly known, 'so that, with all accuracy in the computation of achromatic objectives, their perfection is doubtful and seldom comes up to expectations'.[191] Moreover, the materials were inadequate, especially English flint glass, 'which is never entirely free from striae'. He was initiated into the secrets of glass manufacture by his colleague, P. L. Guinand, an able French-Swiss glassworker. After twenty years of experiments Guinand had succeeded, in 1799, in fabricating faultless discs of flint 10 to 15 cm. in diameter; and afterwards he was able to produce even larger discs, up to 30 and 35 cm., by welding good small pieces in softened condition. His skill and craft traditions were preserved and maintained afterwards in the French glassworks.

Fraunhofer investigated deeply the refraction of light of different colours in various types of glass by means of exact measurements with a theodolite. He was continually hampered by the difficulty that the red or yellow light to which his measures related could not be exactly identical and recovered. At last, on experimenting with the prismatic solar spectrum, he discovered the fine black lines crossing it, which afterward were called by his name, 'Fraunhofer lines'. 'By means of many experiments and variations,' he wrote in 1817 in a paper for the Bavarian Academy, 'I have now become convinced that these lines and streaks belong to the nature of the solar light and are not caused by diffraction or appearances.'[192] He counted more than 500 lines, and the strongest, which he marked by the letters A to H, could now be used as precise marks for definite kinds of light. By measuring their indices of refraction for each kind of glass he was able to compute exactly the combination of lenses producing the most perfect achromatism. The optical industry here made the transition from skilled craft to scientific computation. The practical result, however, was not greater simplicity. By his experiment Fraunhofer found that the refrangibility of the different colours was not strictly proportional in the different kinds of glass; hence it was not possible to bring together all colours exactly into one focus; there remained small differences, producing the 'secondary spectrum'. Moreover, the theory of lens combinations was too complicated and difficult to supply the best forms through mere computation. Nevertheless, through a combination of practical intuition and theoretical knowledge, he succeeded in making excellent objectives for telescopes, up to 24 cm. (9 inches) in diameter. These objectives produced small, sharp, circular stellar images over the entire field of view, showing not only the faintest stars as well as larger mirror telescopes do but also the most delicate details on the surface of the planets and the moon.

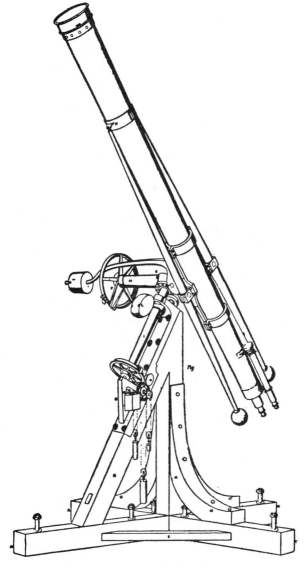

Fig. 30. Fraunhofer's telescope

The mounting of the telescopes was also improved. In the Fraunhofer or German mounting (fig. 30), the tube—in the first smaller telescopes of wood, later of metal—was attached at right angles to the end of a 'declination axis', which itself could turn about a fixed polar axis. All the parts and axles were accurately worked in steel and brass, perfectly fitting and so exactly balanced that the large instrument could be moved by the lightest pressure of a finger. Moreover, it was provided with graduated circles, to read hour-angle and declination, and with screws for fine motion. In every respect this type of telescope constituted a higher stage of perfection and easy handling compared with the unwieldy telescopes of Herschel, with their coarse mounting and moving contrivances of poles, chords and pulleys. The progress here apparent in astronomical instruments corresponds to the transition from the coarse iron rods of the first eighteenth-century steam engines—still to be seen in the South Kensington museum in London—to the carefully finished and polished factory-made machines of the nineteenth century. Another important—nay, indispensable—improvement introduced by Fraunhofer in his telescopes was the automatic following of the stars by the motion of the telescope. Moved by a driving clock, the declination axis, with the telescope, slowly turns about the polar axis, so that the star seems to be entirely at rest in the field of view. This nineteenth-century type of telescope was called a 'refractor', in contrast to the reflector employing a concave mirror. It was the result of the refined skill of devoted craftsmen applying the technical perfection of metal and glass industry based on scientific methods.

Fraunhofer, though a pioneer, was not the only constructor in this field; Steinheil in Germany and Troughton in England were also improving the older types of telescope. Moreover, the reflector still maintained itself because, owing to its wide aperture, it had a large light-gathering power. After twenty years of experimenting, William Parsons (afterward Lord Rosse) at Birr Castle near Parsonstown in Ireland, in 1842–45 built a telescope with a mirror 6 feet (182 cm.) in diameter, with which he discovered the spiral structure of some nebulae. Yet the refractors remained the favourite instruments. The workshops succeeded in increasing their size continually. After the 24-cm. telescope made in 1820 by Fraunhofer for the Dorpat Observatory, considered a giant at that time, had proved its high quality in Struve's double-star work, several German and Russian observatories were fitted out with this type of instrument. In 1839 Fraunhofer's successor, Merz, constructed a refractor of 38 cm. (15 inch) aperture and 22 foot focal distance for the new lavishly-built observatory which, by order of Tsar Nicholas I, was erected as a central scientific institute on the hill Pulkovo, south of St Petersburg. A similar instrument was constructed for Harvard Observatory at Cambridge, Massachusetts, when it was founded in 1843–47 by the citizens of Boston.

In the following years many observatories in Europe were provided with this kind of telescope. In Germany progress in instrument-building slackened, and after 1870 it was surpassed by Grubb in England. In the meantime the art of grinding lenses had found a new home in America, when Alvan Clark in Washington began to make lenses of unprecedented dimensions. His first product, exceeding all predecessors, was an objective of 18 inches (43 cm.) for Chicago; it became famous because his son, while testing the perfection of the images in 1862, discovered the companion of Sirius. When his name was established through the quality of his product, European as well as American observatories ordered telescopes with Clark lenses, every succeeding lens surpassing the predecessor in size. Thus in 1871 Washington acquired a 26 inch (66 cm.) lens, with which Asaph Hall in 1877 discovered the satellites of Mars, and Pulkovo in 1885 acquired one of 30 inches (76 cm.). Other constructors followed Clark closely; Grubb made a 25 inch for Cambridge in England, and the Henry brothers of Paris, astronomers themselves, in 1886 made a 76 cm. refractor for Nice and one of 83 cm. for Meudon.

The English tradition, contrary to the Continental European custom, that pure science was not a government affair, was adhered to in the United States also. The great observatories were not founded by universities but mostly by private persons. Here in the last part of the nineteenth century it was the millionaires who, by seizing and monopolizing the natural riches of a large continent, had gradually become masters of big business and increasingly dominated the economic and social life of the nation. Some of them came forward as patrons of art and science; they endowed universities and founded museums, libraries, laboratories, and, of course, observatories, which were provided with the most costly instruments. Since fame was usually the motive, a giant telescope exceeding all earlier telescopes was the coveted object. Thus it was in the last will of a rich Californian speculator, James Lick; after his death a committee of expert trustees, freely disposing of the money, ordered a 26 inch Clark refractor as the chief instrument of a well-equipped astronomical institute, the Lick Observatory, built in 1888 on Mount Hamilton near San Francisco. Mantois of Paris had furnished the glass discs; soon afterwards he made another set of glass discs, still slightly larger, which were ground by the Clarks into a 40 inch (102 cm.) objective. For this telescope the Yerkes Observatory was founded in 1897 at Williams Bay, Wisconsin, as an institute of the University of Chicago. These giant telescopes were constructed on the same model as the former smaller ones of Fraunhofer; they show the same slender, graceful build, with a length about twenty times the aperture, and the same type of mounting. They were executed with the same care and precision but on a four times larger scale. Everything in them is more colossal, hence heavier; yet they are equally easy to handle, masterpieces of constructive ingenuity. The comparison of these giants with their prototypes of the beginning of the nineteenth century illustrates the progress of that century in all departments of mechanics and science: in the casting of glass, the grinding of lenses, the construction of metal mountings, the control of mechanisms, the organizing of big institutes. America had now taken the lead in technical progress and, therefore, also in practical astronomy.

This great development in astronomical optics depended on the progress of glass technique, but, conversely, the increasing demands of scientific research stimulated this technique. In France the workshop of Mantois and the glass factory of St Gobain, later combined, succeeded in casting ever larger and better glass discs. In Germany the Carl-Zeiss Werke at Jena, founded by Ernst Abbe, gained fame for the excellence of its optical instruments based on thorough theoretical foundations laid by Abbe himself, as well as by its social structure. It was connected with the Schott glassworks, where, in continuous experiment, attempts were made to prepare new kinds of glass with special qualities, as desired by the scientists, e.g. to remove the secondary spectrum. Though not competing with the construction of giant telescopes, the Zeiss Works has provided observatories all over the world with new instruments ingeniously devised.

The limit of size, however, was now reached. Lenses of 40 inches diameter are so thick at the centre that their absorption of the stellar light begins to be effective, and they come near to causing deformity by their own weight. If larger glass discs were constructed, they could not be used in the same way.

Again we have to retrace our steps and return to Fraunhofer's workshop. His chief merit is not that he improved telescopes as instruments for viewing the heavens but that he made of them the most perfect measuring instruments. The small distances of close stars or the dimensions of small planetary discs can, of course, be measured more accurately than the co-ordinates of single objects; many sources of error common to adjacent points disappear in their relative place. For such measurements micrometers are used. Fraunhofer provided each of his telescopes with a filar micrometer, which, by careful construction and accurate finishing of every part, was now modelled into an instrument of the highest precision. A perfectly worked screw of small pitch moves a wire relative to a fixed cross-wire, and the distances are easily read at the divided screw-head. By turning the entire micrometer about the axis of the telescope, the fixed wire is brought into the desired direction of the two stars, and this direction can also be read. Thus the relative position of the two objects is given in polar co-ordinates, i.e. as position angle and distance.

For the use of such a micrometer it is, of course, necessary that the telescope must follow the stars exactly, so that they appear to be completely at rest in the field of vision; then the wires can be pointed precisely upon the centre of the stellar images. William Herschel had not been able to do this, because his telescope did not automatically follow the stars; so, in measuring a double star, he could only bring the wires to such a distance that, by estimate, it seemed to be equal to that of the stars travelling through the field. This explains the remarkable increase in accuracy now attained. The second decimal of a second of arc now was no superfluous luxury or product of averaging; it was the natural unit of which the errors of measurements were only a small multiple. With all the later refractors up to the American giants, the filar micrometers constituted an essential, constantly-used accessory that gave them their real utility.

Another type of instrument for exact measurements, too, was developed by Fraunhofer. In the eighteenth century Bouguer had measured the diameter of the solar disc by putting two equal objectives beside one another at such a distance that the two images of the sun formed by them were in exact contact; then, knowing the focal length of the objectives, he could derive the sun's apparent diameter by measuring the distance between their centres. Ramsden improved the instrument, called a 'heliometer', by taking half-lenses, which produce the same complete solar image. In moving them alongside one another the two images could be brought to coincidence as well as at great distance, and small distances could be measured as well as large ones. Fraunhofer took up the idea and put the two halves of a good objective into frames displaced by a screw, whereupon the distance was read by means of microscopes on an exactly divided scale of millimetres. Retaining the name 'heliometer' he constructed a first-rate precision instrument for the Königsberg Observatory, able to measure far larger distances than the filar micrometer, up to several degrees. It was not ready until 1829, three years after his death. In using it for the first determination of a stellar parallax, Bessel subjected it to a thorough theoretical and practical examination, so that it acquired a birthright in astronomy and was used, up to the end of the century, in a number of important researches on stars and planets, sun and moon, where moderate distances had to be measured.

Then, however, it met with a competitor in this field of astronomical practice. From the middle of the nineteenth century photography was brought increasingly to bear upon the heavenly bodies. The first photographs of the sun and the moon showed, in short exposures, an abundance of detail that would have demanded hours and months of observing in direct drawing and mapping. Since the existing objectives, achromatized for visual rays, did not give sufficiently sharp images in the photographically active blue and violet light in consequence of the secondary spectrum, Lewis M. Rutherfurd, at New York in 1864, for the first time constructed a photographic objective of 29 cm. aperture. It provided him in the next year with the first usable photographs of the starry heavens. Every progress in photographic methods found its application in astronomy, as in 1871, when the wet collodion process was replaced by the far more sensitive dry silver-bromide-gelatin plates, that photographed faint and even telescopic stars in a short exposure. It soon appeared that the nice, perfectly round, stellar images on the plate, though rather large, were more easily and accurately measurable than the vibrating, often irregularly jumping stars in the telescope; the exposure had averaged the irregularities. Thus a new astronomical practice appeared: to use the telescope only for taking photographs and then to measure them at leisure by day.

Besides the gain in accuracy, the photographic method presented other advantages. First, all the stars of a field were pictured in one exposure, so the few clear and favourable nights might be exploited to the full. Their measuring, moreover, could be postponed until the stars were needed. Secondly, there was the increase in size and visibility of the images that could be achieved by simply lengthening the exposure time, so that the effect of larger apertures on the brightness of the stellar images was enhanced. Thirdly, in using lenses with great angular aperture—i.e. a short focal distance relative to the aperture, a great surface brightness was produced. Extended faint, nearly invisible nebulous objects, which by no optical means could be made brighter, now by long exposures revealed for the first time the beauty of their intricate structure.

Astronomy profited from the increasing perfection of optical techniques developed on behalf of practical photography since its discovery in 1839. Laborious theoretical computations, as made by Seidel and Petzval, combined with the practical inventiveness of constructors like Chevalier, Steinheil, Brashear and Zeiss, gradually created a number of increasingly more perfect types of optical systems, consisting of different combinations of lenses. It was the struggle with the problem of how, for any point of an extended object, all emerging rays of all colours could be united into one point, thus forming a sharp image situated in one flat field. The demands for great brightness, an extended field, and faultless depicting could not all be met at the same time; according to the purpose, one or another demand dominated; thus a wealth of different types has been invented and constructed, first of simple triplets, gradually improving to systems of 6 or 8 lenses. Portrait objectives of large angular aperture for the use of amateur photographers are found all over the world in thousands of cameras. With the same angular aperture of 1 : 4 or 5, in larger dimensions, they have been made especially for observatories, providing a new type of instrument that offered new aspects of the celestial objects.

If the positions of the stars are to be determined accurately by measuring them on a photographic plate, there is the difficulty that an ordinary achromatic objective consisting of two lenses depicts the object in a somewhat distorted way. This defect can be avoided by using a well-computed system of more lenses. Before this was fully realized the brothers Paul and Prosper Henry in Paris succeeded, with self-made objectives, in photographing stars down to the fourteenth magnitude. Full of enthusiasm for this result, an International Conference convoked in Paris in 1889 resolved, all too hastily, to construct by international co-operation an atlas of the entire sky down to the fourteenth magnitude and a catalogue of exactly measured star places down to the twelfth magnitude; this was the so-called *Carte du ciel* enterprise. Since the plates were 2° square only, the error of distortion was rather insignificant; but the number of plates needed now was so large that after tens of years the end was not yet in sight. Schlesinger demonstrated in 1928 that on larger plates of 5° square, provided that they are taken with the right optical systems, the star positions can be determined with the same precision as with meridian circles. Indeed, the Astronomische Gesellschaft had already resolved that its AG catalogue, started in the seventies, should be repeated not by meridian circle observations but by means of photographic plates.

With all these new practical appliances, the striving for more light, i.e. for larger apertures of the instruments, persisted. Lenses of more than 40 inches diameter were unusable; here the limit had been reached. To gather more light a return was made to the reflector at the end of the nineteenth century. A mirror not traversed by the light rays but reflecting them at the surface can be supported at the back to prevent it from bending by its weight. By giving it a parabolic shape, the rays are collected exactly in the focus. To be sure, the field presenting perfect images is small; but when single stars are investigated, as in spectral work, this does not matter. Moreover, it is possible to get sharp images over a larger field by interposing an appropriate correcting lens between the mirror and the photographic plate as was devised by Ross.

Since the beginning of the twentieth century, reflecting telescopes of increasing size have been built; instruments of 40 inch aperture became common, and soon apertures of 60, 70 and 80 inches came to surpass them. Foremost among them is the 100 inch Hooker telescope at the Mount Wilson Observatory, devised and ground by Ritchey and put into use in 1919. It won fame for its size, for solving the multitude of construction problems and still more for the amount and high quality of the research performed with it. The experience acquired in this construction induced the plan of an all-surpassing enterprise, the construction of a 200 inch telescope. The story of its gradual realization, the overcoming of difficulties and setbacks before the telescope was erected on Mount Palomar, reads like an epic of modern technique.

Modern development of astronomical instruments would not have been possible without the growth of techniques in nineteenth-century industry, which revolutionized the entire aspect of society. The same precision with which the skilled artisans of a century ago produced instruments of high perfection was now an essential part in the construction of the increasingly colossal machinery of modern giant telescopes. As in other industrial machines, great size is here united with precision in detail. In the vital parts, to secure the exact shape of the glass surface of lens and mirror, control to single thousandths of a millimetre is demanded, for which modern techniques affords its aid. The fact that heavy masses can be moved by a finger is no longer regarded as remarkable. The glass and metal masses of hundreds of tons are moved by electric motors. The astronomer, hardly visible in his cabin near the focus is, so to speak, the small brain of the huge steel organism and directs all the motions of the mammoth machine by the mere pressing of buttons. Technical precision in electrical control of gigantic instruments is the material basis of modern astronomy.

CHAPTER 33

DISTANCES AND DIMENSIONS

A KNOWLEDGE of distances and dimensions is the basic necessity for a good concept of the universe. At the end of the eighteenth century a rather rough idea of the sun's distance and of the dimensions of the stellar system had been acquired, expressed by the parallax of the sun. The dimensions of the stellar system, however, expressed in the yearly parallax of the stars, were entirely unknown.

The increased precision of nineteenth-century technical implements imposed as a first task what might be called a 'geodetic' space survey. Here man was faced with the same kind of problem as when, confined to the few square feet at the top of a hill or tower, he has to determine the distance of distant towers or mountains. Parallaxes had to be measured. Reliable distances and the structure of the outer world could not be ascertained until it was possible to measure parallaxes to hundredths of seconds.

The nineteenth century gave expression to its higher standards in a more elaborate treatment of the best and most important data of observation bequeathed by the preceding century. Bessel gave a new reduction of Bradley's observations of star places. Johann Franz Encke (1791–1865), astronomer at Gotha, undertook a new reduction of the transit of Venus in order to derive the solar parallax. In 1788 Duke Ernest II had founded an observatory on the Seeberg near his residence of Gotha; at this small court science stood in high favour, just as did literature in the neighbouring court at Weimar. For a short period Gotha was a centre of rising science in Germany; in 1796 an astronomical congress assembled here, and a first review of astronomy, called *Monatliche Correspondenz*, was published by F. X. von Zach, the director. Encke worked here after 1813 and took up the reduction of the Venus transits, before he was called to Berlin in 1825 to modernize the observatory there.

In this reduction, published in 1822 and 1824, he applied a new principle to the mathematical treatment of natural phenomena, chiefly developed by Gauss in the preceding years. In former centuries the astronomer selected from among his observations those that seemed the

339

best; this made him liable to bias or inclined to select such data as showed a possibly unreal agreement. It could not seem unreasonable that what agreed in the end was deemed the best. We remember how with Tycho Brahe the cycle of differences of right ascension around the sky resulted in a sum total differing from 360° by only a number of seconds, whereas every separate difference was uncertain by more than half a minute. In the seventeenth century scientists like Huygens and Picard realized that the average of a number of equivalent measurements would be better than one of a couple selected from them, and in the eighteenth century this averaging came more and more into use, all the more so since the concept of chance or probability of errors as a quantitative character had gradually become clearer. The notion of 'laws of chance', already applied by Huygens ('plays of luck'), by Jan de Witt, and by Halley ('mortality tables'), acquired its pure theoretical form a century later in the theory of errors developed by Laplace and Legendre and in Gauss's quadratic-exponential law of errors. It afforded to the computers a method of dealing with a series of observed data according to rules which excluded any arbitrariness.

Thereby a new attitude was brought into being, typical of the nineteenth-century scientist towards his material: it was no longer a mass of data from which he selected what he wanted, but it was the protocol of an examination of nature, a document of facts to which he had to defer. His method of working he found prescribed in the formulas of computation derived in 1804 by Gauss as the 'method of least squares'. By the condition that the sum total of the squares of the remaining errors shall be a minimum, the 'most probable' value of the unknown quantity is found. By making this condition for each of the unknowns of the problem, the most probable values for all of them can be solved directly from the equations. This new method of computing was enthusiastically greeted by the astronomers, who were always struggling with the problem of how to derive the best result from an abundance of data. The 'best' was now defined as the 'most probable'; all hesitation and doubt gave way to solid certainty. Moreover, the deviations themselves that remained in the observed data, called their 'errors', allowed the derivation of a 'mean error' and a 'probable error' as objective indicators of the remaining uncertainty in the results.

This method, afterwards used in all astronomical research, was applied by Encke to the observations of the transits of Venus. These observations gave the moments of ingress and egress. If the relative position of Venus's and the sun's centre in two co-ordinates is known for the middle time and also the velocity of relative displacement, the moments of ingress and egress can be computed as observed from the centre of the earth and, by adding the influence of the relative parallax, for the observing station. The differences between the observed and the computed moments were due not only to the errors of observation but, in addition, to the error in each of the assumed co-ordinates (the velocity was known with sufficient exactness) and to the error in the assumed parallax. Each observation at any place on earth gave a relation between these three unknowns; thus they could be solved by treating all these relations as one body of data. Moreover, Encke could expect a substantially better result than the former computers, because now for a number of stations far to the east or the west, like Hudson Bay, Tahiti, Orenburg, Peking and others, the essentially relevant longitude differences with Europe had been better determined. Thus he got a result for the solar parallax which, after an additional correction in 1835, was given in the form 8.57116″ ±0.0371″.

This form shows how astronomy in these early days of the new method of computation revelled in the delight of being able to compute exact results. According to the probable error after the ±, there is an equal chance that the error of the result is larger, or that it is smaller than 0.037″. It is clear, then, that the third and fourth decimals in the result are not only valueless but even senseless, since the second decimal is already uncertain. Modern scientists would have written more soberly 8.57″ ±0.04″. It will appear that thereby the real uncertainty was still underrated.

Instead of the many different values derived from various combinations of data, we now had one established value derived from their totality; this value has been accepted and used for half a century. With the known radius of the earth, 6,377 km., this solar parallax fixed the distance of the sun—more exactly, half the major axis of the earth's orbit—at 153.3 million km. (95 million miles). It is the astronomical unit of length in which are expressed all the distances and dimensions in the solar system; the sun's radius now is found to be 714,000 km. The sun's mass relative to that of the earth is also connected with it; according to Newton's formula, the third power of the ratio of the orbits of the earth and the moon, i.e. of the ratio of the parallaxes of the moon and the sun, divided by the square of the ratio of their periods of revolution, gives the ratio of the masses, for which 356,000 is found.

These are numbers only, of more or less figures. To understand what they meant for man, who now realized the size of the universe, we have to look into the popular literature that spread scientific interest and knowledge among broad masses of the people in the middle of the century. Here we can read that if a big gun on the sun fired a shot at us we would not perceive it until 25 years later we were struck down by the ball. In a widely read booklet, *An Imaginary Journey through the Universe*, a German writer, Adolph Bernstein, in an amusing way dealt first with the enormous diameter of the sun and then with its volume, computed to be 3,500 billion (3.5×10^{15}) cubic miles (German 'geographical' miles of 7.5 km. or 4.5 statute miles). Then in a chapter entitled 'All Respect for a Cubic Mile', an unsuccessful attempt was made to fill a box of one cubic mile with all the towns and all the people on earth, and equally to count the number of these boxes, which would take millions of years. Against this delight in large figures, F. Kaiser, the restorer of astronomy at Leiden, and a successful popular writer, remarked that the greatness of the universe did not consist in its size but in its order as established by the rule of universal law.

In the first half of the new century the problem of the fixed stars was also taken up. The astronomers of the eighteenth century had failed repeatedly in their attempts to find a yearly parallax of a star. Fraunhofer, however, had now provided such excellent instruments that better results could be expected. Stars presumed to be near us had to be tried first, so that a large parallax might be expected. In his discussion of Bradley's observations, Bessel had met with a fifth-magnitude star in the Swan (61 Cygni) that had an extremely large proper motion of 5.2″ yearly—surely an indication of proximity. In the years 1837–40 he repeatedly measured with his heliometer its distance from two small stars at 8′ and 12′ distance. Since his measurements were very accurate (with the mean error of an evening result only 0.14″), the parallactic circle it yearly described appeared with great clarity. The parallax was found in 1838 to be 0.31″ and in 1840 to be 0.348″. Hence the star was at a distance of 590,000 astronomical units.

At the same time Struve at Dorpat attacked the problem by means of the filar micrometer attached to the 24 cm. Dorpat refractor. He chose the star Vega (γ Lyrae) because of its brilliance, its considerable proper motion (0.35″ yearly) and its situation near the pole of the ecliptic, so that its yearly orbit was an unforeshortened circle. His observations in 1835–38 afforded a parallax of 0.26″. At the same time, Henderson, the director of the Cape Observatory, and his successor Maclear, observed the first-magnitude southern star α Centauri. On many grounds this star was suspected to be very near; it had a large yearly proper motion of 3.7″, and it was a binary describing in a short period a large orbit, which might be supposed to appear large only because of its proximity. The observations were made by means of an ordinary mural circle, thus being far less accurate; but they afforded (1839–40) a large parallax of 0.91″. Later, more exact measurements have reduced this value to 0.76″, still the largest of all stellar parallaxes; so α Centauri remained our nearest neighbour in the world of stars, at a distance of 270,000 astronomical units.

Thus the old problem posed by Copernicus was solved, and the distance of three nearby stars was determined. 'I congratulate you and myself that we have lived . . . to see the great and hitherto impassable barrier to our excursions into the sidereal universe, that barrier against which we have chafed so long and so vainly . . . almost simultaneously overlapped at three different points. It is the greatest and most glorious triumph which practical astronomy has ever witnessed.'[193] Thus spoke John Herschel, as President, to the Royal Astronomical Society when he explained why it had awarded its gold medal to Bessel. And he added that, considering the many false announcements of stellar parallaxes in former years, there might be a certain hesitation as to the results of Struve and Henderson. Bessel's results, however, showing clearly the regular increase and decrease in the measured distances in a yearly period, could not leave the least doubt as to the reality of the parallax found.

It is easily understood that these examples incited the astronomers to follow the same path and to determine parallaxes of other, preferably rapidly moving, stars. Now, however, disappointments supervened. The results of different experienced observers for the same star, though derived from carefully made series of measurements, showed far larger differences than were expected after the agreement within each series. Such was the case, for example, with a star of the seventh magnitude in the Bear, called 1830 of Groombridge's catalogue, which had a proper motion of 7.07″, greater than Bessel's star; its parallax, measured with the Königsberg heliometer, was found to be 0.182″, but with the filar micrometer of the Pulkovo Observatory, to be 0.034″, a fifth of the former figure. For 61 Cygni, Bessel's final result of several years of continued measuring was 0.35″; but Otto Struve at Pulkovo found 0.51″. Obviously, there were large systematic errors present in all these measurements, the origin of which could be only imperfectly guessed: yearly differences in temperature, differences in aspect through atmospheric dispersion, personal errors in pointing stars of different brightness, the comparison stars being usually much fainter than the object star. Thus the determinations of parallax became the severest test of infinite and minute carefulness in the arrangement of measurements and the elimination of sources of error. The heliometer remained the favourite instrument, and it was chiefly David Gill (1843–1914), first in Scotland and afterwards in Cape Town, who, from 1870 on showed how with well-devised handling this instrument could give reliable results. He and his assistant Elkin measured the parallaxes of a number of southern stars, and this work was continued by Elkin at the Yale Observatory at New Haven, Connecticut, in measuring a number of reliable parallaxes of northern stars.

The results obtained for the first dozens of parallaxes allowed some general conclusion to be drawn. Average values had already been derived in the forties by Peters at Pulkovo; for the average parallax of a second-magnitude star he found the small amount of 0.017″, and for fainter stars it must be still smaller. But it became increasingly clear how little such an average meant. Among the brightest stars measured by Elkin, some large parallaxes occur: 0.76″ for α Centauri, 0.38″ for Sirius, 0.32″ for Procyon, 0.24″ for Altair; but also very small values, like 0.028″ for Betelgeuse and 0.008″ for Rigel. The latter, of the same apparent brightness as α Centauri, with the parallax taken as stated, must be about a hundred times farther away, hence 10,000 times more luminous. Conversely, large parallaxes were found for very faint stars that could be singled out from the mass of faint stars as near stars only by their rapid proper motion, such as Lalande No. 211885 of the seventh magnitude, with 0.40″, and Kapteyn's star of the eighth magnitude with 0.32″ parallax. The latter, being at the same distance as Procyon, must have only $\frac{1}{1000}$ of its light power. So there is such a large diversity of luminosity among the stars that one can be millions of times brighter than another.

While the knowledge of distances in the world of the nearest stars thus gradually progressed, the determination of their fundamental unit, expressed by the solar parallax, had entered into a dramatic epoch. The quiet assurance, that through Encke's result we knew it to $\frac{1}{200}$ of its amount, was shaken about the middle of the century by a series of blows. Among the perturbations of the moon's course caused by the sun, there is a term of 125″, called the 'parallactic equation' (already mentioned in Chapter 30), which depends on the ratio of the solar and lunar parallaxes. From this ratio in 1857 and 1863, Hansen, in his theory of the moon, derived a value of 8.92″ for the solar parallax. At the same time, Leverrier found a solar parallax of 8.95″ from the mass of the earth, derived by means of its perturbing action upon Venus and Mars.

Another criticism came from the side of physics. Formerly the velocity of light had always been derived from astronomical data, the constant of aberration fixing the ratio of the velocity of light and of the earth, combined with the velocity of the earth. The former, 20.44″, according to a new determination by Otto Struve and the latter, 30.56 km./sec. from Encke's solar parallax, determined it at 308,000 km./sec. However, methods were now devised to measure the velocity of light by direct physical experiments, first (in 1849) by Fizeau with a toothed wheel, then (in 1862) more accurately by Foucault with a rotating mirror. These afforded a far smaller value, 298,000 km./sec. The aberration constant could not be in error so much as 0.1″ or $\frac{1}{200}$ of its amount; so Foucault immediately pointed out that the adopted solar parallax must be wrong and must be increased to 8.8″.

Distrust now arose concerning the Venus transits, and attention was being directed towards the old method of directly measuring the parallax of Mars in favourable opposition. Better results than in former centuries might now be expected, since the accuracy of observation had so greatly increased. At the perihelion opposition of 1862 a number of corresponding measurements of the declination of Mars were made at northern and southern observatories; a preliminary discussion afforded 8.96″ and 8.93″, and it was believed that the true value would be near 8.90″. Moreover, a new discussion of the Venus transit of 1769 was undertaken in 1864 by Powalky. Because he had at his disposal more and better longitudes than Encke and interpreted some observations in a different way, he got a different result, viz. 8.83″. So Encke's result had to be dropped entirely and the world of astronomy could without bias face the question of where between 8.80″ and 8.90″ the true solar parallax was to be found.

Astronomers began to prepare for the next transits of Venus which would take place on December 8, 1874 and December 6, 1882. They did not restrict themselves to Halley's method of observing solely the moments of ingress and egress, which was proposed at a time when there was little accuracy in direct measurements. The full benefit of the occasion could be hoped for only by measuring the position of Venus on the solar disc during the entire transit. Expeditions were sent to all parts of the world. Ten German expeditions were distributed over the continents, all provided with similar heliometers, considered the best instruments for this purpose. Moreover, photographic telescopes were used especially by English and American expeditions, so that the position of Venus upon the sun could be found by measuring the plates later. An enormous amount of work was expended on the reduction; the results of the German expeditions were published twenty years later in five bulky volumes, and the work of the English and Americans was hardly less. The results were rather disappointing: the English contact results were 8.76″ (Airy), 8.88″ (Stone), 8.81″ (Tupman); from the American photographs Todd derived 8.88″; from the German heliometer measurements of 1874, Auwers found 8.88″ ±0.04″. Not only did the values from different sources diverge but the mean errors added showed a lack of inner agreement too. It can be understood, indeed, that in measuring positions of an object against the luminous background of the sun, with the light rays traversing the heated and vibrating air layers, we are in a worse condition than when using the same instrument at night in observing the stars.

345

When in 1877 Mars again had a perihelion opposition, David Gill took his heliometer from Scotland to Ascension Island, near the equator, to derive the parallax by measuring the position of Mars between the stars, in the east in the evening, in the west in the morning. Whereas collaboration of northern and southern observers was needed for the north–south effect of the parallax, the east–west effect could be measured by one observer, so that personal differences had far less effect. The stars could be made to coincide with the centre of the planet's disc very accurately, and the result, 8.78″, was considered to be very reliable.

Earlier the idea had already been expressed—first by Galle in 1872— that greater accuracy and especially greater freedom from systematic errors were to be expected if, instead of Mars, a starlike small planet was used. Objects must then be chosen whose perihelion came nearest to the sun and the earth. These minimum distances to the earth for the chosen planets Iris, Victoria, and Sappho are 0.83 or 0.84 astronomical units, which are certainly large compared with 0.37, the distance of Mars, so that the parallaxes to be measured were far smaller. It was expected, however, that what was lost in this respect would be gained by the absence of systematic errors. Comparison stars of equal magnitude were carefully selected and determined; a broad programme of meridian and heliometer measurements of the positions of the planets relative to the stars was drawn up. The observing campaigns in 1888 and 1889, in which a number of northern and southern observatories took part, came up to expectation; the result of a careful investigation and discussion of errors by Gill was 8.802″.

The confidence that we were now on the right track and that the solar parallax was near 8.80″ could not be shaken by the deviating values from the Venus transits, where so many points of distrust were present. It was strengthened by the physical result; the velocity of light was determined by Michelson and Newcomb, using Foucault's method, to be $299,860 \pm 60$ km., certain to $\frac{1}{5000}$ of its value; combined with an aberration constant of $20.47″ \pm 0.02″$, it afforded a solar parallax of 8.80″, with an uncertainty of no more than 0.01″.

Better was still to come. In 1898 Gustav Witt at Berlin discovered a small planet, No. 433, later called Eros, the orbit of which is not situated between Mars and Jupiter but, in its perihelion, comes within the orbit of Mars, near the earth. Only in such a near approach was it possible to discover the tiny body; computation showed that in the opposition 1900–1901 it would approach the earth to 0.27 astronomical unit, and in the opposition 1930–31 still nearer, to 0.17. By a piece of good luck, the astronomers were presented here with a celestial body more likely than any other to procure for them the coveted solar parallax by a value of the highest precision. The opposition of 1900–1901, in a complete discussion of all the data by Hinks, gave 8.807″±0.003″ from the photographic and 8.806±0.004″ from the visual measurements.

The twentieth century added new results of similar high quality. Stars in the ecliptic show yearly periodic variations in radial velocity because the earth is alternately approaching them and moving away from them. Hough in 1912, by means of spectrographic measurements, determined the orbital velocity of the earth, from which a solar parallax of 8.802″ was found. From the strong perturbations of Eros caused by the earth, Noteboom in 1921 derived the mass of the earth, which provided a parallax of 8.799″. In 1924 Spencer Jones at Cape Town derived for the parallactic inequality of the moon a value of 125.20″, from which a solar parallax of 8.805″ followed. Then De Sitter at Leiden made an exhaustive discussion of all the mutually related astronomical constants, from which the solar parallax came out at 8.803″±0.001″. This does not mean that the third decimal was now secured. Spencer Jones in 1929 made a new discussion of a still more complete mass of star occultations by the moon, which resulted in a parallactic inequality of 125.02″±0.033″ and a solar parallax of 8.796″±0.002″.

For the even more auspicious opposition of Eros in 1930–31 a great campaign of meridian and micrometer observations and photographic plates was started, in which nearly forty northern and southern observatories took part with their best equipment. The reduction took ten years, and in 1942, during the war, Spencer Jones, then at Greenwich, published the result. It was 8.790″±0.001″. Considering the extent of the collaboration, the large amount of observational data, the perfection of the methods used, the watchful elimination of sources of error, the careful discussion, we may say that another determination of the same or of a higher quality is not to be expected in the near future.

This does not mean that now all is well and that astronomers in this respect are free from worry. This latest best value again deviates more from the former best values than could be expected from their mean errors. There must still be some hidden sources of systematic error. Eros is not a well-formed globe; it has been suspected of being either an irregular, elongated, or a double body; to have a result free from such irregularities, photographs at northern and southern stations should have been simultaneous. So the indication 0.001″ should not be taken too literally. Yet how far was our knowledge advanced!

The history of the determination of the solar parallax is one of the most striking examples of what has been called the 'struggle for the next decimal'. One decimal more means enormous progress, the diminution of the possible error to one-tenth of its former amount. At the end of the seventeenth century it was known that the solar parallax was about 9″, with some seconds' uncertainty, say $\frac{1}{8}$ of its amount. After the Venus expeditions of the eighteenth century its value was held to be between 8.5″ and 9″, with an uncertainty of some tenths, $\frac{1}{30}$ of its amount. Toward the end of the nineteenth century, after some hesitation, its value could be said to be near 8.80″, with some hundredths of a second uncertainty only, i.e. $\frac{1}{300}$ of its amount. And now we may with some confidence add a third decimal, with the certainty that it is accurate to some thousandths of a second, $\frac{1}{3000}$ of its amount.

From the solar parallax and the equatorial radius of the earth (6,378 km.) the astronomical unit is now found to be 149.7 million km., 92 million miles, $\frac{1}{57}$ less than Encke's value. It is the unit not only for the distances but also for the dimensions of the celestial bodies. To find them, their apparent diameters must be accurately measured. This became possible in the nineteenth century by means of instruments from Munich. Since the diffraction phenomena of light rays passing substantial wires prevent the use of filar micrometers, the heliometer is the approved instrument for measuring diameters, by bringing the two images of a luminous disc into contact. So Bessel was the first to derive usable results; after him, Johnson and Main at Oxford, Kaiser at Leiden (with a more simple double-image micrometer, carefully handled), and Hartwig at Bamberg worked along the same lines. The semidiameter of Mercury was found to be 2,380 km. (by Kaiser), of Venus and Mars 6,372 and 3,370 km. (by Hartwig); so with the earth they form the inner group of smaller planets. For Jupiter and Saturn, Kaiser found an equatorial radius of 70,550 and 59,310 km., 11 and 9 times larger than the earth; the polar diameter is considerably smaller, the flattening being $\frac{1}{171}$ for Jupiter, $\frac{1}{62}$ for Saturn. The semidiameter of the sun, for which a special mode of measurement was needed, amounts to 696,400 km., 109.2 times larger than the earth.

Parallel to these developments went the connected problem of the stellar distances. Towards the end of the nineteenth century, when photography was used more and more in astronomy, especially for the determination of positions, the first attempts were also made to use it for stellar parallaxes, but in the beginning with poor results. It appeared that systematic errors vitiated the results more than they did in the case of visual measurements. Diverse causes could soon be traced in the photographic or observational processes. The object star for which the parallaxes were desired was usually much brighter than the comparison stars, which mostly appeared on the plate as faint little discs. If, for some reason, part of the light fell at a spot beside its right central place, its impression for the faint star was too weak to leave any trace, but the image of the bright star was somewhat disturbed or even displaced. Such a deviation might be caused by refraction when during the exposure the field came nearer to the horizon. The observer in 'following' with the solidly connected visual telescope kept the star exactly at the cross-wire; but, because of atmospheric dispersion, the photographic image gradually deviated from the visual image. This explanation of the systematic errors pointed to the means of avoiding them: by removing the difference in brightness and taking the plates when the altitude of the stars did not change, i.e. in the meridian. The example was set by Frank Schlesinger, who in 1903 began to take parallax plates with the 40 inch refractor of the Yerkes observatory, continuing it, from 1905 on, with a photographic refractor at the Allegheny Observatory. The brightness was equalized by having a screen with adjustable free sectors rapidly rotating before the place where the image of the bright object was formed, so that it was illuminated for a few moments only. The sectors were opened to such a width that this image corresponded to the average size of the comparison stars. Since, moreover, the plates were taken symmetrically about the meridian, between one hour before and one hour after the meridian passage, and the long focus of the instrument gave a large scale, the results at once proved to be highly accurate and reliable. The accidental errors of his parallaxes were no more than 0.01″, and they decreased in the course of the work.

This example was soon followed by other observatories. A regular co-operation on stellar parallaxes ensued with six other observatories, Greenwich, Cape Town, and four American, all of which had long-focus refractors or other superior instruments. All were working with the same standard of accuracy; when the results of different observatories for the same star were compared, nothing was found of the former large differences. The differences that were now at most some few hundredths of a second, testified to the new high standard of accuracy. The number of reliable parallaxes increased in the first decades of the twentieth century from tens to hundreds and at last to thousands; Schlesinger's *General Catalogue* of 1924 contained 1,870 stars; and this number was steadily increasing. All stars easily visible to the naked eye and a number of telescopic stars with large proper motions now have their parallaxes measured. The knowledge which earlier had been limited to a few stars has now grown out into a survey of the thousand stars forming our nearest universe. Their distances can now be computed; the unit generally used is the 'parsec', the distance for which the parallax is one second, whereas in popular books the light-year, nearly ⅓ parsec, is often found. Of the stars surrounding us at not too great a distance, the relative situation can now be represented in a spatial model—of course, not at greater distances, because they are too uncertain. If a parallax is determined at 0.01″ with an uncertainty of 0.01″ also, it means that it is probably between 0 and 0.02″, so that its

349

distance probably is beyond 50 parsecs—how much, we do not know. Negative values for parallaxes also occur; it means either that, because of accidental errors, a small positive parallax has been depressed below zero or that the parallax of the object star is smaller than the small parallaxes of the comparison stars. Though such negative parallaxes have no meaning for an individual distance, they should not be omitted in statistical researches where averages are formed.

The fundamental problem of the parallaxes and distances of the fixed stars has thus been solved. Originally aiming at a demonstration of the truth of the heliocentric system, it was also inspired by the desire to acquire knowledge of distances in the world of stars. When a few dozen stellar parallaxes had been determined, they could already be used for a statistical treatment of the nearest surrounding stars. With the thousands of parallaxes now at our disposal, a deeper insight into the structure and character of the sidereal world may be expected.

CHAPTER 34

CELESTIAL MECHANICS

WITH the title of his work, *Mécanique Céleste* ('Celestial Mechanics'), Laplace had set forth a programme of theoretical astronomy. Mechanics is the science of forces and motions; through Newton's law of gravitation the forces were known by their dependence on the positions; then came the task, through integration of the differential equations expressing this dependence, of computing the general characteristics of the orbits of the bodies. The special dimensions of the orbits and the constants determining them had to be derived from the observations.

Astronomy now became the science of patient computation. Besides the practice of untiring observation and measurement, indefatigable computation now took its place as equal partner in practical astronomy. It consisted of two kinds of work: first, the theoretical solving of the mathematical problem, resulting in a body of formulas, which taxed the mind and ingenuity of the most brilliant mathematicians; and then the practical elaboration of this theory in numerical calculation, laid down in endless rows and tables of figures. Astronomy at this time was the only science of nature in which exact practical computation was an important activity; computing was the daily task, so that in astronomy techniques and methods of computing were devised and improved through painstaking practice. Besides the astronomer who observed the celestial bodies in the telescope was a man of equal merit, the astronomer at his desk, who with his pen followed the celestial bodies in their course; they were, of course, often embodied in one person.

Soon there was a lot of work to be done first in the field of the simple two-body problem of motion under the attraction of the sun alone. Comets appeared as faint, small nebulae, but were nevertheless assiduously traced by diligent comet-hunters; their orbits across the planetary system had to be computed. In the eighteenth century they also had been computed, but with difficulty. After the first primitive indications by Newton various mathematicians had occupied themselves with the problem, and Laplace had given formulas for the computation of a parabola through successive approximations; but the procedure was cumbersome and unsatisfactory. The practice of the astronomers who had to deal with them led to the solution of the problem before the turn of the century. In 1797 Wilhelm Olbers (1758–1840) published his treatise *Ueber die leichteste und bequemste Methode die Bahn eines Cometen aus einigen Beobachtungen zu berechnen* ('On the Easiest and Simplest Method to Compute the Orbit of a Comet from several Observations'). Olbers, a physician at Bremen with a busy practice was at the same time a practical astronomer who at night observed the stars and the comets and computed their orbits. Though an amateur, he was highly esteemed among the astronomers. He had devised for himself a method of computing orbits and practised this for many years, not suspecting, in the simplicity of the times, that it was anything special, until his friends persuaded him that it was worth printing. During the entire nineteenth century his method was used by successive generations of astronomers, young and old; some of them succeeded in devising technical improvements in minor points only.

But there was more to it than the comets and their parabolic orbits. In the structure of the planetary system the gap between Mars and Jupiter had often caught attention. Kepler had employed it to insert his tetrahedron. In the eighteenth-century notions about the origin of the planetary system such a gap did not fit. This showed most clearly when the succession of distances was reduced to the mathematical form of a series, the so-called 'law of Titius' or of Bode—Titius had published it in an inconspicuous footnote of a translated book, and Bode had dug it up and made it public in 1772. It renders the size of the orbits of the planets from Mercury to Uranus by the numbers 4; $4 + 3 = 7$; $4 + 2 \times 3 = 10$; $4 + 4 \times 3 = 16$; etc., up to $4 + 64 \times 3 = 196$; but $4 + 8 \times 3 = 28$ had to be omitted: there was no planet known at that distance.

So attention was directed to it, and the opinion was expressed that it should be tracked. But it was found by chance. On the first day of the new century, January 1, 1801, Piazzi at Palermo found among the stars he was observing a faint one of the seventh magnitude that moved. He first called it a starlike comet, without nebulosity; but soon it was realized that it must be a new planet, with a period of revolution of about four years; he afterwards named it 'Ceres', the tutelary deity of Sicily. The discovery was made known by letters to foreign astronomers; but communication in those days was slow because Italy was full of warring armies, and before others could observe it, the planet had been overtaken by the sun. Piazzi had made a number of observations, but he withheld them in order to publish the orbit computed from them at the same time. He worried over this task without much success, because no regular method existed of computing an orbit from an arc as small as a few degrees. When the planet was sought at the computed place, after the period of invisibility, it was not there; it was lost and could not be

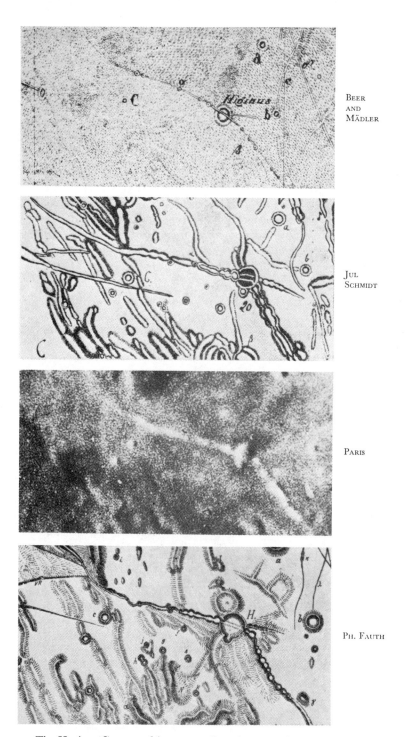

BEER
AND
MÄDLER

JUL
SCHMIDT

PARIS

PH. FAUTH

11. The Hyginus Crater and its surroundings (pp. 372–4)

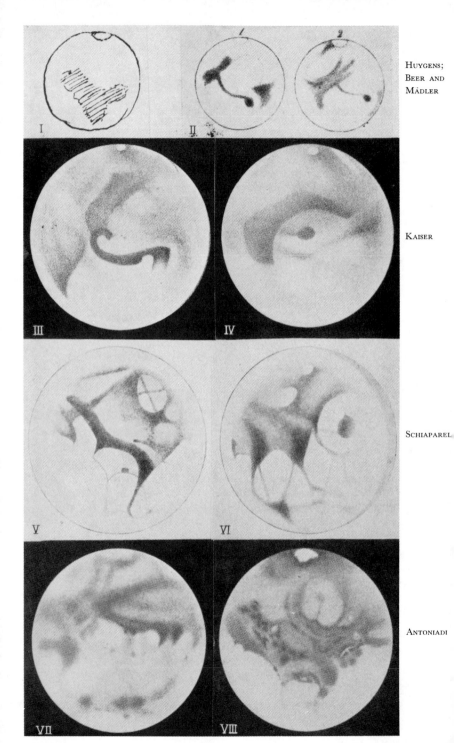

12. Drawings of Mars (p. 376)

recovered. Then the young mathematical genius, Carl Friedrich Gauss (1777–1865), who happened to be working on the problem of planetary orbits, turned his attention to the lost planet. From Piazzi's observations he determined the orbit by his new method, and the planet was found before the end of the year at the place predicted by him, quite different from the former computations.

The success of a skilfully devised theory gave Gauss's method of computing orbits its prominence during the entire nineteenth century. It was expounded at length in his *Theoria motus corporum caelestium*, which appeared in 1809, in form and method the most perfect textbook on motion in unperturbed orbits. Whereas Olber's method supposed the orbit to be a parabola, determined by five elements, Gauss's method makes no supposition about the form and character of the orbit; the character of the conic section expressed by the eccentricity comes out as one of the six computed elements.

This method soon found further application. In the region where he looked for Ceres, Olbers, in April, 1802, discovered another moving star of the seventh magnitude, a second small planet, called 'Pallas', having an orbit of nearly the same size but strongly inclined to the ecliptic. Two more planets were added in the next years—in 1804 the planet Juno, discovered by Harding; and in 1807 the planet Vesta, discovered by Olbers. Gauss's method in each case, after a short period of observation, produced the elements; by means of these elements the new planets were secured and their course computed in advance.

Thus the gap was filled, but in quite an unexpected way: by four tiny planets instead of by one large one. They were more than a thousand times fainter than Mars or Jupiter, looked exactly like stars, and represented another type of celestial body. They were called 'planetoids' or 'asteroids' or simply 'minor planets'. They have orbits of nearly equal size, 2.3–2.8 times the earth's orbit, and periods of revolution between 3.6 and 4.6 years.

For nearly forty years their number remained four. Astronomers, however, were convinced that there must be more. Therefore, to discover them in an easier way, the Berlin Academy organized the construction of star maps covering the zodiacal zone and containing all the stars down to the ninth magnitude. This was made possible by the co-operation of about twenty astronomers, each taking a field of 15° square, inserting the stars of the catalogues according to their co-ordinates and subsequently the bulk of the other stars by eye-estimates at the telescope. Thus furthered, in 1845 there ensued a regular stream of new discoveries which increased steadily. In 1852 the number of minor planets had risen to 20; in 1870 it reached 110. Of course, they were increasingly smaller bodies, and their brightness on discovery decreased to the ninth magnitude, then lower and lower, to the tenth, eleventh, and twelfth magnitudes. The Berlin star maps were now no longer of any avail; the chief discoverers, such as C. H. F. Peters at Clinton and Palisa at Vienna, had to construct new maps to include the fainter stars, and these maps were published. Then Palisa simplified the work by using photographs taken with long exposures by Max Wolf at Heidelberg, with a wide-angle objective, and published reproductions of them. Thus he was able himself to add nearly one hundred new asteroids to the list.

A new and still more efficient photographic method was introduced about 1891 by Max Wolf. When he photographed a large star field with an exposure of several hours, the camera exactly following the stars, the motions of the minor planets during this time produced small streaks, which, merely by visual examination, could at once be distinguished among the thousands of pointlike stellar images. These trails when measured afterwards, gave good positions of the planets for the central moment of exposure. In this way Wolf rapidly increased the number of minor planets; in 1900 it had risen to 450, and he himself discovered (up to 1927) a good 500. Other observatories joined him with similar or better instruments. Through a variation of the method, by having the telescope follow the estimated course of the planets so that they produced small points while the stars drew streaks, still fainter objects were hunted up. In this way they increased still more rapidly; in 1938 their number had reached 1,500. Whereas the first four discovered were substantial bodies of 500 or 800 km. in diameter, the last hundreds tracked had diameters of only 50, 40, or 30 km.

This unexpected increase confronted observers and computers, as well as theoretical astronomy at large, with increasing difficulties. From the very first, the computation of orbits meant a lot of work. The computation of an elliptical orbit from three observations— roughly some days after the discovery, but more precisely from observations a month later—could be performed rapidly enough by Gauss's method. But in order to predict the positions in later years with sufficient certainty, a large number of observations, extending over the entire period of visibility, was first needed, and then the computation of the most probable elements from all these observations had to follow. Moreover, the work was never finished, because every succeeding year brought a new opposition, with new observations. If however this work were neglected, the predicted result would be more and more in error, so that the planets could not be found again among all the little stars, and if seen by chance later could not be identified; general confusion would result. Now and again it was said that we should disregard all the small fry and drop them; but where is the limit? It was the same here as with the entire technical development of the nineteenth century; people were dragged along in endless labour which allowed no slackening. In the early years, about the middle of the century, the enthusiasm and perseverance of young scientists and the charm of astronomical computing sufficed to satisfy the needs. When the numbers increased alarmingly, when, moreover, the ever smaller bodies lost their salient individuality although names were assigned to them, this supply of work began to fall short. Now the work was concentrated more and more in computing offices, where official duty and routine, organization and mechanized computing methods, combined to cope with the ever increasing flood, though with a decimal less in accuracy. The Berlin Computing Office afterwards transferred to Dahlem and christened 'Copernicus Institute', among other duties took over the major part of this crowd of celestial Lilliputians.

If only the regular orbit computation had been the only task! All these little bodies, however, were subjected to the attraction not only of the sun but also of the major planets. So their orbits were continually changed through perturbations. These perturbations are even larger than those of the older planets, first because the minor planets come nearer than any other to Jupiter, the great perturber of the solar system, and also because their orbits often have great eccentricities and inclinations. Yet the perturbations must be computed. If we should neglect them, the computed orbit within a few years would be far too incorrect; the errors of the computed positions would be so large as to render the entire computation useless. Nor could we think of developing the perturbations, once and for all, in a general algebraic way, as Laplace had done for the seven major planets. Whereas in the latter case, where the high order terms rapidly decrease, the theory of one planet may already demand exertion over dozens of years, the number of terms for a minor planet would be endless. Hansen in later years, in 1856, indicated an approximate method and applied it to some of the first discovered asteroids. However, all the astronomers who first encountered the problem—Gauss, Encke, Olbers, Bessel—agreed that there was only one practical way to take account of the perturbations. This was the same method used since Clairaut for the comets: to follow the planet in its course continually, from place to place, from week to week or month to month, computing for every moment the perturbing forces, motions and displacements, and to see where they next brought it. It is a never-ending work, each year requiring as much time as did the preceding one. Since, however, there was no other way, this method of 'special perturbations' has been built up, mainly by Encke, into a handy, fixed and simple routine scheme that has been used by all good computers of the first dozens of planetoids, to give a solid basis to the derivation of orbits. Finally, however, when the number grew into hundreds, even the greatest ardour and patience could not master the work. The computing offices had now to devise gross methods for rapid and approximate computation of perturbation terms for a number of similar orbits. In the judicious balancing of the opposing demands of feasible work and attainable accuracy, no less ingenuity was needed here than formerly for the mathematical problems themselves. Yet astronomers always have to face the question of whether it pays to derive orbits for all those small and smaller lumps of rock of some tens of kilometres straying through space.

The reward of all these labours took the form of new discoveries, new insight and new interesting problems. Kirkwood, in 1857, when the number of minor planets was hardly sufficient, and again in 1866, pointed out that the periods of revolution were not regularly distributed about their average of 4.7 years. There were gaps at 5.93 and 3.95 years, exactly $\frac{1}{2}$ and $\frac{1}{3}$ of Jupiter's period. When the number of planetoids increased, these gaps stood out still more clearly, and less definite gaps also appeared at $\frac{2}{5}$ and $\frac{3}{5}$ of Jupiter's period. It must be an effect of Jupiter's attraction, which obviously does not tolerate these simple commensurabilities and drives the small body away from this situation. As an unexpected new phenomenon, it confronted celestial mechanics with new problems.

In 1906 four new planetoids were discovered shortly after one another —some more were added later on—with periods of revolution of 12 years, exactly identical with Jupiter's. They travelled in a similar orbit at a distance of five astronomical units from the sun, keeping a distance of nearly 60° in longitude from Jupiter, some preceding, some following; in this way they formed an equilateral triangle with Jupiter and the sun. This had already been anticipated by celestial mechanics; Lagrange had theoretically designated these triangular points as points of equilibrium, where small objects under the combined attraction of the sun and Jupiter are relatively at rest. Further analysis showed that small bodies can oscillate about these points in so-called 'librations', always accompanying Jupiter at nearly 60° distance. These remarkable fellow-travellers were named Achilles, Hector, Patroklus, Nestor and other heroes of the Trojan War; the entire group was called the 'Trojan planets'.

Remarkable in other ways and more important practically are such minor planets as in their perihelion pass to within the orbit of Mars and come near the earth. One of them, Eros, we have already met as an excellent aid in determining the solar parallax. Its oscillations of brightness, not even entirely regular, indicated that it was not a perfect globe but a more or less irregular block, rotating irregularly. Surely, large numbers of such misshapen boulders are swarming through space; but they are visible to us only when they pass the earth at a short distance. In 1932 Delporte at Brussels discovered an object on his plates that could be followed in its rapid motion for some days and that allowed an uncertain orbit to be derived; it had approached the earth to a distance of 0.11 astronomical units, i.e. 44 times the distance to the moon. Shortly afterwards Reinmuth at Heidelberg found on his plate a trail produced by a planetoid passing the earth at 0.06 distance. The extreme in close approaches was an object which in 1937, on October 30th, dashed past the earth at a distance of 0.004, i.e. $1\frac{1}{2}$ times the distance to the moon. Compared to the ordinary minor planets, such bodies are of an even lesser order, measured by single miles, and only visible at such very short distances. The astronomer is faced with the question, down to what size do these boulders have to be treated as planets for which orbits are computed? For mankind they pose the more important question of what will happen if they pass us at far shorter distances or even hit the earth. Their mass is so small that such collisions cannot perceptibly disturb the motion of the earth or moon. However, if they happened to strike a continent, the transformation of arrested energy of motion into heat, making the body explode into gas at thousands of degrees of temperature, would cause catastrophic devastation.

Such collisions, though on a far smaller scale, have in fact occurred. Ancient reports sometimes spoke of fiery stones that fell from the heavens; but in the eighteenth century they were relegated to the realms of fables. When meteorites fell in the French province of Gascogne in 1790, the Paris Academy refused to accept a protocol on this phenomenon, in order not to encourage a 'superstition unworthy of these enlightened times'. The German physicist Chladni, however, demonstrated in a well-documented study that a number of real cases had occurred; and shortly thereafter new instances confirmed his view, one meteorite even setting fire to a farm. Small meteorites are now collected and preserved in museums; bigger ones have been found in far countries, often worshipped as legendary objects. A great crater in Arizona is ascribed to the impact of a gigantic meteorite in prehistoric times. On July 30, 1908, a meteorite struck and devastated a region in Siberia fortunately uninhabited. Though optical atmospheric phenomena were widely perceived all over the earth, their cause remained unknown until, ten years later, a scientific expedition brought the occurrence to light.

From this digression by way of planets to meteorites, we now return to the orbits. Gauss's method had originated from the problem of the planetary orbits. It was applicable of course to any ellipse, even to the parabola, where the computation would produce an eccentricity of one; so it was also used for comets. In the first half of the nineteenth century comets repeatedly turned up, small and faint nebulosities, for some of which, as became immediately apparent, a parabola did not fit. Here, by means of Gauss's method, strongly elongated ellipses of small dimensions and short periods of revolution could be deduced. The smallest orbit was shown by a comet discovered by Pons at Marseilles in 1818, which, however, contrary to custom, has always been named Encke's comet after the computer who during the following years took care of it by computing its course. Its period was 3.3 years only, and it turned out that at former appearances—in 1786, 1795 and 1806—it had been observed but not recognized as a periodical appearance. After Halley, Encke was the second man to predict the return of a comet, and it has since been computed, predicted and observed at every return. Its distance from the sun varies between 0.34 and 4.08 astronomical units, so that it does not reach as far as Jupiter's orbit. To compute the perturbations, there was no way but to follow the comet continuously along its orbit by means of a careful computation of special perturbations. As a reward, this procured an accurate derivation of the mass of Mercury, because in 1835 the comet passed it at close quarters. A second result was the curious phenomenon that the orbit gradually became smaller and the period shorter. Encke attributed this to a resisting medium filling the space between the planets; such a resistance can reveal itself only in such extensive and tenuous objects as comets. Other comets, however, did not show it; and later computers, continuing Encke's work on this comet (first Backlund and then Von Asten), were inclined to ascribe the resistance to separate encounters, in special parts of its orbit, with other comets crossing its track.

Shortly after the first short-period comet, another was discovered in 1826 by Biela; it had been perceived in 1772 and in 1805, and had a period of 6¾ years; we shall return to it later. Gradually, more comets with elliptical orbits were discovered. Their number, however, remained restricted: a dozen with periods of about 5 or 6 years, some few with 13 and 33 years. Halley's comet returned twice, in 1835 and in 1910, and was of course carefully observed and computed. A small comet discovered by Olbers in 1815 was found to have a similar period of 72 years and duly returned in 1887. That cometary periods occur in such groups seemed to be connected with the planets; elliptical orbits of 6, of 13, of 33, of 75 years have their aphelia at the distance of Jupiter, Saturn, Uranus, Neptune. That a comet could acquire a short five-year period through the attraction of Jupiter had already been stated by Laplace; if in its parabolic orbit it comes very near to Jupiter, the strong attraction of this powerful planet can throw it into an entirely different track and keep it captive within the solar system. It was supposed for the other groups that a close encounter here with Saturn or with Uranus or Neptune had produced the orbits of 13, 33 and 75 years. Such orbits will not always be final; if it is not in the meantime displaced by other perturbations, the comet must return to its meeting place with the planet and can suffer a new large change in orbit. Such has been the case with a comet observed in 1770, for which Lexell at St Petersburg derived an elliptical orbit of 5.6 years. It has never been seen again. According to Lexell's computation, confirmed by Burckhardt, the comet had acquired this orbit in 1767, shortly before its appearance, through a near approach to Jupiter; after two revolutions it came back to the same spot in 1779 when Jupiter was there again. And it was thrown into a rounder ellipse that forever kept it out of our sight. Other analogous cases occurred later on.

The essential domain of celestial mechanics did not lie in all these computations of orbits; the perturbations of the major planets and the moon were its most important object. After Laplace had brought the theoretical work of the eighteenth century to a close, the task of deepening the foundations of theory and of bringing the developments to a higher stage of accuracy was left to the nineteenth century. The former task was taken up by such ingenious mathematicians as Jacobi (about 1830) in Königsberg, and Henri Poincaré (about 1880) in France. The exact computation of the perturbations of the planets was the object of many studies on the part of able theorists in the first half of the nineteenth century, such as Cauchy, Bessel and Hansen. The most thorough and complete treatment of these perturbations was the work of Leverrier.

Urbain Jean Leverrier (1811–77) gained fame mainly by his theoretical discovery of the planet Neptune, solely from the perturbations it produced in the motion of Uranus. That the motion of Uranus presented irregularities not accounted for by the attraction of the other planets was first perceived by Alexis Bouvard, a farmer's boy from the Alps, who had come to Paris to study science and who, owing to his talent for computation, became an invaluable aide to Laplace. He computed tables for the major planets, but his tables of Uranus, derived from regular observations in the 40 years after its discovery, could not represent the earlier scattered data when Uranus had been observed as a star. In publishing his tables in 1821, Laplace spoke of 'some extraneous and unknown influence which has acted upon the planet'. When in the following years Uranus again began to deviate more and more from the tables, the opinion became widespread among astronomers that there must be an unknown planet disturbing the motion of Uranus. In the late 'thirties Bessel made one of his pupils, F. W. Flemming, attempt a computation of this unknown planet from Uranus's deviations; but Flemming died when the work had just started. In 1842–43 J. C. Adams (1819–92), a gifted student of mathematics at Cambridge, began to tackle the problem; in September, 1845, he was able to communicate to Airy, the Astronomer Royal, and to Challis, the director of Cambridge Observatory, his results on the orbit and position of the guilty planet. Because both astronomers lacked confidence in the research, and as a consequence of his own modesty, the result was not published, and no attempt was made to discover the planet. In the meantime, urged by Arago, Leverrier set to work on the problem. First he made a thorough revision of the theory of Uranus and published it in November 1845. In June, 1846, he added his results on the orbit of the supposed unknown planet and its position in the sky. When Airy and Challis saw his result, which agreed closely with that of Adams, Challis began to make a search in July and August by registering at different days all the stars in a field about the assigned place, to see whether one among them changed its position. Diverted by other work, he then failed to reduce and compare his observations; otherwise he would most certainly have discovered the planet, for it was among the registered stars. In the meantime, Leverrier, impatient that no observer gave heed to his results, in a letter to Galle, astronomer at the Berlin Observatory, asked him to examine with the large refractor the stars about the place indicated, to see whether one showed a disc. A short time before, the Berlin Academy map of this very region had been received at the Berlin Observatory. On receipt of Leverrier's letter, on September 23rd, the map was at once compared with the sky, and the planet was immediately found as a foreign star of the eighth magnitude not present on the map. It received the name of Neptune.

This course of events made a deep impression on the world of scientists, but no less on the world of educated laymen. From all countries honours were showered upon Leverrier, and the discovery at a desk of a body never seen was the ruling topic for a long time. It was in this mid-century that science came to dominate the world concepts of the middle class in western Europe, and in a spiritual struggle gradually superseded the traditional biblical ideas. A number of popular books on science, by spreading knowledge, furthered the *Aufklärung* ('enlightenment'); welcomed enthusiastically among intellectuals and laymen, they served as an aid in the fight against antiquated political and social ideas and institutions. In such an environment this unexpected demonstration of the power of science and the certainty of its predictions came like a brilliant ray of light to strengthen the fight against darkness. Surely the astronomers were right who pointed out that any of the hundreds of computed perturbations used in the planetary tables, whose exactness was confirmed by subsequent observation, was as strong a demonstration, silently repeated every day, of the truth of science. Yet the brilliance of this discovery, happening in the intellectual atmosphere of the years that led to the stormy 1848, made it an important event in the history of science.

To the astronomers it at once brought new problems and new worries. The difficulty in the research had been, first, that it was of course an inverse problem—to derive not the effect from the cause but the cause from the effect; the derivation of the perturbing body from the perturbations was quite a new adventure. The second difficulty was the uncertainty of the data, owing to the short period of regular observation of Uranus. So as not to make the solution practically impossible by the larger number of unknown elements of the unknown orbit, both Adams and Leverrier had assumed, for the mean distance to the sun, 38 astronomical units, according to Titius's law; the corresponding period was 217 years. Both found a large eccentricity of about 0.10, which, for about 1820, made its distance to the sun 34 astronomical units. They could represent in this way the older observations of Uranus before its discovery, that of Flamsteed in 1690 excepted. After Neptune had been discovered and had been observed for some months, Adams (and also Walker in Washington) found that its orbit was much different; it was smaller, with a mean distance to the sun of 30 units and a period of 164 years, and with a very small eccentricity. The real Neptune, therefore, had occupied quite different places in space and had exerted forces upon Uranus widely different from those of the theoretically computed planet. Hence the Neptune discovered by Galle, said the American computers Peirce and Walker, was another body than the planet computed, and its discovery near the designated place had been pure luck. The European astronomers, Leverrier himself in the van, passionately tried to refute these criticisms and doubts. Peirce and Walker in their further work have contributed most to dispel them, first by showing that the real Neptune, through the perturbations it produced, could entirely represent the observations of Uranus, that of Flamsteed included. Moreover, they pointed out that the problem facing Adams and Leverrier, like so many inverse mathematical problems, allowed of several different solutions, all satisfying the data. By assuming the mean distance to be 38, they had hit on one of the solutions, whereas the real Neptune represents another. Figure 31, in which the arrows represent the perturbing forces working upon Uranus, shows that they mattered only in the period between 1790 and 1850 about the conjunction of the two planets. In these years the real and the computed Neptune, because of the large eccentricity of the latter, stood nearly in the same position relative to Uranus; the remaining small difference in distance could be accounted for by a somewhat larger mass of the computed planet. The large differences in position occurred at times when the perturbing forces were very small.

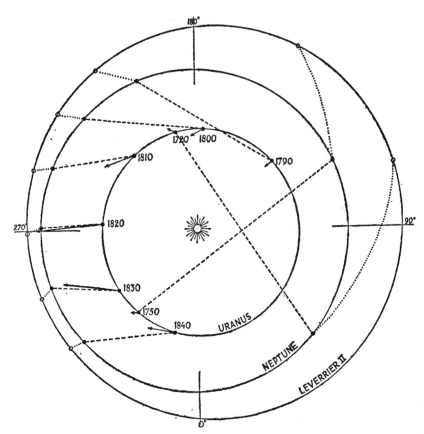

Fig. 31. Perturbing forces of Neptune on Uranus

These researches by the American scientists met with a chilly reception in Europe, and they were mostly ignored as injurious to the honour of the discoverers. The Leiden astronomer Kaiser, in a popular book on planetary discoveries, expressed his surprise and disapproval over such unscientific conduct. He said that evidently the European astronomers had seen the discovery of Neptune chiefly as a means to impress the lay world with the perfection of their science, which in reality, as a human product, it never can attain, and from prejudice rejected everything that could throw doubt upon the agreement between prediction and realization. 'In North America they did not proclaim the miraculous character of the discovery, but they worked all the harder to make it subservient to the benefit of science.'[194] What Kaiser here observed and criticized can be understood when we consider that in the United States there was no need for such a fight for social progress against antiquated social systems and powers as in Europe where, on the contrary, militant natural science was an important factor of a new culture.

Leverrier, since 1853 director of the Paris Observatory, further devoted himself to the perturbations of the planets. Before the Neptune interlude he had already given an accurate theory for Mercury; now he took up all the other major planets. Using new mathematical methods, he could extend the approximations to a far higher order of terms than his predecessors had done. In this way he achieved (a difficult approximation on account of the multitude and smallness of the terms) an accuracy that had formerly been unattainable but which was needed to keep pace with the greatly increased accuracy of the observations. Whereas at the beginning of the nineteenth century astronomers had been content with an agreement of 10" or 20", now the uncertainty decreased to a single second or even less.

As a result of his investigations which, between 1855 and 1877, filled many volumes of the annals of the Paris Observatory, he could state a nearly-perfect agreement between theory and observation. There remained a few discrepancies. The longitude of Mercury's perihelion according to theory must increase 527" each century; the observations, however, of Mercury's transits before the sun since 1631 with great accuracy afforded 565", i.e. 38" more. Moreover, the nodes of Venus's orbit regressed more slowly, and the perihelion of Mars advanced farther (by 24") than theory demanded. The latter deviations were explained when it appeared that the earth's mass, assumed after Encke's solar parallax, was too small by $\frac{1}{10}$. The large deviation of Mercury, however, could not be removed by any acceptable change in the masses. Different explanations were suggested—an undiscovered planet inside the orbit of Mercury; diffused attracting matter about the sun; a small variation in Newton's law, consisting in an increase in the exponent 2 by $\frac{1}{8000000}$. None of them was satisfactory.

When Simon Newcomb (1835–1909) became chief of the computing office for the *American Nautical Almanac*, he had already spent a great deal of time in preparing exact solar and planetary tables for the ephemerides. An able theorist, he succeeded, by using symbolic methods of analysis, in conveniently arranging, organizing and taking into account the entire field of higher terms, thus improving Leverrier's work. With no less practical ability, he succeeded in collecting and reducing uniformly all the old and new series of often unpublished observations made in Greenwich, Paris, Washington and elsewhere, which had been only partly used by Leverrier. He condensed them into conveniently arranged, easily examined tables and used them for the derivation of the best possible elements. This combination of theoretical and practical skill resulted in a definite test of the theory.

It appeared that, for the four innermost earthlike planets, the secular variations of the orbital elements showed entire agreement between the theoretically computed and the practical values, the differences remaining mostly below the probable errors of the latter, though with some few exceptions. For the recession (per century) of the nodes of Venus, the observed and the computed values were $1,783''$ and $1,793''$, a difference of $10'' \pm 3''$; for the advance of Mars's perihelion, they were $1,603''$ and $1,595''$, a difference of $8'' \pm 4''$; for the advance of Mercury's perihelion, they were $575''$ and $534''$, a difference of $41'' \pm 2.1''$. The first two cases mean differences in the positions of the planets of less than $2''$; for Mercury the deviations in position may reach $8''$, more than can be admitted for a result of many accurate observations. This, then, remained the only weak spot in the solid structure erected on Newton's law of gravitation. It would have been an idle attempt to remove this contradiction by artificially invented explanations. We had to wait, as is so often the case in such situations, until new points of view should arise from entirely different quarters. They came from the side of physics, in 1905, through Einstein's principle of relativity.

The theory of relativity is based upon the fact that no absolute motions but only relative motions can be observed. In 1914 the principle of the general theory of relativity was postulated by Einstein: that the true form of the laws of nature cannot depend upon the state of motion assumed for the observer. Newton's law of attraction did not conform to this principle, quite understandably, since Newton had proceeded from the notions of absolute space and absolute time. To satisfy the principle of relativity, Newton's simple formula had to be modified, and the consequence was a movement of the perihelia of the orbits. The difference, depending on the square of the ratio of the planet's velocity and the velocity of light, was so small that only for Mercury, the most rapidly moving planet, was the effect perceptible; it consisted in an advance of the perihelion by $43''$ per century. The only serious discrepancy that had remained between theory and observation found its explanation here in a natural way, through a refinement of theory, without any arbitrary assumption. Should someone have doubts about the certainty of the relativity principle, there was another consequence: light rays passing large world bodies were subjected to its attraction and might be deviating from their straight course, just as would a particle passing with the velocity of light. This deviation, amounting to $1.75''$ for a ray grazing the sun's surface, was established by photographs of the field of stars surrounding the sun during the total eclipse in 1918, taken by two English expeditions; later expeditions confirmed this first result. It gave certainty to the theory of relativity as a whole and so to its explanation of the motion of Mercury.

Though the theory of the planets, especially of the motion of the earth reflected in the apparent solar motion, had practical consequences for the whole of astronomy, it was in the theory of the moon that the vital importance of celestial mechanics lay, because of its importance to practical life and trade. The result of the eighteenth century's work was that Laplace, in his formulas and tables, could represent the course of the moon to less than half a minute of arc. This was far more than the errors of observation; so theory had to be developed further. Laplace had included terms with the third power of eccentricity and inclination. A prize problem was set by the Paris Academy in 1820, asking for the construction of tables of the moon, entirely founded upon theory. In his answer Plana gave a new form of theory, whereas Damoiseau, following Laplace's method, extended it to terms of the seventh order and reached an accuracy to a small number of seconds—yet insufficient for modern needs.

It was owing to the complication, multitude and mutual dependence of the terms that the elaboration of the theory was so very difficult. The ablest mathematicians tried to overcome the difficulty by starting developments according to new theoretical designs. In 1838 Peter Andreas Hansen (1795–1874) at Gotha, lacking good instruments for observation, began to develop theoretical formulas for the motion of the moon. This was the basis of a research that was to occupy him for twenty years. His aim was, by his theory, to represent, with no greater deviations than a second of arc, the complete series of observations of the moon made at Greenwich since 1750. The resulting tables of the moon were published in 1857 by the British Admiralty and were used for more than half a century as the basis of the *Nautical Almanac*. In 1862–64 he added to it an exposition of the theoretical basis and all the data used. Along an entirely different though equally general way Charles Delaunay in Paris at the same time developed a lunar theory and published his results in 1860 and 1867. So very different were their methods that, later on, only by an extensive special research, was Newcomb able to compare their results for each of the terms. Then such a wholesale agreement came to light that theory could be said to have given an unequivocal answer to the question of the moon's motion.

Yet there were some weak points. The rapid advance of the longitude of the moon's perigee of about 40° per year, which had embarrassed Clairaut in the eighteenth century, had remained a difficulty. Through a laborious computation of eight terms, Delaunay had succeeded in approximating its value to $\frac{1}{8000}$; since this margin amounts to 18″ yearly, the uncertainty is far too great to be tolerated. Then George W. Hill, computer at the American 'Nautical Almanac Office' (described by S. Newcomb as 'perhaps the greatest living master in the highest and most difficult field of astronomy, winning world-wide recognition for his country in the science, and receiving the salary of a department clerk')[195] in 1877, by a stroke of genius, solved the problem through an entirely new principle of treatment. Whereas the common method proceeds from the general orbit of the two-body problem, the ellipse with its chance eccentricity, to which the solar perturbative action is added, he proceeded from the orbit in the three-body problem in its most simple form, without chance characters—a circle transformed by Tycho's variation—to which the chance eccentricity is then added. This fitted in so well with the inner essential character of the problem that the successive terms of the development decreased with a factor $\frac{1}{30000}$, so that a short computation of some few terms sufficed to produce the desired quantity to $\frac{1}{100000000000}$ of its amount. Was it now in exact agreement with observation? The question cannot be put in this form. In addition to solar attraction, the moon's perigee is advanced by the flattening of the earth, in a way that depends on the density distribution within the earth. If the computed solar effect is subtracted from the observed value, there remains the flattening effect. The resulting flattening is the mechanical flattening (i.e., the difference between the moments of inertia), not the geometrical flattening of the outer form. It can be used as a datum to derive the distribution of mass inside the earth. Hill's theory has been used by Ernest W. Brown as the basis of his lunar tables, which since 1922 have replaced Hansen's tables in the nautical almanacs.

A second point was the secular acceleration of the moon. Laplace had found that the ancient eclipses and the theoretical computation of the sun's perturbing effect agreed in establishing its value at 10″ per century. By theory Hansen had found 11.47″, and from observations 12.18″; not a great difference. In 1853, however, by more extensive computations including higher terms, Adams found the theoretical value considerably smaller, 5.70″. At first he was contradicted from all sides, chiefly because the former value fitted in so well with the results of observation; but gradually his conclusion was confirmed by Delaunay, by Plana, by the eminent mathematician Cayley, and was finally also recognized by Hansen. This involves the presence of another factor accounting for the other half of the observed acceleration.

This other factor was found in tidal friction. The high-tide wave is dragged along by the earth's rotation; the mutual attraction of the wave crest and the moon retards the rotation of the earth and pulls the moon forward, increasing its angular momentum. This results in an expanding of its orbit and a slowing down of its orbital motion; expressed in days of increased length, the month is apparently shortened and the motion accelerated. It was tacitly assumed that the tidal friction of the lunar tides was able to explain half the observed acceleration. In 1909, however, Chamberlin, geologist at Chicago, deduced that the friction of the oceanic waters against sea bottom and shores was too small to produce such an effect. So this way out seemed to be cut off, until in 1919 Geoffrey Taylor showed that it is especially in shallow inland seas and bays with strong tidal currents that the energy of motion of the earth and moon is lost through tidal friction. According to a rough computation, the Irish Sea could produce $\frac{1}{86}$ of the necessary loss; the Bering Sea later on was found to be responsible for more than half the amount needed, whereas the oceans contributed little. It was of course not possible to compute the exact amount due to all the seas on earth; so the value adopted by Hansen must be considered as an empirically and not a theoretically determined quantity.

In the Greenwich longitudes of the moon there remained, after all other perturbations had been inserted, a slow fluctuation amounting to some tens of seconds. Hansen supposed it to originate from a perturbation by Venus, with a period of 240 years amounting to 21″. Delaunay, by theoretical computation, could find for this term an imperceptible amount only, below 1″. Yet Hansen, because the Greenwich observations clearly indicated it, introduced it into his tables, confident that theory by means of further researches would be able to explain it. So it was now an empirical term, based upon the rather short time of one century; the earlier observations—less reliable, it is true—did not confirm it. What was worse: in the years since 1860, when Hansen's tables had been introduced as a basis for the almanacs, the moon began to deviate increasingly from the tables. In the sixties it was retarded some seconds; in 1880 it was already 10″ behind; and in every following decade the retardation increased. Then Newcomb in 1878 took the matter in hand. He had already begun to collect, as a check on the meridian observations, all the ancient observations of eclipses and star occultations since the invention of the telescope; these he copied from the manuscript data in the archives of the observatories, while he was on a journey across Europe in 1871. The entire collection of observed data could be represented by introducing an empirical term of 17″ with a period of 273 years—practically confirming Hansen. An explanation, however, was still lacking; so it was not to be expected that in the coming years the differences between the tables and the observed positions would cease. Tisserand in the third volume of his standard textbook, *Traité de mécanique céleste*, in 1894 closed his exposition of the lunar theories with the words: 'The theory of the moon finds itself arrested by the difficulty we have just developed; in Clairaut's time also, theory seemed to be unable to explain the movement of the perigee. It will also vanquish the new obstacle now presented; but a great discovery has still to be made.'[196]

The discovery was indeed made and the problem solved, but in an entirely unexpected way. The cause of the deviations did not lie in the field of theory, nor in the moon's motion, nor in the forces of attraction, but in irregularities in the earth's rotation. Newcomb for a moment had considered it, but in his last work on the moon, in 1903, he had dropped the idea as improbable and not sufficiently warranted. Such variations in the period of rotation of the earth must reflect themselves in corresponding apparent irregularities in all rapidly moving heavenly bodies. In 1914 Brown established that in the motion of the sun, Mercury and Venus the same oscillations occurred as with the moon, at the same time but to a lesser amount. It was confirmed by Spencer Jones's detailed studies at Greenwich in 1926 and 1939. When Willem de Sitter (1872–1934) at Leiden investigated the motion of Jupiter's satellites and worked out a complete theory of all the perturbations in this sub-system, he found in 1927 the same irregularities in their motion. So the conclusion was clear; the moon was not guilty; it was the earth. The empirical term by Hansen and Newcomb could now be dropped; in its place came more or less sudden leaps in the earth's time of rotation. About 1667 it was lengthened by 0.0011 second; about 1758 it was shortened by 0.0006; about 1784 by 0.0017, and in 1864 by 0.0027. This caused the alarming deviations from Hansen's tables. Then followed a lengthening of 0.0017 in 1876, of 0.0034 in 1897, and a shortening of 0.003 in 1917.

It is a remarkable fact, though not surprising, that the simple conditions of world structure given by primitive experience and assumed throughout the centuries to be the results of cosmological or mechanical principles could not be maintained against modern refined research. Such was the case with the invariable position of the rotation axis within the earth, as well as now with the invariable period of the earth's rotation. Their close approximation was an important aid in establishing a simple and harmonious astronomical world concept. The great precision of the nineteenth-century measurements showed the deviating—

though very little deviating—reality. Theoretical mechanics, in the hands of Euler, had demonstrated that only as a highly exceptional case could the rotation axis keep an invariable position in a rotating body. In 1885 S. C. Chandler, in a series of latitude determinations at Cambridge, Mass., found progressive changes of 0.4″ in half a year, which could not be ascribed to any error, so that 'the only alternative seemed to be an inference that the latitude had actually changed. This seemed at the time too bold an inference to place upon record . . .'[197] The same thing had already occurred in a series of very accurate observations at Washington in 1862–67; the observers could find no error, but did not venture to explain the periodic deviations as real variations in latitude. In 1888 Fr. Küstner published the results of accurate measurements he had made in Berlin in 1884–85, in order to derive the constant of aberration; he explained the differences found as real variations in latitude. To test the explanation, a second observer was sent to Honolulu at the other side of the North Pole; so it was ascertained that the pole had moved. It described a small orbit of some tens of feet around the end of the axis of maximum inertia. Euler had computed that such an orbit in a rigid earth is completed in 305 days; the search for variations in latitude in this period had always prevented the discovery of the real variations. The real period appeared now to be about 430 days, besides an additional yearly oscillation. Newcomb soon explained the difference by the consideration that the ocean waters, obeying centrifugal force, continually try to adapt their figure to the momentary axis of rotation and in this way slow down the displacement of the axis. The yearly displacement of water masses between arctic and antarctic regions, by melting and freezing, as a cause of the yearly oscillation of the poles, points to other displacements of masses within the earth as a cause of polar variations in general. Now that the period of rotation has turned out to be inconstant, we have to appeal to the same field of events: geological changes within the earth. Since the total moment of momentum is constant, an acceleration of the rotation means a shrinking, its retardation means an expansion, in the earth's mass. It must be added that redistributions of mass to account for the observed leaps demand, from a geological point of view, the transportation of enormous quantities of matter.

For the moon tables in the almanacs the situation now looked precarious. There is no possibility of predicting or computing beforehand, with indefinitely increasing accuracy, the course of the moon measured by the common days of the earth's rotation. Unpredictable changes in the earth's rotation, dependent on unknown geophysical events, cause irregularities in our counting of time which appear in the observed places of the moon. That great ideal of perfect tables of the moon, which navigation needed, on which for several centuries so many of the ablest theorists and observers had spent their best efforts and which seemed within reach—was it to remain unattainable?

Navigation no longer needs the moon. For the centuries-old problem of the longitude at sea a complete solution has meanwhile been found in quite a different way, by wireless time signals. In 1886 the physicist Heinrich Hertz at Bonn made his first experiments with electrical oscillations propagated as waves. In 1895 Marconi introduced wireless telegraphy by means of these radio waves. In 1913 international collaboration was established to send out time signals every exact hour, Greenwich time; they can be observed by ships at any point of the ocean. Once every hour the precise standard time is given, which, compared with the local time, gives the longitude. The problem of the longitude at sea was an episode in the history of astronomy, but highly important for the progress of science now closed. It greatly stimulated celestial mechanics as an important branch in the general theoretical knowledge of mankind. In its stead we are now faced with phenomena of our own earth, and this may be a starting point for new geophysical study.

CHAPTER 35

PLURALITY OF WORLDS

A SILENT force in astronomical research, among scientists and especially among wide circles of interested laymen, was the desire to acquire knowledge about other worlds as an abode of other men. What for authors of antiquity was mere playful fancy, had grown, since Copernicus and Bruno, to be a serious though hesitant opinion that possibly on the moon and the other planets conscious beings might live, gifted with intelligence and reason. Now the study of the planets, of their surface and their conditions, acquired a background of more profound interest.

As long as telescopes were imperfect, this study had to be primitive. At the end of the eighteenth century a diligent amateur, Amtmann Schroeter of Lilienthal near Bremen, using his reflecting telescope, made hundreds of drawings of lunar mountains and of markings on the planets. Compared with the standard of the period that followed, they were rather coarse. This higher standard was set by the refractors from Munich; with their sharp images and easy handling, they opened a new way for this study.

That the moon should be inhabited entirely suited the rationalist mode of thinking of the eighteenth century. In one of the first of the papers he presented to the Royal Society, William Herschel had expressed his conviction that 'lunarians' existed; but on Maskelyne's advice he cancelled this sentence as insufficiently founded. Afterwards the question was raised by Gruithuisen at Munich, who said in 1816 that he had seen clouds on the moon and had recognized fortifications and other human buildings in some lunar formations. Such fancies, however, were soon cut short. In 1834 Bessel established that from occultations of stars the moon's diameter was found to be not perceptibly smaller than from direct measurements. This means that the light rays from the star grazing along the moon's edge were not deviated by atmospheric refraction. The refraction in the horizon, which for the earth's atmosphere amounts to about 2,000″, can be no more for the moon than a few seconds, so that its atmosphere must at least be $\frac{1}{2000}$th of the density of the earth's. This would be insufficient for the respiration of man or animal.

Why does the moon have no perceptible atmosphere? If the moon originated from the earth, we might expect that it had got its share of air. Can it have had air and later lost it? Such a loss of atmosphere could be conceived in terms of the kinetic gas theory. In 1870 Johnstone Stoney argued that a small percentage of gas molecules, which have a velocity much larger than the average, might escape into space, so that gradually such a gas would disappear entirely from the attracting body. The limiting velocity for particles to escape from the earth is 11 km./sec., from the moon it is 2.4 km./sec., whereas at 0° C. the mean velocity of the molecules of hydrogen is 1.6 km./sec., of water vapour 0.53, of oxygen 0.4. That hydrogen is absent in the earth's atmosphere proves that a velocity of escape only seven times the mean velocity allows for its disappearance. That water vapour is present proves that a ratio of 21 is too large to allow it. Later this theory was refined by Jeans, Milne and others. Applied to the moon, these data tell that all water vapour and all oxygen from the moon must have diffused into empty space.

For the astronomers there simply remained the task of making a careful topography of the moon. A fascinating programme to observe another world from far away and to picture it! A small telescope already shows such an immense wealth of minute details that a total picturing would require many years of patient and devoted efforts. First, in 1824, came the work of Lohrmann, himself a land surveyor; so, instead of publishing the original drawings as Schroeter had done, he worked them into topographical maps, in which the mountains were represented as on geographical maps. Only a few maps were published: eye disease prevented the observer from completing them. In 1828 a Berlin banker, Wilhelm Beer, and a young astronomer, J. H. Mädler, began their common study of the moon in a small observatory erected by the former, provided with a Munich telescope of 4 inch (108 mm.) aperture. Following the example of Tobias Mayer on a larger scale, they first laid down a network of precisely determined points as a basis for the topography; with the micrometer the positions of over a hundred well-identified primary and many more secondary points were measured, as well as the heights of a thousand mountain tops by means of their shadows. They formed the frame of topographical maps picturing the entire surface of the lunar disc on the scale of one metre in diameter, in four sheets (plate 11). It was published in 1836 after eight years' work in 600 nights of observation. Throughout all this work they saw no change in the mountainous formations, nor any change suggesting atmospheric phenomena. Stark and rigid stood the lunar surface, with its pitch-black shadows, a dead world of rock and stone.

Though Beer and Mädler's map of the moon was far more detailed and complete than any former representation, it did not contain more than was visible through any small telescope. So the task remained for astronomers and amateurs to make more detailed studies with their larger telescopes and to publish them as more detailed maps. Outstanding was J. F. Julius Schmidt (1825–84), who in Athens, with small instrumental means but a fine climate, applied himself most diligently to all kinds of observing work that is often left to amateurs. His drawings of the moon were sufficiently complete to combine them into a map of the moon two metres in diameter, which was published in 1878 (plate 11). Neison in England in 1876 published a book on the moon, with extensive descriptions and a number of detailed maps.

A solid basis was thus laid to deal with the question of whether small changes did occur on the moon. Julius Schmidt himself, in 1866, pointed out one case of unquestionable change: the small crater Linné, which formerly had been drawn by Lohrmann, by Beer and Mädler, and by himself, had disappeared, and in its place a large whitish spot or a shallow hole was seen. Other more dubious assertions of changes were published occasionally, resulting in much controversy, because it could never be decided whether the maps were sufficiently reliable in such details and especially whether an object missing on earlier maps was really new or had been overlooked.

For a body so rich in detail as the moon, photography meant invaluable progress. A single photograph picturing the entire disc at once replaced hundreds of drawings that would have taken months and years at the telescope; moreover it was trustworthy as a document. The first good photographic picture of the moon was made in 1850 by W. C. Bond, the director of Harvard College Observatory. This was soon followed by one by Warren de la Rue in England, who, from 1852 on, worked with a reflecting telescope provided in 1857 with clockwork motion. He obtained pictures of 28 mm. diameter, which were so sharp that they could be enlarged 20 times. Henry Draper in America, working with a self-made reflector, in 1863 took moon pictures of 32 mm. diameter which could be enlarged to 90 cm. All this was experimentation and could not increase our knowledge of the moon. Great progress was brought about here by the introduction of dry bromide plates in 1871, which reduced the exposure time to a few seconds or less. The exposure could be short because the pictures were taken in the focal plane of the objective, without enlargement, which would have diminished the brightness too much. Progress was now achieved by using long-focus telescopes. The great 36 inch refractor of the Lick Observatory, with its 57 foot focus, supplied moon photographs of 14 cm. diameter. The Paris *equatorial coudé* (2-elbow-telescope) was regularly used by Loewy and Puiseux after 1894 to take moon

pictures of 18 cm. diameter. Enlargements of special parts of these
negatives were reproduced and published, with the Lick pictures on a
scale of 1 metre, with the Paris atlas on a scale of 2.6 metres to the
moon's diameter. At this size the limit was reached at which the silver
grain precluded the visibility of finer detail. More delicate detail was
reached by photographs made with special care with the 40 inch
refractor of Yerkes Observatory, and with the 100 inch reflector at
Mount Wilson. They were smaller in number and did not cover all the
phases of the moon.

A photographic atlas of the moon differs from a visual atlas, in that it
gives the direct aspect of the moment with all its shadows; it is not a
topographic map constructed by the astronomer out of a number of
drawings at different phases. For every landscape, instead of one map,
an entire series of reproductions at different phases is needed, without
ever giving completeness. Moreover, in comparing different repre-
sentations of the same landscape (as, for instance, the crater Hyginus
with surroundings, plate 11), we see that the photographs cannot
compete in wealth of minute detail with the visual work of careful
observers with smaller telescopes. Such was the work of Ph. Fauth, an
amateur at Landstuhl, who since 1895 has published many maps of
special lunar landscapes. This experience showed that visual work
should not be abandoned; many amateurs with good telescopes, up to
4 or 6 inch aperture, continued their study of the details of special
objects, chiefly to check the occurrence of small changes. Conversely,
photography keeps its place because of its documentary value. Observers,
like Krieger (in his maps of 1898–1912) and Goodacre, often made use
of the photographs as basic work, in which to fill all further detail at the
right spot.

The development of lunar topography in the nineteenth century was
confronted with two problems. The first was: Is the moon's surface
indeed rigid and invariable, or does anything happen there? In other
words: What is the amount of reality and what is the character of the
observed small differences appearing sometimes in a landscape invariable in its gross features? A close collaboration between different visual
observers was necessary to answer this question. After-effects of earlier
lunar forces might possibly produce or wipe out craterlets, small holes
or hills. W. H. Pickering later reported narrow haze or cloud tracks
and snowlike spots around craterlets, waxing and decreasing with the
amount of sunshine, and ascribed them to thin water-vapour exhala-
tions from rifts in the soil. He also paid close attention to the dark parts
seen within some circular walls at full moon, when there are no shadows.
He described them as growing and variable 'colouring', i.e. darkening
of parts of the surface, followed later by bleaching. Photographs con-

firmed the changes in their general character, but, because of their coarse grain, could not test the minute details. To explain them, Pickering assumed a primitive, but crowding form of life, such as low vegetation performing its life-history of growing and fading in the 14 days of sunshine, fed by carbon dioxide coming from rock fissures; and perhaps herds of freely moving primitive organisms. 'We find here next door to us a living world, with life entirely different from anything on our own planet.'[198] This was his conclusion in 1921. According to these reports, studied in minute detail, the moon would not be the entirely dead world that at first sight it appeared to be.

The second problem relates to the origin of all these remarkable features, so different from those on earth, such as the great circular walls, craters, rifts, mountains and large plains (called 'seas'), that constitute the solid lunar surface. It is a problem of geology, here more correctly to be called 'selenology'. The essential problem is why the active geological forces on the moon produced forms so entirely different from those on earth. The answer was sought in two different directions. In 1873 Proctor suggested that the lunar craters had been produced by the impact of large meteorites breaking through the thin crust, after which strong lava waves built the surrounding circular walls. Similar impacts upon the earth must have been weakened by the resistance of the air and their effects were effaced by later climatic and organic influences. In later times the same idea was expressed by the well-known geographer, Kurt Wegener. More generally, however, the origin is sought in the inner forces of the moon itself; physicists and astronomers often tried to imitate the formation of its surface by experiments with molten matter. In connection with the Paris photographs of the moon, Puiseux in 1896 gave an elaborate theory of the formation of the lunar features. As had been done before by the physicist Ebert, he appealed to the tides in the body of the moon, generated by the earth, as the primary determining forces. Because of the larger mass of the earth, these were far greater than those generated by the moon in the earth. This could explain why the effects in both cases are so different.

When in the nineteenth century interest in the moon lessened because it was uninhabitable, attention turned mainly to Mars as presenting the best chances in this respect. The study of the planet's surface, however, lies on quite a different level from that of the moon; it is more difficult and more restricted. In its most favourable perihelion oppositions, Mars presents a disc of 26″ diameter only, where 1″ represents 260 km. What at the end of the eighteenth century was seen in the reflecting telescopes by Herschel and Schroeter, besides the sharply-cut white polar spots, were only vague, ill-defined shades, wherein clear shapes could rarely be recognized. They were taken to be mostly fogs and clouds. With the introduction of the Munich telescopes, another epoch opened. In 1830, with a favourable autumn opposition, Beer and Mädler began to make drawings with their four-inch. They were the first to recognize a definite picture of darker and lighter parts. The difficult nature of the work is evident from their words: 'Usually some time had to elapse before the indefinite vague mass at first seen dissolved itself into clearly distinguishable forms.'[199] Since these forms periodically came back in nearly the same characteristic aspects, it was evident that they were not passing cloud formations but topographical markings on the planet's surface. For the first time, as a result of their work there appeared a world map of Mars in two hemispheres. In order to include the north polar regions, it was necessary to extend the observations over additional oppositions, from 1832 to 1839; because Mars in aphelion shows a small disc, they made use, from 1835 on, of the 9 inch refractor of the Berlin Observatory. Now that the markings could be identified the time of rotation of Mars could be derived by comparing their earlier and later drawings. From the observations in 1830 and 1832 it was found to be $24^h\ 37^m\ 23.7^s$.

Thus a foundation had been laid for later progress. Many observers made and published drawings of Mars (plate 12); their often very dissimilar aspect shows how difficult was the precise delineation of the markings and how strongly dependent it was on personal style. The opposition of 1862 showed some progress; the drawings of Lockyer in England and of Kaiser at Leiden—the latter condensed into a globe of Mars—both working with a 6 inch telescope, show more details. Secchi at Rome, in 1858 and after, working with a 9 inch telescope, made drawings in colour, which show subtle hues of green in the dark parts and of yellow in the light regions. Kaiser identified ancient drawings of Huygens and Hooke (from 1666 and 1667) with some main features of his globe and derived a rotation time of $24^h\ 37^m\ 22.6^s$ to $\frac{1}{10}$ second.

All these new observations showed that, on the basis of a fixed and permanent topography, a number of minor variations of detail occurred. Sometimes details were washed out by hazy cloudlike appearances; dark spots bleached, or light spots temporarily darkened, and also the precise figures and borders showed changes not entirely attributable to different directions of view. The observers generally agreed in calling the yellow or ruddy parts land or desert, and the dark, mostly greenish, hues water, perhaps shallows or vegetation. They assumed the white polar caps, which changed in size periodically with the seasons, to be snow and ice. From these data it was concluded that life-conditions on Mars are not greatly different from those on earth: the temperature somewhat lower, the atmosphere thinner and drier, the water scarcer, conceivable as a further-developed stage in planetary evolution.

The perihelion opposition of 1877 brought new progress and a new surprise too: the discovery of the canals of Mars. Schiaparelli (1835–1919), working at Milan with a 9 inch refractor, at once raised areographic research to a higher level by making micrometric measurements, first of the small southern polar cap in order to fix precisely the position of the rotation axis, and then of 62 well-identified points as a basic network. He perceived long dark lines, narrow and straight, chiefly traversing the yellow northern half of the planet. He called them 'canals', without asserting them to be water; the scale of a planetary image is so small that the finest visible line represents a width of many miles, and some of these 'canals' were broader than that and represented considerable actual width.

In the 1879 and later oppositions these observations were continued, confirmed, and extended. Then, from 1881 on, a new phenomenon was recorded, the doubling of the canals. From then on, Schiaparelli's results were an object of continual controversy on the part of his colleagues. Some of the drawings of other astronomers began to show, hesitatingly, a few canals. While the reality of straight canals did not seem impossible, for their doubling no natural cause could be imagined. So the habitability of the planet from a mere fancy became a practical argument of discussion. Is Mars inhabited by intelligent beings, and are we justified in invoking their intelligence as a causal explanation of the phenomena? In that case even the doubling could be made understandable as deriving from the need for navigation or irrigation. Yet, with regard to the doubling, the general attitude was mostly sceptical. Perrotin of Nice and his assistant Thollon were the only ones who succeeded, in 1886, in seeing double canals with the 15 inch refractor. Yet it was curious that, at first, these observers saw nothing of this for weeks and then at last produced drawings entirely similar to Schiaparelli's. The question was posed as to how far bias could influence the delineation of objects at the limit of visibility. Or was it possible that the physiological effect of a minute, unsuspected astigmatism in the eye had played a role? Against these criticisms, Schiaparelli stuck to the reality of the forms he had reported. In 1892 he published a survey of his researches in the German review *Himmel und Erde*, accompanied by a coloured map of Mars in two hemispheres, so rich in detail that one could not fail to be impressed by the great progress of our knowledge of the planet. However, a second map with all the double canals threw the reader back into doubts about their reality and significance.

It was at this time that the giant 36 inch refractor was installed at the Lick Observatory. The Lick astronomers, Holden and Keeler, in 1888 directed it on Mars, expecting that their powerful instrument would solve the doubts. It was a disappointment, not so much because their drawings showed nothing of the canals and their duplication, but chiefly because they were so poor in detail that they looked like products of an earlier stage of research. It had often been asserted that the great telescopes are not superior but rather inferior to small ones for the study of planetary surfaces, as might be seen by comparing Beer and Mädler, Lockyer and Kaiser, with Herschel, Schroeter and Lord Rosse. Faint differences in shade, extending over wide surfaces, are more difficult to distinguish than when contracted to small dimensions. Moreover, air striae in the broad bundle of light entering a large objective will smooth the image more than in the narrower pencil of a smaller instrument. On the other hand, intricate complexes of objects cannot be distinguished by low powers and are conceived in simplified forms. In the case of the Lick astronomers, lack of practice certainly played a role. The observation of Mars was to them a short interruption in important other work on the stars, whereas only by continuous and assiduous occupation with the planet's surface as his sole object can the observer succeed in discovering all the minute details—though at the same time he clings more firmly to a personal style in interpreting and drawing them.

The problem of the possible existence of Martians exerted its fascination. The well-known French popularizer of astronomy, Camille Flammarion, in 1892 published a large monograph, *La planète Mars et ses conditions d'habitabilité*; the title indicated and the contents clearly expressed his conviction that next to the earth there travels a brother-world, with life-phenomena of its own and of a nature as yet unknown to us. He wrote: '. . . the considerable variations observed in the network of waterways testify that this planet is the seat of an energetic vitality. These movements seem to us to take place silently because of the great distance separating us; but while we quietly observe these continents and seas slowly carried across our vision by the planet's axial rotation, and wonder on which of these shores life would be most pleasant to live, there might at the same moment be thunderstorms, volcanoes, tempests, social upheavals and all kinds of struggle for life. . . . Yet we may hope that, because the world of Mars is older than ours, mankind there will be more advanced and wiser. No doubt it is the work and noise of peace that for centuries has animated this neighbouring home.'[200] He further says: 'The present inhabitation of Mars by a race superior to ours is very probable.'[201] Then he indulges in praise of the beauty and greatness of the conquests of modern astronomy: 'It is the first time since the origin of mankind that we have discovered in the heavens a new world sufficiently like the Earth to awaken our sympathies';[202] and he thinks of a far distant future in which the peoples of both planets, united, will proceed to greater common works. He wrote: 'The Earth becomes a province of the Universe,' and 'we feel that unknown brothers are living there in other fatherlands of Infinity.'

For the present, mankind on earth was restricted to a lot of stories about the Martians and to the assiduous and continuous study of science. Percival Lowell, a wealthy New England aristocrat, enthusiastic, gifted and well-versed in many fields, in 1894 founded an observatory at Flagstaff, in the bright desert climate of Arizona, at a height of 6,000 feet, especially to study Mars and its inhabitants. He used first an 18 inch Clark refractor, later one of 24 inches. A number of drawings, with about 180 mostly new canals discovered by him and his assistants, was published. He emphasized the changes in the aspect of continents, seas and canals, varying periodically with the Martian seasons, and explained them by the yearly movement of the water to and fro, from the melting towards the growing ice cap. The canals, serving as expedients for irrigation to ensure the most economical use of the scanty water supply, disclose control by intelligent beings.

More fruitful and dependable was the work of the French astronomer Antoniadi, who from 1909 onwards made studies of Mars with the 83 cm. (33 inch) refractor of the astrophysical observatory at Meudon, near Paris. With this instrument, surpassing in aperture all European refractors, and in a climate distinguished by steady telescopic images, Antoniadi was able to demonstrate that under such conditions and in the hands of a good and trained observer, large instruments surpass the smaller ones for the study of the planets. What Schiaparelli had drawn as straight or double canals was now dissolved into minute details of small spots, sometimes arranged in regular series, also often irregularly, or simply boundaries of regions of different shades. For twenty years he was engaged in drawing and describing the features of Mars; his work was notable progress and enrichment after Schiaparelli, just as the latter had surpassed his predecessors. Antoniadi sometimes saw white clouds, which he took to be composed of water, and at other times yellow dust clouds, obliterating the markings; but the prominent phenomena to him were the colouring and discolouring according to the season—the dark greenish or bluish parts, probably the lowest regions, which by drying up in summer became yellow, brown or red, each in its own manner. The movement of the water from pole to pole, here melting and evaporating, there condensing and freezing, he likewise considered to be the basis of the visible changes. Antoniadi's work opened the way to further progress by showing how the increase in telescope sizes and the observers' skill can advance the study of the planets if aided by perfect atmospheric conditions.

Could not photography also open new ways here? At the close of the nineteenth century many photographs of planetary surfaces were made; but what they presented was at most a blurred copy of the drawings. Taken directly in the chief focus, the images were very small, of some few millimetres in extent, and the grain coarsened the details. In recording enlarged images, the details, owing to longer exposure, were effaced by atmospheric vibrations, whereas an observer's eye could wait for the short still moments when minute details stand out sharply. Photographs of Mars taken by Barnard with the 40 inch telescope in 1909, by Slipher at the Lowell Observatory in 1922, by Wright and Trumpler with the 36 inch Lick telescope in 1924, and by Ross with the 60 inch telescope at Mount Wilson, clearly show the main shape of the dark markings, as well as the polar caps. Of the slender lines of the canals nothing of course was visible. Yet in other respects these photographs were most instructive. Wright took photographs on plates sensitized for limited ranges of colour. The images in infra-red and red light showed the surface topography clearly, though this was totally absent from the plates sensitive to violet and ultra-violet light; besides the white polar spots, these showed vague shadows only. By comparison with earthly landscapes, Wright could establish that this was an effect of the Martian atmosphere, entirely transparent to red light, while reflecting blue light diffusely. This reflected light altogether exceeded the blue light from deeper layers.

Remarkable progress was reached in recent years by an ingenious method by Lyot to eliminate the grain in the photographic emulsion. By rapidly taking a number of small focal images and combining them in one enlarged picture, the grains are smoothed into a continuous background. These figures show almost the same amount of detail as the best drawings; it may be expected that they will surpass them after further improvement has been made to the new method.

In the nineteenth century the other planets gave rise to much less sensational discoveries. On Mercury no distinct spots were perceived until Schiaparelli in 1881 saw a system of vague and dim streaks, which always kept the same position relative to the terminator, the boundary between the dark and the illuminated parts of the disc. He concluded that the planet always keeps the same hemisphere towards the sun, as the moon does towards the earth, so that it rotates about its axis in 88 days, its period of orbital revolution. This was later confirmed by Lowell at Flagstaff, but was opposed by Leo Brenner, who worked on Martian canals with a small telescope in the beautiful climate of Lussinpiccolo in Istria. The question was definitely settled by Antoniadi, who in 1924–29 made a large number of drawings with the great Meudon telescope. He confirmed the rotation time of 88 days; his drawings showed a system of vague broad bands in which he could recognize most of Schiaparelli's streaks. They are fixed markings on the planet's surface, of course visible only on the illuminated half, since the opposite hemisphere always remains dark. Moreover, now and then he saw extensive and variable white patches, which he supposed to be clouds of dust. Water clouds cannot exist on Mercury because its low gravity would cause water vapour to escape; for dust clouds, however, a small amount of atmosphere must be present.

No astronomer observing Venus ever saw anything more than vague and ill-defined differences in shade, not suitable for ascertaining a period of rotation. Schroeter at Lilienthal said he had seen one of the horns of the Venus crescent sometimes blunted—as also a horn of Mercury—which he explained by shadows of high mountains carried around in a rotation period of about 24 hours. Herschel, however, did not see this. In 1839 Vico at Rome thought he recognized definite forms and derived a rotation period of $23^h 21^m 22^s$, by comparing them with drawings of Bianchini, made a century earlier. On the other hand, Schiaparelli found a rotation time of 225 days, so that Venus, like Mercury, would always turn the same face to the sun. The result raised doubts and contradictions; Villiger at Munich in 1898 pointed out that an entirely smooth matt-white globe illuminated sideways shows gradual differences of shade, which by contrast give the impression of faint markings keeping their place relative to the shadow boundary. Observers always agreed that with Venus we do not see the planet's surface but only the upper side of a dense layer of clouds. That Venus has an atmosphere was established first at the transit of 1761, when the part of the dark disc away from the sun was surrounded by a luminous ring. Moreover in the nineteenth century it was often seen that the horns of the crescent extend over more than 180°.

Observation of the planet Jupiter in the nineteenth century afforded nothing so exciting and new as with Mars or Mercury. It confirmed the chief features already established in the past. They were the two dark belts parallel to and separated by a brighter equatorial zone, all showing small spots and irregularities that indicate a rapid rotation in about 10 hours. Since we see only the upper side of a cloud envelope, the moving spots can only roughly disclose a rotation period; it was found to vary at different latitudes, $9^h 55^m$ in the dark belts, $9^h 50^m$ in the equatorial zone. Since the big planet shows a wealth of ever changing details in small telescopes, it was an interesting object for astronomers with limited equipment. During the entire nineteenth century, a number of observers assiduously made drawings of the spots in the cloud bands, without learning more about them than that they were steadily changing, appearing and decaying. Larger instruments could give information about the different colours, mostly yellow or ruddy in between greenish parts. The only sensational news was the appearance, in 1878, of a large spindle-shaped red spot; it could be found on drawings as far back as 1859 and was indicated even in 1831. After its discovery it remained visible for many years, gradually losing its red colour, with a changing rotation period of nearly $9^h 55^m$.

Saturn, with a greater flattening than Jupiter's, hence a certainly rapid rotation, showed far less interesting surface features. There was a faint dark belt along the equator, sometimes accompanied by other fainter belts at higher latitudes. At times irregular spots appeared therein that offered the possibility of determining a period of rotation. Thus in 1876 Asaph Hall found $10^h 14^m 24^s$ for the equator; in 1894 Stanley Williams found $10^h 12^m 13^s$ at the equator and $10^h 14^m 15^s$ above 20° of latitude; whereas in 1903 a spot at 36° latitude, observed by Denning and Barnard, gave $10^h 38^m$ to 39^m. So here also we see only a cloud layer.

Uranus and Neptune are of course far more difficult objects; the flattened figure (flattening about $\frac{1}{12}$) of the greenish-tinged Uranus indicates a rapid rotation about an axis situated nearly in the ecliptic. Drawings made by Lowell showed some spots, the reality of which is suspect.

As objects equivalent to our moon, the satellites of other planets must be mentioned. Their smallness and resultant lack of atmosphere ruled them out as abodes of life; but they completed the world aspect for their primaries. With the improvement of the telescopes in the nineteenth century, more and more of them were discovered. Venus and Mercury remained without satellites, though deceptive reflex images in the telescope were often announced as moons. In 1877 Asaph Hall, with the Washington 26 inch telescope, discovered two moons of Mars, certainly the smallest known of their kind, with estimated diameters of 9 and 12 miles, visible only because of the proximity of Mars to the earth. The period of revolution of the innermost, $7^h 29^m$, shorter than the rotation period of the planet, offered a distinct problem in the theory of tidal action.

In 1892 Barnard, with the 36 inch Lick refractor, discovered a fifth moon of Jupiter in close vicinity to the planet, a small object perhaps 100 miles in diameter. Outside the system of the four large Galilean satellites, four small satellites were discovered between 1904 and 1914 on photographic plates; this number was increased later. Owing to their remarkably deviating orbits, strongly inclined and eccentric, some even retrograde, strongly perturbed by the sun, they are interesting objects of celestial mechanics. Thus the question of whether they could be former planetoids captured by Jupiter was posed and answered, in the affirmative for some of them, by Moulton.

To the seven satellites of Saturn known at the end of the eighteenth century an eighth was added in 1848, discovered nearly simultaneously by W. C. Bond with the Harvard refractor and by Lassell with his reflecting telescope. A very small ninth satellite, found photographically in 1898 far outside the realm of the others, resembles the outermost Jupiter satellites in having a retrograde motion and great eccentricity. Uranus, which had two satellites discovered by Herschel, received two more discovered by Lassell in 1851, all very faint objects, with their orbits nearly perpendicular to the ecliptic. It was again Lassell, who, shortly after the discovery of Neptune, discovered its moon, which also has a retrograde motion in a strongly inclined orbit.

With respect to size the satellites join the planets to form one continuous series. The third and fourth Jupiter moons, with a diameter of 5,000 km. (3,200 miles), are equal to Mercury; the Neptune satellite is somewhat larger; our moon and the sixth of Saturn are somewhat smaller; so there may be just a trace of overlapping. Then the others follow in a decreasing series down to the Mars satellites running parallel to the similar series of planetoids. Whereas the preceding centuries distinguished the sun, the planets, and the satellites as three different types of size, the discoveries of the nineteenth century combined the latter types into one series, continued in the still smaller meteorites. The sun as a solitary, quite different, body stands separated by a gap from the ten times smaller Jupiter.

Nineteenth-century astronomy did not restrict its findings to the outer surfaces of the celestial bodies. It could also penetrate into their interior, because it was chiefly a theory of attraction, and attraction proceeded from all the matter in the interior. The first new datum derived was the mean density of this matter. When the masses of the planets relative to the earth's mass are known, whether from the motion of their satellites or, with more difficulty, from the perturbations they exert, measurement of their diameters reveals their volume, hence their mean density, relative to the density of the earth.

The mass and the mean density of the earth have been the subject of a series of the most subtle investigations extending over the entire nineteenth century, based on the measurement of the hardly perceptible attraction exerted by bodies in laboratories. The result was a mean density of 5.50, larger than the densities of rocks and minerals constituting the crust of the earth. Hence its interior must have a far higher density, such as is found only for metals.

Now the absolute densities of the other celestial bodies could also be stated. For the moon it is 3.3; for Mercury, Venus, and Mars, 3.8, 4.9, and 4.0, of the same order of magnitude but somewhat smaller than with the earth, decreasing parallel with the planet's size. For the major planets the mean density is found to be far smaller: for Jupiter 1.3, for Saturn 0.7, for Uranus 1.3, for Neptune 1.6. That Saturn has a smaller mean density than water can be understood only if a large part of its apparent volume consists of gas. The surface of its solid or fluid body is situated far deeper, and what we see as its surface is the upper side of a cloud mass floating in an extensive atmosphere. The same, in a lesser degree, must hold for the other major planets.

Celestial mechanics could give still more information. Clairaut had derived a formula by which the visible flattening of a planet depended on the period of rotation, combined with the increase in density toward the centre; thus the inequality in the distribution of matter in the interior was revealed by visible phenomena. Celestial mechanics met here with seismology, which had organized all over the earth the exact registering of the small vibrations and seismic waves pervading its body. This branch of geophysics had derived analogous conclusions, but in a different form. Wiechert in 1897 found the earth to consist of two clearly distinguished parts, a metal core, chiefly iron, with a density of 7.8, surrounded by a layer of rock, chiefly silicates of iron and magnesium, with a density of about 3.3. In 1934 Jeffreys expressed the plausible opinion that the same holds for the other earthlike planets and that their decrease in mean density with volume is due to the iron core diminishing in size until it is entirely lacking in our moon. From the flattening of Jupiter and Saturn it was derived that the increase in density with depth was greater here than with the earth, so that, as with Jupiter, a considerable part of the volume beneath the visible cloud surface must be occupied by atmospheric gas.

The knowledge about the conditions on other worlds that was acquired merely by using telescopes was not very satisfactory. Happily support came from another source; in this realm of study, astronomy had no longer to rely solely on its own forces. The progress of physics in the nineteenth century provided new instruments and new methods of research that were applied to the celestial bodies with increasing success. They consisted chiefly of photometry, spectrum analysis, thermometry and polarimetry.

The principles and methods of photometry—the measuring of quantities of light—are so simple and obvious that they could have been used in earlier centuries; but interest was lacking, as well as the restless urge to investigate everything. Some first attempts were made in the eighteenth century, by Bouguer, who in 1729 made measurements, and by Lambert, who in 1760 gave a first theory of the diffuse reflection of light by even matt surfaces. The fraction of the incident light reflected diffusely by a surface was called by Lambert its *albedo*, the Latin word

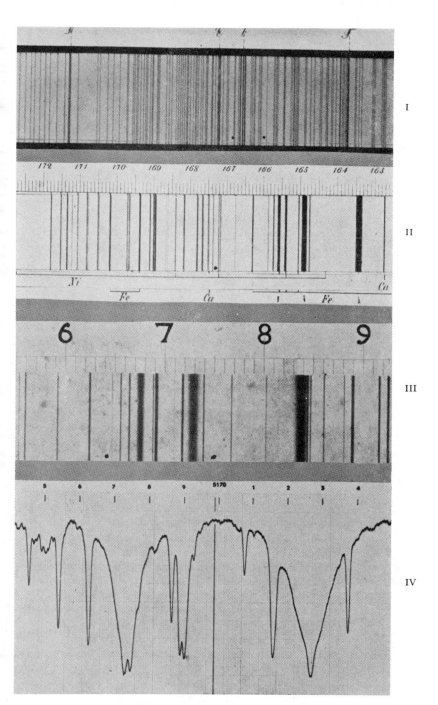

13. Sections of the solar spectrum. I. Fraunhofer; II. Kirchhoff:
III. Rowland; IV. Utrecht. The wave-lengths increase to the
left on I, and to the right on II, III, and IV. The dots mark the
limit of the part reproduced on the next illustration (pp. 406–9)

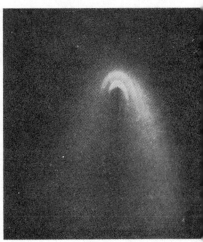

14. Donati's Comet, 1858 (p. 424)   Donati's Comet, enlarged

Morehouse's Comet, September 29, 1908 (p. 425)

for 'whiteness', since his book was written entirely in Latin; this Latin word has remained the usual scientific term. Practical photometric work in astronomy began with John Herschel, who in 1836, during his stay at the Cape of Good Hope, measured the brightness of a number of stars. He compared the star with the pointlike image of the moon formed by a small lens; by varying the distance between the lens with the light point and the eye, he could make it equal to the star. The most fundamental law of optics then says that the intensity decreases as the inverse square of the distance. Because the results of different days were discordant, he supposed that unknown systematic sources of error had vitiated his measures. The real cause was shown afterwards by J. C. F. Zöllner (1834–82): Herschel had computed the brightness of the moon on different days by means of formulae from Lambert which did not fit here. In reality, Herschel's measurements were quite concordant. In 1861 Zöllner devised a photometer in which the brightness of an artificial star was made equal to the real star. The artificial star was formed by a pinhole in front of a flame, and it is diminished at an exactly known rate by an interposed polarizing apparatus consisting of two Nicol prisms. This astrophotometer has found wide application in astronomy.

The same principle of varying the intensity by the use of polarizing apparatus was used by Zöllner in determining the brightness of the moon in different phases. He discovered that this brightness at full moon had a sharp maximum; before and after the exact moment of opposition the brightness rapidly increased and decreased, almost uniform with the phase-angle (the angle at the moon between the directions to the sun and the earth). According to Lambert's theory, the intensity should show a flat top like a hill, slowly varying near the top and more rapidly only at larger phase angles. Zöllner soon understood the cause of the discrepancy: Lambert's theory was devised for an even surface, whereas the surface of the moon is full of irregularities. Directly before and after opposition, when the phase angle is small, shadows become visible and diminish the light; they occupy a part of the surface, uniformly increasing with this angle. The decrease in brightness (expressed in stellar magnitudes) for 1° of phase angle, the so-called 'phase coefficient', for which Zöllner found the value 0.025, may be taken as a measure of the irregularity of the surface. Herschel's results, now used as measurements of the moon's brightness by means of known stars, entirely confirmed Zöllner's results. In 1923 Barabashev measured the brightness of apparently flat parts of the lunar surface; they showed the same dependence on phase angle as the total light, an indication that they were full of small irregularities, holes and knobs, and might be like pumice stone.

The comparison of the brightness of the full moon and of the sun was far more difficult, because the difference is so enormous. Zöllner found a ratio of 618,000. G. P. Bond shortly before had found 470,000 by a different method. Several physicists and astronomers afterwards found widely varying values, and the usually adopted average of 465,000 is uncertain by a tenth of its amount. For the albedo of the moon, Zöllner derived the value 0.17. Later researches resulted in a lower value, 0.07, corresponding to rather darkly coloured stone.

Zöllner, the founder of astrophotometry, also measured in 1862–64 the light ratio of the full moon and some bright stars, as well as the brightness of the planets. He detected some main characteristics; for Mars he found a considerable phase coefficient; for Jupiter, none. From the brightness of the planets relative to the sun, he derived an albedo of 0.27 for Mars, 0.62 for Jupiter, 0.50 for Saturn, 0.64 for Uranus, and 0.47 for Neptune, hence a rather strong reflective power for the major planets.

The Zöllner photometer was used from 1877 to 1893 by Gustav Müller and Paul Kempf at the Astrophysical Observatory at Potsdam, in extensive accurate measurements of the brightness of the planets. They may be condensed in the results for the two characteristic quantities, the phase coefficient and the albedo. For Mercury the former is 0.037 (larger than for the moon), the latter 0.07; for Venus they are 0.013 and 0.59. In these simple numbers the difference in character between the two planets is more distinctly expressed than by any description. The large phase coefficient and the small albedo for Mercury indicate a surface directly seen, unimpaired by an atmosphere, similar to the moon, dark rock with shadow-producing inequalities. The high reflective power of Venus, which makes it our brilliant evening star, indicates that we do not see its solid surface but the white upper side of a cloud layer; this is confirmed by the small phase coefficient. For Mars, Müller and Kempf found a small albedo, indicating that we see its not very dark surface, and a small phase coefficient, indicating that this surface is rather smooth, without large irregularities. The large albedo 0.56 and the small phase coefficient 0.015 of Jupiter show the same conditions as for Venus. For Saturn and Uranus it is difficult, owing to the small phase, to determine a phase coefficient, but the large albedo indicates similar conditions.

For the ring of Saturn photometric measurements have been of special importance. Laplace had demonstrated that flat solid rings could not keep in equilibrium when freely floating in space. In 1859 the famous physicist James Clark Maxwell, the founder of the theory of electricity, dealing with the same problem, pointed out that the ring would be broken by the tensions of attraction and centrifugal force. He suggested that Saturn's rings consist of numerous small bodies which, like a ring of meteorites, freely circle the planet. This theory was sustained by the rift observed between the outer and inner ring, which Kirkwood explained in the same way as the gaps in the ring of planetoids, by a commensurability with the motion of the largest among Saturn's satellites. Another confirmation consisted in the faint 'crape ring' discovered by Bond in 1850, a semi-transparent extension of the ring beyond its inner border.

Müller's photometric measures of Saturn showed that when the ring was invisible, Saturn had a very small phase coefficient. When, however, the ring was widely open, the total brightness increased to $\frac{5}{3}$ times the former amount (the ring giving more light than the ball), and now the phase coefficient rose to 0.044, due to the ring; so that, for the ring alone, it must be still greater. Seeliger at Munich gave an explanation similar to Zöllner's for the moon. If we look at the particles of the ring from exactly the direction from which the solar rays are coming, every particle covers its own shadow upon the particles behind. However, as soon as the two directions deviate by a small angle, the borders of the shadows will appear. Thus, the photometric measurements prove that Saturn's rings do indeed consist of small meteoric bodies.

Turning now to the other physical methods named above, we note that polarimetry is a special refined extension of photometry. The rate of polarization of the diffusely reflected light, positive (parallel to the plane of incidence) or negative (perpendicular to that plane), depends for different substances in a different way on the angle between incident and reflected light (called 'phase angle' above). This method of research could acquire practical importance only when instruments for its measurement became sufficiently accurate. Bernard Lyot in Paris succeeded in improving them so much that the rate of polarization could be measured to 0.1 per cent. In 1924 and thereafter he measured the polarization of the light reflected by the planets and the moon in its dependence on the phase angle. For the moon this dependence had a special irregular course, partly positive, partly negative; of all earthly matter investigated, it was volcanic ash only that showed exactly the same dependence; this gave direct information about the character of the lunar surface.

For Mercury the curve representing this dependence was of exactly the same shape. For Venus it was entirely different; nor did it wholly agree with the light of white clouds. It corresponded rather to light reflected by a fine haze, and it was even possible to recognize the size of the droplets. So looking at Venus we do not see the upper side of a cloud layer but a deep atmosphere, non-transparent, through a thin whitish haze. For Mars the polarimetric curve had some likeness to that of the moon, but it fitted a sand surface better. With Mars, variations and disturbances occur for many weeks, bearing the character of dust clouds. Dust storms from visual observations had also been supposed to occur, indicating a dry atmosphere. The study with polarimeters has in this way afforded with unexpected certitude important knowledge of the physical nature of planetary surfaces. For Jupiter and Saturn, with their small phase angles, only a difference between the equatorial and polar regions could be discerned.

Measuring the heat of distant bodies was always far more difficult than measuring their light, because man does not possess for heat perception so sensitive an organ as is the eye for light perception. It was not until the nineteenth century that physical instruments were constructed of such sensitivity that the heat of celestial bodies other than the sun could be measured. The thermopile of Melloni was applied to the moon by its inventor in 1846, and in a larger investigation in 1869–72 by Lord Rosse. Part of the radiation received from the moon is reflected solar radiation, which consists mainly of visual wavelengths below 0.7 micron. The other part is direct radiation of the heated lunar surface, which, as low-temperature radiation, consists chiefly in long waves above 1 micron. By interposing a sheet of water, which entirely absorbs the long waves, the two can be separated, so that the temperature of the moon can be derived from the intensity of the long-wave radiation, Lord Rosse found that 14 per cent of the lunar radiation was reflected solar radiation, 86 per cent was moon radiation proper; and he derived that the temperature of the illuminated moon surface was 300° C above that of the parts in the dark. His results were confirmed in 1874 by Very's careful measurements at the Allegheny Observatory; in full sunshine the moon's surface was hotter than 100° C; in the dark it fell to little above absolute zero. In 1886 Boeddicker, assistant to Lord Rosse, observed that during a total eclipse of the moon the temperature fell rapidly to far below zero; it shows that the heat is absorbed superficially and is rapidly lost.

The use of Crookes's sensitive radiometer and the rapidly increased sensitivity of modern thermo-elements made it possible, by putting them in the focus of giant telescopes, to measure temperatures of the planets. By interposing sheets of different absorbents like glass and fluorite, it was possible to make a more detailed division in the dark radiation between longer and shorter wavelengths, in order to derive the temperature. Nicholson and Pettit at Mount Wilson Observatory observed the lunar eclipse of 1927 with a radiometer and found that the temperature fell from 70° C to −120° C at the end of the eclipse. From the measurements of Mars made by Coblentz and Lampland in 1926 at the Lowell

Observatory, Menzel derived that the temperature at noon is about, or some degrees above, 0°, the dark parts somewhat warmer than the yellow plains. In the morning the temperature rapidly rises from −100° C; the southern pole, when emerging from the darkness of the winter night, is −100° and in the course of the long summer day rises to above 0°. Former theories that Mars, as a result of the twice times smaller solar radiation, should be covered with a thick sheet of ice, the cracks in which appear to us as canals, are refuted by these measurements. Our earth after all seems to be the better place to live.

For Mercury a temperature of 400° C was measured. The illuminated part of Venus showed a temperature of 60° C and the dark part −20°; the rather high value of the latter shows that it cannot be permanently devoid of sunshine, so that the rotation period cannot be as long as 225 days. The temperature of Jupiter was unexpectedly as low as −130° C; for Saturn it was −150° C; and for Uranus it was still lower, −200°. Because the rapid changes in the cloudy features on Jupiter suggested turbulent processes in its atmosphere, it had often been supposed that the surface of the planet below was hot and, because of its size, had not yet sufficiently cooled. Jeffreys in 1923, assuming radioactive matter to be the source of the observed interior heat of the planets, had expressed his doubts of this idea. The temperatures measured agree with the supposition that the solar radiation, weak on account of the great distance, is their sole effective source. Here, as well as in the case of Venus, we have to remember that the temperatures thus measured hold for the highest layers of the atmospheres, the outer surface of the clouds.

Spectrum analysis was the most powerful aid that physics offered to astronomy in the nineteenth century. Soon after its discovery it was applied to the stars and the planets, first by Huggins in England and Secchi in Rome, soon followed by Rutherfurd in America and Vogel in Germany. What information could the spectra of the planets afford? Their light is reflected sunlight that has twice passed through their atmospheres; so their spectra must show the same Fraunhofer lines as the solar spectrum, but increased by the planets' atmospheric absorptions. If the reflecting surface is coloured, the distribution of intensity over the different colours will deviate from the pure solar spectrum. When accurate determination of wavelengths later became possible (from about 1890), chiefly by the use of photography, velocities in the line of sight, e.g. velocities of rotation, could be measured.

Such was the case with Jupiter. If the slit of the spectrograph is placed on Jupiter's disc along its equator, its eastern and approaching limb has the Fraunhofer lines displaced towards the violet, its western and receding limb towards the red; the lines are inclined, and this inclination is a measure of the rotation period. Keeler, at Lick Observatory in 1895, made spectrograms of Saturn in this way; the spectrum of the disc showed such inclined lines, indicating an equatorial velocity of 10 km. per sec., a confirmation of the rotation period of 10¼ hours. At both sides of this disc spectrum, two bands showed the spectrum of the ring; they had the absorption lines inclined in the opposite direction. This corresponded exactly with the velocities of small satellites describing free orbits about the planet: 20 km. at the inner border, 16 km. at the slower outer one. Thus the spectrum of Saturn afforded new confirmation of Maxwell's meteoric theory, more directly conspicuous than the photometric proof.

These were not new results. For Uranus, Slipher at the Lowell Observatory in 1911 could determine the unknown rotation period of 10.7 hours from spectrograms; this was in accord with the strong flattening of $\frac{1}{12}$ derived from measures of the diameter. Astronomers hoped also that the spectrum of Venus would determine whether the rotation took place in about one day or in 225 days. It was a delicate problem, because in the case of the one-day rotation the equatorial velocity, as with the equal-sized earth, is only 0.4 km. Slipher at the Lowell Observatory, by careful study of spectrograms taken in 1903, could not find any trace of an inclination of the spectral lines; this meant that the velocity could be at most 0.02 km. or that the rotation period was at least some weeks and certainly not in the region of a day.

For the moon, which has no atmosphere, the only difference from the solar spectrum to be expected would be due to the colour of the surface. Wilsing at Potsdam, by spectrophotometric measurements, i.e. by measuring the relative intensity of different colours in the moon's and the sun's spectra, found that the light of the moon is more yellowish, as light reflected by ash or by sand.

For the planets, the main object of study was the constitution of their atmospheres. The absorption of the earth's atmosphere produces in the observed solar spectrum and in all celestial spectra three strong absorption bands in the red, denoted by the letters A, B, and α, produced by oxygen, and many other groups produced by water vapour. If oxygen or water vapour exists in a planetary atmosphere, it must show in the spectrum as an enhancement of these bands. On the question as to whether this is the case in the spectra of Mars and Venus, there has been much controversy. It was an extremely difficult observation not only to see whether such bands could be ascertained and measured in the faint, hardly visible red but, moreover, to decide whether they were stronger than in the moon's spectrum. In the 'seventies the observers Huggins, Maunder (at Greenwich), Janssen (at Meudon) and Vogel (at Bothkamp) agreed that oxygen and water vapour were present in the atmospheres of Mars and Venus; and in 1894 they confirmed their former conclusion. Certainly some preconceived opinion of a natural similarity in the planetary atmospheres must have played a role. Gradually, when better telescopes and spectroscopes were used with greater resolving power, doubts arose. In 1894 Campbell stated that with the powerful Lick instruments he could not perceive any difference between the bands in the spectrum of the planet and of the moon when they stood at equal altitude.

Not until the spectra could be photographed on plates sensitized to red light could accurate decisions be made. Even then they were not easy. When Venus in greatest elongation is approaching or receding from the earth, the series of lines constituting these bands must be displaced relative to the same lines produced by the earth's atmosphere. Slipher could find no trace of doubling or displacement of these lines in the Venus spectra taken in 1908. Clearly, oxygen and water vapour are not present in the highest atmospheric layers of this planet. At Mount Wilson this result was confirmed; here another group of absorption lines was found which appeared to be due to carbon dioxide. For Mars the decision proved to be far more difficult. Slipher in 1908 had taken Mars spectra also; by measuring the intensity of the bands photometrically, Very found that the oxygen bands were 15 per cent stronger than in the moon's spectrum and that there was more water vapour than in the dry Flagstaff air. Campbell however, in 1910, could find no difference from the moon. In 1925 Adams and St John at Mount Wilson took plates with large dispersion to see whether the separate lines of these bands were displaced in comparison to the intermingled true solar metallic lines. Their result was that water vapour in the Martian atmosphere was 3 per cent of the Mount Wilson atmosphere, and oxygen was 16 per cent. Thus the question seemed to be settled. When, however, new sensitizers had been discovered, by means of which the oxygen band B could be photographed with great dispersion, Adams and Dunham in 1933 made a new study of the problem; now the absence of any trace of Martian oxygen could be stated with certainty. The quantity of free oxygen in the atmosphere of Mars must be less than 1 per cent, perhaps even less than 0.1 per cent of what it is on earth. A trace of carbon dioxide was found by G. P. Kuiper in the Martian atmosphere. So the result of all this atmospheric research is that neither Mars nor Venus is inhabitable for beings whose life-energy depends on the use of oxygen.

What, then, is the meaning of the greenish colour of the dark parts of Mars, called 'seas' and 'canals' but interpreted as vegetation? Kuiper succeeded in photographing their spectrum and found the infra-red absorption bands characteristic of cholorophyll were absent; as this is also the case with some mosses and lichens on the earth, a very primitive order of vegetation on Mars is not excluded.

The spectral investigation of the major planets brought results no less surprising. In the 'sixties Secchi and Huggins had already perceived in the Jupiter spectrum an absorption band in the red, wavelength 6180, not known from any substance on earth. In the 'seventies besides this one a number of other unknown dark bands in the red and yellow were perceived. Real progress came when modern sensitizers had opened to our view the extreme red and infra-red. Slipher then established that all major planets exhibited the same absorption bands, in such a way that they are faintest with Jupiter and increase in intensity with increasing distance from the sun. Their origin remained unknown until, in 1932, Rupert Wildt at Göttingen pointed out that a number of them were present in the spectrum of methane, the simplest of hydrocarbons, and others were due to absorption by ammonia. It was confirmed by exact comparisons made by Dunham at Mount Wilson. H. N. Russell deduced from a theoretical discussion of chemical equilibrium, that in the absence of oxygen the atoms of nitrogen, carbon and hydrogen under such low temperatures as below $-100°$ must combine chiefly into these carbon-hydrogen (CH_4) and nitrogen-hydrogen (NH_3) compounds as their most stable molecules. The increase in intensity from Jupiter to Neptune is then connected with decreasing temperature. The structure of the two largest planets, Jupiter and Saturn, was thus described by Jeffreys in 1934: a core of rock minerals is enclosed by an envelope of ice and solid carbon dioxide of a density of about one; they are surrounded by an extensive atmosphere of nitrogen, hydrogen, helium and methane, in which float clouds of crystallized ammonia.

Thus the problem of the plurality of worlds has acquired a new aspect in the last century. Three centuries ago the similarity between the dark celestial bodies and the earth had to be strongly emphasized to give a firm emotional basis to the new heliocentric concept of the world. The conviction of a purpose in creation and the idea that the planets also are destined to be the abode of living beings contributed to its gradual acceptance as a harmonious world view. Nineteenth-century astronomy, however, has demonstrated their dissimilarity with convincing power. It has revealed a far greater wealth and variety in nature than the awakening insight of simpler times could suspect. The ingenuous faith in a purpose had to give way to the scientific view that the conditions of other worlds are determined by natural influences and forces, different for each planet according to size and position. The problem of the reactions and the equilibrium of different atoms and molecules in Jupiter's atmosphere, resulting in various forms and colours, loses nothing of its charm by the exclusion of any idea of living organisms.

Yet this means a fundamental change in our concept of the world. The dream of a plurality of inhabited worlds, the dream of other men living on neighbouring kindred globes, is over. It is true that as yet we know nothing of possibilities in inaccessibly distant stellar systems; but as far as our own solar system is concerned, no other mankind exists than on earth. We do not feel the solitude, divided as we still are into hostile peoples who look upon one another as foreigners and try to exterminate one another. When, however, mankind on earth has become united, the consciousness of our being the sole humanity in our solar system—separated from other systems by impassable distances—will exert a determining influence upon our life-concept.

That the earth occupies this exceptional place among the planets is due to the presence of free oxygen in its atmosphere. Oxygen as a main constituent of a planetary atmosphere is an anomaly; since this element is easily bound by many other elements, it will be rapidly absorbed and fixed in rock minerals and other compounds. It would also have disappeared on earth had it not been continually supplied by the photochemical processes in the plant cells, the dissociation of carbon dioxide by chlorophyll under absorption of sunlight. So the kernel of the problem confronting us here is: how did chlorophyll originate on earth? What special conditions—elsewhere absent, or perhaps present in the future for the atmosphere of Venus with its content of carbon dioxide—brought it about first that protein molecules formed in the early tepid oceans and merged into living matter? What conditions caused the forming of this special compound that could utilize absorbed solar energy for the splitting of carbon dioxide molecules into carbon for themselves and oxygen for the future animal world, and so could open the road leading to man? The study of this fundamental problem as yet has hardly been started. The development of astronomy in the last century has brought it to the fore as a problem specific for the earth.

CHAPTER 36

COSMOGONY AND EVOLUTION

W HAT was the origin of the world? In the earliest dawn of civilization and even before, in the stage of barbarism, this problem
already occupied the minds of men. In myths and legends, man gave
various answers, according to the conditions of his life and the state of
his knowledge. Such cosmogonies do not belong to astronomy but to
folklore and religion, or, as in later days in Greece, they were the work
of poets and philosophers. They were based on the simplest ideas of
world structure, in which heaven and earth were its two equivalent
halves. They usually proceeded from an original chaos into which the
gods, by their act of creation, brought order and structure. A creation
out of nothing was beyond their imagination.

In the Middle Ages and the following centuries cosmogonic ideas
were determined by the Christian doctrine laid down in the Book of
Genesis. The scientific contests of these centuries dealt with the structure
and not with the origin of the world. This changed with the eighteenth
century, the century of rationalism. The Kant and Laplace theory of the
origin of the solar system out of primeval nebula has been dealt with
above. Here, for the first time, primitive legend gave way to scientific
theory. Theory could not go beyond the science of the time. The only
science, apart from astronomy, that had developed solid foundations
was mechanics, the theory of forces and motions. Eighteenth-century
cosmogony therefore could do no more than apply mechanics to the
limited problem of the solar system. A true all-encompassing scientific
cosmogony was for the first time possible through the development of
physics in the nineteenth century.

Atomic theory at the beginning of this century awakened in the scientists the consciousness that matter is indestructible. It implied the notion
of an eternity of the world in the past and the future, dominated by
natural laws, in which there was no room for an act of creation. Thus
the concept of 'cosmogony'—though the word remains in use—was
superseded by the concept of 'evolution'. The science of the evolution of
the universe is intimately connected with the development of astrophysics.

The origin and growth, with increasing rapidity, of astrophysics, was the most important renewal of astronomy in the nineteenth century. Gradually it outstripped the old astronomy of position and motion. Indeed it is the real astronomy, since it is the knowledge of the world bodies themselves. It deals with their physical nature, whereas the science of their motions deals with their outer behaviour under the compulsion of gravitational forces. Most important now are not the small and dark secondary bodies which can only be seen when they are near, but the large self-luminous millions of stars, which, as the sources of light and heat, fill the universe with radiant energy. Among them our sun, the basis of our life, is most important as the foremost object of scientific research.

In this development astronomy was no longer autonomous as in former centuries, when it had to seek its own way. It now has other sciences to lean on, especially physics; the name 'astrophysics' indicated that it is the application of physics to the stars. So it had to wait until, about the middle of the century, physics had developed first the theoretical principles and laws valid for the entire world of phenomena, and secondly the practical methods for studying the distant bodies. The first condition was realized by the rise of the mechanical theory of heat, including the doctrine of energy and entropy, the second by the discovery of spectrum analysis.

The rise of theoretical physics, especially the theory of heat, was intimately connected with the development of the steam engine as the technical basis of nineteenth-century industry. Practical and theoretical study of the working of the steam engine was a matter of first importance for increase in knowledge as well as increase in productivity. Here it was seen every day how the engine produced power and performed work while heat was expended. The obvious conclusion was that heat was transformed into work or working power. From the experience that steam engines had a greater output when the steam had been heated to a higher temperature, Carnot derived another explanation: the work effect is produced because heat (at the time considered as a special weightless matter) descends from high to low temperature, just as in a water-mill it is produced when water descends from a higher to a lower level. On the other hand, machines in operation showed numerous cases where work used to vanquish frictions or resistances turned into heat. Thus in the mind of the physicists the concept of energy was born as the capacity of performing work, which in later days was increasingly to dominate the realm of physics. In the law of conservation of energy, Julius Robert Mayer in 1842 stated that energy is indestructible; it appears in many different forms—heat being one of them—which in physical processes pass into one another, the total amount always remaining the same. At the same time James Prescott Joule at Manchester determined by various experiments how many units of work were equivalent to one unit of heat: 1 calorie, the quantity of heat increasing the temperature of 1 kg. of water by 1° C, is equal to 425 kilogram-metres, raising 425 kg. by 1 metre. In Germany, Helmholtz in 1847 demonstrated the transformation of energy (called *Kraft*, 'force') in all different physical phenomena.

The truth in Carnot's apparently inconsistent view came to light when subsequently (1850–54) Rudolph Clausius in Zurich and William Thomson in England brought it into line with the conservation of energy. Energy in the form of heat can pass into the form of work only when at the same time a quantity of heat is allowed to descend from higher to lower temperature. Of itself heat goes from warmer to colder bodies, through conduction or radiation; of itself mechanical energy passes into heat. This is the automatic positive course in nature. The reverse negative processes can take place only when compensated by a simultaneous positive process. This law, called the 'Second Law of Thermodynamics', was brought into mathematical form by Clausius through the introduction of the concept of *entropy*, which name was given to the 'transformation value' of the heat contained in a system; this 'value' of heat is larger as its temperature is lower. The Second Law states that the entropy (of a closed system or of the world) can only increase. Through the automatic processes the entropy increases; all opposite changes (from lower to high temperature or from heat to work) must be compensated for by simultaneous positive processes. These two laws of thermodynamics—that the total energy in the world remains constant and that the transformations of energy take place in one special direction—became highly important for astronomy.

Immediately after the first law of energy had been established for the phenomena on earth, it was extended to the heavens and applied to the problem of the sun. This was only reasonable. All the transformations of energy on earth from one to another of its different forms—energy of motion, potential energy of gravity, heat, chemical and electrical energy, life-energy—which form the contents of all phenomena, have their source in the energy poured out upon the earth by solar radiation— the tidal motions excepted. Every discussion of the conservation of energy has to include the sun and to pose the problem of to what other forms we have to look to for the origin of solar energy. Practically, it presented itself in the astonishing question of how the sun can maintain its radiation in undiminished power, notwithstanding the continuity of its enormous losses through radiation.

Mayer had an answer directly to hand: heat proceeds from mechanical energy. The earth is continuously battered by meteors, and so to a greater intensity is the sun. The energy of motion of the meteors when they are arrested is transformed into heat. Qualitatively this seemed to be a satisfactory explanation; when, however, a numerical test was applied later on, it failed. The yearly solar radiation amounts to 2.9×10^{33} calories; the energy of 1 gm. of matter coming from afar and falling into the sun is equal to 4.4×10^7 calories; hence, to make up for the radiation, a mass of 6.5×10^{25} gm. yearly has to fall upon the sun, one thirty-millionth of the sun's mass (19.8×10^{32} gm.). It seems little, but it is much, far too much. By such a regular increase in the sun's mass, the planets would gradually revolve more rapidly and in ever smaller orbits. As a year is thirty millions of seconds, the earth's time of revolution would be shortened by something of the order of 2 seconds per year. This is impossible; a thousandth of that amount would have been detected centuries ago.

Helmholtz in 1853 explained the constancy of the solar radiation in a different way, by framing his theory of contraction. Contraction produces heat. When the sun contracts so that all its particles come nearer to the centre, their potential energy of attraction decreases and an equivalent amount of heat is generated. The yearly radiation is covered by a contraction of 75 metres, i.e. one eighteen millionth ($\frac{1}{18000000}$) of the diameter. It is so little that even after a thousand years it would be imperceptible to our instruments. So the contraction theory offered a satisfactory explanation of the problem of the solar heat. Moreover, it could be connected with the Kant and Laplace theory of the origin of the solar system through the contraction of an original nebula. In the eighteenth century this theory had to serve to explain the planets and their circular orbits. Now attention was directed to the contraction of the matter of the nebula—except the insignificant fragments forming the planets—into the small body of the sun. This contraction from a widely extended nebula into the sun's present volume must have produced an amount of heat sufficient to keep up the present radiation for 18,000,000 years. Thus, for the first time, physical astronomy afforded an estimate of the time scale in which the age of the solar system and of the earth must be expressed.

Proceeding along the same line of thought, the American mathematician James Homer Lane pointed out in 1871 that the sun had indeed acquired its high temperature through this contraction, so that an original high temperature of the nebula was not necessary. He demonstrated that, free in space, a sphere of gas able to expand or to contract can remain in equilibrium when its temperature changes in inverse ratio to its radius. If it contracts in losing energy by radiation only part of the generated heat is needed for the radiation, and the remainder increases the temperature. It sounds like a paradox to say that by losing energy a body should become warmer instead of colder. The same paradox is demonstrated in celestial mechanics, that a body, subjected to resistance in its orbit, increases its velocity in contracting its orbit; here the decrease in potential energy of attraction is only partly needed to neutralize the resistance, and the remainder is used to increase the velocity. Lane's law now offered a theory of how the sun had acquired its heat by regular evolution: a cool, widely extended mass of thin gas develops by its radiation into a more and more contracting solar sphere of increasing temperature. Of course this development does not continue indefinitely; Lane's law is based on the validity of the simple gas laws. When the density has increased so much that they no longer hold, Lane's inverse proportionality of temperature and radius loses its validity. Gradually the contraction becomes more difficult and smaller, the energy of contraction becomes less and is not sufficient to increase the temperature or even to compensate for the radiation. Then the sun will begin to cool down and finally turn into a small, dark and cold body. Lane's results afforded a first theory of continuous evolution of the stars, which was to dominate the ideas of the astrophysicists for many years.

It was not by chance that the idea of evolution here appeared in a form entirely different from the ideas of the eighteenth century. We saw that the eighteenth century knew of development only as preparation for a remaining final condition. In the human world this was the growth from barbarism and ignorance towards a society based on nature and reason; we have already pointed out that the nebular theory of Kant and Laplace appeared as the cosmological image of this world concept. The nineteenth century brought the idea of a continually progressive social development. The Industrial Revolution, starting in England at the beginning of that century, gradually spread over the adjacent countries, France and Germany, over the United States, and ever wider over the earth. All trades were involved, all the life-conditions of man were fundamentally changed, and in half a century the aspect of our world was transformed more profoundly and more rapidly than in all preceding centuries together. Thus the human mind became accustomed to seeing the world as a continual process of development in which no end and no goal could be discerned.

This new view expressed itself in various scientific and philosophic theories. Hegel's philosophy had already presented the world as a 'dialectic' process of unfolding of the Absolute Idea. In biology the doctrine of the development from lower to more highly organized forms, after Charles Darwin's publication in 1859 of his *Origin of Species*, found wide recognition. The idea of progressive development found its physical expression in Clausius's Second Law of Thermodynamics: all autonomous processes in nature go in one direction; the entropy of the world can only increase, never decrease. Physicists sometimes expressed this law by saying that the entropy of the world tends to a maximum. This maximum would be reached when all mechanical energy had been transformed into heat, and the temperature would be the same everywhere. The final world would consist of a dispersed nebula with nothing but molecular motions. In this alleged final condition no account is taken of the powerful world force of gravitation; on the contrary, we saw that according to Lane's results a dispersed isothermal nebula must condense into a small, dense and hot sphere: intensification instead of smoothing of the temperature differences.

The idea of evolution now took increasing hold on astronomy; everywhere in the universe it discovered processes of development. George H. Darwin, the son of the great biologist, in a series of theoretical studies after 1879, dealt with tidal friction as an important agent of evolution. It was generally accepted that the earth and the moon had originally been soft and deformable through their tide-raising mutual attractions. If there were no friction and the earth could immediately take the equilibrium figure, the high-tide humps would daily go around the earth, with the moon always exactly in zenith (or nadir). By inner friction this course was retarded; the humps followed the moon at some distance, dragged after it, or, expressed in another way, the rotating earth dragged along the humps from beneath the moon over a certain distance. Then the attraction between the high-tide humps and the moon had a retarding influence upon the earth's rotation and, simultaneously transferring rotation momentum to the moon, widened its orbit and increased its period of revolution. Darwin traced this process back into the past and found that originally the daily rotation and the monthly revolution had been equal, between 3 and 5 hours; in this condition the moon and earth continually faced and almost touched each other. When originally the earth then shrank by cooling and accelerated so that the moon fell behind, tidal friction, considerable on account of the close proximity, began to operate in driving the moon away and retarding the rotation. Day and month both increased, the latter most, until at last 29 rotations came in one revolution. Later on, the moon's increase slowed up, so that now this number is $27\frac{1}{3}$ only. This will lead to a final state in which day and month will be equal, but now amounting to 55 of our present days; at a great distance, the two bodies will once more always show the same side to each other. What is treated here is a simplified theoretical case, because in practice the tidal action of the sun will disturb this course of things. The question of how, assuming this theory, a moon can revolve more rapidly than the planet rotates (as in the case of the inner Mars satellite) was answered by

Moulton: while the solar tides retard the rotation of the planet, the satellites of Mars are so small that they practically raise no tides and are therefore not subjected to tidal influences either.

More important than this distant future was the initial condition to which Darwin was led by theoretical reckoning into the past. It indicated that originally the moon and the earth had probably been one body that had split into two unequal parts. Here the conclusions coincided with the results of an investigation by the great French theorist, Henri Poincaré, in 1885 on the equilibrium and stability of rotating fluid bodies. In the eighteenth century, in 1740, Maclaurin had deduced how the flattening of a rotating fluid sphere increases with the velocity of rotation. In 1834 Jacobi, at Königsberg, demonstrated that a flattened ellipsoid of rotation, such as we have with Jupiter and with the earth, is not the only equilibrium figure of a rotating fluid; an ellipsoid with three considerably different axes is another. Poincaré now investigated the conditions of stability of these different forms. When the rotation (e.g. by shrinking) gradually becomes more rapid, we have first an ellipsoid of rotation with increasing difference between the polar and the equatorial diameters. With more rapid rotation, this form loses its stability and two unequal equatorial axes appear. With still more rapid rotation, this three-axial ellipsoid also becomes unstable; then it is constricted asymmetrically into the shape of a pear, indicating that it is going to split into two unequal parts. If the period of the solar tides in the earth, corresponding to the rotation time, is equal to the period of the body's proper vibration time, which is estimated at between 3 and 5 hours, then the deformations will increase unhampered and a catastrophic development may set in. In this way Darwin explained how the original body of the earth may have split into two unequal bodies; the resulting system is the starting point for his discussion of evolution through tidal friction.

These investigations provided the first response based on exact scientific analysis to the old mysterious and impressive cosmogonic problem of the origin of the present world. Afterwards Jeans, in a series of researches on cosmogony published in 1922, confirmed these results and extended them over more general cases of gaseous masses of different distributions of density. An interesting result was that, with a rapidly rotating mass of thin gas, the surplus matter flows out at two opposite points of the equator, forming two spiral arms. Could it give a clue towards explaining the spiral nebulae?

Most noteworthy, however, is the fact that all this nineteenth-century research on cosmogony is not applicable to the problem that is of foremost interest to mankind; it is silent regarding the origin of the planetary system. Repeatedly critical remarks pointed out that a system like ours could not have originated in the way expounded by Kant and Laplace. The chief difficulty is that 60 per cent of the total moment of momentum (what in a popular way might be called the 'quantity of rotation') of the solar system is supplied by the orbital motion of Jupiter, which possesses no more than one-thousandth of the mass of the system, and only 2 per cent by the rotation of the sun. It is not easily conceivable that in the original division the rotational momentum could thus be partitioned between the outer ring and the condensing central mass. Nor was it possible that such an amount of rotational momentum could be transferred by tidal friction from the sun to the circulating small body. The astronomer Moulton and the geologist Chamberlin, at Chicago, who put these arguments forward strongly, concluded that the bulk of the rotational momentum must have been brought into the system from outside. So in 1906 they formulated the theory that the planetary system had originated from the passage of another star close to the sun. By its attraction eruptions of solar matter had taken place, which, because of the sideways attraction of the passing star, acquired orbital motions constituting the final rotation momentum. In the surroundings of the sun it condensed into small bodies, called 'planetesimals', which afterwards, by their collisions, coalesced into planets and thereby produced the high temperature in the solid interior of the earth.

This encounter theory of the origin of the planetary system, with many variations as to details, met with the approval of Jeans, Eddington and many other astronomers. Jeans pointed out, and Eddington agreed, that such close encounters of stars, because of their large mutual distances, must be so extremely rare that perhaps no other case has happened among the millions of stars of our galactic system. 'The calculation shows that even after a star had lived its life of millions of millions of years, the chance is still about a hundred thousand to one against its being surrounded by planets' (Jeans).[203] Then it follows that possibly our planetary system is unique in the stellar universe and that hence the earth, as an abode of living beings, is unique in the world; 'not one of the profusion of stars in their myriad clusters looks down on scenes comparable to those which are passing beneath the rays of the sun' (Eddington).[204]

This startling conclusion surely cannot be considered as a final dictum of science; if one supposed cause is found to have so infinitely small a probability that it could happen once only in millions of millions of cases, numbers of other apparently improbable possibilities remain. That it was proclaimed notwithstanding its hypothetical character as a probable conclusion must be considered not only as a result of scientific reasoning but also, and even more, as the expression of a new concept of the world. From the very first, the special and unique place of man in the world was closely knit with religion. When Giordano Bruno proclaimed his doctrine of a plurality of inhabited worlds, he put himself in opposition to the accepted church doctrine. In the middle of the nineteenth century the doctrine of a multitude of planetary systems, all inhabited by intelligent beings, formed part of a world concept often expressed in materialist and rationalist forms, strongly antagonistic to the dominant religious creeds. Afterwards, in Europe, the materialist and rationalist ideas lost much of their prestige against the increasing influence of religion, doubtless in connection with the social crises, catastrophes and world wars, engendering uncertainty of life and the future. It is not surprising that in such times the rejection of former materialist doctrines also involved a critical mood towards belief in a plurality of inhabited worlds.

CHAPTER 37

THE SUN

WHAT is the sun? The old belief that the sun was a globe of pure fire and light was shaken when the telescopes revealed the existence of dark spots in 'the eye of the world'. By analogy with molten metal, they were compared with scoriae. Gradually their pitch-black appearance engendered the idea of a black interior of the sun surrounded by an ocean of fire. Lalande took the solar spots to be mountains protruding above this ocean. In 1774, however, Alexander Wilson at Glasgow observed that regular round spots, on approaching the sun's border, show the surrounding half-shade ring to be broad at the outer side and narrowing at the inner side, like the inclined edge of a hole. He concluded that solar spots are holes in the brilliant shell through which we see the dark interior.

William Herschel adopted this idea and in a paper dated 1795 extended it to the supposition that the dark solar surface could quite well be inhabited by living beings. They had to be protected against the heat from above by an interposing layer of clouds; classicists may be reminded here of Aristotle's world structure, with the fire above the air. Herschel produced good arguments against the objection that the heat from above would scorch the dark surface beneath. He said: 'On the tops of mountains of sufficient height . . . we always find regions of ice and snow. Now if the solar rays themselves conveyed all the heat we find on this globe, it ought to be hottest where their course is least interrupted. Again, our aeronauts all confirm the coldness of the upper regions of the atmosphere . . .' To explain this phenomenon, he assumed that the sun's rays produced heat only by contact with 'the matter of fire' contained in the substances that are heated; then we had further to assume that the lower atmosphere and the dark surface of the sun are not 'capable of any excessive affection by its own rays'. Thus he could find similarity with the other globes of the planetary system, which 'leads us on to suppose that it is most probably also inhabited, like the rest of the planets, by beings whose organs are adapted to the peculiar circumstances of that vast globe'.[205] In the first half of the nineteenth century this notion of a dark solar body was generally accepted; it shows how far physical consciousness lagged behind the extent of astronomical knowledge. A more profound insight could not force its way until the concept of energy was established. Then, in the second half of the century, the study of the sun advanced with rapid strides.

At first this study proceeded along the old paths, though with more breadth and perseverance. It was restricted chiefly to observation of the sunspots. In 1826 Schwabe, a chemist at Dessau in Germany, began to notice and register the sunspots regularly, chiefly with a view to discovering an eventual new small planet within the orbit of Mercury. What happened to him he compared to the experience of Saul, who went out to search for his father's asses and found a kingdom. After many years, comparing all his notes in 1843, he found (and published in 1851) that the number of sunspots showed a periodic variation. In 1828 and 1829 and also in 1836–39, the sun was not without spots for a single day, whereas in 1833 and 1843 on half the days of observation no spot was present. The total number counted in 1828 was 225; in 1833 only 33; in 1837, 333; in 1843, 34. So the maxima and minima returned after nearly 10 years. Rudolf Wolf at Bern, later at Zurich, by investigating all the historical data about sunspots, could trace the periodicity through the former centuries; the mean value of the period was $11\frac{1}{9}$ years, but with large irregularities between 7 and 17 years. Numerous attempts have been made to find the cause of this curious phenomenon, mostly by connecting it with the course of the planets, chiefly Jupiter, but without result. The forces producing sunspots must be located in the sun itself.

Still more remarkable was a discovery by Lamont, a Scottish scientist at Munich, published in the same year (1851), that the irregular perturbations of the magnetic instruments and the earth's magnetic field, also in a 10 year period, were alternately stronger and smaller; the aurorae connected with them showed the same periodicity. Sabine in England and Wolf in Switzerland immediately pointed out that both periods coalesced; the magnetic perturbations and the aurorae followed the sunspots not only in their periodicity itself but also in all its irregular variations. Even the appearance of single large spots produced magnetic storms and aurorae on earth. Thus a most remarkable and mysterious effect of solar disturbances upon terrestrial phenomena came to light.

More important than the simple counting of sunspots was the determining of their position and motion. The first object was, as with the planets, to find the rotation period of the sun. One of the foremost workers in this field was Carrington, at Redhill, who determined the positions of sunspots in the years 1853–61. He belongs to the numerous groups of British amateurs who, by the quality of their work, count as fully-fledged astronomers in the sense that William Herschel, South, Lassell and Lord Rosse were also amateurs. England from olden times had produced a class of wealthy landowners and merchants, later on also industrialists and businessmen, who expected nothing from government but had to do everything on their own initiative. They founded libraries and academic chairs and, if attracted by astronomy, built their own observatories and did scientific researches for themselves. Carrington had installed a meridian circle with which at night he made observations for a catalogue of northern stars to complete Bessel's zone work, and in the daytime he observed the sunspots. He found that the period of rotation increased with the distance from the solar equator. Near the equator the period was 25.0 days; at 20° it was 25^d 18^h; at 30° it was 26^d 11^h and it increased to $27\frac{1}{2}$ days at 45°, where the spots are scarce. Hence the spots cannot be fixed parts of a solid solar body. His results were confirmed by similar work (in 1860–73) by Spoerer, a German amateur at Anklam in Pomerania. Both observers perceived another peculiarity. During the years of most numerous spots these came gradually nearer to the sun's equator; their latitude decreased from about 25° to about 10°, and then they became extinct about latitude 5°. At the same time (the time of minimum number), at a high latitude of 25°–30°, the first numbers of a new cycle appear, which in the next years of increasing number expand towards lower latitudes. Thus the periodicity of the sunspots consists in a succession of consecutive series or waves, all starting at high latitude, swelling as they descend towards low latitude and being extinguished there.

Photography, of course, was used to make pictures of the sun immediately after its discovery. Here, contrary to other celestial objects, it was the abundance of light that had to be neutralized by special devices. Now it was possible in a split second of exposure to register all the details of the sun's surface—the spots, faculae and other markings—so that afterwards their number, their total surface, their shapes and movements, could be studied. A spectacular effect was obtained by combining two pictures in a stereoscope, one taken a short time after the other, when the sun had rotated to a small extent; in this way the sun was seen as a globe, with the dark spots as pits and the faculae floating at high level. In 1858 Warren de la Rue devised and erected a photoheliograph at Kew, afterwards transferred to Greenwich and used for the regular photographing and measuring of solar pictures as routine work.

Besides this work, which was used chiefly for statistical purposes, there were refined technical methods for the study of the minute details in the structure of the spots as well as of the fine granulation of the undisturbed photospheric surface. Here it was the French astronomer P. Jules C. Janssen, at Meudon, who in the seventies excelled in his enlarged photographs of the granulation and the sunspots. With the rapid variations in the spots, the finest detail was not so important here as with the planets, except for the granulae. Through a careful comparison of a number of photographs taken in rapid sequence, Hansky at Pulkovo in 1905 was able to determine the average lifetime of the separate granulae at 2–5 minutes; then they dissolved and were replaced by others.

Up to the middle of the century only the sun itself was observed by the astronomers. In 1842 for the first time, at a total eclipse of the sun visible in Southern France and Northern Italy, their attention was directed to those luminous phenomena around the dark disc that make total eclipses of the sun the most wonderful sight and the most important source of knowledge: the far-extending glory called the 'corona' and the small pink 'prominences' protruding at different points outside the dark limb of the moon. They had already been observed earlier, in 1733, by Wassenius at Gothenburg in Sweden, and described as red clouds in the moon's atmosphere. They were mentioned even in the medieval Russian monastery chronicles, where, e.g. on May 1, 1185, is written: 'The sun became like a crescent of the moon, from the horns of which a glow similar to that of red-hot charcoals was emanating. It was terrifying to men to see this sign of the Lord.'[206]

The eclipse of 1851, visible in Sweden, made it possible to ascertain that the prominences belonged to the sun, not to the moon, and that they are the highest parts of a narrow pink ring (afterwards called the 'chromosphere') surrounding the sun, which also had occasionally been mentioned before. At the eclipse of 1860, observed in Spain, photography was used on a big scale to ascertain the objective reality of all these phenomena. From now on, at every total eclipse astronomers travelled to the zone of totality, in whatever distant country it might be, to make, in the few minutes available, the various observations that by startling discoveries extended our knowledge of the sun.

Then in the years 1859–62 spectrum analysis arose, chiefly through the work of Kirchhoff and Bunsen. It was, so to speak, in the air. Several scientists, such as Stokes, Foucault, Ångström and others, had perceived that Fraunhofer's double line D in the solar spectrum coincided with a bright-yellow double sodium line. The conclusion that sodium must be present in the sun was obvious. Stokes described the process in this way: the particles that had absorbed light of this special wavelength from a light source were afterwards able to emit it. It was Gustav Kirchhoff (1824–87), eminent theoretical physicist, who gave a solid scientific basis to spectrum analysis. He demonstrated that for any wavelength the ratio of emission (quantity of emitted light) and absorption (the fraction absorbed from incident light) is the same for all bodies and is equal to the emission of a 'perfectly black body' (supposed to absorb 100 per cent). The latter is a continuous function of wavelength and temperature. So a hot gas that absorbs one special wavelength and thereby produces a strong black line in the spectrum of a light source will emit the same wavelength strongly, whereas adjacent wavelengths, which are little or not at all absorbed, are emitted little or not at all. The significance of the Fraunhofer lines in the solar spectrum was now at once clear; they indicated what absorbing particles were present in the sun. Kirchhoff measured, on an arbitrary scale, the position of some thousands of Fraunhofer lines and established their coincidence with lines emitted by diverse chemical elements, such as hydrogen, iron, sodium, magnesium, calcium, etc. He concluded that these elements were present in the sun's atmosphere, absorbing their special wavelengths from the continuous spectrum emitted by the solar body. Ångström in 1868 replaced Kirchhoff's arbitrary scale by the natural scale of wavelengths; he expressed them in a unit of one ten-millionth of a millimetre, afterwards called by his name; for the visible colours the wavelengths in this scale are numbers of four figures (red, 6,500; green, 5,000; violet, 4,000).

Spectrum analysis miraculously realized what shortly before had been declared forever impossible: establishing the chemical composition of distant inaccessible bodies. The French philosopher of positivism, Auguste Comte, in 1835 in his *Cours de philosophie positive*, in order to emphasize that true science is impossible if it is not based on experience, wrote of the celestial bodies: 'We understand the possibility of determining their shapes, their distances, their sizes and motions, whereas never, by any means, will we be able to study their chemical composition, their mineralogic structure, and not at all the nature of organic beings living on their surface.' And some pages later: 'I persist in the opinion that every notion of the true mean temperature of the stars will necessarily always be concealed from us.'[207] It is apparent here, just as it was with Descartes and with Hegel, that philosophy must stumble when it tries to prescribe and predict the results, or even the methods, of science; its task is to use them as materials for its own theory of knowledge, epistemology.

The new discoveries brought new ideas on the nature of the sun. It was no longer possible to believe that the sun's interior was a dark and cold body. Kirchhoff considered the sun to be a red-hot sphere, solid or liquid—because its spectrum is continuous—surrounded by a less hot atmosphere containing the terrestrial elements in a gaseous state, which produce the Fraunhofer lines. He considered the sunspots to be cooler clouds in this atmosphere; he was well aware that relative darkness was bound to a lower temperature and that the often-heard opinion that darkness was due to a smaller power of emission was refuted by physical law. His ideas on the sun's interior were corrected in 1864 by Secchi and John Herschel, who assumed it to be gaseous also; they ascribed the continuous spectrum to droplets floating like a cloud layer deep in the atmosphere. Experiments showed that strongly compressed gases also emitted a continuous spectrum; moreover, the physicist Andrews in 1869 discovered that any matter above a certain 'critical temperature' cannot exist in a fluid but only in a gaseous state. Yet the cloud theory was generally adhered to. 'It seems almost impossible to doubt that the photosphere is a shell of clouds,' was written by Young in 1882 in his book *The Sun*.[208] Father Secchi, as well as the French astronomer Faye, explained the sunspots as openings in the cloud layer which at such places was volatilized by an uprush of hotter gases expelled by pressure from below. More complicated accessory explanations were needed to meet the objection that in that case the spot would appear hotter than the surroundings. Faye was especially struck by whirling phenomena observed in the spots, and he compared them with tornadoes on earth.

To establish what elements are present on a luminous celestial body is, in principle, very simple; it demands nothing but the observation of exact coincidences of spectral lines or the equality of exactly measured wavelengths. Since instrument making and glass techniques were highly developed, excellent prism spectroscopes for various purposes could soon be constructed. The introduction of photography not only afforded the same advantages as in other domains of astronomy but, moreover, gave access to a new field of wavelengths invisible to the eye —the ultraviolet part of the spectrum between 4,000 and 3,000 Å (Ångströms). Thus spectrographs increasingly superseded spectroscopes.

A prominent achievement in this development was the construction of concave gratings, in 1887, by Henry A. Rowland at Baltimore. A concave metal mirror produces a sharp image of a slit without the need of lenses. On such a mirror a grating of fine parallel lines, 25,000 per inch, was engraved, which produced a series of diffraction spectra of great dispersion and high resolution. Rowland worked for many years on the perfecting of an engraving machine capable of automatically cutting the fine grooves at exactly equal distances. The reward of this painstaking work consisted in spectra which remained unequalled for dozens of years. With these gratings Rowland photographed the solar spectrum and published it in 1888 as an atlas of maps, on a constant scale of 1 Å = 3 mm. so that the entire spectrum between 3,000 and 6,900 Å forms a band of 40 feet in length. It contains more than 20,000 Fraunhofer lines in all intensities, from barely visible traces up to heavy dark bands. By measuring the original photographs Rowland was able to publish in 1896 a catalogue of all these lines, with their wavelengths given to three decimals (hence in seven figures) and their estimated numerical intensities, a standard work used henceforth by every astrophysicist. From the lines he could ascertain the presence of 36 terrestrial elements in the sun. In a later revision of his catalogue by St John at Mount Wilson in 1928, this number was increased to 51. The most important later progress was the publication, in 1940, by Minnaert and his co-workers at Utrecht, of a photometric atlas of the solar spectrum. Instead of by separate lines, the entire spectrum is pictured here by a continuous intensity curve, so that, in addition to the place, the curve also shows the width, the character and the distribution of intensity (the profile) of every line. In the figures in plate 13 the progress of our knowledge of the solar spectrum can be seen.

We have now to revert to the nineteenth century for the first application of the new method of spectrum analysis to the special solar phenomena. The occasion of the first total eclipse after the introduction of spectral research was of course seized upon to satisfy the curiosity as to the nature of the newly discovered prominences and corona. Numerous observers with their spectral apparatus went to India for the eclipse of August 18, 1868. In their results they all agreed, with small differences in detail only, that the spectrum of the prominences consisted in some few bright lines and that, hence, the prominences were glowing masses of gas. The brightest lines were the red and green emission lines of hydrogen (designated Hα and Hβ) coinciding with Fraunhofer's C and F, and, moreover, a yellow line, first taken for the sodium line D, afterwards seen to be different and denoted D_3. It was not known from any terrestrial spectrum and was hence ascribed to an element present only on the sun and called 'helium'.

The observed emission lines were so brilliant that one of the observers, Janssen from Meudon, immediately understood that the darkness of an eclipse was not needed to make them visible. The next day he placed the slit of his spectroscope just outside the sun's limb and without difficulty could observe the emission lines in full daylight. For some weeks he could study the sun 'during a period equivalent to an eclipse of 17 days', as he wrote in a report to the Paris Academy. 'I have made charts of the prominences which show with what rapidity (often in a few minutes) these immense gaseous masses change their form and position.'[209] At the same time, Lockyer was working in England on the same lines; two years before, supposing that the prominences consisted of glowing gas emitting bright spectral lines, he had already contrived that they could be made visible if the continuous spectrum of the sky obliterating them could be weakened by strong dispersion. Had not the instrument-maker left him waiting a long time for the spectroscope he had ordered, he would have made his discovery long before the eclipse. Now the discoveries of both astronomers were presented to the Paris Academy on the same day in 1868.

The situation had now changed considerably. To study the spectrum of the prominences, it was not necessary to wait for an eclipse; it could be done on any bright day. Moreover, having plenty of time now, the observer could derive the extent and shape of a prominence by moving the slit. This was done still better, first by Huggins, by opening the slit widely; if the continuous background was sufficiently weakened by strong dispersion, the entire prominence could be looked at and followed during all its rapid changes. Several observatories soon put the regular observation of prominences on their programmes of work. They were found to be most frequent in the years of sunspot maxima and at the same latitudes as the spots, though they occurred less frequently in high latitudes, up to the poles, where the spots are absent. As to form and character, two main types could be distinguished: the quiet prominences, floating like pink clouds in the atmosphere, and eruptive prominences, shooting up like fountains of fire to great heights and then dissolving or being sucked into the pits of the sunspots. Sometimes quiet clouds were suddenly torn into a mass of fragments as if dispersed by a violent storm. The same velocities of hundreds of miles per second were also indicated by distortions, i.e. local displacements of the lines in slit spectra. All these fascinating scenes of events on a scale of thousands of miles presented new problems; they helped to sustain the idea of eruptions as the cause of sunspots.

At the eclipse of 1870 in Spain, Young discovered in the faint continuous spectrum of the corona a narrow green emission line of a wavelength first given as 5,315 but afterwards found to be 5,303; as it was found not to occur in any known spectrum, a second unknown solar element was assumed, called 'coronium'. Another important discovery was made by Young; having set the slit of his spectroscope nearly tangent to the sun's limb, at the moment of the eclipse he saw, as in a flash, the flaring-up of an innumerable number of emission lines; after one or two seconds they disappeared, when the moon covered the thin layer of only 1″ width (representing 500 km.) that emitted them. It seemed as if all the Fraunhofer lines were for a moment reversed into bright lines, because the gas layer absorbing them was seen sideways without the background of the sun; so it was called the 'reversing layer'. In order to observe it without an eclipse, Young sought the steady air on the top of Mount Sherman; here he saw that the spectrum consisted of the same numerous metal lines as the solar spectrum, but mostly with different relative intensities. The brightest of these lines also appeared in the eruptive prominences. In the quiet prominences, in addition to the hydrogen lines and the helium line D_3, a small number of unknown lines were seen; when Ramsay in 1895 had discovered helium in terrestrial sources, so that its entire spectrum could be investigated, they appeared to be other helium lines.

When photography was applied to these phenomena, Huggins in 1875 discovered the harmonic series of ultraviolet hydrogen lines, in continuation of the four lines Hα to Hδ. At continuously decreasing separations these lines merge at last at a limiting wavelength, below 3,700 Å. On the photographs it appeared, moreover, that the two violet calcium lines coinciding with H and K of Fraunhofer, with wavelengths 3,968 and 3,934 Å, surpassed all other chromospheric lines in brightness; indeed, H and K themselves are the strongest absorption lines in the Fraunhofer spectrum.

For the purpose of photographic observation of solar eclipses the appropriate instrument was devised and constructed by Lockyer. Joseph Norman Lockyer (1836–1920) was also an amateur, an official at the War Ministry, who could devote only his spare time to astronomy, yet was esteemed a first-rate scientist. He constructed the 'prismatic camera' simply by placing a prism before the camera lens. A self-luminous object is then pictured in as many images as it emits separate wavelengths, each image showing by its shape the distribution of its atoms emitting this line. A photograph taken at the exact moment of the 'flash' shows all the metal lines of the reversing layer as short narrow arcs, whereas the hydrogen lines H and K present the prominences in their true form; the green corona line is pictured as a faint and broad luminous ring, sharply cut off at the inner edge by the limb of the dark moon. In 1893 the first imperfect photographs had been obtained— showing, because the exact moment of the flash was missed, only the arcs of hydrogen, helium, H and K. After 1896 prismatic cameras were in regular use at every eclipse because of the vast amount of information they afforded. Mitchell in 1905 perfected the method by using a Rowland grating instead of a prism.

The mysterious coronal lines remained an important object of study at solar eclipses, because all attempts to make them visible in full daylight had failed. The curious fact here was that different observers at different eclipses reported new and different coronal lines not perceived before. A few of the brightest lines appeared regularly: besides the green line, a red and a violet line; but for the others there remained doubt as to whether they were real and whether the coronal spectrum was variable. For all these lines the origin was unknown; perhaps there were more 'coronium' elements.

The regular photography of prominences led to a new method of research. The bright lines of hydrogen or calcium were visible not only outside the sun's border but also at different strongly disturbed places on the solar disc, mostly in the vicinity of sunspots. They appeared here as bright, narrow emission lines in the centre of the broad, dark absorption lines and were considered to be due to luminous gaseous masses in the upper layers of the atmosphere. In 1890–91 Deslandres at Paris and George E. Hale at Chicago, independently and in somewhat different ways, constructed an instrument, called a 'spectroheliograph', to picture these high emissions. The slit of a spectrograph is made to slide over the sun's image; a firmly connected second slit behind the prism, allowing only the narrow emission line to pass, slides over the photographic plate. Such pictures of the sun in the light of one wavelength were made first in Chicago and at the Yerkes Observatory; to conduct a more profound study, Hale founded the 'Solar Observatory' at Mount Wilson, where later the restricting word 'solar' was dropped. The regular study of these pictures, most in the light of the calcium K line and the hydrogen $H\alpha$ or $H\gamma$, revealed an abundance of remarkable structures, especially about the sunspots, reminiscent of spiral arms or whorls and vortices, sometimes connecting two adjacent spots by curves reminiscent of iron filings over two magnetic poles.

The spectrum of the sunspots was also an object of many researches. Though sunspots appear black by contrast, they radiate a strong light. In 1866 Lockyer found, as a cause of the relative darkness, the broadening of most Fraunhofer lines and the appearance of numerous additional fine lines. Their detailed investigation had to wait for the use of more perfect spectrographs. It was after 1920 that photographs with such great dispersion were taken at Mount Wilson that a catalogue of spot lines could be published (in 1933) by Charlotte Moore, which in their completeness approached Rowland's solar spectrum. Careful comparison showed interesting differences; the so-called 'high-temperature' lines were weaker, the low-temperature lines were stronger in the spot spectrum, whereas numerous fine lines belonging to molecular bands appeared. Both gave clear evidence that the spots are regions of low temperature.

Most metallic lines in the spot spectrum with great dispersion show a reversal in their centre, a bright line separating the dark line into two components. They were considered to be of the same character as the bright emissions in the centre of the hydrogen and H and K lines, i.e. as due to high layers of hot gases. In 1908, however, Hale discovered at Mount Wilson that their origin was entirely different, viz. the magnetic doubling of the lines through the Zeeman effect. The two components were circularly polarized in opposite directions, showing that strong magnetic fields were acting in the sunspots. It was remarkable that pairs of spots following one another in the solar rotation, which in the Hα light often show opposite directions in the vortex structure, also showed opposite magnetic fields. More remarkable was it that the sequence of polarity in such a pair in the northern hemisphere was opposite to the sequence in the southern hemisphere. It was suggestive of the opposite direction of rotation of tornadoes on earth at both sides of the equator. When the whirling charged particles in the two spots forming a pair had opposite directions, Hale considered them as opposite ends of one vortex tube situated in the deeper parts and producing spot phenomena where it ended at the surface. The many new ideas and problems raised by these phenomena acquired a still more curious and mysterious aspect when, after the sunspot minimum of 1912, it appeared that the polarities of the northern and southern hemispheres had interchanged. After the minimum of 1922, a new interchange took place. So it might be said that the real period of the sunspots, especially in their magnetic and rotational phenomena, was not 11 but 22 years.

Gradually a wealth of knowledge on solar phenomena had been acquired by a long series of patient researches and startling discoveries. Yet all this could still only be called a prescientific period of solar physics. Half a century after the discovery of spectrum analysis, astrophysics was in the same stage as was the old astronomy before Kepler and Newton. It consisted of an abundance of data and facts but without the basis of a solid theory. New theories of the sun were now and again devised by prominent observers to explain new facts just discovered, but they were based on the imperfect general ideas of the time. There were the more general theories of keen outsiders, like August Schmidt in Germany, who in 1891 explained many phenomena, such as, for example, the sun's sharp border, by strongly curved light rays. Proceeding from the same idea, the Utrecht physicist W. H. Julius gave an explanation of the Fraunhofer and the chromospheric lines as consequences of anomalous dispersion. In all these attempts the rigid certainty of unfailing principles was lacking.

This was due not to astronomy, which could not make experiments with its objects, but to physics. Astrophysics could not become a real science until physics had developed the phenomena of radiation into a perfect theory. This was accomplished at the beginning of the twentieth century. It was possible only by such a fundamental revolution in the principles of physics that—as was then often said—to a physicist of 1890 not only were the new physical laws entirely incomprehensible but their terms even formed a foreign language which he could not understand.

This development began with the establishment of the general laws

413

of radiation. In 1870 they were so utterly unknown that Secchi put the temperature of the sun's surface at some millions of degrees, whereas the physicist Pouillet found 2,000°. Both results were based on almost the same experimental value of the sun's amount of radiation, but in the one case the radiation was assumed to increase proportionally, in the other exponentially, with the temperature. In 1879 the Austrian physicist Stefan, from accurate measurements over a large range of temperatures, deduced that the total radiation was proportional to the fourth power of the absolute temperature. In 1884 Boltzmann gave a rigid theoretical demonstration of this law. Following along the same paths, W. Wien in 1893 deduced that the radiation of a perfectly black body was given by one single function of the wavelength, which with changing temperature shifted in such a way that the wavelength of its maximum was inversely proportional to the temperature. These two laws, applied to observational data, established that the temperature of the sun's surface was nearly 6,000°. In 1906 they found their final completion in the radiation formula of Max Planck, which expressed the black radiation as a combined function of wavelength and temperature. More than by its own contents, this formula became famous because Planck had to assume for its derivation that radiation energy is emitted and absorbed not in a continuous stream but in discrete portions. His theory of the 'quantum' of action placed its stamp upon all later physics.

At the same time the knowledge of the regularities in the monochromatic radiations of the atoms, which we observe as the emission line spectrum of each element, also made great progress. In 1885 J. J. Balmer at Basle published his famous formula for the wavelengths of the four visible hydrogen lines. They can be exactly expressed by the value 3645.6 multiplied by the ratios $\frac{9}{5}$, $\frac{16}{12}$, $\frac{25}{21}$, and $\frac{36}{32}$, hence by the fraction $n^2/(n^2-4)$, in which n is taken as 3, 4, 5 and 6. This formula was confirmed by the ultraviolet lines which Huggins and Draper had photographed in the spectra of the star Vega and of the solar chromosphere. They correspond to the larger values of n from 7 on increasing indefinitely. The wavelengths decrease with decreasing intervals, so that the lines crowd ever nearer together, until they coalesce in the limiting wavelength 3645.6. After this lucky find, analogous but more complicated formulas for other spectra were devised, first by Kayser and Runge and then in a more promising form by Rydberg. The abundant lines in these spectra were full of regularities and numerical relations; their grouping in doublets, triplets and other multiplets was connected with the position of the element in the periodic system of elements. Too little was known, however, of atomic structure to reduce these spectrum relations to atomic properties. The spectra were, according to a saying of those days, an answer of nature to which we did not know the question. In the light entering our instruments the distant celestial bodies transmit messages telling of their condition; but the messages are in code, and, as long as there was no key, they could not be deciphered.

Niels Bohr's atomic model, constructed in 1913 on the basis of the structure derived by Rutherford in 1911 from his experiments, was the key enabling the physicists to break the code and to decipher the light messages. In the next dozen years the rapid progress in theory and experiment made the entire field of spectral structure and corresponding atomic structure into an assured domain of science. Every line emission or absorption is produced by the transition between two atomic states of different energy; its wave number (the reciprocal of the wavelength) is the difference between two 'terms', corresponding to the energies of these two states. Thus hundreds of spectral lines could be reduced to some few tens of terms.

Based on this theory of atomic spectra, the Fraunhofer lines could give information on the interchange of energy of the atoms and hence on the physical conditions in the sun's atmosphere. The solar spectrum, of course, could only give information on the outer layers, the only ones from which light enters our instruments; but this information was complete. The state of these layers could now be ascertained. In 1905 Karl Schwarzschild (1873–1916), a pioneer here as in so many fields, gave a theory of the sun's atmosphere based on the principle of radiative equilibrium, according to which the temperature at any point is a result of the radiation it receives from all sides as the main mechanism of heat interchange at these high temperatures. Here the cloud theory of the photosphere has disappeared completely; temperature and density regularly increase in going downwards. Building on this foundation Milne from 1921 on and Eddington in 1923 gave a thorough theoretical treatment of such an atmosphere as the gradual thinning-out of the dense and hot deeper layers. The continuous spectrum, as well as the Fraunhofer lines, now appeared as the combined product of all the deeper and higher layers of atoms; the distinction between photosphere, reversing layer and atmosphere no longer corresponded to real qualitative but only to practical quantitative differences.

The structure of a Fraunhofer line now also became an object of investigation. What was formerly called a 'line', with one wavelength, in reality had width and structure. Theory had shown that the atomic processes existed not only in taking up and emitting radiation (as Stokes originally had assumed) but also in collisions, in which energy of thermal motion was transferred to or from the atoms (corresponding to Kirchhoff's law). Adjacent wavelengths participated in these processes, though to an amount decreasing with distance. After the example given by Unsöld's theory in 1927, the entire intensity profile of a Fraunhofer line could be derived in the following years, so that the conditions in the solar atmosphere—temperature, pressure, ionization and the abundance of the different elements—could be derived from the line intensities. A remarkable unexpected result was that the abundance of hydrogen atoms surpasses a thousand times the abundance of all the metal atoms.

The origin of the continuous spectrum now began to present difficulties. About 1870 it was easily explained by the cloud theory, and about 1910 again by the radiation from deep gases of high density. However, since atomic theory had shown their radiations to be the result of jumps of energy giving line spectra, these explanations became doubtful. In 1939 Rupert Wildt pointed out that hydrogen atoms are able to bind and to release a second electron; in the absorbing and emitting of the needed small amount of energy, all the wavelengths of the visible solar spectrum are involved. These processes are rather infrequent; yet they play an important role because hydrogen atoms are in enormous abundance compared with other atoms. It is a curious fact that the incomparable beauty of the deep hues shading off in the continuous solar spectrum—the source of all light and colour on earth— has its origin in such an accidental and secondary process.

Astronomers could now (about 1930) be satisfied that the nature of the sun was known in its main outlines and that only details had to be added. Nature, however, in her inexhaustible richness, ever again surprised us by unexpected phenomena. We mentioned how photographing the sun with a spectroheliograph had become everyday routine at Mount Wilson. Photographing, with all its documentary value, is in some respects a new work of registering; one misses the direct contact with the happenings of every moment. Hale, himself the inventor of the spectroheliograph, was not satisfied with it, because he missed the charm of seeing what happened; so in 1926 he constructed a 'spectrohelio-scope', on the same principle, adapted to visual observation. This new method of observing led, in the following years, to the discovery of sudden explosions occurring on the sun, bright flashes of hydrogen radiation, mostly in the disturbed spot regions, which lasted about ten minutes and after half an hour had disappeared. It was then remembered that Carrington in 1859, without spectroscope, had observed a short-lived brilliant starlike flare-up lasting only five minutes, which he ascribed to a large meteor falling into the sun. In 1933 it was perceived that unexpected radio fadings coincided with these flares on the sun, a new case of solar disturbances influencing the phenomena on earth. But what was their cause?

Moreover, some phenomena which did not fit into the general picture had been detected. The presence of a line of ionized helium in the

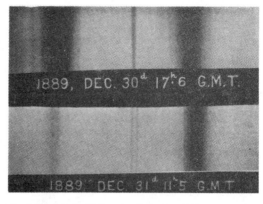

Doubling of the K line in β Auriga (the broad lines to the left and right are the 5th and 6th hydrogen lines) (p. 435)

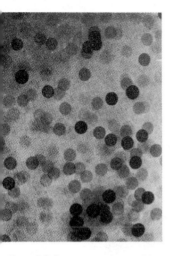

Part of Orion, extrafocal exposure (p. 441)

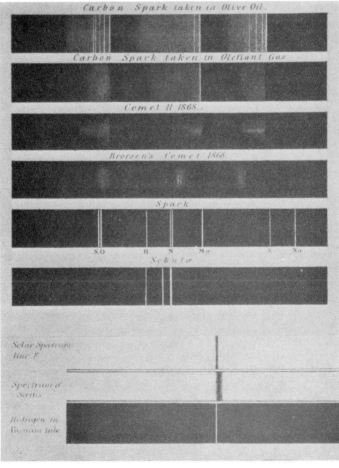

Spectra by W. Huggins (pp. 426, 451)

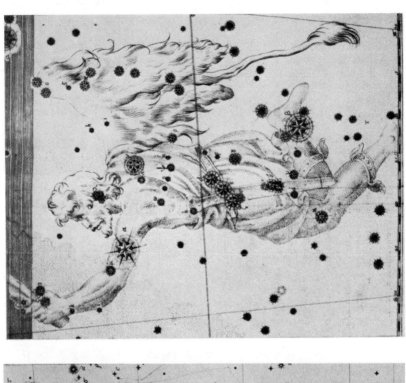

Constellation of Orion from Bayer's *Uranometria* (p. 215)

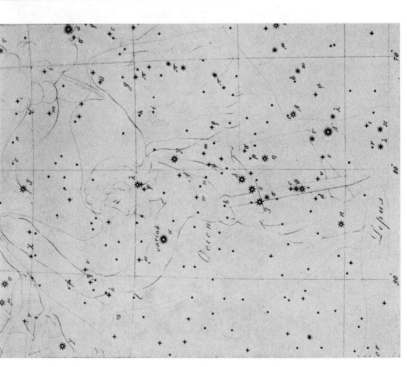

16. Constellation of Orion by Argelander from *Uranometria Nova*

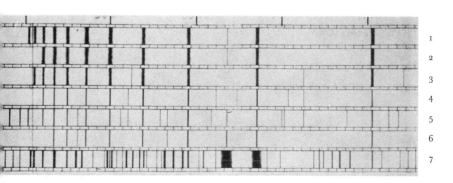

7. Huggins' first ultra-violet stellar spectra (*Philos. Transact.* 1880)
   1. Vega; 2. Sirius; 3. η Ursae; 4. Spica; 5. Altair; 6. Deneb;
   7. Arcturus (p. 452)

Types of Star Spectra according to A. Secchi. 2nd type (the sun,
Pollux); 1st type (Sirius, Vega); 3rd type (α Henc. β Pegasus); 4th
type (152 Schjellerup) (p. 450)

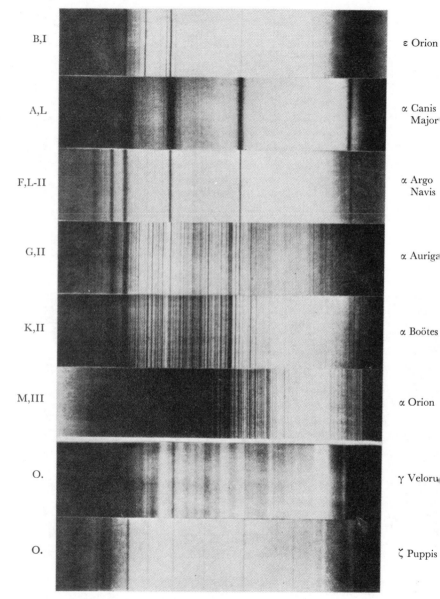

B,I	ε Orion
A,L	α Canis Major
F,L-II	α Argo Navis
G,II	α Auriga
K,II	α Boötes
M,III	α Orion
O.	γ Veloru
O.	ζ Puppis

18. Harvard types of stellar spectra (p. 453)

chromospheric spectrum indicated as its cause the presence of radiations of far greater ionizing power than could be produced by a surface temperature of 6,000°. The emission lines of the chromosphere were too broad to correspond to this temperature and indicated a greater velocity of the particles, i.e. a higher temperature, of the order of 30,000°. In the continuous spectrum of the corona, which is reflected sunlight far outside the sun, the Fraunhofer lines were invisible; the possibility was suggested that they could be smoothed out by rapid molecular motion. In 1930 Lyot had succeeded in solving the difficult problem of observing the coronal lines in daylight by carefully excluding all disturbing light coming from diffraction, impurities and atmosphere. The lines could now be studied at ease, and they appeared to be so broad that the atom velocities producing them demanded temperatures of millions of degrees. It seemed incredible, until in 1941 came the unravelling of the mystery of the 'coronium' lines. Following an indication by Grotrian, the Swedish physicist Bengt Edlén established that the strongest among them were emitted by 9, 10 and 13-fold ionized iron atoms, 11 and 12-fold ionized calcium, and 11 to 15-fold ionized nickel atoms. To strip off all these strongly attached electrons, temperatures of many hundred thousands or millions of degrees were needed. Thus it has been ascertained that the sun, with its surface at 6,000°, is surrounded up to a great distance by a widely extended atmosphere of extremely thin and hot gases. It contributes hardly anything to the total radiation of the sun but adds to it an appendage of strongly ionizing radiations of very short wavelength. How this heating from above influences the stormy processes in the prominences remained an object of further research.

A complete answer to the question, What is the sun? has to include knowledge of its interior. Important progress was made on this subject in the same years about 1920, on the basis of the above-mentioned discovery of the exact laws of radiation. In the years 1878–83 the German physicist A. Ritter had already tried by theoretical deductions to derive results about the interior of the sun. Since, however, the phenomena of radiation constituted at the time an unknown chapter of physics, the right basis to his work was lacking. In 1907 R. Emden constructed a general numerical theory of gaseous spheres in space, applicable to the sun and the stars. He again knew of no other mechanisms of heat transfer than convection and conduction; so the results were unsatisfactory. Yet he could give a good explanation of the granulae visible at the sun's surface, considering them as vortex elements in ascending and descending streams. Where Ritter and Emden had failed, Eddington succeeded when our knowledge had matured. In 1916 Eddington (1882–1944) started his series of theoretical researches, extending Schwarzschild's earlier work on radiative equilibrium to the inner layers of a star and making use of the numerical computations of Emden. Now it was possible to compute exactly the physical conditions of matter at any point of the interior as a function of its distance from the centre: its temperature, density, pressure, ionization and coefficient of absorption. The values found looked fantastic: at the centre of the sun a temperature of 18 million degrees and a pressure of 9,000 million atmospheres; yet they were products of exact computation. This does not mean that they are final and absolute; in such researches there remains a certain arbitrariness in the assumed basic data which must be introduced in a simplified way. Matter, under these extreme conditions not directly observable and manageable, because they far exceed our experimental possibilities, now became an object of science because its effects present themselves in the very existence of the sun and the stars.

CHAPTER 38

PASSING LUMINARIES

ARISTOTLE had placed fire as the highest sphere of the unsteady earthly world; all the transitory luminous phenomena not admissible into the celestial realms of eternal ether were given a place there above the air. Here his intuition was not so very different from what later science taught. All that was above he called meteors, the higher things; this name was later restricted to the phenomena which were popularly called 'shooting stars'.

For many centuries shooting stars did not attract the attention of scientists; they were considered to be a kind of lightning. To learn something about their character, one has first to answer the question about their place. At what height do they flare up and become extinguished? In the last years of the eighteenth century, in 1798, two young scientists, Benzenberg and Brandes, students at Göttingen, tried to answer these questions by simultaneous observation at two different stations. Identical times of observation, after carefully co-ordinating their watches, ensured that they observed identical objects; the different apparent courses as projected among the stars made it possible to compute the heights of the objects. The observers soon perceived that these heights were far greater than they had suspected, not some few hundreds or thousands of feet but tens, even hundreds of miles, so that they had to make their observing points much farther apart. Shooting stars appear and disappear far above the cloud region of the atmosphere, in and even above the layers called in modern times the 'stratosphere'. Though their duration was difficult to estimate, it was apparent that with such distances the real velocities amounted to tens of kilometres per second and were comparable to planetary velocities.

So the meteors were shown to be objects which from planetary space penetrate into our atmosphere, where, because of the resistance of the air, they lose their velocity and are burned up. Benzenberg supposed that they might be little stones erupted in former times by the volcanoes of the moon, which after various wanderings through space at last hit the earth. That they came from the farther depths of the solar system was demonstrated in 1833 by a magnificent meteor shower. For more than five hours during the night of November 12th–13th, shooting stars came in a dense display, like a fiery hailstorm of bright and small stars, an overwhelming spectacle especially in North America. They all seemed to radiate from a single point in the sky, in the head of the Lion; this radiation point (or 'radiant') moved along with the stars, so that it could not belong to the earth. Olmsted, at New Haven, immediately explained the radiation from one point as an effect of perspective, where in fact the objects are in parallel motion. Since at the same date a year before a large number of meteors had also been seen, Olmsted supposed that a cloud of small bodies described an orbit about the sun in half a year, meeting the earth after exactly a year at the same place in space. It was then remembered that Alexander von Humboldt, at the start of his travels over South America, in the town of Cumana had observed such a meteor shower in 1799 on the night of November 11th–12th. In his description Humboldt mentioned that old people there had told him that about 30 years earlier the same phenomenon had been seen. In the years after 1834 the November meteors (called 'Leonids' after their radiant) became scarce; Olbers suspected that the full display of Leonids would come back after another 34 years, in 1867. Erman at Berlin supposed that the Leonid meteors described a common orbit of somewhat more than a year, which intersected the earth's orbit where the earth was on November 13th; the meteors were dispersed over the entire orbit, not equally, but mostly heaped about one place. For the dense swarm to meet the earth on November 13th every 33 years, its time of revolution might be $1 + \frac{1}{33}$ or $1 - \frac{1}{33}$ years as well as 33 years. Since every year about August 10th a great number of shooting stars are seen, for which now a common radiant was also found, in the northern part of Perseus (hence called 'Perseids'), an analogous explanation was given; but here the meteors were more evenly distributed over the entire orbit.

In 1864 H. A. Newton of New Haven, through historical investigation, found that the Leonid meteor showers had been seen also in former centuries but that, most curiously, the date had been earlier in earlier centuries—as early as October 31st in the year AD 902. This meant that the point of intersection with the earth's orbit gradually advanced in longitude, nearly 1° in 70 years, J. C. Adams applied the formulae of celestial mechanics to this case; he found that perturbations could entirely explain the displacement for an orbit of 33 years but not for an orbit with a period of about one year. So it was established, in 1867, that the Leonid swarm described an orbit about the sun in 33 or 34 years; since, moreover, for the moment and place of meeting the direction of the swarm relative to the earth was given by the radiant, all the elements of the orbit could be derived.

This, however, was not an entirely new discovery; the year before, Schiaparelli at Milan had derived, in a different way, the orbit of the August meteors. It had long been known that shooting stars are more frequent after midnight than in the evening. The reason is at once clear; in the earth's orbital motion the parts having morning are at its front, those having evening are at its back. The former catch up all the particles at rest or coming to meet it. The ratio of the number of shooting stars in the first and in the second half of the night allowed Schiaparelli to derive the ratio of the mean velocity of the meteors and the earth's velocity. This ratio was found to be $1.4 = \sqrt{2}$, exactly the ratio of parabolic and circular velocity at the same distance from the sun. So Schiaparelli concluded that meteors in general, among them the Perseids, also moved in nearly parabolic orbits. Now, computing size and position, i.e. the elements of the orbit of the Perseids, from the radiant he found them to correspond to the orbit of a bright comet that had passed its perihelion on August 23, 1862 (designated 1862 III), for which a period of 119 years had been derived.

A new and wonderful insight sprang from these results; they meant that meteor swarms and comets, properly speaking, are identical. Meteor swarms at a great distance look like comets, and comets arriving in our close vicinity are found to consist of a swarm of meteors. In former years panic had been caused by the prediction that a comet would cause a catastrophe if it collided with the earth. Now the comforting conclusion was that it would produce only brilliant fireworks, because the little stones would burn up in the envelope of air that protects us and the earth.

In the meantime the Leonids had made a new display in 1866, as predicted by Olbers and Newton, this time in Europe, nearly as brilliant as in 1833; and the meteor shower was repeated in 1867 and 1868, on November 13th–14th. Schiaparelli and Leverrier computed the orbit and found that it was nearly identical with the orbit of Temple's comet 1866 I, for which Oppolzer had computed a period of revolution of 33 years. Though the orbits coincided, the objects did not; the meteors that hit the earth in November were stragglers, running ten months after the comet, which had passed the intersection point in February.

Gradually other identities between cometary and meteoric orbits were found. The meteors appearing rather regularly about April 20th, with a radiant in Lyra (the Lyrids), have the same orbit as comet 1861 I. More remarkable were the numerous meteors observed about December 6th by Brandes in 1798 and by Heis in 1838. They radiated from a point near the star γ Andromedae and were connected with Biela's comet mentioned on p. 358. The comet had been discovered in 1826 and had a period of revolution of 6.8 years. It had split into two cores in 1846 and had disappeared after 1852. It was assumed that the former split had been the beginning of a wholesale dissolution into small invisible fragments. But meteors from Andromeda were now often seen, not in December but at the end of November. Computation of its perturbations by the planets showed that the node of its orbit regressed rapidly, so that the dates of meeting came earlier. Because the comet would come near to the earth in 1872, Galle predicted an abundance of Andromedid meteors for November 28th of that year. And, indeed, on the evening of November 27th a fiery rain, lasting for many hours, burst over Europe. Thirteen years later, after two more revolutions, the spectacle was repeated with equal splendour; the comet itself remained invisible, following the observed swarm by 60 days according to the computation. At the next appearance in 1892, on the nights of November 23rd–26th, only a small number of meteors was seen.

Thus those dozens of years of exciting celestial fireworks came to an end. In 1899 many observers on November 14th kept watch for the Leonids, but no meteor shower appeared; only a somewhat greater number of shooting stars than on other nights. Computations by Berberich and by Downing disclosed that the perturbations through the great planets had displaced the orbit, so that the swarm now passed the earth at twice the former distance. So meteor showers from the Lion can never again be expected. Swarms of meteors become visible as showers of shooting stars only under special conditions; and numbers of them may describe their orbits through the solar system unperceived by us. Of course, perturbations may change other orbits so as to make them intersect the earth's orbit and produce a new meteor shower.

Observation of shooting stars during the entire nineteenth century was a regular field of work for amateurs. It consisted partly in corresponding observations at two stations, to derive the height of their appearance and their extinction; partly it was one-man work to find out radiants. Denning at Bristol, after many years of observing, in 1899 published a catalogue of 3,000 radiants, but they only partly represented real swarms; conspicuous separate fireballs were among them. Olivier in 1920 made a list of 1,200 radiants, half of which he assumed to be real swarms. Every night among the many isolated meteors there are some belonging together in groups. All the swarms in space are gradually dispersed along, as well as outside, their orbits and become less marked, while the number of solitary objects increases. That the space between the planets is full of small bodies condensed in the vicinity of the sun is manifest in the zodiacal light, mentioned first by Cassini in 1683 as a special phenomenon. Its brightness decreases with increasing angular distance from the sun; but in 1854 Brorsen pointed out that at 180° distance, just opposite the sun, the brightness increases in the so-called

Gegenschein ('counterglow'), which, as a faint nebulous patch, wanders along the ecliptic. Zöllner explained it by means of the sharp maximum of brightness which every planet shows at opposition.

Meteor tracks sometimes appear on stellar photographs, but then the exact instant is uncertain. Photography was used at the Harvard Observatory to determine the most difficult datum—the duration and velocity of the phenomenon. For this purpose the track was interrupted every tenth of a second by a rapidly rotating screen before the camera. The velocities found, the result of atmospheric resistance, are employed to study the conditions in the highest layers of the earth's atmosphere.

The connection found in the nineteenth century between meteors and comets had disclosed that the meteors come from the vast spaces outside the planetary system and that the comets consist of extensive thin clouds of smaller and larger meteoric bodies. Is this all that can be said?

We have already met with comets, but chiefly as objects of fear and objects for computation of orbits. From olden times comets impressed man as imposing celestial luminaries, mostly appearing as big stars with an appendage which we call its 'tail' but which in former times was mostly described as hairs(the word *kometes* means 'hairy') or a beard. Their uncommon aspect, sometimes a bright band of light spanning half the sky, sometimes a faint ghostly torch or only a wisp, combined with their sudden appearance and arbitrary course between the stars, made them, especially in perturbed times, awe-inspiring phenomena. Chiefly at the end of the Middle Ages and in the centuries that followed they were, apart from regular astrology, considered as heralds of calamities and signs of divine wrath. The appearance of Halley's comet in 1456, a few years after the fall of Constantinople, with its scimitar-like shape, was considered as announcing the Turkish peril for Europe. Even after Tycho had shown that comets were celestial bodies in world space, comet fear remained. It did not abate until, after Newton and Halley had computed their orbits and predicted their return, the prediction was verified in the rationalist climate of the eighteenth century. Yet an instinctive uneasiness remained, which, in the nineteenth century when large numbers were discovered, became the fear of a catastrophic collision between the earth and a comet. Most of them, certainly, were innocent faint nebulae; only the bright ones developed the tails which were feared so much.

Where did these tails come from? As early as 1531 the 'Imperial Mathematician', Bienewitz (latinized 'Apianus') remarked that comet tails are turned away from the sun. Kepler in 1618 said that the luminous matter was driven away from the comet's head by a repulsive power from the sun. With the great comet of 1811 Olbers perceived with his telescope a starlike nucleus, surrounded at some distance by a parabolic envelope passing into the tail. In a spraying fountain the separate jets, curving downwards, also show a parabolic envelope; so he supposed that matter ejected from the nucleus in different directions and repelled by the sun, in streaming into the tail showed the aspect of the luminous envelope. He pointed out that different repelling forces might produce tails of different forms: 'The longer and straight tail [of the comet of 1807] must consist of particles repelled more strongly by the sun than the matter forming a curved tail.'[210] And on the nature of the repelling force, he said: 'It is difficult to restrain oneself from thinking of something analogous to our electrical attractions and repulsions.'[211]

The study of these luminous phenomena was continued at every appearance of a great comet. When Halley's comet returned in 1835, Bessel observed a fan-shaped jet of luminous matter directed towards the sun, oscillating in a period of $4\frac{1}{2}$ days, obviously driven sideways by solar repulsion. Bessel developed a theory of the shape of the tail and computed the repelling force of the sun as compared with its gravitational attraction. That the tail is formed continuously by new rapidly outflowing matter was shown by the brilliant comet of 1843, which in two hours made a turn of 180° near the sun's surface, the tail remaining directed away from the sun. In the head of Donati's comet, a beautiful appearance in the autumn in 1858, several luminous envelopes were visible, which proceeded from the nucleus, expanded while moving outward, and then dissolved; sometimes luminous fans connected them with the nucleus (plate 14). Two phenomena always stood out clearly; the streaming of luminous matter out of the nucleus on the sunward side and the repulsion of the matter by the sun. The Russian astronomer Th. Bredichin in Moscow, in a number of papers from 1862 on, continued Bessel's work on the repulsive force. He found that long, straight tails are produced by a repulsion 15 to 20 times greater than the attraction; the best-known curved tails, like a scimitar, develop if the repulsion is about once or twice the attraction, whereas in the case of heavy particles, for which the attraction far surpasses the repulsion, short, strongly-curved tails appear.

The physical character of the repulsive force could not be derived from the observations and remained a matter of speculation. Most investigators spoke of electrical forces, since other cases of repulsion were not known. In 1900 the prominent Swedish chemist Svante Arrhenius, author of a couple of remarkable astronomical works, pointed out that light pressure, theoretically derived by Maxwell, and afterwards confirmed by subtle experiments, could play a role. A theoretical discussion by Schwarzschild in 1901 showed that this pressure, for particles smaller than one micron, could amount to 18 times the gravitation at most. That stronger repulsive forces had sometimes been observed, Arrhenius explained by assuming the particles to be porous like smoke puffs, combining great surface with small weight.

Photography was used here with unexpected results, after the right method had been found. Though an amateur with a common portrait camera in 1858 had made a good photograph of Donati's comet in seven seconds, Draper in 1881 with his telescope needed an exposure of 2½ hours to get a good picture of Tebbutt's comet 1881 III. And not until a professional photographer at Cape Town in 1882 had once more made a beautiful picture of Cruls's comet, did astronomers realize that it was not the linear aperture, but the angular aperture of the lens that mattered. A comet, unlike a pointlike star, is a faint luminosity extending over a great surface; in order to have a great surface intensity acting upon the plate, the aperture must be large relative to the focal distance. Once this was realized, astronomers began to use portrait lenses and doublets in photographing all kinds of faint extensive celestial luminosities. At the Lick Observatory pictures of Swift's comet were made in 1892 by E. E. Barnard and of Rordame's comet in 1893 by J. W. Hussey. With increased exposure time, comet tails, formerly smooth and ghostly, hardly visible phantoms, now appeared on the plates as brilliant torches with rich detail of structure, with bright and faint spots, never before seen or even suspected. Such photos, taken of every succeeding bright comet (like Morehouse's comet in 1908 and Halley's in 1910, often reproduced in scientific and popular reviews) gave a new impulse to the study of comet tails (pl. 14). Comparison of pictures on consecutive days showed that the bright condensations moved away from the head. Their velocity could be derived; Heber D. Curtis, also a Lick astronomer, found for Halley's comet that the velocity increased with the distance from the nucleus, from 5 to 10 up to 90 km. per second. The same phenomenon—the repulsion of cometary matter by the sun—formerly derived by theoretical discussion from telescopic structural changes in the comet's head, was now made directly visible in small-scale images of the total object.

The origin and nature of the comet's light could be ascertained only by spectrum analysis, which was developed in the same sixties in which comets were recognized as meteor swarms. It was immediately applied and used for every new comet. As faint nebulae at great distances, they showed only the reflected solar spectrum. When, however, in approaching the sun they became bright and formed a tail, a new spectrum appeared. Donati, observing Temple's comet in 1864, perceived a spectrum of three faint luminous bands, yellow, green and blue.

Huggins in 1868 succeeded in identifying these bands—which were sharply cut off at the red side—with the bands emitted by ethylene vapour and other hydrocarbon compounds made luminous by electrical discharges, described earlier by Swan (pl. 15). Huggins said the comet's light was emitted by luminous carbon gas. Meteors heated in a laboratory give off gases consisting of carbon monoxide and dioxide, hydrogen and hydrocarbons; so it could be understood that swarms of such stones, heated in approaching the sun, produced atmospheres about their nucleus consisting of the same gases.

Improvement of the spectral apparatus and its application to all the comets in the ensuing decades led to a continual increase of knowledge. Photographic spectra showed a number of ultraviolet bands, chiefly belonging to the spectrum of the carbon-nitrogen compound cyanogen. A new and unexpected phenomenon was presented by Wells's comet 1882 I; when it approached the sun, the hydrocarbon band spectrum disappeared and was replaced by the bright yellow line of sodium. The same sodium line, now accompanied by several iron lines, was shown by the brilliant September comet, 1882 II, which was visible in the daytime at the sun's limb and grazed its surface in perihelion passage; here the metal lines disappeared and gave way to the hydrocarbon bands when the comet had receded to a greater distance from the sun. Also later comets, 1910 I and 1927 IX, when near to the sun, showed the sodium emission; it was a regular phenomenon belonging to a great intensity of solar radiation.

The amount of matter involved in these phenomena was determined by Schwarzschild and Kron in an exhaustive study of some few photographs of Halley's comet, taken by chance, but standardized. They measured the surface intensity of the tail $\frac{1}{2}°$ behind the head. By assuming that the light here was not reflected sunlight but solar energy absorbed and emitted as new radiation (called 'fluorescence'), they deduced that 150 gm. of matter flowed per second from the head into the tail. The density of this luminous matter then was 10^{20} times less than that of our air, i.e. one molecule per cubic centimetre, scarcely more than is assumed for empty outer space. It can hardly be called a 'gas', since in a gas the billions of collisions of the particles, through interchange of energy, produce average conditions of equilibrium. In these comet tails, however, every molecule or atom, once ejected and propelled by solar repulsion, runs its course without being disturbed by mutual collisions.

The processes indicated in these older researches by the vague names of 'luminescence' and 'fluorescence' acquired determinate precision after 1914, through Bohr's atomic theory of spectral lines. Every atom, because of internal changes of energy, emits or absorbs certain determinate wavelengths. The rapid increase in knowledge of atomic line spectra was soon followed by an analogous increase in knowledge of the molecular band spectra. Molecules consisting of two or more equal or different atoms have far more additional freedom of internal change by absorbing or emitting energy in small and slightly different bits; so the spectrum shows an abundance of close lines, which in small instruments appear as bands. The unravelling of molecular spectra since 1920 found its application in the study of the ever-increasing number of various bands disclosed in cometary spectra by the improved spectrographs. This study revealed that the Swan spectrum had nothing to do with hydrocarbons; it was radiated by the carbon molecule C_2. Next to it the cyanogen molecule, CN, contributed most to the light of the comet's head. The fainter light of the tail is chiefly due to ionized carbon monoxide (CO^+). Fainter bands are due to such molecules was CH, CH_2, OH, NH, and (in the tail) N_2^+. Chemically, such molecules would be called free radicals, unable to exist in a gas in equilibrium where disintegration of compounds is neutralized by recombinations. In the extremely thin comet matter, collisions and recombinations are practically absent, so that what has been dissociated or ionized by solar radiation remains in this condition. When the swarm of meteoric blocks and stones comes near to the sun, they give out gases as they do when heated in a laboratory: hydrogen, nitrogen, carbon dioxide, hydrocarbons. When these molecules absorb the right wavelengths out of the solar radiation, they are split; CO_2 produces CO and CO^+; C_2N_2 produces CN; hydrocarbons produce C_2 and CH. They form the nebulous envelope of about 100,000 km. diameter called the 'comet's head', and by radiation pressure they flow into the tail.

Modern atomic theory has attributed a new character to this radiation pressure. Instead of the pressure exerted by the total solar radiation upon globular 'small particles' treated by Schwarzschild, we are now dealing with atoms and molecules pressed by only such special wavelengths as they are able to absorb. In absorbing this radiation coming from the sun, they get an impulse directed away from the sun; the opposite back-push, when they emit the same radiation, is, on the average, directed evenly to all sides and has no effect upon the motion. In 1935 Karl Wurm computed that in the case of the CO^+ molecule these impulses and emissions are strong and numerous; they explain the strong intensity of its spectrum in the tail and produce a great repulsive force, 80 times greater than gravitation, in agreement with the observed velocities.

It is a curious thing that such striking effects come from this simple mechanical basis: swarms of blocks and stones, not so very different from earthly matter, in their elongated orbits from far spaces coming near the sun are able, in consequence of simple, recently discovered physical processes, to produce those mysterious luminous phenomena that filled earlier generations with fright and awe and fill men today with surprise and wonder.

CHAPTER 39

PECULIAR STARS

WITH the nineteenth century, supremacy in astronomy switches over from the solar system to the world of fixed stars. Its realm widens with a big leap; its horizon broadens millions of times. In former centuries the stars served only as the fixed background behind the play of planetary motions in the foreground. True, their positions were carefully measured, but it was chiefly to supply a basis for the motion of the planets and the moon. Even in the first half of the nineteenth century the careful mapping of telescopic stars was done to promote the discovery of minor planets.

Now the fixed stars became increasingly the object and purpose of astronomical research for their own sake. After the pioneer work of William Herschel, the new age had to study the now opened field with more thoroughness and greater accuracy. Though it was known that the stars were distant suns akin to our sun, they were not entirely similar. The dissimilar objects, of course, attracted most attention, because here new things could be found; such peculiar objects were double stars, variable stars, and star clusters.

Double stars in the world view of that time were so peculiar that their existence in great numbers was at first doubted. People were accustomed to one sun in a system. Moreover, they were convinced that one sun, providing light and heat to the surrounding inhabitable planets, was the most suitable world system. What purpose could the Creator or Nature have with two, often unequal and differently coloured, suns? How would planets move about two such centres of attraction? Though we will never be able to see them, they yet presented an interesting theoretical problem which in later times was treated, by means of extensive numerical integrations by Ellis Strömgren at Copenhagen. The problem of the motion of the two component stars was of more direct importance. Nobody doubted that Newton's law of attraction applied also to the remote world of stars; it could be settled by observation of the double stars. Two bodies attracting each other must describe ellipses (or, more generally, conic sections) about their common centre of gravity, and the orbit of one star relative to the other has a similar figure. Seen obliquely, this (relative) orbit also appears like an ellipse, but one of different form and with different position of the focus. So it was the task of the astronomers, by means of accurate measurements of the relative positions of the components, to derive, first, their apparent orbit, and therefrom the real orbit.

Herschel had published lists of hundreds of double stars and in 1803 found relative motion of the components for about 50 of them. His pioneer work, however, was too coarse for the demands of a good determination of orbits. Double-star astronomy became the first field of application for the refined nineteenth-century instruments, with their higher standard of accuracy. It began when F. G. Wilhelm Struve (1792–1864) at Dorpat (now Tallin), who in 1819 had already measured double stars, in 1824 introduced the new 9 inch refractor, the biggest Fraunhofer instrument, to double-star work. First a survey over the new field was needed; a more complete list, containing about 3,000 double stars, was gathered together and was published in 1827. In the next ten years the distance and position angles for each of them were repeatedly measured with Fraunhofer's perfect filar micrometer. The high standard of accuracy is shown by the use of two decimals of a second in all the results. There were, or came, more workers in this field; John Herschel, South and Dawes in England, Bessel and Mädler in Germany, and Kaiser in Holland, either worked contemporaneously or followed his example; afterwards his son, Otto Struve, his successor as director of the new Pulkovo Observatory, carried on the work. Thus the motion within each pair could be followed year by year over increasingly larger parts of the orbits, and soon complete orbits could be derived for the most rapid binaries. In 1850 about 20 orbits had been computed; the smallest period of revolution (for ζ Herculis) was 31 years in an ellipse measuring 2.4″.

This kind of work continued throughout the entire century; the mere passage of time brought increase of knowledge. Nearly every astronomer with a good filar micrometer considered it his duty to take part in this work, and some among them, Dembowski in Naples and Burnham in Chicago, devoted all their time to double-star work. This general participation proved to be highly appropriate, because, firstly, smaller instruments in the right hands produced results nearly as good as big telescopes and, secondly, all results are subjected to systematic personal errors of unknown origin, which are diminished in the average of many observers. What, then, is the advantage of big instruments? It is their greater separating power. Stars which in smaller instruments appear single, or at most somewhat elongated, are seen with larger apertures as two well-separated stars which can be measured. Also faint companions are made visible by larger instruments. The construction of ever bigger telescopes involved new discoveries, especially the separation of close pairs. First Otto Struve, with the Pulkovo 15 inch refractor, extended his father's catalogue. Then it was chiefly S. W. Burnham, who in 1873, with a small 6 inch Clark telescope, had already been able to discover, on account of his sharp vision, duplicity of stars not previously suspected. Afterwards, having at his disposal the Lick 36 inch and the Yerkes 40 inch, he added a thousand new objects to the lists, which were all carefully measured. They were mainly very close pairs, separated by less than 1″; this is often due to actually small distances involving small periods of revolution. The shortest period was $5\frac{1}{2}$ years for δ Equulei, with a distance of 0.3″. In all, 5 per cent of the stars examined were found to be double.

After discovery and measurement, computation had to follow. The astronomers were faced with the problem of computing double-star

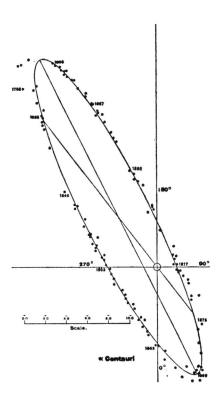

Fig. 32. Revolution of the double star α Centauri

orbits, while at the same time they were greatly occupied with planetary and cometary orbits. Naturally the solution was sought for along the same lines. An ellipse is determined by five points and can bep comuted from five measured relative positions. After Savary in 1828 had first derived and applied a method of computation chiefly by trial and error, Encke in 1830 developed a system of formulae somehow analogous to that of planetary orbits. Though often used, it was gradually found less satisfactory. The cases were too different; a small error of, say, 0.1″ on a distance of 4″ distorts the apparent course more strongly than an error of, say, 3″ does a 10° motion of a comet. So the shorter graphical method, especially for close pairs, gradually superseded the apparently more rigid algebraic method. Every measurement of distance and position angle, plotted as polar co-ordinates, determined a point on the graph; the totality of these points presents a clear picture of the apparent orbit, and an ellipse is drawn as well as possible through the assemblage of points. To construct or compute the true ellipse, of which the apparent ellipse presents a projection, is a simple geometrical problem. It is a different matter when in a large, slowly described orbit every short part of the performed arc is well determined by the average of a large number of observations; in that case it is worth while to make use of exact algebraic formulas and to calculate the most probable elements of the orbit by the method of least squares.

This was still more true after Ejnar Hertzsprung in Potsdam between 1914 and 1919 applied photography to the measurement of double stars. By taking a large number of exposures on one plate and measuring them carefully, he obtained results that at once surpassed the visual measures by an entire decimal; the uncertainty of one plate was no more than 0.01″, and the results were given in three decimals. The method was restricted to such wide pairs that the images were well separated. To eliminate the errors proceeding from a different brightness of the components, a coarse grating was put before the objective, which produced smaller diffraction images on both sides of the central image. With this accuracy, an exact computation of the orbit became possible, even when only small arcs had been described. Moreover, the reality of irregularities due to a perturbing third body could now be strictly decided. The measurements of the famous double star 61 Cygni clearly show deviations due to the attraction of an invisible body that perhaps might be called a big planet, since its mass is about $\frac{1}{60}$ of the sun's mass. Thus the photographic method has introduced a new epoch into double-star astronomy.

In the quiet progress of double-star astronomy in the nineteenth century, an exciting episode was the discovery of 'dark stars'. In 1844 Bessel showed that the bright star Sirius and also Procyon, for both of

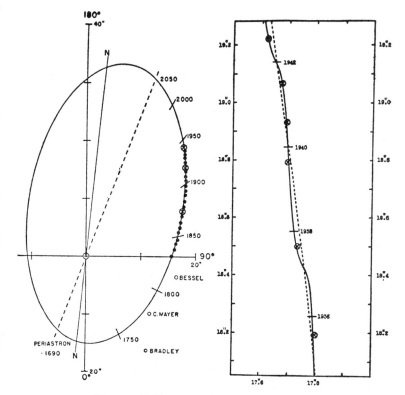

Fig. 33. Orbit of the double star 61 Cygni

Left: Ellipse computed from measurements made between 1830 and 1940.
Right: Part of the orbit between 1935 and 1942, from the photographic measurements of K. A. A. Strand

which there existed a large number of good meridian observations of position, presented irregularities in their motion. In 1825 and 1833 Pond at Greenwich had asserted that this was the case for a large number of stars, and he was convinced it was a result of their mutual attraction. Bessel, however, demonstrated that such attractions must be imperceptible on account of the enormous distances. But with Sirius the matter was different; the irregularity was real. In considering various possibilities, he concluded that an invisible body of great mass must exist in the vicinity of Sirius which, by its attraction, caused the irregularities. It might seem a piece of irony that the brightest of our stars should be subject to a dark body, reversing the relation held until then as the only natural order of things. In reality, however, we had two equivalent bodies, one luminous and one dark, which, by their mutual attraction at short distances, formed a peculiar specimen of binary. The observed motion of Sirius was its orbital motion about their common centre. The orbit derived by C. A. F. Peters in 1850 had a period of revolution of 50 years and a half-major axis of 2.4″. Then it happened in 1862 that Alvan Clark, in order to test the quality of his just finished 18 inch objective, had it pointed at Sirius, and that close to it a small speck of light was perceived, hardly visible in the glare of the mighty star. When the discovery was made known and the object was seen also in other telescopes, measurements showed it to stand relative to Sirius exactly in the direction of the centre of the orbit, at a distance two times greater. It was Bessel's dark star. So it was not entirely dark, only very faint, a star of the eighth magnitude, nearly 10,000 times fainter than Sirius. Yet, since it was twice as far from the centre of gravity as Sirius itself, its mass was as much as half the mass of Sirius. Though not a dark star in the strict sense, this Sirius B, as it is called, remains a remarkable object because of this combination of great mass and small luminosity— how remarkable appeared later on. Sirius has since been measured regularly like any other binary. The dark star for Procyon was discovered in 1895, by Schaeberle with the 36 inch Lick telescope, as a still fainter star of the thirteenth magnitude.

Binaries are highly important in astronomy, because they are the only objects showing the effect of attraction, so that only by them can we acquire knowledge of the masses of the stars. The formula deduced by Newton as an extension of Kepler's third law—that the ratio of the cube of the major axis and the square of the period of revolution is proportional to the total mass—can be applied to binaries as soon as the parallax is known. The parallax is needed to derive the major axis in astronomical units from the observed major axis in seconds of arc. For Sirius, half the major axis (7.5″, i.e. being 20 times the parallax 0.37″) is 20 astronomical units; with the period of 50 years, a total mass of $20^3 : 50^2 = 3.2$ times the sun's mass is found. So, in the search for stars interesting for parallax measurements, binaries were preferred. Since for Sirius the distance of the chief star from the centre of gravity was known to be $\frac{1}{3}$ of the distance of the two components, it followed that the separate masses of the components were $\frac{1}{3}$ and $\frac{2}{3}$ of the total amount. The same computation could be made for some other bright binaries when a great number of meridian observations disclosed the motion of the bright component. The masses thus found were no more than 10 times greater or smaller than the sun's mass, though their luminosities may diverge a million times.

A new kind of binary was discovered in photographing stellar spectra. In 1889 Antonia C. Maury at Harvard Observatory perceived that in

434

the spectrum of ζ Ursae Majoris the K line, the only clearly visible narrow line, sometimes was double, sometimes single; still more regularly the same happened with β Auriga, where the line on successive days was alternately double or single (plate 15). 'Double' means the star has two different radial velocities, i.e. it consists of two components whose spectra shift periodically in opposite directions because they revolve about their centre of gravity. For β Aurigae the period is four days. Such 'spectroscopic binaries' have been found in great numbers; they form a special class, because they can be discovered only when their orbital velocities are large, amounting to tens and hundreds of kilometres per second, whereas their orbits are so small that their duplicity cannot be observed with the largest telescopes.

At the same time, about 1890, Vogel and Scheiner at Potsdam were photographing spectra of bright stars with a new excellent spectrograph, to determine their radial velocities. They perceived that with some stars, such as Spica in Virgo and Algol in Perseus, this velocity periodically increased and decreased. So these stars also were spectroscopic binaries, different from the other kind because the second spectrum was invisible. This does not mean that the other star is dark, but only that it is so much fainter that its absorption lines are blotted out. Spectroscopic binaries of this kind, with single periodically displaced lines, are far more numerous than those with doubling lines. In the following years Campbell at Lick Observatory discovered a large number of them; in his catalogue of 1924 more than one thousand are listed. He estimated that one-third of all stars belong to them; though single stars like our sun form the majority, yet duplicity is a common condition. The theoretical deductions on the division of rapidly rotating bodies and the effect of tidal friction might well be applied here; the wide pairs with periods of hundreds of years, however, can hardly be explained in this way.

The above-mentioned star Algol in Perseus was already renowned in astronomy, because Goodricke in 1782 had discovered its regular variations in brightness; always, after a period of 2 days 22 hours, its light decreases for 4 hours to a minimum and then increases in the same time to its normal constant brightness. Algol was one of the 'variable stars', i.e. variable in brightness, of which a small number had been discovered by chance and observed more or less regularly. A regular, uninterrupted engagement with these objects began with F. W. A. Argelander (1799–1875). He introduced the method of estimating small differences in brightness in 'steps', thus improving Herschel's former method by using numerical values instead of commas and dashes. His first results were published in 1843 in a paper on β Lyrae; and in the following quarter of a century he continued to observe the known and newly-discovered variables, deriving their periods and light curves. In 1844, in an 'Appeal to Friends of Astronomy', he indicated a number of important objects of research for amateurs, which demanded only small instruments or none, and which the astronomers in their observatories had to neglect because their big and expensive instruments had to be used for more difficult objects. Such were twilight, the meteors, zodiacal light, the Milky Way, the brightness and colour of the stars, and especially the variable stars. His appeal and example stimulated many younger people and pupils, like Eduard Heis at Munster, Julius Schmidt, who in Athens worked on all these objects, and Eduard Schönfeld. Besides these, a number of amateurs occupied themselves with variable stars, of which more and more telescopic specimens were discovered owing to the improved star maps. A catalogue, compiled by Chandler in 1889, already enumerated as many as 225 items. The different types known from the eighteenth century by some bright representatives all increased in membership now. In the standard catalogue published in 1919 by Gustav Müller and Ernst Hartwig, there were 131 variables of the Algol type, which, in reality, are common stars occulted now and then by others. The β Lyrae type, regularly varying in mostly small periods between two equal maxima and two unequal minima, numbered 22 stars. The continuously changing stars of short period, called 'Cepheids' after δ Cephei, accounted for 169 items. The largest number, more than 600, were strongly variable red stars with periods of the order of one year, called 'Mira type' after the first discovered 'wonderful star of the Whale'. Furthermore, there are bright red stars rather irregularly oscillating by small amounts, such as α Herculis, already mentioned as discovered by William Herschel, and Betelgeuse, found to be variable by his son, John Herschel, in 1836.

The cause of the variation in brightness in the case of the Algol stars was obvious. Goodricke's explanation of an occultation by a dark satellite star was a plausible hypothesis. It acquired certainty when Vogel and Scheiner in 1889 found that the visible star Algol recedes from us before the minimum, and approaches after the minimum, and hence during minimum is behind the companion. This explanation holds for all the variables of the Algol type, therefore called 'eclipsing variables'. Though belonging to the variable stars according to observational practice, they belong by their nature to the spectroscopic binaries. That their variation is not physically real but the result of a geometrical situation gives them a special importance. With all the other stars, our sun excepted, we must be content with studying their integrated light. We see them as points—by the telescope magnified into spurious discs—not as real discs, so that the different parts cannot be distinguished. In stellar eclipses, however, one disc gradually covers and uncovers the other; so they are the only stars in which the differences between separate parts of the disc, e.g. centre and edge, can be an object of investigation. The diversity of relative diameter and surface brightness of the components, combined with the situation of the orbit, produces a rich variety in the shapes of the light curve; conversely, these light curves allow the physical and geometrical conditions to be derived. The variable U Cephei, discovered in 1880 by Ceraski at Moscow, which in full light has a magnitude of 6.9, shows a minimum magnitude of 9.2, which remains constant for a couple of hours; hence a small disc enters before or hides behind a large disc. The variable Y Cygni, discovered by Chandler in 1886, had an eclipse every 1½ days, in which it decreased to half the full light. The period at first seemed to be irregular, until in 1891 Dunér at Uppsala found that the interval was alternately smaller and larger; the difference, moreover, gradually decreased and, after passing zero, increased in the opposite direction. He could explain it by two exactly equal stars alternately occulting each other in a period of three days, in an elliptical orbit, in which the major axis revolved rather rapidly. So we have here a perturbation in a stellar system, caused probably by deformation through mutual attraction. Numerous eclipsing variables show two alternating minima of unequal depth, in all intermediate degrees between exact equality and a secondary minimum hardly perceptible if the second star is faint.

Is Algol's dark companion really dark? Several observers tried in vain to detect a secondary minimum, until in 1910 Stebbins succeeded in discovering it by means of his selenium photometer. Its depth was only 0.06 magnitude, corresponding to 6 per cent loss of light.* Here it appears that, for the study of Algol stars, Argelander's method of step estimates is not sufficiently exact to afford values of the brightness suitable to be compared with geometrical theory. Photometric measurements were needed. About 1880 some had been made in Potsdam and at the Harvard Observatory, but only in far too small a number. It was Dugan who, at Princeton Observatory in 1905–10, determined, photometrically, exact light curves for a large number of eclipsing variables, with only a few per cent of uncertainty for each value of the brightness. This accuracy was obtained as the average of extensive series of observations; since all this photometry was based on ascertaining the equality of two stellar images by eyesight, its accuracy depended on the limited power of the eye to detect differences in brightness; this limit is about 5 per cent. Fundamental progress was possible only by substituting a

---

* As explained in the next chapter, 0.06 magnitude means a logarithmic decrease of 0.4 × 0.06 = 0.024, which is the logarithm of 1.059.

physical apparatus for the human eye. Stebbins' selenium photometer, in which light produced a change in the electrical resistance, was the first attempt of this kind and was rewarded by the detection of Algol's secondary minimum.

The derivation of the elements of the eclipse—the relative size and brightness of the stars and the dimensions, form and inclination of the orbit—from the shape of the light curve demands very laborious computations. Formulas were derived by Pickering in 1880 and by Myers in 1895. An effective solution of the problem was given in 1912 by Henry Norris Russell of Princeton, in the form of a practical computation scheme based upon a number of auxiliary tables. They were used by Harlow Shapley in 1913 in a careful determination of the geometrical elements for all the eclipsing variables for which Dugan had provided sufficient data. Because the distribution of intensity over the disc was also an unknown element, Shapley computed the other elements under two extreme suppositions: either uniform brightness over the entire disc or its regular decrease to zero at the limb. He found that in both cases all the points of the light curve could be equally well represented. This means that the coefficient of limb darkening could not be found in this way; the light curves computed for both cases (each with its own elements) coincided well-nigh completely and deviated only at a few places by some few hundredths of a magnitude. For this purpose the photometric results were still too coarse; more precise measurements were needed.

Fortunately, a better method had been developed only a few years before. About 1911 the German physicists Elster and Geitel had improved the photoelectric cell into an extremely sensitive instrument for measuring small light intensities. It was soon used for the stars by Guthnick and Prager at Berlin and by Rosenberg in Tübingen and was found to determine the brightness of a star with a ten times greater accuracy than was possible by visual comparisons: the magnitude differences had to be expressed in three decimals, with errors of some thousandths instead of hundredths of a magnitude. This accuracy, not necessary for common stars, was valuable for variable stars—the two Berlin astronomers detected a small variability in hitherto unsuspected stars—and it was just what was needed for the eclipsing variables. At the Lick Observatory Gerald Kron constructed a highly sensitive instrument of this type and in 1937–38 determined the light curve of the Algol star 21 Cassiopeiae with a probable error of not more than 0.002 of a magnitude. Thus he was able to derive the coefficient of limb darkening at 0.58 ±0.04, the first reliable value next to that of our sun.

Eclipsing variables may be useful in still another way. The duration of the eclipse relative to the period determines the dimension of the stars relative to the size of the orbit. The diameter, to the third power, determines the volume; the radius of the orbit, to the third power, determines the mass; so, together, they determine the mean density of the stars without needing parallaxes or true dimensions.

A sensational case was the third-magnitude star ε Auriga. Since 1843 Julius Schmidt had suspected it of showing small irregular variations. Argelander and Heis in 1847–48 saw it as a fourth-magnitude star; but in the following years it showed doubtful fluctuations only. In 1875 it was again seen to be faint, but only for a short time. When this was repeated in 1901 and a variable radial velocity had been found in the meantime, Ludendorff at Potsdam realized that it was an eclipsing variable of unusual dimensions: it had a period of 27 years, in which the star decreased for seven months, remained constant for ten months and increased in seven months to normal brightness.

The diversity in the phenomena of eclipsing variables is increased still more by variations of the full light between the eclipses in some of them. First, because in the case of a small distance the fainter star, strongly illuminated by the other, increases the total light on both sides of the secondary minimum, and second, because the stars by their mutual attraction are elongated and show their larger surfaces between the eclipses. Such stars form a transition to the β Lyrae type, which, in continuous variation in each period, shows two equal maxima and two unequal minima. Or, expressed in another way, the β Lyrae stars are extreme specimens of eclipsing variables. Periodical variations of radial velocity had indeed been established by Lockyer in 1893 and by Belopolsky at the Pulkovo Observatory in 1892; the components, different in brightness and spectrum, revolve with great velocity (of 155 km.) at so short a distance from each other that they must nearly touch. For β Lyrae, considered as a double star, elements were computed by Myers in 1896 and by Stein in 1907. Yet, with its many curious spectral changes, it remained one of the puzzles of astronomy.

Periodical variations of radial velocity were also established for δ Cephei in 1894 by Belopolsky, and it was soon found to be a property of all these short-period variables. In 1907 Sebastian Albrecht at the Lick Observatory remarked that the light curve and the velocity curve not only coincided in that the most rapid approach took place at maximum brightness, but even their shape was similar in minor details. Nobody at the time had any doubt that these Cepheids were spectroscopic binaries; but there were difficulties. The irregular shape of the light and velocity curves—often a rapid increase and a slow, sometimes interrupted, decrease—was hard to account for, even with a large eccentricity. That the temperature at maximum brightness was also higher than at minimum had been demonstrated by Schwarzschild in

1899 by photographic measurements, which showed a range in brightness $1\frac{1}{2}$ times larger than was observed visually, indicating a colour more blue at maximum than at minimum. Tides produced by the attraction of the companion were invoked, or a heating at the front during rapid motion through a resisting medium was thought to produce the maximum. In this case, however, the resistance would greatly change the period, which, in reality, was almost constant. The moderate orbital velocity combined with a short period demanded, in the case of a binary, a very small mass, less than $\frac{1}{100}$th of the sun's mass. The large size of the star derived from other data meant that the centre of gravity was situated deep in its interior, so that in its motion it would scarcely be justified to distinguish a front.

Then the pulsation theory was advanced; though suggestions along this line had been made by others, it was Shapley who in 1914 exposed and defended it in all its consequences. The alternate approach and recession measured in the spectrograms are the effect of the alternating expansion and contraction of the star. These periodic changes in dimension and volume are adiabatic changes, i.e. the heat needed for or produced by the change of volume is given off or taken up by the stellar gas as decrease (during expansion) or increase (during contraction) of temperature. There is something odd about stars that cannot find their equilibrium but alternate in endless pulsation between too large and too small, too cold and too hot. Yet this explanation for the short-period variables was soon accepted by the astronomical world, especially when Eddington in 1918 had developed a theory of pulsating stars, based upon his researches on the constitution of the stars in general. Yet in this theory, too, there remained difficulties not directly solvable, especially as to why the observed maximal brightness and temperature came, instead of at the moment of smallest volume, at exactly a quarter of a period later, when the star was already rapidly expanding.

With their theoretical importance, the necessity for a more accurate determination of the light curves of the Cepheids was felt. During the entire nineteenth century the light curves had been derived by astronomers and amateurs by means of step estimates following Argelander's example. These, however, could not guarantee a uniform scale over the entire range of variation. Here photometric measurements had to come to the rescue. Since an accuracy of 0.01 magnitude sufficed and 0.001 magnitude was not needed here, visual photometry could be used. Photography also was tried, when methods of photographic photometry developed, especially in connection with the great star-mapping enterprises such as the *Carte du ciel*. On a photographic negative of a field of stars, the size of the black dots varies with the brightness of the stars. Empirical formulae were derived in about 1890 by many astronomers (Charlier, Scheiner, *et al.*) to find stellar magnitudes from measured diameters. It appeared that bad objectives, with much stray light about the stellar images, gave better results (because of the greater diameters) than plates taken with good objectives; but the indefinite borders of the star discs always made it a coarse procedure. It was an important improvement or, rather, an entire renovation of the basis of photographic photometry, when Schwarzschild in 1899, in the abovementioned research, introduced the use of extra-focal images. By placing the plate at some distance, inside or outside the focus, he obtained, for all the stars, smooth discs of equal size but of different blackness, i.e. density of the silver deposit. Differences in blackness can be distinguished far more accurately than differences in diameter on focal plates (pl. 15).

A Cepheid variable (η Aquilae) was the first object and motive of his method, which has proved most important for this class of variables. In this first research Schwarzschild had to estimate the blackness in a self-made artificial scale of comparison discs. The accuracy was increased when, in the same year, Hartmann at Göttingen had constructed his 'microphotometer' for the measurement of silver densities by equalizing them—in a most exact way, by the disappearance of the border line—to the blackness at some point in a wedge of dark glass. In order to reduce the measured densities to magnitudes, Hertzsprung in 1910 devised how to put a scale on each plate by taking the photographs through a coarse grating placed before the objective. Then diffraction images appear on both sides of the star disc, fainter discs of the same size, of which the ratio of brightness can be computed exactly. Thus the basis was laid for an exact photographic photometry of the stars, and several observatories have since contributed in this way to the derivation of accurate light curves of Cepheids.

The red variables of large period and great amplitude, called the Mira type, remained as the appropriate field for step estimates by Argelander's method. With an extent of variation of 4 to 8 magnitudes, or even to 10, and with the many irregularities in the light variation itself, errors of 0.1 magnitude are irrelevant. Here was a fertile field of work for amateurs with small telescopes, though the greater telescopes of the observatories had to complete the regular record for the faintest minima below the twelfth magnitude. With the increase in the number of objects, the number of workers in this field also increased, and they organized into special societies, in order by co-operation to ascertain the star's behaviour from day to day. A variable radial velocity has also been found in Mira stars, but most important here are the spectral changes, which present many riddles.

The most peculiar among peculiar stars are the new stars, the 'novae'. No innocent, ever repeating, fluctuations here! What they present are catastrophes, alarming man and raising problems of world destiny. Stellar catastrophes stood at the cradle of astronomy—if the tale is true that Hipparchus was induced to make his star catalogue by the appearance of a new star—and were beacon lights in its revolutionary epoch, when the novae of 1572 and 1604 drove Tycho to astronomy and inspired Kepler and Galilei. Then they became scarce; during the entire eighteenth and the first part of the nineteenth centuries, none were mentioned. Was this through lack of attention? Surely there has not been such a display of conspicuous apparitions as that which occurred in the second half of the nineteenth century. It was opened by a modest nova of the fifth magnitude in Ophiuchus, discovered by Hind in 1848. We omit here the star η Carinae in the great nebula of the Ship as forming a separate type; earlier regarded as of the fourth magnitude, it was observed by John Herschel in 1838 as a first-magnitude star; after a relapse, it flared up again in 1843 to a brightness equalling Sirius, and then gradually faded to the sixth magnitude.

In 1866 a second-magnitude nova appeared in the Crown; in 1876 another of the third magnitude in the Swan; in 1892 a fourth-magnitude nova appeared in Auriga; in 1901 a first-magnitude one was found in Perseus; and in later years occurred several of magnitudes 5, 4 or 3, with again a first-magnitude star which appeared in 1918 in Aquila. They all exhibited the same normal, common behaviour: they flared up suddenly, in a single day, to a brightness 10,000 times and more, surpassing their former light; then they began to decrease slowly, sometimes with irregular or periodic variations, until they faded out. Nova Aurigae 1892, as it appeared on Harvard plates taken earlier, had been fluctuating about the fourth magnitude many weeks before its discovery. It suggested to Seeliger the explanation (contrary to the then prevalent opinion of a collision between two stars) that the star was heated by friction during its rapid passage through a nebulous cloud of variable density. Nova Persei in 1901, half a year after its outburst, a faint star already, was seen to be surrounded by a faint nebulous ring at 7″ distance, which slowly expanded. Kapteyn and Seeliger explained it as the glare of the outburst spreading with the velocity of light over the surrounding nebulous masses. Thus it was possible to derive for it a distance of about 200 parsecs.

To be free from the chance discovery or non-discovery by amateurs on the lookout, a 'sky patrol' was installed at the Harvard Observatory, a large-field camera, continually photographing the entire sky every night. Any foreigner can be tracked immediately, and for every newly-discovered nova the previous history can be read afterwards. All the novae in this way were completely registered; most of them were, of course, telescopic in their greatest brightness. On the basis of these records Bailey estimated that 10 to 20 novae brighter than the ninth magnitude appear yearly. On the conjecture that they are all within a distance of 10,000 parsecs and that the number of stars therein is 10,000 million, it is found that, on the average, each star has a chance to become a nova once in a thousand million years, certainly less than its probable term of life. The possibility that every star—our sun included —once or oftener during its existence should be subjected to an all-destroying fiery blaze provided a new, rather disquieting, aspect for our little world's future. It might be, however, that the conditions for nova outbursts as a kind of instability are present only in special stars; cases are known of repeated outbursts of the same star. Milne, in his general theory of stellar structure, in 1928 also gave a theory of collapsing stars, where inner instability results in a catastrophe. It is clear, however, that only a profound knowledge of the processes in the interior of the stars will shed light on the origin of nova outbursts.

CHAPTER 40

COMMON STARS

THE common stars differ among themselves in apparent brightness—to which fact we owe the picturesque aspect of the starry heavens—and also in colour. Ptolemy had already expressed the differences in brightness by forming six classes, which he called the first to the sixth magnitudes, the last at the limit of visibility. From the number in each of these classes—15, 45, 208, 474, 217 and 49—it is clear that the two faintest classes were very incomplete. The use of the word 'magnitude' for brightness does not mean simply that when a star is bright it is automatically called a big star, but it includes the assumption that bright stars have greater size than faint stars. Tycho Brahe assumed the size of first-magnitude stars to be 2', of third-magnitude stars 1'. Because the eye easily distinguishes smaller differences in brightness than one magnitude, Ptolemy added to some of the stars the designation 'smaller' or 'larger'.

In the two centuries of revolutionized astronomy the problems of position and motion so engrossed men's minds that little attention was given to brightness. On the celestial maps, in the tradition of antiquity, the human or animal figures of the constellations were prominent, while the stars themselves were often inconspicuous. The transition to a more rational celestial cartography came through the work of Argelander. Moving to Bonn in 1837, he spent the years while the observatory was being built, when he was without instruments, in constructing an atlas which was to depict all the stars visible to the naked eye in their true magnitude as established by careful intercomparisons. In this atlas, published in 1843 and called *Uranometria Nova* (pl. 16), as a modern substitute for Bayer's *Uranometria*, the numbers of the stars for each magnitude from the first to the sixth are 14, 51, 153, 325, 810, 1,871—in total, including variables and nebulae, 3,256. The regular increase shows the incompleteness of the former work for the faint stars. In the catalogue added to the maps the brightest and faintest stars of each class were specially indicated, as with Ptolemy; he designated the sub-classes, e.g. of the third magnitude, by 3; 2,3 and 3,4.

His example was followed by Eduard Heis at Munster, who, because of his sharp eyesight, was able to distinguish fainter stars, which he indicated by 6,7. His *Atlas coelestis novus*, appearing in 1872, contained 5,421 stars, far more than Argelander's. Since both astronomers lived in Germany, the southern part of the sky below 40° of southern declination was omitted from their work. This hiatus was filled when the American astronomer Benjamin A. Gould (1824–96), expelled from the observatory at Albany through a quarrel with the 'Board of visitors', was called to Cordoba in the Argentine to organize an observatory. The extreme clarity of the sky allowed him and his assistants to add a seventh magnitude in the *Uranometria Argentina*, atlas and catalogue, which appeared in 1879. By careful intercomparisons of the stars, the precision could be increased and their brightness, by adding one decimal, expressed in tenths of a magnitude.

This greater precision was not entirely new. The steps used by Argelander for variable stars corresponded to nearly 0.1 magnitude. The signs (commas, dashes and points) used by William Herschel in his sequences of stars also denoted small fractions of a magnitude; but his observations, though published, were left unreduced. John Herschel, continuing his father's work during his stay at the Cape of Good Hope (1834–37), arranged sequences of bright southern stars and in this way could express their magnitudes with two decimals added.

It must be remarked that the addition of decimals fundamentally changed the character of 'magnitude'. From a quality, a class, an ordinal number, a basis of statistics, it has turned into a quantity, a measure, an amount that can be divided into fractions, a basis of computation. We cannot very well speak of a star of the 2.78th magnitude; but we can say that its magnitude is 2.78.

What difference in brightness is expressed by 1 magnitude? The opinion of William Herschel—also accepted by his son—was mentioned in Chapter 31; they supposed that a star of the first magnitude, if removed to twice or three times or seven times its distance, would appear as a star of magnitude 2, 3 or 7; this means that their brightness can be represented by $\frac{1}{4}$, $\frac{1}{9}$, $\frac{1}{49}$. At the same time, he estimated that the light radiated by a first-magnitude star was 100 times that of a sixth-magnitude star. In 1835 Steinheil expressed his opinion that magnitudes of stars do not indicate differences but ratios of light intensity; he derived the ratio for 1 magnitude to be 2.83. This was in accordance with what later on, in 1869, was formulated exactly by Th. Fechner as the 'psychophysical law': what the eye experiences as an equal difference in brightness is not a constant difference but a constant percentage of the quantity of light. He was led to this conclusion by the observation that in a cloud the differences between white and grey parts remain equally distinct if it is viewed through a dark glass. In a more general way he could express his law by saying that the psychic impression is as the logarithm of the physical action. Fechner's law implies that between successive magnitudes there is a constant ratio, which must be derived from photometric measurements.

John Herschel's measurements of stars, in 1836 at the Cape, have already been mentioned; because he compared them in his 'astrometer' with a starlike image of the moon, whose light made faint stars invisible, he was restricted to 65 bright, mostly southern stars. About the same time Steinheil constructed a photometer, by which extra-focal different-sized images of two stars could be made equal in brightness by varying the size of the discs. Since equality of surface brightness can be judged very accurately, especially by seeing the border line disappear, such measurements are very precise. Steinheil's photometer was used by L. Seidel in 1852–60 in measuring 208 bright stars; the results were very accurate indeed. Because such star discs for stars smaller than the third magnitude are too faint to be measurable, this instrument could not compare, for the great number of fainter stars, with photometers measuring the point-like stars themselves, such as the Zöllner photometer.

In these first researches it was evident that stars called 'first magnitude' because they surpassed a certain limit differed too much to form one really complete class; everyone can see that Procyon surpasses Regulus by an entire magnitude and that Sirius surpasses Procyon by far more. In taking them all together, Steinheil had found too great a ratio. Between the less bright magnitude classes, smaller ratios, between 2.2 and 2.5, were found; taking the latter value, a difference of five magnitudes would correspond to a ratio of $(2\frac{1}{2})^5 = \frac{3125}{32}$, or nearly 100. In 1850, Pogson at Oxford proposed that this value be adopted as correct by definition; this meant that artificial magnitudes derived from measured light intensities should be used. If five magnitudes correspond to an intensity ratio of 100, with logarithm 2.0, then one magnitude corresponds to a difference in logarithm of exactly 0.40, i.e. a light ratio of 2.512. This came into general use, so that now the results of photometric measurements are expressed in these artificial magnitudes. The magnitude defined in this way is $2\frac{1}{2}$ times the negative logarithm of the intensity, negative because the brightest stars have the lowest magnitude values. Since for the stars from the second to the fifth magnitude these numbers are taken to correspond nearly to the traditional magnitudes, the brightest first-magnitude stars get negative values: Aldebaran, 1.0; Procyon, 0.5; Vega, 0.1; Canopus, −1.0; Sirius, −1.4. Continuing the scale for the still brighter planets, Jupiter when at its brightest is about −2½ and Venus −5. There is no reason to exclude sun and moon from this easy method of expressing brightness; the full moon then is about −11, and the sun has a stellar magnitude of −26.7.

Two great researches to determine the brightness of the stars were carried out in the last quarter of the nineteenth century, one at the *Astrophysikalisches Observatorium* in Potsdam, the other at the Harvard College Observatory in Cambridge, Mass., both extending down to magnitude 7.5. In Potsdam the measurements were made by Gustav Müller and Paul Kempf in 1885–1905 by means of Zöllner photometers of different power, with a carefulness in plan and performance in accordance with the best German traditions; the mean error of a resulting magnitude, based usually on two measurements, was 0.07 magnitude only. At the Harvard Observatory the director, Edward C. Pickering (1846–1919), devised a 'meridian photometer', in which each star during its meridian passage, by means of a birefringent calcite prism with Nicols, was equalized to one of the pole stars. The measurements were made by Pickering, who could pride himself on having made more than a million settings in photometric work. Begun in 1879 with the brightest stars, it gradually extended over the fainter classes. It was done less carefully, with more mistakes and gross modes of reduction, and shows a greater mean error (0.10 to 0.15) than did the Potsdam results. But the Potsdam catalogue comprised the northern hemisphere only, whereas Pickering, who had a clear feeling of the importance of complete, though less precise, data, extended his work over the entire sky. Whenever a programme had been finished at Cambridge, an observer with the same instrument or an identical one was always sent to the southern hemisphere, mostly to Arequipa in the Peruvian Andes, a site with a brilliant climate. On account of their completeness, most later researches made use of the Cambridge catalogues and not of the better Potsdam results.

For completeness' sake the Oxford catalogue of magnitudes of the naked-eye stars (1885), measured by Pritchard with the 'wedge photometer', must be mentioned. The brightness was given by the thickness of dark glass needed to extinguish the star. In accuracy and extent it is inferior to the others. More important was the determination of the photographic brightness of the stars—called, therefore, not photometry but 'actinometry'—made in 1904–08 by K. Schwarzschild at Göttingen. Here he continued his former Vienna work with extra-focal star discs; again working with surface densities of silver deposit; to get exactly equal surfaces for all stars, he devised a *Schraffier-Kassette* ('hatching plateholder') that extends the light of each star over a small square. Like all work done by Schwarzschild, it was a model of careful ingenuity and exceeded all visual catalogues because of its greater accuracy; the mean error of a result was only 0.02–0.04 magnitude. Only the first part, however, was completed, comprising 3,500 stars down to the magnitude 7.5 between 0° and 20° of declination; thus it could serve as a model but could hardly be used in general researches.

What we measure in this way, the brightness of a star, is a characteristic for us, but not for the star itself; it is apparent brightness. It determines the aspect of the starry heavens, which, as mankind's great experience of nature, unchanged through all centuries, we have in common with our barbarian ancestors and our classic predecessors. Every modern astronomer, however, has been aware that stars apparently connected into one time-honoured constellation, like Orion or Cassiopeia, do not really belong together but could be at widely different distances, one behind the other. This was so self-evident that it caused a certain shock when Proctor in 1869 and, more definitely, Klinkerfues at Göttingen in 1878 concluded from the proper motions that, among the seven chief stars of the Great Bear which form the Wain, the five in the midst belong together and, like a shoal of fishes or a procession of wagons over a field, run after one another in the direction of the Eagle. A second surprise was Hertzsprung's discovery, in 1909, that Sirius keeps pace with that procession like a fellow-traveller. More such processions have been discovered.

The apparent magnitudes, which make the brightest stars appear to us like one family, depend greatly on the distances; from the parallaxes it was soon deduced that, for example, Sirius, Vega, and α Centauri owed their brightness to their vicinity; Canopus, Rigel, and Betelgeuse to their enormous real light power or luminosity that makes them a special class of stars. As soon as the astronomers had the photometric magnitude as well as the parallax of a star at their disposal, they could compute its luminosity. It was expressed by the 'absolute magnitude'—the magnitude a star would have if placed at a fixed standard distance. Though Kapteyn and others made the logical attempt to introduce 1 parsec for this standard distance, practical use led to the adoption of 10 parsecs, with a standard parallax of 0.1". Then the absolute magnitude (M) is found from the apparent magnitude (m) by adding 5 times the logarithm of 10 times the parallax. Thus for Sirius $(m = -1.4, p = 0.37")$, M was found to be 1.4; for α Centauri $(m = 0.1, p = 0.76")$, $M = 4.5$; for Rigel $(m = 0.3, p$ at most $= 0.01")$, $M = -4.7$ (or brighter), which is 6 magnitudes brighter than Sirius. On the other hand, taking faint nearby stars, we find for Bessel's star, the binary 61 Cygni $(m = 5.6$ and $6.3, p = 0.30")$, $M = 8.0$ and 8.7; and for the runaway 'Barnard's star' $(m = 9.7, p = 0.54")$, $M = 13.3$, again 100 times fainter. The extreme values of absolute magnitude were found to differ by 18 or 20, corresponding to a ratio of 16 or 100 million in light power. When the sun was added, with an absolute magnitude of 4.8, it appeared that

19. Sir Arthur Eddington (p. 417)

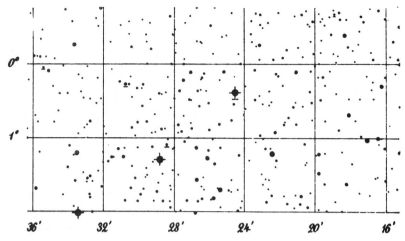

20. Telescopic stars in the Belt of Orion
    Above: According to Argelander's atlas (p. 444)
    Below: According to the Franklin Adams maps (p. 469)

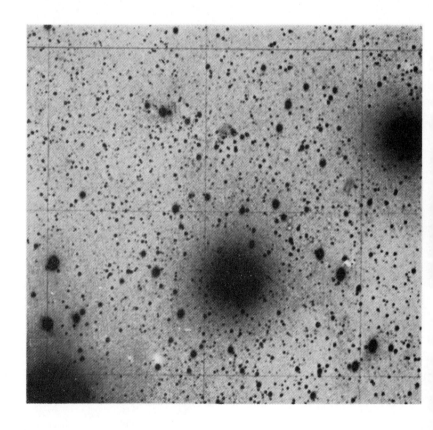

Milky Way; Cygnus (p. 475)

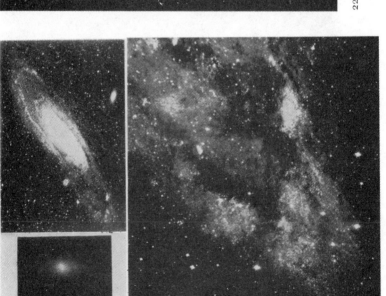

22. The Andromeda Nebula
Top left: Drawing by Kaiser
Top right: Photograph by L. Roberts (p. 486)
Bottom left: Photograph of the south-western outer edge. Mt. Wilson
Right: The Andromeda Nebula photographed in 1901 by

it is in the middle between stars 10,000 times brighter and 10,000 times fainter. Jeans denoted them by the descriptive names of 'lighthouse' stars and 'glow-worm' stars.

Whereas differences in the light power of the stars cannot be seen but only computed, differences in colour can be observed directly and have been known since Ptolemy. Struve was struck by the strongly—partly subjective—contrasting colours in double stars, which he described as red and blue or green. Zöllner constructed his photometer to serve simultaneously as a colorimeter, by inserting a plate of quartz into the polarized light pencil of the artificial star. The colours produced by rotating the quartz plate, however, were entirely different from the real stellar colours; the latter form a one-dimensional series from red through orange and yellow to white and bluish white; green shades, appearing by contrast, do not really occur among them. They are the colours appearing in an incandescent body heated to increasing temperatures; the colour differences in stars have always been considered as indicating temperature differences. When reddening occurs through absorbing nebulae or atmospheres, the colours remain within the same series.

This was clearly recognized by Julius F. J. Schmidt, who observed the stellar colours and expressed them—as also had Herman J. Klein— by one number, estimated on a scale on which 0 designated white, 4 yellow, and 6 to 10 shades of red of increasing depth (or saturation). His colour estimates, however, comprise only a moderate number of bright stars; the interest in star colours at the time was not very great. When making their photometric measurements, the Potsdam observers also estimated the colours; they expressed them not by a number but by letters indicating the names white, yellow (*gelb*), red: W, GW, WG, G, RG, GR, R, the scale being refined by adding + or −. Such letters indicating yellow, orange, and red hues (W, Y, OY, O, OR, R) were also used in 1884 by Franks, who did not restrict his colour estimates to a one-dimensional series but, by adding numbers for the degree of saturation, used a two-dimensional scheme. A most important contribution to the knowledge of stellar colours was given by H. Osthoff, an amateur in Cologne, who in the years 1883–99 estimated the colour on Schmidt's scale for a thousand stars. By careful practice he attained such accuracy that his results have a mean error of only 0.2 units. An equally extensive work in colour estimates of mainly reddish stars was performed by Fr. Krüger. Later on, K. Graff constructed a colorimeter in which a wedge of red glass gives to an artificial star all colours from white to red; the equality of colour must of course be estimated by eyesight.

Progress in science usually consists of replacing estimates of quality by measurements of quantity. Thus, instead of colour estimates, came the determination of 'colour equivalents', which are quantities directly connected with colour. When the photographic magnitudes of the stars could be measured accurately, the difference between visual and photographic magnitude, from 0 for white stars to 2 magnitudes and more for red stars, could be used as a 'colour index'. Another colour equivalent, introduced by G. Comstock, was the 'effective wavelength', the weighted mean of the wavelengths active in producing the photographic image. Comstock determined it by putting a coarse grating before the objective and afterwards measuring the distance from the short sideways spectra to the central stellar image; thus for white and for red stars he found 4,100 and 4,400 Å. All these different colour indications are expressions of the same basic fact: the distribution of intensity over the spectrum, which, according to Planck's formula, with increasing temperature shows an increasing preponderance of the shorter wavelengths.

The spectra themselves, however, fully disclose the diversity in the nature of stars. Our real knowledge of the nature of the stars begins with the discovery of spectrum analysis. Our science of these brilliant bodies that fill the universe with their radiation has been created by nearly a century of study of the stellar spectra.

Fraunhofer in 1817 perceived, and in 1823 described in detail, the fact that in the spectrum of Castor and Sirius dark lines presented themselves clearly different from those in the solar spectrum. He was, however, much too busy with his optical work to give more attention to this fact, and he died too soon to come back to it. After Kirchhoff had laid the basis of spectrum analysis and shown the meaning of the dark lines, spectroscopes were attached to telescopes to produce sufficiently bright stellar spectra for study. Besides Rutherford in America and Donati in Italy, there were William Huggins in his private observatory at Tulse Hill in England, and Father Angelo Secchi at the Specola Vaticana at Rome, who most thoroughly cultivated the new field. Huggins studied the spectra of a number of bright stars, identified the dark lines of sodium, iron, calcium and magnesium, and stated in 1863 that the same elements are present in the stars as in the sun and on the earth. Thus the unity of the material constitution of the entire universe, previously only assumed, was demonstrated with certainty.

Secchi, during 1863–68, examined the spectra of over 4,000 stars and found that they could be classified into four types, with some very few intermediate or deviating forms (plate 17). The first type, comprising more than half the number of visible stars, all white or bluish, showed four strong absorption lines, one in the red, one in the green, two in the blue-violet, all due to hydrogen, and moreover some very faint other lines. The second type, consisting of yellow stars (Capella, Arcturus, Aldebaran), showed a spectrum resembling that of the sun: numerous narrow lines identical with the solar lines but somewhat stronger in the last-named stars. The third type, the red stars (Betelgeuse, Antares, α Herculis, Mira), moreover, showed dark bands sharply cut at the violet side and gradually fading towards the red.* His attentiveness in this observation is shown by his discovery (in 1868) of a fourth type, faint and fiery-red stars, showing different bands fading towards the violet side and corresponding in wavelength to the emission bands of electric discharges in hydrocarbons. He also perceived some few exceptional stars, such as γ Cassiopeiae and the variable star β Lyrae, which showed bright emission lines, as also did the nova appearing in 1866 in Corona.

Huggins in 1864 directed his spectroscope to the planetary nebula in Draco and discovered that the spectrum consisted of only one bright and two fainter green emission lines; one of the latter was the green line of hydrogen (Hβ), and the other two were thought to be due to nitrogen. The same was the case with the great Orion Nebula. Thus was answered the question posed by William Herschel: whether there are nebulae that do not consist of a number of stars but of a 'shining fluidum'. There are; the spectrum shows them to be a thin, glowing or luminous gas. The cause of the luminosity could not be exactly indicated; electrical discharges in a tube with rarefied gas could serve as an instance of what was indicated by the term 'luminescence'. Such gaseous nebulae are few in number; most non-resolvable nebulae, among them the large Andromeda Nebula, showed a continuous spectrum, so that they must be clusters of densely-packed small stars.

Huggins also was the first to determine, according to Doppler's principle, the radial velocity of a star through the displacement of its spectral lines (cf. figure in plate 15); his example was soon followed at Greenwich Observatory. The measurements, however, were so difficult that the results were often uncertain by tens of kilometres. An essential improvement was made when, in 1887, Vogel and Scheiner at the Potsdam Observatory constructed a spectrograph with proper provisions against a flexure and changes of temperature. Now the results had such greater accuracy that the orbital velocity of the earth was clearly shown by a periodic change in the measured radial velocity of the stars. Thus the validity of Doppler's principle—previously doubted by some physicists—was demonstrated, as well as the yearly motion of the earth, for whomever might need it.

Secchi's four types of spectrum have remained in use as a main classification. Vogel called them classes I, II, IIIa and IIIb and considered them as phases in a natural development; through cooling, class I

---

* In his publications Secchi adorned his pictures of the spectra with their real colours which we have left out of our reproduction.

stars change into class II and class II into class IIIa or IIIb. Subdivision Ib for white stars with narrow lines, as well as Ic and IIb for stars with bright lines, disturbed the regularity; moreover, the question was raised as to whether it was safe to identify a formal classification with an evolutionary theory. A classification of all the stars brighter than the seventh magnitude was started on these lines at Potsdam. When, however, the zone between 0° and 20° of declination had been finished, the work was not continued; for in the meantime photography of stellar spectra had advanced and replaced the laborious peering at faint colour bands by an easier and more reliable working method.

Here Henry Draper of Virginia and Huggins in England were the pioneers. Draper had ground a mirror of 72 cm., and, by placing a quartz prism (quartz does not, as does glass, absorb the ultraviolet rays) before the focus, he succeeded in 1872 for the first time in photographing a stellar spectrum (of Vega). In 1879–82, working with a 28 cm. refractor and a Browning spectrograph, he photographed the spectra of fifty stars. Soon after Draper, Huggins succeeded in photographing stellar spectra (pl. 17). His attention was drawn to the remarkable regular series of absorption lines in the spectrum of Vega, Sirius and other white stars; since they joined and continued the hydrogen lines in the blue and violet at decreasing mutual distances, he did not hesitate to ascribe them to hydrogen. They were the same lines that had been photographed shortly before in the chromospheric spectrum and that shortly afterwards could be represented by Balmer's formula.

After Draper's death in 1882, his widow bequeathed all his instruments and a sum of money as a 'Henry Draper Memorial Fund' to the Harvard Observatory to continue his work on stellar spectra. Seldom has a moderate amount of money been so well spent for science. Pickering used it to provide the objective of a wide-angle telescope with an objective prism, a large glass disc ground into a prism with small refractive angle. Thus the images in the focal plane became small spectra, instead of round star points; one plate contained the spectra of all the hundreds of stars of a large field at the same time. They were classified according to their aspect and were indicated by letters A, B, F, G, K, M, N—representing really different classes of spectra. A and B corresponded to Secchi's first type; F, G, K, to the second type; M was identical with the third and N with the fourth type. The first *Henry Draper Catalogue* in 1890 gave the spectra of 10,000 stars, among which, in the case of faint spectra and hardly visible lines, many erroneous allocations still occurred and had to be corrected afterwards. The work was then continued with larger instruments; the 10 inch Bache telescope was provided with two objective prisms of some few degrees refracting angle, which used separately served for faint stars and combined for bright stars. By taking the same kind of plates at Arequipa, the work was extended over the entire southern sky.

The spectra of the bright stars taken with the larger dispersion were investigated by Antonia C. Maury. She distributed them over 24 classes, designated I–XXIV; this more detailed distinction, however, did not meet with general acceptance. In a number of bright stars in Orion, whose spectra were classified in classes I–IV, corresponding to the B stars of the *Draper Catalogue*, she perceived a number of characteristic lines, which for the moment were called 'Orion lines'; they had also been found by Vogel and Scheiner in some of their stars. After terrestrial helium had been discovered by Ramsay in 1895, it appeared that all these 'Orion' lines were helium lines. In each of the classes, moreover, she distinguished stars with broad, with very broad, and with narrow lines by adding the letters 'a', 'b', 'c' to the Roman numbers. The lines in the sub-class 'c' were not only narrower than in 'a' but also different in relative intensity; she added a list of lines that are strong in the 'c' stars and pointed out that these lines probably constituted a separate characteristic group. The importance of the 'c' stars was shown afterwards.

Pickering's chief aim with all these plates was to extend the survey of stellar spectra down to the ninth magnitude, a 'mass production'. Such a work never could have been accomplished by means of slit spectra of all the separate stars. Slit spectra, because they are not blurred by air vibrations, are necessary for refined study of the spectral lines, whereas the spectra taken with the objective prism, less sharp through unsteady air, are used for classification. The classification in this case was the work of Annie J. Cannon, who acquired a special ability in distinguishing minute gradations. She used the system of letters of the first *Draper Catalogue*, which in this way, in spite of its queer origin, became dominant in the field of stellar spectra (pl. 18). It appeared that the different classes indicated by these letters formed a continuous series with gradual transitions: from the B stars (with helium lines) and A stars (with broad hydrogen lines), through F, G (the exact solar type), and K, to the M stars (with bands). The transition forms were indicated by numbers forming a kind of decimals: Bo, B5, B8; Ao, A2, A5, A8; Fo, etc.; the estimates could not very well be made more precise than ¼ of a class; exact decimals appeared in later times as the result of measurements. All but one single per cent of the totality of stars could be classed in this series, especially after side branches of R–N stars (with different band systems) were added at the end and O stars at the beginning. Thus, after many years of work, the great *Henry Draper Catalogue*, giving magnitude and spectrum of 225,000 stars, was completed and published in 1918–24 in nine volumes of the Harvard Annals.

The above-named letter O was introduced for some rare peculiar spectra, showing bright, mostly broad, emission lines on a continuous background; the strongest had wavelengths 4686 and 4650. The first stars of this type—fifth and sixth-magnitude stars in a bright galactic patch in the Swan—had already been discovered visually in 1867 by the Paris astronomers C. Wolf and G. Rayet; hence this kind of stars was often designated as 'Wolf-Rayet stars'. On the plates of the southern sky Pickering detected a second-magnitude star in the Ship, γ Velorum, as the brightest of this type, sometimes called the 'fifth type'. Still more important was the discovery, in 1896, of the peculiar spectrum of another bright star in the Ship, ζ Puppis. Between the hydrogen lines it showed other lines of nearly the same intensity, alternating with them so regularly that Pickering could represent their wavelength by Balmer's formula when, instead of the whole numbers 3, 4, 5 . . ., the values $3\frac{1}{2}$, $4\frac{1}{2}$, and $5\frac{1}{2}$ were inserted. So this 'Pickering series' was ascribed to hydrogen in some unknown abnormal condition. Rydberg pointed out that then the line with wavelength 4686, exactly $\frac{5}{7}$ times that of the first Balmer line Hα, also belonged to this spectrum. Since it is the brightest line in the O stars and since, morover, the lines of the Pickering series here are faintly visible, a bridge was built to the Bo stars, where they occur as faint absorption lines. Thus the O stars found their place preceding the B stars.

The series of spectral classes indicated by these letters, from O and B to M and N, was a series of decreasing temperature. The idea that it was a series of consecutive stages of development, due to cooling, as formerly expressed by Vogel, was a self-evident conclusion. However, there were difficulties. Cooling means decrease in intensity; and small bodies cool more rapidly than large ones. Monck, in 1892, pointed out that the frequent occurrence of double stars consisting of a bright red and a faint blue-white component contradicted this scheme of evolution.

An entirely different theory had formerly been proposed by Lockyer. Lockyer not only enriched astrophysics with many practical instruments and new discoveries but also became renowned through original, sometimes fantastic, new theoretical ideas, laid down most extensively in his *Chemistry of the Sun* (in 1887) and *Inorganic Evolution* (in 1900). Through his observations of spectra in the sun and in the laboratory he had gained the conviction, first expressed in 1878, that the atoms of our chemical elements were not really elementary particles but that, under high temperature or strong electrical discharge, they broke up into simpler constituents, which he called 'proto-elements'. They were characterized by special spectral lines, which were enhanced when going from the arc spectrum of an element to the high-tension spark spectrum; he called them 'enhanced lines', and they are often denoted

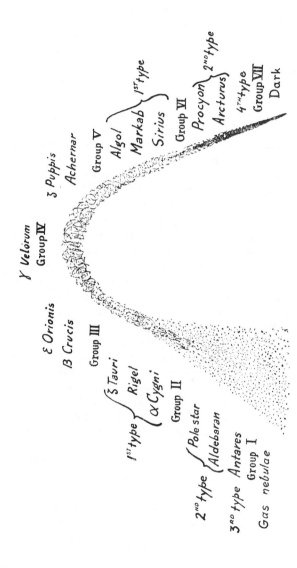

Fig. 34. The evolution sequence of the stars according to Lockyer. (Names of typical stars are added, as found in his theory of 1900)

as 'spark lines'. They were, as he remarked in 1900, the same lines that Miss Maury had found to be strong in her 'c' stars. In his many photographs of stellar spectra Lockyer found that among stars of the same spectral class some had very strong enhanced lines, others had them faint; the first, in which the proto-elements were more abundant, he considered to be in a more primitive state. Thus he was induced to assume an ascending and a descending line in the evolution of the stars, in agreement with the theoretical results of Lane. The series of ascending temperatures began with the cold nebulae and proceeded along stars contracting with increasing temperature, passing through the third, second and first type stars with strong spark lines: Antares (M according to the Harvard classification), the Pole Star (F8), Deneb (A2), Rigel (B8) to the brightest of the Orion belt stars: ε Orionis (Bo). After Pickering's discoveries, he added, as the highest summit of temperature, ζ Puppis and γ Velorum, considering the Pickering lines as the spectrum of proto-hydrogen. The series of descending temperatures from the first to the fourth type, under continued contraction, was characterized by the faintness or absence of spark lines: Algol (B8), Sirius (Ao), Procyon (F5), Arcturus (Ko), down to the N stars. This classification of the stellar spectra, with its theory of evolution, found little acceptance among astrophysicists, chiefly because Lockyer had linked it to his 'meteoric hypothesis', which assumed meteor swarms to be the original state, from which the increasing heat engendered by collisions led to the nebulous and then to the stellar state.

What value and what truth there was in this theory was shown when, in the next dozen years, the growing number of parallaxes and proper motions presented new data on the luminosity, diameter and density of the stars. When their enormous diversity in luminosity was connected with the diversity in spectra, their mutual dependences became clearer. That the A stars, and still more the B stars, far surpass the sun in luminosity was easily conceivable, because the temperature regularly decreased from B over A to F and G. Then, however, from G to K and M the average light power increased, notwithstanding the decreasing temperature. This contradiction was cleared up by Ejnar Hertzsprung, who in 1905 remarked that among the yellow and the red stars there were two sorts, one sort very bright, having large diameters and surfaces; the other faint, of small size and luminosity. He called them 'giant stars' and 'dwarf stars', names that have since come into general use and are indicated by adding 'g' and 'd' before the spectrum; our sun is dGo (o = zero). The ruddy K and M dwarfs, numerous in space but faint, are visible only when situated in the nearest surroundings of our sun, so that they form a small minority, recognizable by large motion and parallax among the bulk of the remote giants, which, though outnumbered by the dwarfs in a volume of space, are visible over a thousand times larger spaces and thus determine the average large luminosity.

At the same time Hertzsprung found that Miss Maury's 'c' stars were distinguished by imperceptible parallaxes and small, nearly imperceptible, proper motions, so that they must be situated at a great distance and hence must have a high luminosity; 'they may perhaps correspond to the whales among the fishes'[212] as he said in a zoological metaphor, rightly emphasizing the differences in kind; 'supergiants' they have since been called. Since Miss Maury had found the 'c' characteristics to be present in the Cepheid variables, these, too, must be, as already mentioned, gaseous spheres of enormous size.

The treatment of all the data on luminosity, density and spectrum by Hertzsprung and by H. N. Russell in Princeton resulted in the famous Hertzsprung-Russell diagram, a graphic representation of the relation between spectrum and absolute magnitude, best known in the often reproduced figure drawn by Russell in 1913. The co-ordinates are absolute magnitude (M) and spectral class; every star is given by a point, and values deduced from an average of a number of very small parallaxes are represented as open circles. In this diagram it is seen that the stars are situated along two belts, one horizontal at $M=0$, with giant stars in all classes a hundred times brighter than the sun; the other inclined, following a sequence from A stars with $M=0$ decreasing to $M=3$ for the F stars, $M=5$ for the G stars(the sun among them), to $M=7$ and 10 for K and M stars. The inclined belt, since it contains the greater majority of stars, was later on mostly called the 'main sequence'. Red dwarfs of the M type, of extremely small luminosity $(M=10-15)$, continue the main sequence at its lower end. White giants of the B and O type $(M=-1$ to $-5)$ continue it at its upper end. Sparse supergiants of all types are situated above the belt of common giants.

The wide gap between the two kinds of yellow stars could be demonstrated by Russell in a striking instance. Among the numerous eclipsing variables, almost all of the B and A types, two were found of the G type —W Ursae, with a period of 8 hours and W Crucis, with a period of 198 days. The densities, computed as explained above (Chapter 39) were found, for the former, to be twice, for the latter, $\frac{1}{600000}$ the sun's density. The former is a common dwarf, the latter must be an enormous ball of thin gas, i.e. a supergiant.

A queer object, not fitting in with either of the two sequences and represented by a lonely point at the lower left-hand side of the diagram, was found in a small star of the tenth magnitude in Eridanus. With a smaller adjacent star, it forms the binary o_2 Eridani, with a period of 200 years. It has a large proper motion and a large parallax of 0.20″; hence it is a dwarf with a luminosity 400 times smaller than the sun.

From its orbital motion a mass of 0.4 times the sun's mass was derived; mass and luminosity correspond to the red dwarfs of the M type. It was found, however, to be a white star of spectral class A, the first specimen of a 'white dwarf'. The companion of Sirius is also a faint star with a moderate mass, to which special attention has already been given. In 1915 W. S. Adams at Mount Wilson succeeded in photographing its spectrum, notwithstanding the glare of Sirius' light in which it is drowned; it contained nothing but broad hydrogen lines; hence it must be an A type (or, at most an F star). It was a second white dwarf, occupying a special place in the diagram beside the Eridanus star. This isolation from all the other stars symbolized their entirely mysterious character—small luminosity, moderate mass and strong radiative power. Either it must have a very small size with an enormous density of 60,000, or it must be a large dark sphere with a small piece of surface, radiating white-hot—both equally absurd.

Russell in 1913 immediately identified his diagram with Lane's theory of evolution. The branch of ascending temperatures was represented by the horizontal belt of giant stars; because the decreasing size here was compensated for by increasing temperature, the luminosity remained constant when the star developed from a red to a white giant. On the descending branch temperature and size both decreased and the star rapidly declined in luminosity along the inclined belt of the main sequence. He pointed out that Lockyer was thus justified in respect of the main question of an ascending and descending branch; but his allocation of the third type to the former and the fourth type to the latter had been too rough. Now the spectral differences also fitted into the picture; Lockyer's enhanced lines characterizing the ascending branch were identical with Miss Maury's 'c' lines characterizing the biggest supergiants.

The correlation between the luminosity of the star and the intensity of special lines in the spectrum, so clearly exhibited by the 'c' stars, appears also, though in a lesser degree, when we compare common giants with dwarfs. When these qualitative differences could be brought into the form of a quantitative relation, it would become possible to derive the luminosity of a star from the intensity of its spectral lines. Kohlschütter and Adams at the Mount Wilson Observatory developed this method in 1914–18 by taking spectra of nearly all the accessible F, G and K stars. For every spectral class or subclass, all the available data of parallax and proper motion, transformed into absolute magnitudes, were used to derive empirical tables of mutual dependence; they served to derive the absolute magnitude for all the other stars. Comparison of this absolute magnitude with the apparent magnitude at once gave the parallax of the star. Thus, in an unexpected way, parallaxes

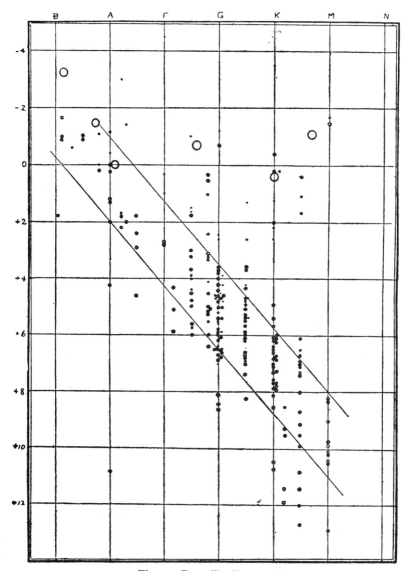

Fig. 35. Russell's diagram

far too small to be measured directly could be found and used in studies on the structure of the stellar system; they were reliable to about $\frac{1}{5}$ of their amount. Such 'spectroscopic parallaxes' were derived at Mount Wilson for more than 1,400 stars of the second type, and the example was soon followed by several other observatories.

In the same years physical theory had progressed to such an extent that it was possible to give a theoretical basis and a complete explanation for all these empirical discoveries and methods. When Bohr in 1913 expounded his theory of the atom, he could at once add an explanation for the Pickering series in the O stars. It was due not to hydrogen but to ionized helium. When the helium atom loses one of its two electrons, it acquires the same structure as the hydrogen atom, with the spectral lines in a series of the same structure but, owing to its twice as great nuclear charge, twice as close together. This implied that the O stars, notwithstanding their generally more yellowish colour, must have a higher temperature than the B stars with their lines of un-ionized helium. Moreover, it was now clear at once that it was ionized atoms of other elements which emitted and absorbed the lines met with as Lockyer's enhanced lines, our spark lines, and the 'c' lines of Miss Maury. So Lockyer had been right when he said that the atoms were split up into smaller particles by high temperature. But he was wrong in supposing these particles to be more primitive proto-elements; the normal atoms split into ionized atoms and free electrons.

Ionization, single and multiple, now appeared as the main factor, besides simple absorption and emission, which dominates the many radiations of the stars. Ionization theory became fertile for astrophysics, when in 1920 the Bengal physicist Megh Nad Saha derived the ionization formula, expressing the rate of ionization of any atom as a result of temperature and pressure. Now the intensity of a spectral line became a quantity calculable and dependent on the physical conditions of the stellar atmosphere that produced it; conversely, these physical conditions in a star can be derived from the line intensities in its spectrum. Thus the ionization formula became the basis of a quantitative treatment of stellar spectra. Saha could at once explain the differences in the consecutive spectral classes indicated as O, B, . . . K, M, by differences in temperature. Formerly the names 'helium stars', 'hydrogen stars', 'metal stars', implied a difference in the chemical constitution of their atmospheres, to which the differences in temperature were added as an independent datum. Now it was evident that chemically equal atmospheres, if increasing in temperature, had to show the sequence of M, K, G, F, A ,B, O spectra. Moreover, Saha's formula, indicating the increase in ionization by decrease in pressure, explained the great intensity of the enhanced lines in the voluminous giant stars with their small density. By the theoretical work of H. N. Russell at Princeton, who adapted the ionization formula to astrophysical practice; of E. A. Milne at Oxford, who developed the theory of stellar atmospheres; and the practical work of Cecilia H. Payne at Harvard Observatory, who, from 1924 on, applied theory to the rich collection of Harvard spectra; by these the study of stellar spectra from a qualitative became a quantitative affair. For normal stars the spectrum was found to depend on two parameters, the effective temperature and the gravity at the star's surface. The former measures the stream of heat energy pouring out from the surface of the star; the latter determines the gradient of density in the atmospheric layers, small in giant stars, large in dwarf stars. The intensity of the spark lines, which empirically seemed a mysterious effect of the luminosity of the stars, was now found to be an effect of the small density gradient in the giant stars, i.e. of the small surface gravitation due to the large size of the star. Therefore, what was said above on the binary stars, that they were the only objects which allowed the derivation of the mass of a star, is not exactly true. Fundamentally, the spectral lines are also able to do this, provided that theory and measurements are sufficiently accurate.

Temperature now appears to be the main factor determining the spectrum. What are the real temperatures of the stars? The first attempts to determine them were based on the intensity distribution in the spectrum. The empirical rule that stars hotter than the sun have the blue part stronger and the red part fainter than the sun, and the opposite for cooler stars, was refined to exact quantitative relations through Planck's formula; it allowed the computation of temperatures from measured intensity ratios. Wilsing and Scheiner at Potsdam, in 1905–10, measured visually the intensity relative to a glowing black body of known temperature as a standard of comparison for a number of wavelengths in the spectra of 109 stars of different spectral classes. Rosenberg in Tübingen in 1914 published measurements on photographic spectra of 70 stars, compared with the sun. The results were startling in their lack of agreement; for the classes Ao and Mo the Potsdam observers found 9,300° and 3,100°, whereas the Tübingen photographs gave 28,000° and 2,600°. It soon became clear that the radiation of a star proceeding from many layers of different density and temperature was different from the theoretical black-body radiation according to Planck's formula. Each result had to be corrected, and, with some range of uncertainty, a value of 10,000°–11,000° for Ao and 3,000° for Mo could be adopted.

The helium stars of class B were hotter than the A stars; their temperature could not be found in this way, because space absorption, which is strong on account of their great distance, makes their colour more yellowish. Here the spectral lines came to the rescue. By observing the appearance and disappearance of the lines of once, doubly and trebly ionized silicon in the sequence of A, B and O stars, Cecilia Payne in 1924 could establish a scale of temperatures for these stars: for B5 to B0 they increased from 15,000° to 20,000°. The first O stars (O9 and O8) had 25,000°–30,000°–; farther on, the increasing strength of the lines of ionized helium, as well as the lines of equally difficult ionizable oxygen, nitrogen and carbon atoms (among them the strong Wolf-Rayet line 4650), indicated still higher temperatures, but she could not exactly determine them.

This was done by Zanstra in 1925 by an entirely different method. The radiation of a high-temperature star, of 30,000° or more, consists almost entirely of far-ultraviolet wavelengths invisible to the eye and not accessible even to photography, because all wavelengths below 3,000 Å are absorbed by an ozone layer high in our atmosphere. According to Planck's formula, the radiation of such a star is strongest at wavelengths of about 1,000 Å; the wavelengths above 3,000 and 4,000 Å, through which we photograph or see the star, constitute the small extreme border part of its radiation. The invisible wavelengths of great intensity below 911 Å consist of large quanta and exert a strong ionizing power upon the hydrogen atoms in surrounding space. In recombining to normal atoms these ionized hydrogen atoms emit the lines of hydrogen, among them the Balmer series; our plates and eyes are sensitive to these wavelengths, and we see this surrounding matter as a faint nebula (a planetary or a ring nebula), emitting hydrogen and other lines. Thus the gaseous nebulae are explained; Herschel's 'shining fluid' is an extremely rarefied gas of hydrogen and other atoms illuminated (in a complex way) by a high-temperature star. Wright at the Lick Observatory had established in 1918 that the central star visible in planetary or ring nebulae was an O star. The light seen around the star, produced by its strong invisible radiation, is far stronger than what we see directly as starlight. This difference was the basis of Zanstra's deductions; the higher the temperature of the star, the greater this difference between nebula and central star. Thus for nearly (or entirely) invisible stars in bright nebulae he could derive temperatures from 34,000° to 40,000°— in a single case up to 70,000°.

The matter did not rest here. In 1926 the Californian physicist Bowen had solved the riddle of 'nebulium', the fantastic element invented to explain the unknown lines in the spectrum of nebulae, among them the strong green line first discovered by Huggins, which did not occur in any terrestrial spectrum. They are so-called 'forbidden lines' of ionized oxygen and nitrogen, radiated in transitions so difficult and scarce that in ordinary gases and atmospheres they are always forestalled by other transitions producing the common lines. Where the latter transitions are absent, in consequence of the extreme tenuity of matter and the faintness of radiation obtaining in the nebulae, the 'forbidden' transitions have a chance. Bowen pointed out that the smallest wavelengths in the stellar radiation consist of quanta so large that they are not entirely needed to ionize hydrogen and hence leave a large amount of energy to excite the oxygen and nitrogen atoms; these wavelengths are strongest in stars of the highest temperature. This means that the 'nebulium' lines relative to the hydrogen lines are stronger, the higher the temperature of the central star. Wright had already arranged the nebulae, according to the relative line intensities, as more or less excited; now it appeared that this was indeed a temperature sequence. Low in this sequence stand the bright extended nebulae in Orion and in the Ship, both excited by groups of O stars (in the former the quadruple star θ^2 Orion is called the 'trapeze'); they have a 'moderate' temperature of 30,000°. For the invisible stars of the most strongly excited nebulae, temperatures above 100,000° have been found.

Turning now to the other side of the scale of temperatures, we have 3,000° for a red Mo star like Betelgeuse. The characteristic bands discovered by Secchi and in 1907 found by A. Fowler to be produced by titanium oxide dominate the spectrum in cooler sub-divisions. For such low temperatures the maximum and main amount of radiation is infra-red invisible heat radiation; only a border of the smallest wavelengths is visible as red light. In the variable Mira stars the strong absorption bands do not permit the exact finding of the temperature by means of the gradient of spectral intensity. For the deep-red N stars, 2,000° is adopted; for Mira and analogous stars, Pettit and Nicholson at Mount Wilson, by radiometer measurements, derived temperatures at their maximum brightness of 2,400°–2,000°, which at minimum decrease to 1,300° or 1,400°. Though such stars radiate much infra-red heat energy, they are almost invisible. With plates sensitized to infra-red radiation, fairly bright stars have been photographed that are entirely invisible on ordinary panchromatic plates; their temperature must be below 1,000°, so that here we may suspect the transition to really dark stars. Whether they have a corresponding small mass we do not know.

Thus the range of surface temperatures of the stars extends from, say, 1,000° to 100,000°, where the extremes at both ends are almost, or entirely, invisible. This, of course, is due to the specialty of our organ of vision, that is especially adapted to the radiations of the sun. Red stars, which appear very bright to us, must also apparently be gigantic globes. Can we indicate how large they appear to us? We always speak of the stars as light points, indicating that they do not show as real discs; what it means can be readily computed. If a G star like Capella has the same surface temperature as the sun, its apparent brightness must be smaller in direct proportion to its apparent surface. Its brightness is 40,000 million times smaller than the sun's; hence its apparent diameter must be 200,000 times smaller than the sun's, i.e. 0.01″. This is indeed too small a disc to be recognized, because the diffraction disc (for a wavelength of 5,000 Å, i.e. $\frac{1}{2000}$ mm.) with a telescope of 40 inch or 1 metre aperture is 0.10″, ten times the real disc of the star. To reduce the diffraction disc to 0.01″, an aperture of 10 metres of 400 inches would be needed.

In 1890 it dawned upon the American physicist A. A. Michelson that such a small stellar disc could still be made perceptible. When the light of a star is caught by two mirrors at 10 metres distance and reflected into one telescope, the interference of the two light pencils produces upon the diffraction disc of the star a pattern of dark and bright lines, at a distance of 0.01″, corresponding to the distance of the pencils of 10 metres. This, however, holds for absolutely parallel light, i.e. a point-like star image. If the star is a very close double or has a real disc of 0.01″ diameter, the alternation of dark and bright lines would be smoothed out, and the line pattern would disappear. Though the basic idea was soon applied to the separation and measurement of close double stars, it was not realized until 1919, in the practical construction of an 'interferometer' at Mount Wilson. Anderson and Pease succeeded in determining the diameters of some few bright red stars: Betelgeuse, 0.045″; Antares, 0.040″; Arcturus, 0.022″; Mira, 0.056″. Only red stars were taken into account; white stars, in order to present such diameters, would have to appear hundreds of times brighter than they do. The measurements, though they constitute no new discovery, are important as a triumph of physical theory over technical limitations.

All that we can know of the stars through observation relates to their outer surface. What about their interior?

When dealing with the sun, we mentioned Eddington's researches on its interior. Here, however—so much wider was the scope of modern research—the sun was a chance object only, and the research dealt with the stars generally. Eddington's book, which in 1925 collected all his work in this field, had the title *The Internal Constitution of the Stars*, and his numerical data were taken from the best-known stars (plate 19).

The basis of Eddington's theory of the stars was formed by the concept of radiative equilibrium; with the high temperatures of millions of degrees obtaining here, radiation is the main and practically the only mechanism of heat transfer within the stars.

In the treatment of the conditions for thermal as well as for mechanical equilibrium, three new points of view were introduced. First, the radiation pressure, theoretically deduced long before in Maxwell's theory of electricity; almost imperceptible in experiments on earth, it can carry in the stars an appreciable part of the weight of the matter, larger or smaller, depending on the intensity of the outward stream of heat. Secondly, the high degree of ionization; by these fierce radiations entire shells of electrons are torn from the atoms, down to the limit where the electrons are subject to be either torn away or caught up in recombination. In an endless play of ionization and recombination of the atoms and electrons and alternate absorption and emission of radiation, the energy is transported from the interior to the outside. Thirdly, the production of energy in the interior, which keeps up the outward stream of energy that the star is sending into space. These three kinds of phenomena determine the state of matter and energy in the stars: temperature, pressure, density, absorption coefficient, ionization of all kinds of atoms, as functions of the distance from the centre.

The most important result of Eddington's work was the mass-luminosity law, stating that the luminosity of a star—save for small, calculable deviations dependent on the spectral class—is entirely determined by its mass, according to a relation derived theoretically. In former years, in a study of the masses of binary stars, astronomers had perceived, with some surprise, that stars of the same mass but of different spectral class had almost the same luminosity. What had been found as a casual coincidence in a small assemblage of data Eddington showed to be a general law of fundamental importance. It gave a new outlook on the evolution of the stars and the meaning of the Hertzsprung-Russell diagram. If it is true that for constant mass the luminosity is also constant, then the inclined belt of the main sequence from the bright A stars to the faint M stars cannot be a line of evolution by cooling. The theory of evolution which the diagram was supposed to illustrate cannot be right and must be replaced by a new one. The stars, in keeping their mass, must follow a horizontal line in the diagram, from the right to the left when increasing in temperature and then back when getting cooler; the main-sequence belt represents the crowding at the reversal of the change. The smaller the mass, the lower is the line of evolution situated in the diagram and the lower the maximum temperature.

Eddington had derived the mass-luminosity law on the assumption that stellar matter, as might be assumed for the giant stars, behaved like a perfect gas. When in 1924 he compared his results with the practical data of a number of stars, chiefly to see what deviations would be shown by the dense dwarf stars, he found to his surprise that the latter entirely conformed to the law: they behaved as if they consisted of thin gas. He soon understood the cause; the atoms in the deep layers of the star,

having lost their outer shells of electrons in their high degree of ionization, occupy so small a volume that, even with high density, they freely
run their course unhampered by one another. An unexpected consequence followed immediately: these stripped atoms can be so densely
packed that they form matter of density 60,000, as was computed for the
white dwarfs but deemed impossible. That the white dwarf Sirius B
was really a very small star—of the size of Uranus—of great mass,
hence with a large gravity potential at its surface (20 times that of the
sun), was confirmed in the next year by Adams. Einstein had deduced
from the theory of relativity that light emitted in such a gravity field
had its wavelength changed; Adams, by taking the spectrum of Sirius B,
in fact found the wavelength of the lines changed to the computed
amount. The certainty that matter of that 60,000 density really existed
set new problems to physics; the name 'degenerate matter' indicates
how a new department of physical theory had thus been opened.

Continuing Eddington's work, Milne in 1928 investigated more
general models of stellar structure, and found some of them becoming
unstable and collapsing to a small volume. In such a stellar catastrophe
a large amount of gravitational energy is liberated and turned into a
heat eruption, an explanation, at the same time, of a nova outburst and
of the origin of small O stars of great density such as we find in planetary
nebulae; perhaps they are remnants of former novae.

Eddington's work has opened up the interior of the stars to science.
The science of astrophysics, i.e. the study of the inner as well as the outer
nature of the stars, takes increasing precedence in astronomical investigation. Its latest development is so intimately connected with modern
astrophysical problems that it can hardly be treated in an adequate
manner as a piece of history.

CHAPTER 41

THE GALACTIC SYSTEM

Do those millions of stars visible through our telescopes and occupying the wide spaces around us form a coherent system? William Herschel had accepted this as self-evident and had connected the system with the visual phenomenon of the Milky Way. From then on, the Milky Way, or rather its central circle, was the basic plane in all researches on the stellar system; its structure was referred to co-ordinates based on this plane, galactic longitude and latitude.

In a primary survey Herschel had traversed these spaces to the farthest limits in a rush; the task of the nineteenth century, while penetrating step by step, was to take complete possession of them. Such research had to go in two directions: investigation of the arrangement of the stars in space and discovery of the laws of their motions.

Spatial arrangement had to begin with surface arrangement over the celestial sphere, taking stock of positions and magnitude. Neither in the mapping of telescopic stars for the search of minor planets nor in the meridian determinations of their exact position was this the deliberate aim of the work. For a real inventory, the precision of meridian work was not needed, and, on the other hand, the catalogues were too incomplete, and the magnitudes treated as secondary were too crude. That they could yet be of some use for this purpose was shown by W. Struve in 1847 in a study of the spatial distribution of the stars based on Bessel's zone catalogue; he derived the rate of crowding of faint stars near the galactic plane compared with their low density at great distances from this plane.

Argelander in Bonn was the first to realize the importance of a good inventory of the stars. His careful and complete cataloguing of the magnitudes of all the naked-eye stars in his first Bonn years has been mentioned above. Now he extended this task to the telescopic stars. He recognized that the method of mapping followed thus far, by alternately looking at the sky and the map and inserting the stars by eyesight, was laborious and inaccurate. So he devised a better working method. The observer, with the telescope motionless, saw all the stars of the same declination successively pass through the field of view. Every time a star passed a line directed north-south and carrying a scale, he gave a signal, the time of which was noted by the recorder sitting in front of the clock; this afforded the right ascension. The observer at the same time noted and called out the scale reading of the passage (affording the declination), as well as the estimated magnitude. Thus a step or zone, in declination as narrow as the field of the telescope but widely extending in right ascension, could be finished in one session. Each part of the sky was covered twice by such zones, so that there was a check and greater accuracy by averaging the two values. Owing to a well-considered limitation of the precision of positions (whole seconds of time in right ascension and o.1′ in declination) and a wise restriction to ninth magnitude stars (by using a small telescope of only 76 mm. aperture), it was possible to complete the entire northern hemisphere in seven years of work, from 1852 to 1859, during which the enthusiasm of the observers (the assistants Thormann, Krüger and Schönfeld) was able to carry on unabated. From this catalogue, called *Bonner Durchmusterung*, which contains about 324,000 stars (the faintest called 9.5), an atlas of star maps was constructed that surpassed all former maps in completeness and reliability. Both catalogue and atlas in all later years have become indispensable aids in astronomical work. Even now, far within the twentieth century, this old Bonn catalogue has not yet become superfluous for statistical researches.

The name *Durchmusterung* ('survey') was given by Argelander, deliberately to emphasize the coarseness of the positions for the sake of a complete inventory of the stars. It became the international designation for later similar catalogues. First his work was continued by Eduard Schönfeld, his successor at Bonn, to 24° southern declination, in order to include the entire ecliptic with its minor planets. In 1885 Juan Thomé, Gould's successor at the Córdoba Observatory in Argentina, began to extend it over the southern sky, starting with 22° southern declination. He took a somewhat larger telescope, showing stars to the tenth and eleventh magnitudes, so that his programme was larger; when the government grants were later cut so that he had to make all reductions himself, when, moreover, irrigation works in this dry province caused greater cloudiness, his work slowed down considerably. At his death, in 1908, he had only reached the parallel of 62°, though about 579,000 stars had already been observed. And it was not until 1930 that new observers completed the work to the South Pole. Moreover, because of the attempt to keep the estimates adapted to photometric measurements, the *Córdoba Durchmusterung* lacks the homogeneity that is so valuable for statistical discussion.

The Bonn scale of magnitudes is also not entirely homogeneous, since it is based on estimates on a merely mental scale. Exact photometric magnitudes, first provided by the great catalogues of Potsdam and Cambridge, reached only to magnitude 7.5; these have, of course, been used in later statistical work. Photometric measurement of all the hundreds of thousands of *Durchmusterung* stars was an impracticable task, at least visually; photographically, it would be fairly possible, thanks to modern technical methods. In order to make the *Durchmusterung* magnitudes usable for statistical purposes, Pickering measured all the stars within narrow belts of declination 10′ wide, situated at regular intervals, at 0°, 5°, 10° . . ., of declination. By means of these stars the scale of estimated magnitudes could be reduced to the photometric scale for all the separate declinations.

In quite another way J. C. Kapteyn (1851–1922) made an inventory of the southern sky; he used a complete set of photographic plates taken by Gill in Cape Town in 1885–90. To avoid the troublesome reduction of rectangular co-ordinates on the plate to the spherical co-ordinates right ascension and declination, he used a small theodolite which was pointed at a plate placed at a distance equal to the focal distance of the photographic telescope. By setting the theodolite and the plate in the right way it was exactly as if the star field was looked at in Cape Town when just setting. Then the spherical co-ordinates of every star could be read directly on the graduated circles of the instrument, with errors of some seconds of arc only, more accurately than had been done in the visual Bonn catalogue. The magnitudes were derived from the measured diameters of the stellar images by means of empirical formulas. In this way, through ten years of measuring in a small laboratory room at Groningen, the *Cape Photographic Durchmusterung* was achieved, containing about 454,000 stars between 19° southern declination and the South Pole, down to the eleventh photographic magnitude.

The mapping of the millions of fainter stars was possible only by photography. Measuring and cataloguing them all was not necessary, since the maps themselves as reproductions were always at hand. Thus the Harvard Observatory distributed an 'atlas' of all stars to the twelfth magnitude in the form of boxes of glass plates which are copies of the original plates. The slowly progressing co-operative work of the Carte du Ciel has already been mentioned. Making use of the excellent modern optical systems, which produce star images as sharp black points even to the edges of the plate in extensive star fields, Franklin Adams, an English amateur, provided for the needs of the astronomers in a still better way. In 1902–05, first in England and then in South Africa, with a doublet of 10 inch aperture, he photographed the entire sky on 206 plates of 15° square; they were photographically reproduced in maps, where the smallest star points of the fifteenth magnitude can be seen only with a magnifying glass (plate 20). Cataloguing them all with well-reduced magnitudes would be an impossible task.

The *Durchmusterung* catalogues are the basis for statistical studies on the distribution of the stars; first the apparent distribution over the celestial sphere and then the real distribution in space. This field of research was opened by the extensive theoretical and practical work of Hugo Seeliger at Munich, in a series of studies from 1884 to 1909. By means of Pickering's measurements, he derived corrections to the Bonn magnitudes. They varied not only with the declination but also with the star density; in fields full of stars the estimated magnitudes are fainter, and their limit is met at a brighter level—as if the eye were dazzled by their total light, and in the crowding more faint stars are omitted than in poor regions. Working now with fairly correct magnitudes, Seeliger could establish different regularities in the distribution. The number of stars increased in a ratio of 2.8 to 3.4 per magnitude. That every succeeding class comprised three times more stars than the preceding one was a remarkable result; for if space were filled equally with stars, every following class should be four times as numerous, since every next limiting sphere at a distance of $\sqrt{2.5}$ times greater has a volume four times larger. Thus Seeliger concluded as a first result that the space density of stars (number of stars per unit volume) decreases outward at a determinate rate. Of course this is a strongly schematized picture of the universe.

The density, however, is not equal in all directions; the surface density increases towards the Milky Way. This increase is hardly perceptible for the naked-eye stars, and becomes increasingly greater with fainter classes. For Herschel's stars it is far stronger than for the Bonn stars. The same fact can be expressed in another way: the rate of increase of the number of stars with magnitude is small at high galactic latitudes, large at low galactic latitudes, and largest in the Milky Way. This means that in the Milky Way the decrease in star density with distance is small and that it is rapid towards the galactic poles.

These are general qualitative results which do not take into account the diversity of stellar luminosities. This diversity is expressed by the luminosity law, which gives the number of stars (in unit volume of space) as a function of the luminosity. Seeliger brought all these relations into mathematical form and developed formulae expressing the number of stars counted for every apparent magnitude in its dependence on the density distribution and the luminosity law. He demonstrated that, if the star numbers should exhibit a constant rate of increase with magnitude, the resulting decrease in space density could be computed without needing the luminosity law. The counted numbers, however, did not conform to this supposition; the rate of increase became smaller for fainter classes. Seeliger explained this by assuming that we here arrived at the limit of the stellar system. It is clear, however, that knowledge of the luminosity law is necessary for better results —more generally, that we cannot get complete insight into the structure of our stellar system by mere statistics of star numbers, without making use of the data of parallax and proper motion.

Is there any regularity or law to be discovered in the motions of the fixed stars that can lead us, as with the planets, to the knowledge of spatial structure and controlling laws? A regularity had been detected from a small number of stars by William Herschel: the stars seem to converge towards a certain point in the sky, because the solar system moves in the opposite direction. In the first part of the nineteenth century there was much controversy over this question; Bessel, using an indirect method, could find no solar motion from his more extensive material. But Argelander, then at Åbo in Finland, in 1830 demonstrated with a larger number (390) of proper motions, carefully determined by himself, that Herschel had been right and that the solar system did move towards the indicated apex.

The nineteenth century, in increasing the number of good meridian observations, produced an increasing number of catalogues of proper motions. The most important was, in 1888, the new reduction of Bradley's stars by A. Auwers, made by using modern values for all reduction elements. The Auwers-Bradley catalogue, containing the reliable proper motions of 3,200 stars, remained for many years the basis of all researches on stellar motions, until in 1910 it was superseded by L. Boss's *Preliminary General Catalogue* of 6,188 stars. All computations on the basis of these and other catalogues of proper motion confirmed the results of Herschel and Argelander; the positions found for the apex were all situated in the vicinity of the point 270°, +30° (i.e. right ascension, 270°; declination, +30°). With one exception: when Kobold treated this entire material by Bessel's method, he found an apex far more to the south, near the equator. Bessel's method was based on the situation of the arrows indicating the proper motions, in such a way that forward and backward directions were not distinguished. How it could give a deviating result, indicating some other regularity, remained unknown for a time.

When spectrographic determinations of radial velocity had been made in sufficient number, they were also used to determine the solar motion; besides the direction towards the apex, they could also provide the linear velocity. Campbell derived them in 1901 from 230 stars and again, ten years later, from 1,180 radial velocities measured at the Lick Observatory; the apex found was 268°, +25°, and the velocity 19 or

20 km./sec., comparable to the earth's orbital velocity of 27 km./sec.

So a regularity in the motions of the stars had been found, a general drift towards the anti-apex at 90°, −30°. It was, however, not a motion of the stars themselves but a reflection of the solar motion. Attempts to find systematic regularities in the stellar motions themselves were made in the nineteenth century; thus in 1848 Mädler thought he had found a general motion about a centre of gravity in the Pleiades; but he was mistaken. When freed from the 'parallactic motion' due to the sun's motion, the remaining 'peculiar motions' showed only an irregular chance distribution. Yet another advantage was gained from all this work; because the sun in a century moves a distance 420 times larger than the distance from sun to earth, the parallactic motion of a star in a century is 420 times greater than its parallax. In this way the distance for remote stars could be found, not for single stars, because of their peculiar motion, but for groups, in which, on the average, the chance motions of the separate stars were eliminated.

This method has been used on a large scale by Kapteyn when in the 'nineties he started his researches on the stellar system. He thus derived the main distances of the stars of the third, fourth, and fifth magnitude and found that their rate of increase was less than $\sqrt{2.5}$, which was to be expected from their brightness. This meant that every succeeding magnitude class consists of stars whose average luminosity is smaller than that for the preceding class. Here the astronomers were confronted with the full difficulty of the problem: the distribution of the stars over their apparent brightness is a combined effect of two causes—their different luminosity and their different distance. The problem of how to find the unknown laws of both from their combined effect cannot be solved by any mathematical deduction.

The way in which Kapteyn solved the problem was an admirable piece of practical ingenuity. In 1901 he first computed, from the most reliable sources, average parallaxes for specimens of stars within certain limits of magnitude and proper motion. They were used to derive an empirical formula for the parallax as a function of magnitude and proper motion; such a parallax, expressing the distance, gave at the same time the luminosity. This formula was applied to all the Bradley stars and such fainter stars as had reliable proper motions; their lack of completeness had to be corrected through statistical counts. All these stars could then be distributed in world space by placing them between definite limits of distance, i.e. within successive spherical shells of space. In every shell the stars of different luminosity could be counted; thus Kapteyn could construct a table giving the number of stars between certain limits of distance—hence also for a unit volume of space at different distances—and certain limits of luminosity. The latter afforded the luminosity law in table form. The table showed that the number of stars increased rapidly with decreasing luminosity, but this increase slowed down more and more for the fainter absolute magnitudes until it ceased for stars a hundred times fainter than the sun, for which the frequency reached a maximum. The numerical values could be expressed by a quadratic-exponential function, in the same way as Gauss's probability law of errors.

We are meeting here a new kind of astronomy, which may be called 'statistical astronomy'. It appears where we have to deal not with stars individually but with hundreds or thousands or even millions of them. Then the nature of the problems has changed with the object; we do not ask which stars, but how many stars have certain characteristics (colour, spectrum, duplicity) or certain values of the parameters (temperature, density, luminosity, magnitude). Counting supplies the measuring. The positions (in the sky or in space) do not matter, but the densities of distribution (over the sky or over space). Statistical laws of distribution are the objects and the working instruments of the astronomer who is dealing with thousands and millions of the heavenly host.

With the knowledge of the luminosity law, the derivation of the spatial distribution of the stars became much easier. Kapteyn took up this problem first, in 1908, by determining the number of stars of every magnitude—or, more precisely, the number per square degree above a certain limit of magnitude—as a function of galactic latitude. He did not restrict himself to reducing the counted numbers of bright stars and *Durchmusterung* stars to statistical data; it was now possible to go down to the fifteenth and sixteenth magnitudes, since photometric measurements of comparison stars for very faint minima of variable stars, made at Harvard and by Parkhurst at the Yerkes Observatory, had extended the scale of reliable magnitudes to this faint limit. It was clearly shown by these figures that the rate of increase in the number of stars per square degree for the fainter magnitudes lessened greatly in these classes. It could also be rendered fairly well by a Gaussian curve, different, of course, for different galactic latitudes. Schwarzschild in an ingenious analysis showed how the density function (space-density dependence on distance) could be directly derived from the star-number function (dependence of the number of stars on magnitude) and the luminosity function, if all three could be represented exactly by Gaussian probability functions.

The distribution derived by Kapteyn consisted in a density nearly constant around the sun up to a distance of 100 parsecs, then continuously decreasing rapidly towards the galactic poles, slowly in the galactic plane. The surfaces of equal density were strongly flattened ellipsoids of revolution, so that we may speak of an 'ellipsoidal distri-

bution' of the stars. In the next dozen years computation at the Groningen Astronomical Laboratory gave more accurate numerical results: a
density of $\frac{1}{16}$ the central density around the sun was found in the galactic
plane at 3,500 parsecs and towards the poles at 660 parsecs.

That Kapteyn's stellar system had the figure of an ellipsoid of revolution was not a result but a presupposition; since in this first approximation differences along galactic longitudes were neglected and averaged,
and only the variation with latitude was considered, the result was
identical for all longitudes. Thus the Kapteyn system was a strongly
schematized and smoothed picture of the stellar universe. It was
schematized also in the variation of density with distance; because the
luminosities of the stars were widely different, what was visible in the
broad figure of the luminosity curve, all short-range variations of space
density with distance, were smoothed out and barely perceptible in the
distribution of the stars over the apparent magnitudes.

The Milky Way was spoken of here as a luminous belt around the
sky or sometimes as a plane or layer of stars in space. Now it must be
viewed as a proper object of study. What really is the Milky Way?
Exactly speaking, it is a phantom; but a phantom of so wonderful a
wealth of structures and forms, of bright and dark shapes, that, seen on
dark summer nights, it belongs to the most beautiful scenes which nature
offers to man's eyes. It is true that its glimmer is so faint that it disappears where the eye tries to fix upon it—it is perceived only by the rods,
not by the cones of the retina, hence is seen only by indirect vision; yet,
when all other glare is absent, it gives an impression of brilliant beauty.

It was mentioned earlier (p. 158) that Ptolemy in his great work
gave a description of its course and irregularities. It is a remarkable fact
that in all later centuries no attempt was made to repeat and improve
his work and to depict the phenomenon as it appeared to the eye. Star
maps mostly showed a worthless picture of a uniform, sharp-bordered
river. Probably the reason is that what always remains the same does
not attract the attention. John Herschel, when he was making observations at the Cape of Good Hope in 1834–38, was the first to be so strongly
struck by the uncommon sight of the southern part of the Milky Way
not visible in Europe that he was induced to make a sketchy picture of
it. After Argelander, in his appeal to amateur astronomers, had pointed
to the Milky Way as one of the objects of study, it was taken up by
Julius Schmidt in Athens and by Eduard Heis in Münster. The latter, in
his atlas of the northern sky, published in 1872, gave a drawing of the
Milky Way. Gould had his assistants Thomé and Davies do the same
for the southern Milky Way on the maps of the *Uranometria Argentina*. In
both cases the Milky Way was an accessory feature inserted on maps

where the stars were the chief objects; this is also shown by the lack of well-elaborated fine details. More detailed drawings in two publications devoted to the Milky Way as sole object appeared at the end of the nineteenth century shortly after each other—one in 1892 by O. Boeddicker, assistant to Lord Rosse at Parsonstown, Ireland; the other in 1893 by C. Easton, a Dutch amateur at Dordrecht. These pictures, as well as later ones, though presenting considerable differences in the mode of conceiving and depicting the details, agreed in showing how intricate an object it is. Where a superficial view showed only a broad luminous band, attentive study revealed a sequence of irregular clouds and patches, connected by light streams of various intensities, separated by and mixed with dark fissions and interruptions. Yet there was a regularity visible in its general appearance, in that at one side of the sky it is far brighter than at the opposite side. The brightest light patches are seen in Sagittarius, at galactic longitude (reckoned from the inter-section point with the equator in the Eagle) of 330°. On both sides the brightness decreases, though rising again to secondary maxima in the Swan (at 40°) and in the Ship (at 260°); in Perseus (at 120°) only a faint glimmer is visible.

After this difficult visual work, the application of photography was a revelation (plate 21). In 1869 several astronomers—H. C. Russell in Sydney, Max Wolf in Heidelberg, and E. E. Barnard at the Lick Observatory—began to photograph the Milky Way—as well as comets, with lens systems of great angular aperture, mostly 1 : 5, to have a great surface brightness. After about three hours' exposure there appeared on the plates the bright cloud forms of the Milky Way in great intensity and with a wealth of detail that never could have been detected by the eye, no matter with what optical aid. In 1891 Max Wolf published beautiful pictures of a bright nebula near Deneb in the Swan that, from its peculiar form, was called the America Nebula. Barnard used increasingly better instruments, but finally a 10 inch doublet of 1 : 5 obtained with the aid of a grant by Miss Catherine Bruce. He employed it to photograph all the parts of the Milky Way visible in the United States; his pictures are reproduced in an atlas published in 1927 after his death. Still more perfect was the Milky Way atlas of Frank Ross, expert constructor of optical systems, who devised one that pictured fields more than 20° in diameter, with the stars shown as sharp, fine points up to the edge. It is to be regretted that complete-ness of this atlas failed by the omission of the most southern part, between longitudes 240° and 300°.

What all these photographs revealed was, first, that the real Milky Way, the bright clouds and streams, was made up of hundreds of thousands of very faint stars, from the thirteenth, fourteenth, and fifteenth magnitudes downwards. They were situated at great distances, between 1,000 and 10,000 parsecs, where Kapteyn's stellar system faded out in more and more thinly populated outer parts. The small densities computed for his schematized universe were deceptive averages of large, nearly empty spaces and restricted realms of great density. In those outer spaces large condensations of stars occur, similar to, or even larger than, the central parts up to 500 parsecs of Kapteyn's system, now mostly called the 'local system'. What we see as the Milky Way surrounds it at far greater distances. So it was not inappropriate that Proctor in 1869 conceived the Milky Way as a sinuous ring surrounding our solar regions at a great distance. Easton, in a series of studies in 1894–1900, transformed it into a spiral structure, with the great Cygnus cloud, between β and γ Cygni, as its nucleus and the sun situated in one of the spiral arms. Such theories were hypotheses of what might be. The problem of what really is could only be solved by deriving spatial densities from extensive counts of stars.

The second conspicuous phenomenon in the Milky Way photographs, giving them their most picturesque aspect, were the dark features, empty spaces almost without stars, often sharply defined. They appeared in all dimensions between very large and very small, in the most freakish shapes, such as patches, canals and lanes interposed between the bright star masses. Barnard gave much attention to them and in 1919 published a catalogue of 182 dark objects, afterwards increased to 352—objects, indeed, because they must be dark clouds or structures of absorbing matter extinguishing the light of the stars behind them. The biggest were already known from visual observations: the famous Coalsack near the Southern Cross, some dark spots in the Swan, and an almost black, starless space south of θ Ophiuchi. In former times they had been considered as empty spaces, interrupting the star-filled spaces and separating the single star clouds, just as did the great dark rift separating the two branches of the Milky Way from Centaurus to the Swan. The small black spots on the photographs, however, could not be explained in this way.

Outside the Milky Way also and at its borders, regions with a lack of stars were found. On Schönfeld's atlas of the Southern Bonner *Durchmusterung* a large irregular part of Ophiuchus with a great deficiency of stars is seen. On the Franklin Adams maps a number of irregular, mutually connected voids is seen in Taurus. Dyson and Melotte at Greenwich in 1917 published a drawing of them; they concluded that the dark nebulae causing the lack of visible stars were no farther distant than 100–200 parsecs. Later investigations of other astronomers confirmed this result; the great Ophiuchus Nebula, causing a scarcity even of naked-eye stars, is at nearly 100 parsecs or less. For smaller dark spaces between the galactic clouds, distances of 400–500 parsecs have been found. So these dark nebulae may be compared with clouds and threads of smoke or dust, of world dimensions, surrounding us at close distances, and obstructing our view of the far distant galactic star accumulations that can show their full splendour only through the interstices.

Related to the Milky Way are the various kinds of nebulae and star clusters. Leaving aside the gaseous nebulae treated above (p. 451) and also the two Magellanic Clouds near the South Pole that look like small, detached bits of Milky Way, there remain three kinds, all consisting of stars, as is shown by their continuous spectrum. Among the 5,000 nebulae collected in 1864 by John Herschel in his *General Catalogue* and the 13,000, to which, in 1888, Dreyer increased this number from many sources in his *New General Catalogue* (*N.G.C.*), the great majority consisted of so-called 'unresolvable' nebulae, in which separate stars could not be detected with the largest telescopes. They often showed a spiral structure, first discovered by Lord Rosse, so that the entire class was sometimes designated as 'spiral nebulae'. They were found to accumulate around the galactic pole, in the constellation Coma and surroundings, and hence to stand in a kind of antagonism to the stars. A second kind consisted of the nearly one hundred so-called 'globular clusters', where good telescopes showed the outer borders crowded with numberless small stars, which, still more strongly condensed in the central parts, formed an unresolvable bright mass. They are all situated on one side of the sky, in the hemisphere having its centre in Sagittarius. Then, as the third kind, we have the 'open clusters', loose, often irregular, groups of stars, mostly in the Milky Way, from large groups like the Pleiades and Praesepe in the Crab, descending in size and brightness to small clusters of petty stars. Obviously the open clusters are nearest, the spirals most remote.

When Trumpler, at the Lick Observatory, investigated the open clusters and the spectra of the separate stars therein, to derive the distance, he found in 1930 that the relations perceived between the stellar magnitudes and the sizes of the clusters could be explained only by assuming a general absorption in space; such an absorption diminishes the brightness of the stars but not the size of the cluster. His result was confirmed by Van de Kamp at the Sproul Observatory and by others, from the reddening of distant stars. It is well known that small dust particles and gas molecules scatter the light more strongly the shorter the wavelength; hence the blue colour of the sky at daytime and the reddening of far-distant stars, whereby they show redder than corresponds to their spectral class. Such a general absorption in the star-filled galactic layers can explain the apparent accumulation of spiral nebulae at the galactic poles. Photographs taken by Hubble at Mount Wilson show tens of thousands of minute nebulae outside the galactic zone, between the star images, whereas they are entirely missing in the Milky Way, being extinguished by the absorbing matter which extends some-what irregularly over the middle layers of the Galaxy. Probably the separate dark nebulae noted above, among them the extensive absorb-ing nebulae in Taurus and Ophiuchus, are nothing but the nearest and densest parts of the absorbing matter that fills all space between the stars. The occurrence of these formerly unknown irregular absorptions makes the derivation of the space density from the observed numbers of stars far more difficult.

Thus the galactic system appears to consist not only of millions of stars but, in addition, of absorbing matter occupying the space between them. The study of this matter, the nature and the size of the different solid particles of which it consists, was a new and important field of astrophysical research. It is not the only substance filling these spaces. In 1904 Hartmann discovered that in the spectrum of δ Orionis, which has a strongly variable radial velocity, the K line of ionized calcium remained unmoved and hence was produced not in the stellar atmos-phere but in outer space. Other different stars showed the same phenomenon. So ionized calcium atoms roam freely through inter-stellar space like an extremely rarefied gas. Later, in 1919 and thereafter, other atoms, e.g. of sodium and ionized titanium, were noticed. These thin gases do not show the irregular distribution of the solid absorbing particles; instead, they exhibit small radial velocities, indicating currents in world space.

We must now return to the motions of the stars. When Kapteyn started his investigations of the stellar system, he intended to base them not upon the magnitudes but on the proper motions of the stars, because the velocities were far less different than the luminosities. So he tried to establish a law not of luminosity but of velocity. But the proper motions were refractory; they did not fit in with his scheme of formulae; so he had to proceed the other way. For several years the singular behaviour of the proper motions worried him, until he found the solution, which he presented first in 1904 at the astronomical meeting in St Louis, Missouri, and then in 1905 to the British Association meeting in Cape Town: the theory of the two star streams.

If the velocities of all the stars are represented by arrows from one origin, the points of these arrows form the velocity diagram; the density of these points indicating the frequency of the velocities was found to decrease from the centre outward. As a consequence of the solar motion in space these velocities must show a preferential direction of their proper motions, pointing towards the anti-apex. Kapteyn found that in each region the proper motions showed *two* preferential directions, pointing towards two different goals, which he called the apparent 'vertices'. They can be thought to be produced by a combination of the solar motion towards the apex and the motion of the two parts of the totality of stars relative to their common centre of gravity; the latter,

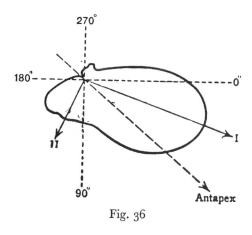

Fig. 36

then, must take place in opposite directions, towards the true vertices opposite each other, found at 91°, +13°, and at 271°, −13°. These were near the directions once found by Kobold by means of Bessel's method. Kapteyn's result was confirmed by a more extensive investigation in 1907 by Eddington, whose diagrams of velocity distribution, because of their curious shapes, became known as 'Eddington's rabbits'. At the same time, Schwarzschild showed by a careful theoretical analysis that the phenomenon could be described quite as well in another way, viz. as an ellipsoidal distribution of velocities in the region examined. The surfaces of equal density in this diagram, according to Schwarzschild's theory, were not spheres but ellipsoids, elongated in the direction of the true vertices, which is the most frequent direction of velocities. Of course it appeared, on further research by different astronomers, that the phenomenon was more complicated; there were diverse groups of stars showing special stream motions. But the main phenomenon was ascertained; a regularity had been found in the stellar motions, and the question now was how to account for it. Kapteyn supposed that the galactic system was a mixture of two kinds of stars rotating in opposite direction, so that its flat shape was explained without rotation of the entire system. The true explanation would afterwards be found to be quite different.

Not all computations of the solar motion resulted in the same apex at 270°, +30°, situated somewhat north of the galactic circle at longitude 23°. If certain star groups have a special motion relative to the main system, it will appear in a different solar motion computed from them. In 1896 Stumpe derived the solar motion from groups of stars separated according to motion and magnitude; he found the apex to be the more displaced to greater galactic longitude the smaller and slower, hence probably the more distant, the stars were—up to 60° galactic longitude for the most distant. For the remotest objects with imperceptible proper motions, only radial velocities could be measured and used. In 1923–24 G. Strömberg at Mount Wilson discussed them and derived the motion and velocity of different kinds and groups of objects relative to the sun. For the far-remote globular clusters, as well as for the spiral nebulae, for which Slipher at the Lowell Observatory had measured a number of radial velocities, he found that they have a common velocity of 300 km. towards a point situated at longitude 250°, somewhat south of the galactic circle. Or, expressed in another way, the sun, relative to these distant objects, moves with this enormous velocity towards another apex at 70° galactic longitude. This motion it has in common with all the stars surrounding it and relative to which it has only the small velocity of 20 km. Another curious phenomenon was presented by a number of 'racer stars', showing abnormal velocities above 60 km. and even up to more than 100 or 200 km. They were all found to move toward one side of the sky, centred about longitude 234°, with none moving to the opposite side. If we should adopt the system of the distant clusters and nebulae as the stationary zero point of the universe, these apparent racers would be the idlers among the rapidly moving bulk of the sun's companions.

The stars of spectral type B, which are almost all situated in the Milky Way, did not show the two streams. They presented another phenomenon, the so-called 'K effect', all wavelengths being too large. This was shown afterward to be a relativity effect of the strong gravity field at their surfaces due to their great masses. In 1922 Freundlich and Von der Pahlen at the Einstein Institute at Potsdam found that, superimposed on this K effect, the radial velocities presented a periodic variation with galactic longitude, a maximum of recession at 0° and 180° of longitude, a minimum (i.e. an approach) at 90° and 270° of longitude. Other stars with small proper motions, hence probably also at a great distance, showed the same fluctuation. Then J. H. Oort in 1927 at Leiden, proceeding from theoretical studies of Lindblad on rotating stellar systems, deduced that in the case of a rotation of the entire galactic

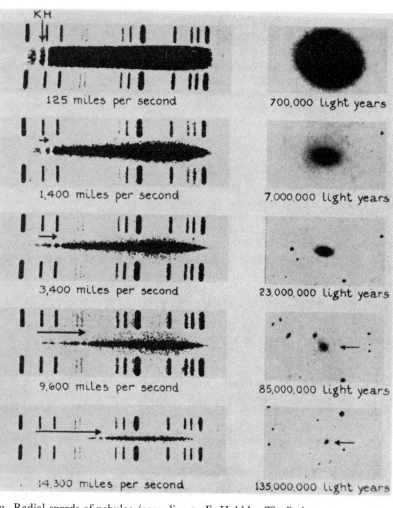

125 miles per second	700,000 light years
1,400 miles per second	7,000,000 light years
3,400 miles per second	23,000,000 light years
9,600 miles per second	85,000,000 light years
14,300 miles per second	135,000,000 light years

23. Radial speeds of nebulae (according to E. Hubble, *The Realm of the Nebulae*). The displacement of the doublet H and K (breaks in continuous spectrum indicated by the arrow above the spectrum) determines the speed of recession (1 mile = 1.6 km; 1 light year = $1/3\frac{1}{4}$ parsecs) (p. 488)

24. The Jodrell Bank Radio Telescope

system about a centre in Sagittarius, where the densest star clouds are situated, just such a periodic alternation of positive and negative radial velocities must present itself. If the sun and the surrounding stars describe circular orbits about this attracting centre as the planets do about the sun, the stars nearer to the centre will have a greater, and those farther from the centre a smaller velocity than the sun. In the first case the preceding stars are escaping from the sun, the following ones overtaking it; in the second case the forerunners are overtaken by the sun, the after-runners left behind. So, as figure 37 shows, they will exhibit radial velocities alternating with their position relative to the sun and increasing with the distance to the sun. Considering the velocity of 300 km. toward 70° galactic longitude, nearly perpendicular to the direction of the Sagittarius star clouds at 330° longitude, as the orbital velocity of the sun and its stellar surroundings, we will have the stars about 0° and 180° longitude receding from the sun, the stars about 90° and 270° approaching it, both perceptible only at distances of some hundreds of parsecs.

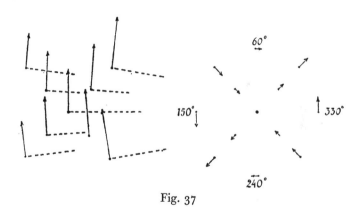

Fig. 37

Thus the galactic system was shown to be a rotating system. Oort worked out its theory in detail; he found its centre to be at a distance of 6,000 parsecs with a central attracting accumulation equal to 60,000 million solar masses, whereas an equal mass is dispersed over the entire system; for a complete revolution, the sun needs 140 million years. When the stars surrounding the sun have motions deviating by chance from exact circles, their velocities relative to the sun, as Oort demonstrated, will show a preferential direction toward the centre. Since this deviates by only 20° from the preferential direction derived by Schwarzschild (Kapteyn's vertex), star streaming, though with a small irregularity, here found its explanation. Lindblad computed the distance to the centre to be 9,400 parsecs, and the sun's period of revolution to be 200 million years. J. S. Plaskett and J. A. Pearce at the Victoria Observatory found from their spectra of B stars that the interstellar gas takes part in the rotation.

The main features of the stellar system to which our sun belongs, its shape and its state of motion, are now established as far different from what had been found in Kapteyn's pioneering investigations. We may call it the 'galactic system'; what we see as the wonderful shapes in the shining light band of the Milky Way, what appears on the photographs as brilliant star clouds, consists of the thousands of millions of stars of this system. But, partly covered and obscured by gigantic dark streaks of absorbing matter, the dense central masses are visible to us only in small parts in the bright galactic clouds of Sagittarius. Our sun is not situated near the centre but far away, somewhere in the border parts—as with the Copernican revolution, another inroad on human pride as the world's centre—so that towards the outside we see only the faint galactic shimmer in Perseus. Whether there is a local condensation about the sun or whether the 'local system' is an appearance only, produced by the absorbing matter, still remained uncertain.

The establishment of the galactic system is not the end, but rather a beginning of research, specifying a task. Just as many centuries were needed after the establishment of the solar system for the investigation of its contents, structure and details, its laws and characteristics, so it is now with the stellar system. Many observational data have been collected, many remarkable results on distribution and motion of special classes have been found—e.g. on the B stars, their clustering as a kind of frame for the system, the cepheids, 'c' stars, and nebulae. The elucidation of all the problems of structure presented by the system of stars and by the dark matter, as well as of their origin and development, is a tremendous task for future research.

CHAPTER 42

INTO ENDLESS SPACE

O̲ur solar system, so large compared with the earth and larger still compared with our surroundings and ourselves, is now included as a petty particle in the greater galactic system, millions of times larger and consisting of thousands of millions of suns, the extent and the dynamics of which we have now come to discover. But the galactic system is not the entire universe. What is found outside it? With this question we enter into a new field, a third platform in the investigation of cosmic space.

Here we are faced with another state of affairs than in the previous chapters. History here is only a preamble to the future. What was discovered, not in the nineteenth but in the first part of the twentieth century, was only a first survey, an opening-up of roads, a dim looming up of the problems that face the scientists. What now occupies our minds is not so much the past in retrospect but the outlook towards the future.

We have already become acquainted with some objects outside the galactic system: the globular clusters, the unresolvable (often spiral) nebulae, the Magellanic Clouds. The globular clusters occupy one hemisphere, with its centre in Sagittarius. The brightest specimen, visible in southern countries, is a small nebulous disc equalling a fourth-magnitude star and named ω Centauri. In northern countries some smaller ones can be seen, just visible to the naked eye as sixth-magnitude stars; they are designated by their Messier numbers, M13 and M92 in Hercules and M5 in Serpens. The Large and the Small Magellanic Clouds, near the South Pole, look like detached pieces of the Milky Way, at 33° and 44° galactic latitude. John Herschel, at the Cape, found them to consist of a dense accumulation of stars, clusters, gaseous and other nebulae, which he catalogued.

In 1895 Bailey at Harvard Observatory, studying a number of photographs of the globular clusters M3, M5, and ω Centauri, discovered in their outer parts a large number of variable stars. Afterwards the same was found for some thirty other clusters. Most of these variables have periods of about half a day, but periods of several days also occur.

It appears that all of them belong to the cepheid class, and their periods and light curves could be determined from a large number of plates. Then such variables were found in the Small Magellanic Clouds also; and in 1912 Miss Leavitt at Harvard discovered in them an exact correlation between their period and their medium brightness (the average of maximum and minimum magnitude). The brighter they are, the slower their pulsations. For a period of 2 days the medium magnitude was 15.5; for 5 days, it was 14.8; for 10 days, 14.1; for 100 days, 12.0. Since the difference between apparent and absolute magnitude is the same for all the stars in such a Cloud, this means that for cepheids generally the period varies in a regular way with the luminosity. This law could not have been found from the cepheids in our galactic system, because they are situated at different unknown distances. Assuming that this law holds for cepheids in general, Hertzsprung in 1913 was able to fix the scale of their luminosities. From the small proper motions of 13 galactic cepheids between the second and the sixth magnitudes, reduced to equal distance, he could derive their parallactic motion and hence their mean parallax and luminosity. He found that a medium absolute magnitude of −2.3 belonged to a period of 6.6 days. Miss Leavitt gave for this period a photographic magnitude in the Cloud of 14.5, corresponding to 13.0 visually; so the apparent magnitudes were 15.3 magnitudes fainter than the absolute magnitudes. This meant that the distance of the Small Magellanic Cloud was 11,000 parsecs. Afterwards it became evident that the scale of magnitudes for the faint southern stars was considerably in error; a repetition of the computation with better data by Shapley in 1918 modified the distance to 29,000 parsecs, larger than the dimensions of our galactic system.

In this way the cepheid variables afforded the yardstick with which to measure distances in celestial space. 'Lighthouse' stars they had been called by Jeans on account of their periodical flaring up; now indeed they were found to be beacons to guide the astronomers into the far depths of space. Shapley used them first in his discussion of the photographs of globular clusters, which he had taken from 1916–17 with the 60 inch telescope at Mount Wilson. He measured the photographic and photovisual magnitudes of the brightest as well as of the variable stars in these clusters; the latter afforded him their distances. He found that the brightest stars were red and were 1½ magnitudes brighter than the numerous cepheids of ½-day period and, hence, that they were supergiants. He found, moreover, that the smaller the size of the clusters, the fainter were the stars and the fainter also was the total brightness of the cluster, as determined formerly by Holetschek in Vienna. This meant that all the globular clusters are built on the same plan and that their different aspects are due to different distances. These distances could now be derived for all the 86 globular clusters: for the brightest and nearest from the cepheids, and for the smaller ones from apparent diameter and total magnitude. The nearest is the bright ω Centauri at a distance of 6,500 parsecs, the most remote is a cluster of magnitude 9.7 at 67,000 parsecs. They occupy a large space, situated on one side of us, with their centre at 20,000 parsecs towards longitude 325° in Sagittarius. In the central plane of the galaxy they are lacking, evidently through the strong absorption in this plane. Though they are situated outside the galactic system of stars, Shapley considered them as belonging to this system as a kind of environment. His work was the first clear indication that the galactic system extended far more towards Sagittarius than had formerly been assumed. In 1921 he even estimated its total size to be 100,000 parsecs, far greater than would correspond to Cort's later result of the centre at 6,000 parsecs. The figure is certainly too large, perhaps because the weakening by the absorbing matter of the light of the remotest clusters has not been taken into account. The two Magellanic Clouds at distances of 26,000 and 29,000 parsecs and with diameters of 4,000 and 2,000 parsecs constitute kinds of satellites of the galactic system.

The next problem was to find out where the unresolvable nebulae were situated, outside or inside the galactic system. The prevailing opinions may be learnt from a discussion which took place in 1921 between Harlow Shapley and Heber D. Curtis on 'The Scale of the Universe'. Curtis held that they were 'island universes', separate stellar systems outside and comparable to the galactic system, which he assumed to be no larger than 10,000 parsecs. He criticized Shapley's exact luminosity-period relation for the cepheids and the distances derived therewith; on this point, however, he was not successful. Shapley on the other hand, pointing to the accumulation of the spirals at the galactic poles, considered them not as distant galaxies but as belonging to our galactic system, itself regarded as larger, a 'continent universe'. The fact that no stars were visible in these not very distant nebulae, though the spectrum was continuous, he ascribed to strongly scattering nebulous matter within them; and he held as a tentative hypothesis that the 'spirals' are not composed of typical stars at all but are truly 'nebulous objects'.[213] His chief argument, however—their accumulation at the poles and their absence in the Milky Way—lost its validity when it was found to be only an appearance produced by the absorption in the galactic plane.

Gradually the number of these nebulae had increased. To the 13,000 of Dreyer's *New General Catalogue* of 1888 the photographs of the sky studied by Max Wolf and by Palisa (for the purpose of finding planetoids) had first added many thousands; the Franklin Adams maps and

Harvard plates increased them still more, with ever smaller items. Were they really small objects, or was the apparent smallness due to their distance? Gradually the latter opinion became dominant. Yet they were not built exactly on the same model. Among the large specimens, round or flattened, many show a smooth brightness distribution, decreasing to the border; others are spirals, either with large spiral arms extending from a small nucleus or with narrow windings closely serried around a broad body. They can be seen in their true shape when viewed from right above (as the oft-pictured beautiful M33 in the Triangle and M99 in the Hounds). Others viewed obliquely appear strongly flattened like the largest among them, the great Andromeda Nebula, M31, the spiral arms of which were first shown on photographs taken in 1890 by Isaac Roberts (plate 22).

In 1885 a new star, a nova of the seventh magnitude, appeared in the bright central parts of the Andromeda Nebula. It was possible that by mere chance an ordinary nova had appeared right in front of the nebula; but a connection was thought to be probable. In later years, when large-scale photographs of its outer parts were made, a number of stars were found that suddenly appeared and soon decreased more slowly to invisibility, just as do our ordinary novae. Their maximum brightness, however, was no more than the fifteenth to eighteenth magnitudes; therefore, if they are stars like our own novae, the nebula must be very far away. In the years 1919–26 Hubble made a large number of long exposures with the 60 inch and the 100 inch at Mount Wilson, especially of the outer parts and the spiral arms, of the Andromeda Nebula as well as of M33. On these photographs the nebulous light was at last resolved into an abundance of extremely small stars. Careful examination revealed, besides 67 novae in the Andromeda Nebula, about 40 cepheid variables in both. Their periods were between 10 and 80 days; their magnitude at maximum was 18–19; the minima were below the limit of visibility on the plates, which also explains the absence of cepheids with short periods. The long-period maxima, however, were sufficient to establish their distance. Comparison with the Small Magellanic Cloud indicated that M33 was nine times more remote, hence at a distance of 260,000 parsecs, and the Andromeda Nebula at 275,000 parsecs. The diameter of the former, then, is about 5,000, and of the latter about 14,000 parsecs.

Thus it was settled that at a distance of some hundreds of thousands parsecs other stellar systems are found entirely comparable to our galactic system. Other galactic systems appearing to us as spiral nebulae occupy the far spaces around. How far? There is no reason to suppose that only our nearest surroundings should be thus favoured; moreover, numerous smaller nebulae are found in our catalogues. At a 10 times greater distance they must look 10 times smaller and 5 magnitudes fainter; so the nebulae down to the tenth or eleventh magnitude must occupy space up to three million parsecs.

These are brightest ones only. Now that nebulae had become objects worthy of study, a systematic search was made. Pickering in former years had large plates taken with long exposures. Shapley, his successor in 1921 at Harvard Observatory, had them closely examined by Miss Ames to discover and catalogue all the minute nebulae hardly distinguishable from the thirteenth- to sixteenth-magnitude stars. In their thousands and tens of thousands they now appeared, the remote galactic systems occupying ever wider depths of space, at distances of 20, 30 and 50 million parsecs. The systematic survey comprising 60,000 of the smallest nebulae, made by Hubble at Mount Wilson, has already been mentioned.

The space explored by astronomy thus increased immensely, and the number of galactic systems therein might be estimated at hundreds of thousands. 'A galaxy of galaxies', they were called by Shapley. They were not distributed evenly or at random, but mostly condensed in groups or accumulations. The condensation about the galactic pole of rather bright near spirals had already attracted attention in the nineteenth century; now fainter groups appeared in different parts of the sky. Within each group there were differences in dimension; Shapley estimated our own galaxy to be one of the biggest. A 'luminosity function' could even be established for such groups and used in the derivation of distances. Evidently in this greater world, too, there is structure.

Now the question arises: Can we go on indefinitely in this way? In infinite space 50 million parsecs does not mean more than 1 parsec or 1 inch. But is space infinite? Since Gauss's speculations on the axioms of geometry, in the first years of the nineteenth century, and Riemann's discussion of non-Euclidean forms of space, in about 1854, most scientists think that we cannot be sure about the absolute validity of Euclid's geometry for our world. There might be a small deviation towards a Riemann geometry, in which parallels intersect at great distances, the sum total of the angles in a triangle exceeds 180°, and space, though unlimited, is not infinite. Analogy with the case of two dimensions, where the plane is replaced by a spherical surface, introduced the term 'curvature of space'; if it is small, the deviations from Euclidean space are perceptible at great distances only. Our galactic system is too small to show any deviation; in the thousand times larger system of galaxies, indications might appear in the apparent distribution, for instance, in a diminishing increase of their number for fainter classes.

However, on account of unexpected new discoveries, the treatment of these questions was abandoned or rather included in new and wider problems. We spoke previously of the spectrographic measurements of the radial velocities of globular clusters and the brightest among the spiral nebulae, from which the sun's orbital velocity of 300 km. was derived. Deviations of hundreds of kilometres remained as peculiar velocities of the separate objects. Slipher at the Lowell Observatory and Pease at Mount Wilson in 1916–17 measured the radial velocities of a number of fainter nebulae to increase the body of data. In 1919 Shapley pointed out that the remaining peculiar velocities were all positive and hence that the nebulae were moving away from us; but he could not make out where to look for an explanation. In 1929 Hubble demonstrated that the velocity of recession regularly increased with the distance of the object (pl. 23). To test this correlation, it was necessary to photograph the spectra of faint nebulae of the fourteenth to eighteenth magnitudes: all the light falling on the 50 square feet of the 100 inch aperture was condensed into a small spectrum $\frac{1}{10}$ inch long. The only lines visible in such a continuous spectrum (a mixture of chiefly second-type stars) were the H and K doublet of ionized calcium and sometimes the G group and Hγ. They stood far to the red of their true places, displaced tens or even hundreds of Ångstrom units, indicating outward velocities of tens of thousands of kilometres. The greatest velocity measured in 1936 was 42,000 km. for a nebula of magnitude 17.9, the brightest in a group in Ursa, whose distance was estimated at 72 million parsecs.[214]

This discovery that the far galaxies are receding with velocities increasing with distance—500 or 600 km./sec. per million parsecs—presents such a curious and strange phenomenon that it upset all our former concepts of the universe. It does not mean that our galaxy is resting in the centre of the universe; it means that this universe of galaxies is expanding uniformly, so that all its members are receding from one another. Since 500 km. per second is identical with 0.52 parsec per 1,000 years, the rate of yearly expansion is $\frac{1}{2000000000}$ of the present distance. If we assume that every galaxy has always kept its present linear velocity, it follows that two thousand million years ago all these galaxies were in a pile, close together.

Here, all of a sudden, we face a date in the remotest past that never could have been suspected. In all more or less fantastic cosmogonic theories, embodying strong extrapolations from present-day processes, a rather regular development was presupposed to be extensible indefinitely into the past. This provoked, however, some uneasiness, of the kind which always arises when we recklessly venture to speak of the infinite. Now, however, we are unexpectedly faced with an initial date, not precisely of a creation, yet of a starting point of the present development, beyond which we cannot see farther into the past.

This is based, it is true, entirely on the assumption that the velocity of each galaxy has always and forever been the same, which is in accordance with our fundamental mechanical 'principle of inertia', but without absolute certainty. And it must be added that far greater periods of time had been assumed for the evolution of the separate stars. Now, however, attention was directed to other phenomena pointing in the same direction. The percentage of radioactive matter in minerals and meteors, compared with the known rate of decay, points to a date of origin of about a thousand million years ago. The open clusters in our galactic system are supposed to have developed from denser accumulations, which, by the attraction of other stars, have gradually been dissolved; this must have been going on during a finite time of the same order of magnitude. The equipartition of energy—the fact that stars with the greatest masses have the smallest velocity—not yet completely realized, can be explained by the mutual attraction of the stars only if in former times they were more closely packed. All these considerations, though they form no strict proof, pointed to the same conclusion: that the present development started a couple of thousand million years ago, when the galaxies and perhaps the stars also were much nearer to one another. Or, expressed in another way: that an originally united system had been torn to pieces by a kind of explosion, whereupon the fragments, with different constant velocities, began their journey toward infinity. Thus in the picture of a uniform past, gradually fading into the distance, appeared a break, a special moment, that brought contrast and pattern into history. It is true that it is full of inconceivable mystery, but it invites further researches.

The discovery of the retreat of the galaxies came at a time already filled with profound discussions on space and time. Einstein's general theory of relativity, formulated in 1916, reduced gravitation to local curvatures, produced by matter, in the four-dimensional space-time pattern. Through the totality of attracting matter, space must have a positive curvature, i.e. a finite content. The formulae deduced by Einstein and by De Sitter showed that a space cannot be in stable equilibrium; according to the theoretical solution give in 1927 by the Belgian scientist Lemaître, the curvature must change continuously. Combined with the observed behaviour of the galaxies, it led to the theory of the 'expanding universe'. In this conception the material elements of the universe, the galaxies, are diverging not through their own motion but because space, in which they are embedded, is expanding—in two-dimensional space the same would happen with points on an expanding elastic ball.

In this way the newly-discovered great velocities of the nebulae found a natural explanation or, rather, a more profound background, as an effect of the qualities of space. Eddington extended this theory in 1931 and connected it with the atomic structure of matter in such a way that he was able by mere theory to compute the total number of electrons and protons in the universe (each 1.3×10^{79}), their total mass (1.08×10^{22} solar masses), and the velocity of the nebulae (528 km./sec. per million parsec) out of the known physical constants. The intricate difficulties inherent in the idea of an expanding universe repeatedly raised the question of whether the red-shift in the spectra of the remote nebulae could not be due to other influences working upon the light rays on their journeys of hundreds of millions of years. Milne, in a number of studies since 1932, has developed a different cosmological theory, in which our freedom to choose the measure of time when extrapolating from the present to an ever more remote past is used to construct a simple structure of space. The observed red-shift then means that in the ever more remote past the atomic vibrations took place ever more slowly, if expressed in the time of Newtonian dynamics.

Thus astronomy faces a host of new problems. Problems not as before purely astronomical, but problems of space and time, of universe and science, involving physics, mathematics and astronomy. Problems of what formerly, in the absence of definite notions, was simply called the 'infinite' and now is found to be the many-sided enigmatic object of a new science, a combination of astronomy, physics, mathematics and epistemology, for which the name 'cosmology' has come into use. Problems to be treated by the most acute theorists, handling the newest discoveries of astronomy, the most fundamental physical ideas, the most abstract mathematical methods, under careful reflection as to the basis of all thinking and knowledge. In this fusion with other disciplines, astronomy, the science of the stars, has been transformed into a science of the universe.

CHAPTER 43

THE LIFE OF THE STARS

IT was mentioned in Chapter 6 that the Babylonian astronomers, in their instructions for computation, denoted the daily displacement of sun and moon by a term signifying the 'life' of the luminary. Proper motion for them was the characteristic of life. To modern science, life of living beings consists in the first place in transformations of energy. The life-processes of every organism form a part of the great cycle of transformations of matter and energy in nature; every activity or life-phenomenon is an interchange of energy with the surrounding world. The source of all—or nearly all—energy circulating on earth is the solar radiation. All the life-processes in the organisms are a weakened effect of the strong radiation which the earth receives from the sun. And the latter is the effect, 50,000 times weakened by great distance, of the stream of energy pouring out from the sun's surface layers which consist of glowing matter at 6,000°. So, metaphorically speaking, we may say that our life is the remotest gentle rippling of the life, i.e. of the energy transformations, in the sun and, generally, in the stars.

If, therefore, we inquire about the source of all life in the universe, our own life included, we inquire about the origin of the energy of the stars. We faced this problem when we dealt with the constancy of solar heat. The answer given there was not satisfactory; the 20 million years allotted to the sun by the contraction theory was far too short for all the geological processes of disintegration, silt deposit, and rock formation, for which geologists claimed some hundred millions of years. The physics and astronomy of the time, however, could not present a more satisfactory answer.

The problem was presented again, but this time with more detailed precision, when Eddington in 1916 started his researches on the internal constitution of the stars. The mathematical equations determining the internal structure expressed the fact that, in a state of equilibrium, the energy radiated outward by a spherical layer of the star must be equal to the energy produced in the deeper layers. The energy produced in the stellar interior was a fundamental datum in the problem; we had to know where it was produced, by what matter and under what conditions. Strictly speaking, nothing was known about it; so Eddington performed his calculations, using two extreme suppositions: uniform production of energy throughout the entire mass, and production by a 'point source' in the centre only, where the highest temperature and pressure prevail. Happily, the two results did not differ greatly.

Could a source be indicated for the newly-generated energy? This was now possible as a result of the revolution brought about by Einstein in the fundamental ideas of physics. The principle of relativity involves the fact that mass and energy are identical; 1 gm. of mass is equivalent to 9×10^{20} ergs. The immensity of this quantity is evident when we consider that the energy produced by common burning (combining 1 gr. of carbon and oxygen) is only $\frac{1}{10000000000}$ part of it. The production of new energy, then, must take place by the annihilation of mass; in radiating, the star gradually decreases in mass; its matter is 'burning up', as we might call it, in a far more absolute sense than in what is commonly called 'burning'. Eddington, in his classic book of 1926 on the internal constitution of the stars, could put forward two suppositions only; either direct annihilation of matter by the coalescence and mutual destruction of a positive proton and a negative electron, so that the charges disappear and the mass is transformed into high-frequency radiation (gamma rays); or, in another way, by transformation of hydrogen into helium. When four hydrogen nuclei and two electrons combine into a helium nucleus, 0.0124 is lost in mass, i.e., $\frac{1}{320}$ of the combined weight, and is transformed into energy. To compensate for the yearly radiation of the sun (30 million ergs per gramme), the transformation into helium of a quantity of hydrogen is needed equal to one hundred thousand millionth part (1 : 10^{11}) of the sun's mass. The resulting lifetime of the sun is sufficient. The question is whether this process actually takes place or, rather, under what conditions it will take place.

Astronomy, to find an answer, had to seek assistance from physics, especially from the new, rapidly-developing nuclear physics. Since in 1919 Rutherford had transformed an atomic nucleus into another atomic nucleus through the impact of extremely rapid particles penetrating into it, many physicists had used this method. Thus detailed knowledge was acquired of all these transmutations, their energy balance and their frequency. The great velocity of the particles (electrons, protons, or alpha-particles) needed to break into an atomic nucleus is found in nature only in the case of extremely high temperatures, of millions and hundreds of millions of degrees. Such temperatures may be expected only in the interior of the stars. The stars are the big world furnaces, which, fed by the energy produced in these transmutations, keep themselves in the state of the intense heat needed.

Thus with the interior of the atoms, the interior of the stars is opening up before the eyes of the scientists. At the outer surface of a star, with temperatures of about 10,000°, we see exhibited in the stellar spectra the outer electrons jumping up and down or torn away from the atom and recaptured, in an unceasing play of absorption and emission of radiation, accompanying the processes of excitation, ionization and recombination. Deeper in the star in the layers whose radiation does not penetrate to the outside, with temperatures of hundreds of thousands of degrees, entire shells of electrons are torn away from the atoms. Only the more strongly bound electrons, nearer to the nucleus, by being alternatingly freed and captured, transport the energy in alternating absorption and radiation. We may call it the 'internal life' of the star, but this life is passive only, a passing-over of the energy from the interior of the star to the outer layers. At last, when in still greater depth we meet with temperatures of millions of degrees, we find these atoms entirely crushed, as naked nuclei unable to hold the disorderly running host of free electrons and to organize them into stable systems.

But this is not all. At still higher temperatures and velocities, within the densest central parts of the star, new processes come forth, now in the atomic nuclei. Some of those most rapidly running particles, protons or alpha-particles, break into the heavier nuclei to form new ones. Mostly these soon disintegrate by ejecting negative or positive electrons or by dividing, thereby emitting or absorbing gamma-radiation. Some of the nuclei thus formed will be stable, and in this way nuclei of heavier atoms are gradually built up. All these processes, which, owing to the progress of experimental physical data, can now be treated theoretically, with their energy balance and their frequency dependent on temperature and density, constitute the true active life of the stars. It is a continuous play of transformation, of constitution and of demolition of nuclei under extreme conditions of velocity, temperature and pressure, for which we have to make use of million-numbers just as when dealing with the dimensions of the universe. There remained the difficulty that the central temperatures in the stars were found to be insufficient for these processes. It was solved by Gamow, who, by means of wave mechanics, showed that if the velocity of the particles is too small to break into a nucleus, yet a certain petty percentage—sufficient for the effect—manages to creep in.

The science of the nuclear transmutations is as yet in its first stages, but the way is now open towards the solution of the many problems of the origin, life and future of the stars. An investigation by Bethe in 1938 led to the discovery of a cycle of processes of forming and splitting of carbon, nitrogen and oxygen nuclei, which results in building helium nuclei out of protons, to such an amount that it can explain the present radiation of the sun. Thus the old problem of the source of the sun's heat has now at last been solved. Moreover, by knowing the source of the sun's heat, we can derive how it depends on temperature and density and can ascertain that it is produced in the deepest centre of the sun. The obstacle that hampered Eddington in his investigations thus has been removed, and the inner constitution of the stars can be computed with greater exactness.

So we begin to know something of the life-processes in the stars. But life is not only an endless repetition of ever the same transformations of energy. Life is development. Just as in any organism life does not consist solely in the interior processes but also in its growth, in the genesis and development of the individual and the species, so it is in the life of the stars. Life is progressive change. It is a directed process with no return, according to the Second Law of Thermodynamics. Life is ageing, is genesis and perishing.

Here we have the old problem of the evolution of the stars, now appearing in a new context. It is the question of how stars develop and what types and forms transform into what other types. Ever again, when new points of view had been opened through observation in the last century, this question was posed and answered in different ways, in later years in connection with the Hertzsprung-Russell diagram. Certainty, however, was lacking. The life of mankind is too short to perceive continuous changes in the stars; we see the multitude of forms, but how one develops into another, and which forms belong together as older and younger evolutionary phases, can be perceived only by the mind and disclosed by theory. This will be possible when the interior atomic processes are completely known.

Now they begin to be known. We see indeed a progressive change in the processes deep in the interior of the stars. Out of the primary substances—the protons—are built helium nuclei and then heavier nuclei. The hydrogen content of the star gradually decreases; thereby the mean weight of the particles changes, with the result that the density and temperature of the successive layers change. Then also the production of energy changes as well as the spectral type. Bengt Strömgren and Eddington have studied the gradual exhaustion of hydrogen atoms as the determining factor in the evolution of the stars. As far as we can see, the star must finally be extinguished when the stock of hydrogen nuclei runs out. It is possible, of course, that other nuclear processes then play a part and complicate the course of development.

Indeed, the observation of astronomical phenomena shows that everything is not moving along the smooth paths of assured stability. Only from the science of the nuclear processes may we expect the knowledge to explain the lack of stability appearing in the pulsations of the cepheids, and still more to explain the far more serious instability that appears suddenly in the stellar catastrophes which we see in the flaring-up of novae.

Spectrum analysis, coming into being at the very moment when many bright novae appeared, could establish what happened. In the first flaring-up, which was just caught in some cases, the star showed an ordinary A- or B-type spectrum. After one day it changed into a spectrum with broad emission lines; a layer or shell of hot gases expanded with a velocity sometimes exceeding 1,000 km., clearly flung away by sudden enormous pressure from within. Expanding continually, the shell cooled, so that the radiation and the brightness decreased, and finally it was dissipated. What remained was a small star of the former brightness, but hotter and more condensed, like an O star. We have already found the expanding Crab Nebula in Taurus to be the afterglow of the Chinese guest star, the nova of 1054.

The cause of the instability which results in such sudden flaming-up remains a problem. And even more remarkable phenomena than the ordinary novae have now come to the front. If the seventh-magnitude nova appearing in 1885 in the Andromeda Nebula was really situated within this distant galaxy, it must have had an immense luminosity, 10,000 times greater than the numerous common novae of this system, which reached the sixteenth and seventeenth magnitude only. Such a star of absolute magnitude −15, when situated at the distance of our nearest stars, would be far brighter than the full moon! Baade and Zwicky, in California, in 1934 put forward the theory that such 'supernovae' really occur, whose brightness is not much less than the brightness of the entire galactic system to which they belong. By systematic search, several cases have been traced of bright novae in small spiral nebulae. They are, of course, far less frequent than ordinary novae; the estimate is one every 500 years in every galactic system. Baade and Zwicky think that, in view of its long visibility and slow decline, Tycho's star must have been a supernova, as well as the Chinese nova of 1054. As yet, the supernovae present to the astrophysicists a number of difficult problems.

We now return to the nuclear processes in the common stars. Production of energy in their interior is linked with transmutations of atomic nuclei, producing heavier stable nuclei also. Thus in the hands of the scientists the problem has broadened. The question was: What is the source of the continuously flowing stream of energy pouring out from the stars? The answer now obtained deals not only with the origin of their energy but also with their own origin. This question was raised now and then: What is the origin of all those different atoms, light and heavy, in a determinate proportion found everywhere, on earth, in meteors and in the stars? The idea now lies near at hand that all the present stable atoms have been formed in long periods of development out of the original bulk of protons, the remainder of which appears in the spectra of the A-type stars, the prominences on the sun, and the water on earth, the basic matter of all organisms, including ourselves—without water, no protoplasm and no life would have been possible. This means that in those cosmic furnaces—the central parts deep in the stars—the entire world was, and still is, fabricated, the materials constituting its matter as well as the radiations constituting its life. It is this life which appears, greatly weakened, at the hot stellar surfaces and, again a thousand times weakened, transformed into the life-energy of living beings.

Another difficulty arises: to penetrate into the heavier nuclei, hence to build the heavier atoms, theory may demand still greater densities and greater velocities of the protons, corresponding to still higher temperatures of hundreds or thousands of millions of degrees; these we do not find even in the stellar interiors. Can the heaviest nuclei, of uranium, of lead, of gold, perhaps have been present as original matter in the world? What, then, does the word 'original' mean here? Or shall we assume that the required conditions once existed in the past and now have disappeared—so that the heavy atoms are remainders, a kind of archaeological remains from conditions long passed? It may be plausible to connect the needed high temperatures with the original condition of closely packed galaxies and stars 2,000 million years ago. But all such ideas are a hesitant groping in a dark past.

So here, too, is an endless field of new problems to be cleared, where only the first sods have been cut. Here also we are working with millions, not for distances now and dimensions, but for intensities of energy. Here our path leads not towards the infinitely large, to study the great structure of the universe, but towards the infinitely small, to study the finest structure of nature, of what in the coarse-grained world of the senses is called 'matter' and 'radiation'.

And again this must be accomplished by a combination of sciences, of theoretical physics and abstract mathematics, tested by observation of matter and the radiation of various celestial bodies. Here, too, astronomy takes part in the elucidation of the essence of the world.

APPENDIX A

ARISTARCHUS' DERIVATION OF THE SUN'S DISTANCE

Aristarchus' seventh proposition is the most important, since it is there that the essential numerical result is derived. The demonstration is interesting enough, in its value to future astronomy, to reproduce it here in brief. In the figure A represents the position of the sun, B of the earth, C of the moon when seen halved. Hence angle EBD=angle BAC=3°. Let the angle FBE(=45°) be bisected by BG. Since the ratio of a great and a small tangent to a circle is greater than the ratio of the underlying arcs and angles, the ratio GE to HE will be greater than the ratio of $\frac{1}{4}$ to $\frac{1}{30}$ of a right angle; *i.e.* greater than 15/2. Furthermore FG : GE=BF : BE=$\sqrt{2}$, greater than 7/5; hence FE/GE is greater than 12/5. Combining it with the first inequality, we find the ratio of FE to HE greater than 15/2×12/5=18; and the ratio AB/BC which is equal to BH/HE, hence a little bit larger than BE/HE is

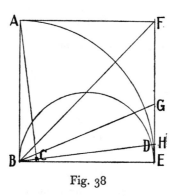

Fig. 38

certainly also greater than 18. Applying, on the other hand, the proposition that the ratio of a great and a small chord is smaller than the ratio of the subtended arcs, upon DE subtending 6° in the half-circle BDE, and the side of a regular hexagon, equal to the radius, subtending an arc of 60°, we find the ratio of $\frac{1}{2}$ BE to DE smaller than 10, hence the ratio of AB to BC smaller than 20.

APPENDIX B

APOLLONIUS' DERIVATION OF THE PLANETS' STATIONS

The computation of the stations of the planets in the epicycle theory was reduced by Apollonius to the geometrical problem of drawing a line intersecting the epicycle in such a way that the sections have a definite ratio. Through the revolution of the entire epicycle towards the left side the planet situated on its circumference moves from point 1 to point 2 over a distance

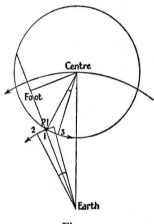

Fig. 39

equal to the angular velocity of this revolution times the distance earth – planet. At the same time the planet on its epicycle moves toward the right side, from point 1 to point 3 over a distance equal to the angular velocity on the epicycle times the radius of the epicycle. The latter displacement from the earth is seen foreshortened in the same ratio as exists between the distances planet – footpoint of the perpendicular and centre – planet. The planet seems to be at rest when the two displacements are seen from the earth to compensate. This will happen when the two distances planet – footpoint and planet – earth have the inverse ratio of the angular velocities. Or in other words: the planet has a station when the distance earth – planet and half the chord in the epicycle have the same ratio as the period of revolution and the synodic period.

APPENDIX C

NEWTON'S DEMONSTRATION OF THE LAW OF AREAS

In his Proposition I Newton derives Kepler's Law of Areas from the supposition that a revolving body is subject to a centripetal force directed to a

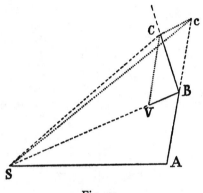

Fig. 40

fixed point. For the demonstration he makes use of equal finite time intervals, after each of which the force gives a finite impulse to the body towards the centre S. During one interval, from A to B, the motion remains the same; then at B it is suddenly changed by the additional motion BV. In the second interval, instead of continuing the motion along Bc = AB the body follows the resulting path BC. Since, geometrically, the areas SAB and SBc are equal and because the impulse BV is directed toward S, the areas SBc and SBC are equal, the area SBC is equal to SAB. This holds for every further interval; every next triangular area is equal to the preceding one. So all the triangular areas described in equal time intervals are equal, and they will lie all in the same place. This holds also when the time intervals are taken ever smaller and their number ever greater in the same rate. Then, finally, we have a continuously acting force and a curved orbit, for which the areas described are proportional to the time used.

APPENDIX D

NEWTON'S DERIVATION OF THE
FORCE OF ATTRACTION

In his Proposition XI (Cajori, p. 56), for the case of an elliptical orbit, Newton derives the centripetal force from Kepler's laws. In the figure of the orbit, where the lines are indicated by small letters, the planet is at P and the

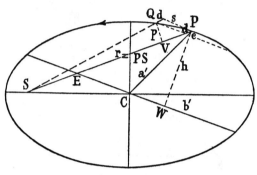

Fig. 41

sun occupies the focus S. The small deviation of the planet's motion from the tangent towards the focus is denoted by e, the motion itself by s. The deviation is proportional to the force and to the square of the time: so the force is

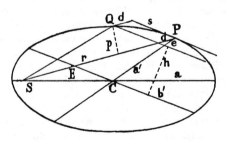

Fig. 42

found by dividing the deviation e by the square of the time interval. Because of the law of areas the time interval may be replaced by the area described by the radius vector. By taking the time interval and the motion s ever smaller

500

the area can be rendered ever more exactly by the triangle SPQ. Newton derives that PE $=a$, half the major axis.

Dropping a perpendicular p from Q to the radius vector and a perpendicular h from P to b' ($2a'$ and $2b'$ are conjugate diameters), we have area SPQ $=\frac{1}{2}rP$; because of the similarity of the triangles PQV and PEW, we have $p : h = s : a$ or $p = h \times \frac{s}{a}$; $h = ab : b'$; hence for the area we find $\frac{1}{2}rs$ $(b : b')$.

The deviation e is related to the small distance d on the diameter by $d : e = a' : a$. For the distance e we have, by considering the ellipse as the projection of a circle, the relation $e \times 2a' : s^2 = a'^2 : b'^2$

so that
$$e = \frac{1}{2}\frac{a'}{b'^2}s^2; \quad d = \frac{1}{2}as^2 : b'^2$$

Then the force is
$$\frac{1}{2}\frac{as^2}{b'^2} : \frac{1}{4}r^2 s^2 \frac{b^2}{b'^2} = \frac{2a}{b^2}\frac{1}{r^2}.$$

Here all the quantities depending on the position of the planet in its orbit have disappeared. So for all the points of the ellipse the force is as the inverse square of the distance to the sun.

REFERENCES

1. F. K. Ginzel, *Handbuch der mathematischen und technischen Chronologie*, II (Leipzig, 1911), p. 21.
2. *Ibid.*, I (Leipzig, 1906), p. 245*n*.
3. *Ibid.*, p. 245.
4. F. X. Kugler, *Sternkunde und Sterndienst in Babel*, II (Münster, 1909), p. 253.
5. Ch. Virolleaud, *L'Astrologie chaldéenne* (Paris, 1908–12) I, *Ishtar*, XXVI, pp. 1–9; also Kugler, *op. cit.*, III (1913), p. 169.
6. L. W. King, *Babylonian Religion and Mythology* (London, 1903), p. 79.
7. Kugler, *op. cit.*, II, p. 259.
8. E. F. Weidner, *Alter und Bedeutung der babylonischen Astronomie*, etc. (Leipzig, 1914), pp. 91–2.
9. *American Journal of Sciences and Arts*, 1855, p. 273.
10. G. Smith, *Assyrian Discoveries* (London, 1875), p. 405; also *Monthly Notices of the Royal Astronomical Society*, XXXIX, p. 454.
11. Thureau Dangin, *Die Sumerischen und Akkadischen Königsinschriften* (1907), pp. 88ff; also *Revue biblique*, LV, No. 3 (1948), p. 403 cylindre A (4–6).
12. Kugler, *op. cit.*, II, pp. 71–2; III, p. 315.
13. Virolleaud, *op. cit.*, Sin XIX, pp. 19–20; also Weidner, *op. cit.*, p. 23.
14. Kugler, *op. cit.*, II, p. 36.
15. *Ibid.*, p. 21.
16. *Ibid.*, p. 77.
17. *Ibid.*, I (1907), p. 64.
18. *Ibid.*, p. 45.
19. *Ibid.*, p. 71.
20. *Ibid.*, p. 77.
21. *Ibid.*, p. 59.
22. *Ibid.*, p. 91.
23. *Ibid.*, p. 97.
24. J. N. Strassmaier and J. Epping, 'Ein babylonischer Saros-Canon', *Zeitschrift für Assyriologie*, VIII (1893), p. 176; X (1895), p. 67.
25. Kugler, *op. cit.*, I, p. 152.
26. *Ibid.*, II, pp. 524–30.
27. Kugler, *Die babylonische Mondrechnung* (Freiburg, 1900), pp. 75–82.
28. O. Neugebauer, *Untersuchungen zur antiken Astronomie*, II ('Quellen und Studien zur Geschichte der Mathematik, etc.,' Abt. B., Band IV, Berlin, 1937), p. 84.
29. O. Neugebauer, 'The History of Ancient Astronomy', *Journal of Near Eastern Studies*, IV (1945), p. 12.
30. Kugler, *Die babylonische Mondrechnung*, pp. 9–10.
31. H. Maspero, *L'astronomie chinoise avant les Han* (T'oung Pao, XXVI, Paris, 1929), p. 288.
32. *Ibid.*, p. 299.
33. *Ibid.*, p. 312.
34. *Ibid.*, p. 336.
35. W. Eberhard and R. Henseling, 'Beiträge zur Astronomie der Han Zeit', *Sitzungsberichte der Preussischen Akademie der Wissenschaften, Phil.-Hist. Klasse*, V (1933) and XXIII (1933).
36. See T'oung Pao, XXXVI (1939), p. 174.
37. *Iliad*, XXII, 26–31.
38. *Odyssey*, V, 271–7.
39. B. Farrington, 'The Character of Early Greek Science, *Proceedings of the Royal Institute*, XXXIII, Part II, No. 151 (1945), p. 291.
40. A. Fairbanks, *The First Philosophers of Greece* (London, 1898), p. 33.
41. See also O. Neugebauer, *The Exact Sciences in Antiquity* (Copenhagen, 1951), p. 136.
42. Plato, *Republic*, VII, A-B, 529C–530B.
43. Plato, *Epinomis*, 990.

44. Simplicius, *Commentary to De Coelo*, II, 12.
45. Plato, *Timaeus*, 40, A.
46. T. L. Heath, *Aristarchus of Samos* (Oxford, 1913), p. 153.
47. Plato, *Timaeus*, 36, C-D.
48. *Ibid.*, 38, C-D.
49. *Ibid.*, 39, C-D.
50. Plato, *Epinomis*, 987.
51. Plato, *Timaeus*, 40, B-C.
52. Aristotle, *De coelo*, II, 13.
53. Geminus, *Elementa astronomiae*, cap. 8.
54. Aristotle, *De coelo*, I, 2.
55. Aristotle, *Physica*, VIII, 10.
56. Simplicius, Commentary to *De coelo*, II, cap. 8.
57. *Ibid.*, II, 13.
58. Proclus, *In Timaeum*, 281, E-F.
59. Aëtius, *Placita*, III, 13, 3 (Diels, *Doxigraphi Graeci*, Berlin, 1879), p. 378.
60. Heath, *op. cit.*, pp. 352–4.
61. *Ibid.*, p. 302; see also Heath, *Greek Astronomy* (London, 1932), p. 106.
62. Ptolemy, *Mathematikè Suntaxis*, III, 1 (Hei. I, 206–7; Man. I, 145) and VII, 2 (Hei. II, 12; Man. II, 12). [*Hei* gives page in edition of J. L. Heiberg (1898, 1903). *Man* gives page in K. Manitius, *Ptolemaeus Handbuch der Astronomie* (1912–13)].
63. *Ibid.*, III, 1 (Hei I, 207; Man. I, 145).
64. *Ibid.*, VII, 2 (Hei. II, 12; Man. II, 12).
65. *Ibid.*, VII, 2 (Hei. II, 15–16; Man. II, 15).
66. See Manitius, *op. cit.*, I Anhang, p. 440–1.
67. Rostovtzeff, *Mystic Italy* (New York, 1927), p. 8.
68. P. Tannery, *Recherches sur l'histoire de l'astronomie* (Paris, 1893), p. 128.
69. Plutarch, *De facie in orbe lunae*, cap. VI, 34.
70. Ptolemy, *op. cit.*, IX, 2 (Hei. II, 211; Man. II, 97).
71. *Ibid.*, IX, 5 (Hei. II, 251–2; Man. II, 120–1).
72. *Ibid.*, X, 6 (Hei. II, 317; Man. II, 172–6).
73. *Ibid.*, X, 8 (Hei. II, 347–8; Man. II, 195–6).
74. *Ibid.*, XIII, 2 (Hei. II, 532–3; Man. II, 333–4).
75. Suetonius, *De vita Caesarum*, I, 40.
76. Ptolemy, *op. cit.*, I, 7 (Hei. I, 24–6; Man. I, 19–20).
77. *Ibid.*, III, 4 (Hei. I, 233–4; Man. I, 167).
78. *Ibid.*, III, 1 (Hei. I, 205; Man. I, 143).
79. *Ibid.*, VII, 2 (Hei. II, 13; Man. II, 13).
80. J. B. J. Delambre, *Histoire de l'astronomie ancienne* (Paris, 1817), Discours préliminaire, XXV, XXXIV.
81. Ptolemy, *op. cit.*, V, 12 (Hei. I, 408; Man. I, 299).
82. *Ibid.*, V, 14 (Hei. I, 421; Man. I, 309–10).
83. Kugler, *Sternkunde*, etc., III, 1913 (Ergänzungen, p. 7).
84. Ptolemy, *op. cit.*, VIII, 3 (Hei. II, 180–2; Man. II, 72–4).
85. Ptolemy, *Tetrabiblos*, I, 2.
86. Al Battani, *Opus astronomicum* (Publ. d. R. Oss. Brera, Milano), Vol. XL, Cap. I, 5.
87. H. Pirenne, *Mahomet et Charlemagne* (Paris, 1937), p. 25 (Engl. edn. George Allen & Unwin, London, 1939).
88. *Die astronomischen Tafeln des Mohammed Ibn Musa al Khwarizmi*, ed. H. Suter (Mémoires de l'Académie, Copenhagen, 1914, p. 17).
89. See J. Bensaude, *L'Astronomie nautique au Portugal*, etc. (Berne, 1912).
90. A. von Humboldt, *Kosmos* (Stuttgart, 1850), III, Cap. VII, 360, note 44.
91. W. E. Peuckert, *Nikolaus Kopernikus* (Leipzig n.d.), p. 33.
92. N. Copernicus, *De revolutionibus orbium caelestium* (ed. Thorn, 1873), I, Cap. 10.
93. *Ibid.*, V, cap. 2.
94. *cf. Bulletin of the Astron. Institutes of the Netherlands*, No. 366, X, 68.
95. Copernic op. cit., V, cap. 15.

96. *Ibid.*, cap. 30.
97. *Ibid.*, VI, cap. 1.
98. *Ibid.*, III, cap. 25.
99. *Die astronomischen Tafeln des Al Khwarizmi*, p. XXIII.
100. *Tychonis Brahei Dani, Opera omnia*, I, p. 153.
101. *Ibid.*, I, p. 156.
102. *Ibid.*, II, p. 16.
103. See L. Prowe, *Nicolaus Coppernicus* (Berlin, 1883), Book I, Vol. II, p. 232.
104. P. Melanchthon, *Doctrinae physicae elementa* (Basel, 1550), Liber I, p. 56.
105. E. Wohlwill, *Galilei und sein Kampf für die Copernicanische Lehre*, I (Leipzig, 1909), pp. 19–20.
106. Tycho, *op. cit.*, VI, p. 217.
107. *Opere di Galilei* ('Ediz. Nationale'), X, p. 68.
108. *Ibid.*, X, pp. 69–70.
109. *Ibid.*, XI, p. 33.
110. F. Plassmann, *Himmelskunde* (Münster, 1898), p. 605.
111. J. Kepler, *Neue Astronomie*, trans. Max Caspar (München, 1929).
112. *Ibid.*, cap. 16, p. 147.
113. *Ibid.*, cap. 18, p. 159.
114. *Ibid.*, cap. 19, p. 163.
115. *Ibid.*, cap. 19, p. 166.
116. *Ibid.*, cap. 40, p. 246.
117. *Ibid.*, cap. 44, p. 267.
118. *Ibid.*, cap. 56, p. 325.
119. J. Kepler, *Opera omnia*, ed. Frisch (Frankfurt, 1853–71), VI, pp. 13–14.
120. *Ibid.*, V, p. 269.
121. B. Farrington, *Francis Bacon* (London, 1951), p. 6.
122. R. Descartes, *Principia philosophiae* (Amsterdam, 1685), III, p. 4.
123. J. S. Bailly, *Histoire de l'astronomie moderne*, II (Paris, 1779), p. 179.
124. *Lettres de M. Auzout sur les grandes lunettes* (Paris, 1735), pp. 57–58.
125. Weld, *History of the Royal Society*, I (London, 1848), p. 46.
126. *Ibid.*, II, p. 465.
127. Christiaan Huygens, 'Systema Saturnium' (*Oeuvres Complètes*), XV, p. 299.
128. *Ibid.*, XV, p. 15.
129. *Ibid.*, 'Cosmotheoros', XXI, 680.
130. *Ibid.*, XXI, pp. 704–6.
131. Fontenelle, *Entretiens sur la pluralité des mondes*, 2e soir (Ed. Dijon, an 2), p. 66.
132. J. Kepler-Max Caspar, *Neue Astronomie* (München, 1929), p. 26.
133. *Ibid.*, p. 28.
134. I. Newton, *Mathematical Principles*, trans. F. Cajori (Berkeley, 1934), p. 397.
135. *Ibid.*, p. XVII.
136. *Ibid.*, p. 13.
137. *Ibid.*, p. 5.
138. *Ibid.*, pp. 2–3.
139. *Ibid.*, p. 41.
140. *Ibid.*, p. 408.
141. *Ibid.*, p. 410.
142. Christiaan Huygens, *Traité de la lumière, et Discours de la cause de la pesanteur*, ed. Burckhardt (Leipzig, 1885), pp. 112–13.
143. I. Newton, *op. cit.*, p. 421.
144. *Ibid.*, p. 422.
145. L. T. More, *Isaac Newton, a Biography* (New York, 1934), pp. 446–50.
146. Newton, *op. cit.*, 395–96.
147. Chr. Huygens, *op. cit.*, p. 118.
148. *Ibid.*, p. 118.
149. *Ibid.*, p. 119.
150. *Ibid.*, p. 121.

151. Chr. Huygens, *Oeuvres complètes*, IX, Letters 2628, p. 523.
152. I. Newton, *Opera* (1779–85), IV, p. 394.
153. *Ibid.*, p. 430–32.
154. *Ibid.*, p. 438.
155. I. Newton, *Mathematical Principles*, p. 456.
156. *Ibid.*, p. 457.
157. R. Grant, *History of Physical Astronomy* (London, 1852), p. 460.
158. Halley, *Philosophical Transactions*, XXX (1718), p. 737.
159. Delambre, *Historie de l'astronomie au 18ᵉ siècle* (Paris, 1827), p. 420.
160. Chauvenet, *Spherical and Practical Astronomy* (Philadelphia, 1885), II, p. 92.
161. See Lynn Thorndike, *A Short History of Civilization* (New York, 1933), p. 476.
162. L. T. More, *op. cit.*, p. 75.
163. *Ibid.* (Letter of February 23, 1668/9), p. 68.
164. Voltaire, *Lettres sur les Anglais* (*Oeuvres de Voltaire*, 1819), XXIV, p. 67.
165. *cf.* F. F. Tisserand, *Traité de Mécanique Céleste*, III (Paris, 1894), p. 64.
166. *Ibid.*, III, p. 76.
167. *Exposition du système du Monde* (*Oeuvres de Laplace*, VI, 1842), IV, ch. 2, p. 228.
168. *Mécanique céleste* (*Oeuvres de Laplace*, V), p. 324.
169. R. Grant, *op. cit.*, p. 60.
170. Clairaut, *Théorie du mouvement des comètes* (Paris, 1760), p. 11.
171. *Ibid.*, p. IV.
172. R. Grant, *op. cit.*, p. 61.
173. *Exposition du système du Monde*, IV, Ch. 5; p. 264.
174. Kugler, III, Ergänzungen (1919–35), p. 352.
175. W. Herschel, 'Account of a comet', *Philosophical Transactions*, LXXI (1781), p. 492.
176. W. Herschel, *Scientific Papers*, I, p. 91.
177. *Ibid.*, I, p. 92.
178. *Ibid.*, I, p. 90.
179. *Ibid.*, I, p. 158.
180. *Ibid.*, I, p. 223.
181. *Ibid.*, I, p. 245.
182. Constance Lubbock, *The Herschel Chronicle* (Cambridge, 1933), p. 197.
183. William Herschel, *Scientific Papers*, I, p. 422.
184. *Ibid.*, I, p. 424.
185. *Ibid.*, II, p. 213.
186. *Ibid.*, II, p. 487.
187. *Ibid.*, II, p. 533.
188. J. S. Bailly, *Histoire de l'astronomie moderne*, III (Paris, 1785), p. 60.
189. F. W. Bessel, 'Ueber die Verbindung der astronomischen Beobachtungen mit der Astronomie', *Populäre Vorlesungen* (Hamburg, 1848), p. 432.
190. J. C. Kapteyn, 'On the Motion of the Faint Stars and the Systematic Errors of the Boss Fundamental System', *Bulletin of the Astr. Inst. of the Netherlands*, No. 14, I (1922), p. 71.
191. J. Fraunhofer, *Bestimmung des Brechungs- und Farbenzerstreuungsvermögens*, etc. 'Denkschr. der Akademie München' (1814–15, Band V), p. 193; also: Ostwald's *Klassiker der exacten Wissenschaften*, No. 150 (Leipzig, 1905), p. 1.
192. Fraunhofer, *op. cit.*, V, 206; also Ostwald, *op. cit.*, p. 15.
193. J. F. W. Herschel, 'President's Report, 1840', *Monthly Notices of the R.A.S.*, V, p. 97.
194. F. Kaiser, *De Geschiedenis der Ontdekkingen van Planeten* (Amsterdam, 1851), p. 139.
195. S. Newcomb, *Reminiscences of an Astronomer* (London, New York, 1903), p. 221.
196. F. F. Tisserand, *Traité de mécanique céleste*, III (Paris, 1894), p. 424.
197. S. C. Chandler, 'On the Variation of Latitude', *Astronomical Journal*, XI (1891), p. 59.
198. W. H. Pickering, 'Erathostenes II', *Popular Astronomy*, XXIX (1924), pp. 404–423.
199. W. Beer and J. H. Mädler, *Beiträge zur physischen Kenntniss der himmlischen Körper im Sonnensystem* (Weimar, 1841), p. 111; also *Astron. Nachr*, VIII (1830), p. 148.
200. Camille Flammarion, *La planète Mars et ses conditions d'habitabilité* (Paris, 1892), p. 592.
201. *Ibid.*, p. 588.
202. *Ibid.*, p. 591.

203. J. H. Jeans, *The Universe Around Us* (Cambridge, 1929), p. 332.
204. A. S. Eddington, *The Nature of the Physical World* (Cambridge, 1929), p. 178.
205. W. Herschel, *Scientific Papers*, I, p. 480.
206. A. N. Vyssotsky, 'Astronomical Records in the Russian Chronicles', *Meddelanden Lund Observatory*, No. 126, II (1949), p. 9.
207. Auguste Comte, *Cours de philosophie positive*, II (Paris, 1835), 8 and 12, 19ᵉ Leçon.
208. C. A. Young, *The Sun* (London, 1882), p. 290.
209. J. C. Janssen, 'Indication de quelques-uns des résultats, etc.', *Comptes Rendus de l'Acad. d. Sciences*, LXVII (1868), p. 839.
210. W. Olbers, 'Ueber den Schweif des grossen Cometen von 1811' (*Monatliche Correspondenz*, XXV (January 1812), p. 12.
211. *Ibid.*, p. 14.
212. E. Hertzsprung, 'Ueber die Sterne der Unterabteilungen *c* und *ac*, etc.', *Astron. Nachrichten*, No. 4296 (1908), CLXXIX, p. 373.
213. H. Shapley and H. D. Curtis, 'The Scale of the Universe' (*Bulletin National Research Council*, II (Part 3, No. 11) (1921), p. 192.
214. See M. L. Humason, 'The Apparent Radial Velocities of 100 Extragalactic Nebulae', *Astrophysical Journal*, LXXXIII (1936), pp. 12, 16.

INDEX

Abbe, Ernst, 334
Abd al-Raḥmân ibn 'Umar, 168
Aberration (of light), 290, 344, 369; spherical (of lenses), 254; chromatic (of lenses), 294–5, 330–1, 336
Abû'l Wefa, 168
Abû-Ma'shar, 166, 174
Abundances of elements, 416, 460, 495
Academies, scientific, 250–1
Achromatic objective, 294–5, 330–1, 336
Acronychian rising, setting, 22
Actinometry, 447
Adams, John Coach, 360–1, 366, 420
Adams, Walter S., 391, 458, 466
Aëtius, 98, 117
Africans and calendar reckoning, 21–2
Age of the universe, 488–9
Agrippa, 145, 149
Agriculture and calendar, 21–6, 82–4, 95, 97
Airy, Sir George Biddell, 326, 345, 360
Albategnius (Al-Battâni), 167–8
Albedo of planets, 384–6
Albert of Saxony, 178
Albertus Magnus, 175, 176
Al-Biruni, 27
Al-Bitrûjî, 170
Albrecht, Seb., 439
Albumazar, 166, 174
Alcmaeon, 100
Aldebaran (star), 32, 446, 450
Alembert, Jean le Rond d', 298, 303
Alexandria, 63, 122–4, 130
Al-Farghânî, 166
Alfonsine tables, 168–9, 180, 189, 198, 206
Alfonso X, 168
Alfraganus, 166
Algol (star), 174, 311, 317, 319, 435–8, 444
Alhazen, 170
Al-Khwârizmî, tables of, 165, 168, 170, 174
Almagest (Ptolemy), 146–59, 166, 175, 180
Al-Ma'mûn, 166
Almanacs, 108, 132, 185, 187; nautical, 279
Alpetragius, 170
Al-Ṣûfî, 168
Al-Ṭûsî, 169
Al-'Urdî, 169
Amateur astronomy, 405, 436, 440
Amenhotep IV (Ikhnaton), 84
Ames, Miss, 487
Ammonius of Alexandria, 163
Anaxagoras, 100, 114
Anaximander of Milesia, 98
Anaximenes, 99
Anderson, J. A., 464

Andrews, Thom., 408
Andromeda nebula, 312, 451, 486, 495; Pl. 22
Ångström, A. J., 406, 407
Anomalistic period of moon, 75, 128, 158
Antoniadi, E. M., 379–80; Pl. 12
Apex of sun, 315, 417, 480
Apianus, P., 423
Apollonius of Perga, 134–5, 180; derivation of the planet's stations, 498
Arabians, 20, 21, 26, 27, 164–70; numeral system, 166, 200
Arago, D. F. J., 360
Aratus, 130, 132
Archimedes of Syracuse, 120–1, 132, 175, 180
Arcturus (star), 32, 157, 163, 450
'Areas, law of' (Kepler), 240, 499
Argelander, Friedrich, 320, 435, 437, 439, 440, 441, 445, 467–8, 471, 474
Aristarchus of Samos: distance and size of the sun, 118–20, 283, 497; heliocentric theory, 116, 121, 134, 147; *see also* 125
Aristotle, 97–8, 223, 227, 228, 231, 233, 247–9; theories of world structure, 113–16, 118, 207; the church and, 175, 230–1; theory of motion, 114, 178, 223, 224, 227, 233
Aristyllus, 124, 126, 150
Arkwright, Richard, 292
Armillas, 91, 93, 129, 137, 151–3, 167, 182–3, 201, 210, 214
Arrhenius, Svante, 424
Arzachel, 168, 197
Aspects of planets, 160
Assyria, Assyrian astronomy, 24, 29, 36–48, 52
Asten, E. von, 358
Astral mythology, 30
Asteroids: *see* Planetoids
Astrolabe, 93, 163, 167–8, 174, 178–9, 185, 201
Astrolabon, 137, 151–3
Astrology: Assyrian and Babylonian, 33–47; Arabian, 166–9; Chinese, 89–92; Greek, Roman, 130–2; Hellenistic, 160–1; Medieval, 176–7; Renaissance, 185–7; Tycho Brahe and, 204–6; Kepler, 235–6; and the Church, 176–7
Astronomical societies, 250–2
Astronomical unit, 341, 348
Astronomische Gesellschaft, 326, 337
Astronomer Royal, 279
Astronomy, origin, 19
Astrophysics, 395–418

A CATALOG OF SELECTED
DOVER BOOKS
IN SCIENCE AND MATHEMATICS

A CATALOG OF SELECTED
DOVER BOOKS
IN SCIENCE AND MATHEMATICS

QUALITATIVE THEORY OF DIFFERENTIAL EQUATIONS, V.V. Nemytskii and V.V. Stepanov. Classic graduate-level text by two prominent Soviet mathematicians covers classical differential equations as well as topological dynamics and ergodic theory. Bibliographies. 523pp. 5⅜ x 8½. 65954-2 Pa. $14.95

MATRICES AND LINEAR ALGEBRA, Hans Schneider and George Phillip Barker. Basic textbook covers theory of matrices and its applications to systems of linear equations and related topics such as determinants, eigenvalues and differential equations. Numerous exercises. 432pp. 5⅜ x 8½. 66014-1 Pa. $12.95

QUANTUM THEORY, David Bohm. This advanced undergraduate-level text presents the quantum theory in terms of qualitative and imaginative concepts, followed by specific applications worked out in mathematical detail. Preface. Index. 655pp. 5⅜ x 8½. 65969-0 Pa. $15.95

ATOMIC PHYSICS (8th edition), Max Born. Nobel laureate's lucid treatment of kinetic theory of gases, elementary particles, nuclear atom, wave-corpuscles, atomic structure and spectral lines, much more. Over 40 appendices, bibliography. 495pp. 5⅜ x 8½. 65984-4 Pa. $13.95

ELECTRONIC STRUCTURE AND THE PROPERTIES OF SOLIDS: The Physics of the Chemical Bond, Walter A. Harrison. Innovative text offers basic understanding of the electronic structure of covalent and ionic solids, simple metals, transition metals and their compounds. Problems. 1980 edition. 582pp. 6⅛ x 9¼. 66021-4 Pa. $19.95

BOUNDARY VALUE PROBLEMS OF HEAT CONDUCTION, M. Necati Özisik. Systematic, comprehensive treatment of modern mathematical methods of solving problems in heat conduction and diffusion. Numerous examples and problems. Selected references. Appendices. 505pp. 5⅜ x 8½. 65990-9 Pa. $12.95

A SHORT HISTORY OF CHEMISTRY (3rd edition), J.R. Partington. Classic exposition explores origins of chemistry, alchemy, early medical chemistry, nature of atmosphere, theory of valency, laws and structure of atomic theory, much more. 428pp. 5⅜ x 8½. (Available in U.S. only) 65977-1 Pa. $12.95

A HISTORY OF ASTRONOMY, A. Pannekoek. Well-balanced, carefully reasoned study covers such topics as Ptolemaic theory, work of Copernicus, Kepler, Newton, Eddington's work on stars, much more. Illustrated. References. 521pp. 5⅜ x 8½. 65994-1 Pa. $12.95

PRINCIPLES OF METEOROLOGICAL ANALYSIS, Walter J. Saucier. Highly respected, abundantly illustrated classic reviews atmospheric variables, hydrostatics, static stability, various analyses (scalar, cross-section, isobaric, isentropic, more). For intermediate meteorology students. 454pp. 6⅛ x 9¼. 65979-8 Pa. $14.95

RELATIVITY, THERMODYNAMICS AND COSMOLOGY, Richard C. Tolman. Landmark study extends thermodynamics to special, general relativity; also applications of relativistic mechanics, thermodynamics to cosmological models. 501pp. 5⅜ x 8½. 65383-8 Pa. $15.95

APPLIED ANALYSIS, Cornelius Lanczos. Classic work on analysis and design of finite processes for approximating solution of analytical problems. Algebraic equations, matrices, harmonic analysis, quadrature methods, much more. 559pp. 5⅜ x 8½. 65656-X Pa. $16.95

INTRODUCTION TO ANALYSIS, Maxwell Rosenlicht. Unusually clear, accessible coverage of set theory, real number system, metric spaces, continuous functions, Riemann integration, multiple integrals, more. Wide range of problems. Undergraduate level. Bibliography. 254pp. 5⅜ x 8½. 65038-3 Pa. $9.95

INTRODUCTION TO QUANTUM MECHANICS With Applications to Chemistry, Linus Pauling & E. Bright Wilson, Jr. Classic undergraduate text by Nobel Prize winner applies quantum mechanics to chemical and physical problems. Numerous tables and figures enhance the text. Chapter bibliographies. Appendices. Index. 468pp. 5⅜ x 8½. 64871-0 Pa. $12.95

ASYMPTOTIC EXPANSIONS OF INTEGRALS, Norman Bleistein & Richard A. Handelsman. Best introduction to important field with applications in a variety of scientific disciplines. New preface. Problems. Diagrams. Tables. Bibliography. Index. 448pp. 5⅜ x 8½. 65082-0 Pa. $13.95

MATHEMATICS APPLIED TO CONTINUUM MECHANICS, Lee A. Segel. Analyzes models of fluid flow and solid deformation. For upper-level math, science and engineering students. 608pp. 5⅜ x 8½. 65369-2 Pa. $14.95

ELEMENTS OF REAL ANALYSIS, David A. Sprecher. Classic text covers fundamental concepts, real number system, point sets, functions of a real variable, Fourier series, much more. Over 500 exercises. 352pp. 5⅜ x 8½. 65385-4 Pa. $11.95

PHYSICAL PRINCIPLES OF THE QUANTUM THEORY, Werner Heisenberg. Nobel Laureate discusses quantum theory, uncertainty, wave mechanics, work of Dirac, Schroedinger, Compton, Wilson, Einstein, etc. 184pp. 5⅜ x 8½. 60113-7 Pa. $7.95

INTRODUCTORY REAL ANALYSIS, A.N. Kolmogorov, S.V. Fomin. Translated by Richard A. Silverman. Self-contained, evenly paced introduction to real and functional analysis. Some 350 problems. 403pp. 5⅜ x 8½. 61226-0 Pa. $11.95

PROBLEMS AND SOLUTIONS IN QUANTUM CHEMISTRY AND PHYSICS, Charles S. Johnson, Jr. and Lee G. Pedersen. Unusually varied problems, detailed solutions in coverage of quantum mechanics, wave mechanics, angular momentum, molecular spectroscopy, scattering theory, more. 280 problems plus 139 supplementary exercises. 430pp. 6½ x 9¼. 65236-X Pa. $14.95

ASYMPTOTIC METHODS IN ANALYSIS, N.G. de Bruijn. An inexpensive, comprehensive guide to asymptotic methods–the pioneering work that teaches by explaining worked examples in detail. Index. 224pp. 5⅜ x 8½. 64221-6 Pa. $7.95

OPTICAL RESONANCE AND TWO-LEVEL ATOMS, L. Allen and J. H. Eberly. Clear, comprehensive introduction to basic principles behind all quantum optical resonance phenomena. 53 illustrations. Preface. Index. 256pp. 5⅜ x 8½. 65533-4 Pa. $8.95

COMPLEX VARIABLES, Francis J. Flanigan. Unusual approach, delaying complex algebra till harmonic functions have been analyzed from real variable viewpoint. Includes problems with answers. 364pp. 5⅜ x 8½. 61388-7 Pa. $10.95

ATOMIC SPECTRA AND ATOMIC STRUCTURE, Gerhard Herzberg. One of best introductions; especially for specialist in other fields. Treatment is physical rather than mathematical. 80 illustrations. 257pp. 5⅜ x 8½. 60115-3 Pa. $7.95

APPLIED COMPLEX VARIABLES, John W. Dettman. Step-by-step coverage of fundamentals of analytic function theory–plus lucid exposition of five important applications: Potential Theory; Ordinary Differential Equations; Fourier Transforms; Laplace Transforms; Asymptotic Expansions. 66 figures. Exercises at chapter ends. 512pp. 5⅜ x 8½. 64670-X Pa. $14.95

ULTRASONIC ABSORPTION: An Introduction to the Theory of Sound Absorption and Dispersion in Gases, Liquids and Solids, A.B. Bhatia. Standard reference in the field provides a clear, systematically organized introductory review of fundamental concepts for advanced graduate students, research workers. Numerous diagrams. Bibliography. 440pp. 5⅜ x 8½. 64917-2 Pa. $11.95

UNBOUNDED LINEAR OPERATORS: Theory and Applications, Seymour Goldberg. Classic presents systematic treatment of the theory of unbounded linear operators in normed linear spaces with applications to differential equations. Bibliography. 199pp. 5⅜ x 8½. 64830-3 Pa. $7.95

LIGHT SCATTERING BY SMALL PARTICLES, H.C. van de Hulst. Comprehensive treatment including full range of useful approximation methods for researchers in chemistry, meteorology and astronomy. 44 illustrations. 470pp. 5⅜ x 8½. 64228-3 Pa. $12.95

CONFORMAL MAPPING ON RIEMANN SURFACES, Harvey Cohn. Lucid, insightful book presents ideal coverage of subject. 334 exercises make book perfect for self-study. 55 figures. 352pp. 5⅝ x 8¼. 64025-6 Pa. $11.95

OPTICKS, Sir Isaac Newton. Newton's own experiments with spectroscopy, colors, lenses, reflection, refraction, etc., in language the layman can follow. Foreword by Albert Einstein. 532pp. 5⅜ x 8½. 60205-2 Pa. $13.95

GENERALIZED INTEGRAL TRANSFORMATIONS, A.H. Zemanian. Graduate-level study of recent generalizations of the Laplace, Mellin, Hankel, K. Weierstrass, convolution and other simple transformations. Bibliography. 320pp. 5⅜ x 8½. 65375-7 Pa. $8.95

CATALOG OF DOVER BOOKS

THE ELECTROMAGNETIC FIELD, Albert Shadowitz. Comprehensive under-graduate text covers basics of electric and magnetic fields, builds up to electromagnetic theory. Also related topics, including relativity. Over 900 problems. 768pp. 5⅜ x 8¼. 65660-8 Pa. $19.95

FOURIER SERIES, Georgi P. Tolstov. Translated by Richard A. Silverman. A valu-able addition to the literature on the subject, moving clearly from subject to subject and theorem to theorem. 107 problems, answers. 336pp. 5⅜ x 8½. 63317-9 Pa. $10.95

THEORY OF ELECTROMAGNETIC WAVE PROPAGATION, Charles Herach Papas. Graduate-level study discusses the Maxwell field equations, radiation from wire antennas, the Doppler effect and more. xiii + 244pp. 5⅜ x 8½. 65678-0 Pa. $9.95

DISTRIBUTION THEORY AND TRANSFORM ANALYSIS: An Introduction to Generalized Functions, with Applications, A.H. Zemanian. Provides basics of distri-bution theory, describes generalized Fourier and Laplace transformations. Numerous problems. 384pp. 5⅜ x 8½. 65479-6 Pa. $13.95

THE PHYSICS OF WAVES, William C. Elmore and Mark A. Heald. Unique overview of classical wave theory. Acoustics, optics, electromagnetic radiation, more. Ideal as classroom text or for self-study. Problems. 477pp. 5⅜ x 8½.
64926-1 Pa. $14.95

CALCULUS OF VARIATIONS WITH APPLICATIONS, George M. Ewing. Applications-oriented introduction to variational theory develops insight and pro-motes understanding of specialized books, research papers. Suitable for advanced undergraduate/graduate students as primary, supplementary text. 352pp. 5⅜ x 8½.
64856-7 Pa. $9.95

A TREATISE ON ELECTRICITY AND MAGNETISM, James Clerk Maxwell. Important foundation work of modern physics. Brings to final form Maxwell's theo-ry of electromagnetism and rigorously derives his general equations of field theory. 1,084pp. 5⅜ x 8½. 60636-8, 60637-6 Pa., Two-vol. set $27.90

AN INTRODUCTION TO THE CALCULUS OF VARIATIONS, Charles Fox. Graduate-level text covers variations of an integral, isoperimetrical problems, least action, special relativity, approximations, more. References. 279pp. 5⅜ x 8½.
65499-0 Pa. $8.95

HYDRODYNAMIC AND HYDROMAGNETIC STABILITY, S. Chandrasekhar. Lucid examination of the Rayleigh-Benard problem; clear coverage of the theory of instabilities causing convection. 704pp. 5⅜ x 8¼. 64071-X Pa. $17.95

CALCULUS OF VARIATIONS, Robert Weinstock. Basic introduction covering isoperimetric problems, theory of elasticity, quantum mechanics, electrostatics, etc. Exercises throughout. 326pp. 5⅜ x 8½. 63069-2 Pa. $9.95

DYNAMICS OF FLUIDS IN POROUS MEDIA, Jacob Bear. For advanced stu-dents of ground water hydrology, soil mechanics and physics, drainage and irrigation engineering and more. 335 illustrations. Exercises, with answers. 784pp. 6⅛ x 9¼.
65675-6 Pa. $19.95

NUMERICAL METHODS FOR SCIENTISTS AND ENGINEERS, Richard Hamming. Classic text stresses frequency approach in coverage of algorithms, polynomial approximation, Fourier approximation, exponential approximation, other topics. Revised and enlarged 2nd edition. 721pp. 5⅜ x 8½.   65241-6 Pa. $16.95

THEORETICAL SOLID STATE PHYSICS, Vol. 1: Perfect Lattices in Equilibrium; Vol. II: Non-Equilibrium and Disorder, William Jones and Norman H. March. Monumental reference work covers fundamental theory of equilibrium properties of perfect crystalline solids, non-equilibrium properties, defects and disordered systems. Appendices. Problems. Preface. Diagrams. Index. Bibliography. Total of 1,301pp. 5⅜ x 8½. Two volumes.   Vol. I: 65015-4 Pa. $16.95
Vol. II: 65016-2 Pa. $16.95

OPTIMIZATION THEORY WITH APPLICATIONS, Donald A. Pierre. Broad spectrum approach to important topic. Classical theory of minima and maxima, calculus of variations, simplex technique and linear programming, more. Many problems, examples. 640pp. 5⅜ x 8½.   65205-X Pa. $17.95

THE CONTINUUM: A Critical Examination of the Foundation of Analysis, Hermann Weyl. Classic of 20th-century foundational research deals with the conceptual problem posed by the continuum. 156pp. 5⅜ x 8½.   67982-9 Pa. $6.95

ESSAYS ON THE THEORY OF NUMBERS, Richard Dedekind. Two classic essays by great German mathematician: on the theory of irrational numbers; and on transfinite numbers and properties of natural numbers. 115pp. 5⅜ x 8½.
21010-3 Pa. $6.95

THE FUNCTIONS OF MATHEMATICAL PHYSICS, Harry Hochstadt. Comprehensive treatment of orthogonal polynomials, hypergeometric functions, Hill's equation, much more. Bibliography. Index. 322pp. 5⅜ x 8½. 65214-9 Pa. $9.95

NUMBER THEORY AND ITS HISTORY, Oystein Ore. Unusually clear, accessible introduction covers counting, properties of numbers, prime numbers, much more. Bibliography. 380pp. 5⅜ x 8½.   65620-9 Pa. $10.95

THE VARIATIONAL PRINCIPLES OF MECHANICS, Cornelius Lanczos. Graduate level coverage of calculus of variations, equations of motion, relativistic mechanics, more. First inexpensive paperbound edition of classic treatise. Index. Bibliography. 418pp. 5⅜ x 8½.   65067-7 Pa. $14.95

COMBINATORIAL TOPOLOGY, P. S. Alexandrov. Clearly written, well-organized, three-part text begins by dealing with certain classic problems without using the formal techniques of homology theory and advances to the central concept, the Betti groups. Numerous detailed examples. 654pp. 5⅜ x 8½.   40179-0 Pa. $18.95

THEORETICAL PHYSICS, Georg Joos, with Ira M. Freeman. Classic overview covers essential math, mechanics, electromagnetic theory, thermodynamics, quantum mechanics, nuclear physics, other topics. First paperback edition. xxiii + 885pp. 5⅜ x 8½.   65227-0 Pa. $21.95

HANDBOOK OF MATHEMATICAL FUNCTIONS WITH FORMULAS, GRAPHS, AND MATHEMATICAL TABLES, edited by Milton Abramowitz and Irene A. Stegun. Vast compendium: 29 sets of tables, some to as high as 20 places. 1,046pp. 8 x 10½. 61272-4 Pa. $29.95

MATHEMATICAL METHODS IN PHYSICS AND ENGINEERING, John W. Dettman. Algebraically based approach to vectors, mapping, diffraction, other topics in applied math. Also generalized functions, analytic function theory, more. Exercises. 448pp. 5⅜ x 8¼. 65649-7 Pa. $12.95

A SURVEY OF NUMERICAL MATHEMATICS, David M. Young and Robert Todd Gregory. Broad self-contained coverage of computer-oriented numerical algorithms for solving various types of mathematical problems in linear algebra, ordinary and partial, differential equations, much more. Exercises. Total of 1,248pp. 5⅜ x 8½. Two volumes. Vol. I: 65691-8 Pa. $16.95
Vol. II: 65692-6 Pa. $16.95

TENSOR ANALYSIS FOR PHYSICISTS, J.A. Schouten. Concise exposition of the mathematical basis of tensor analysis, integrated with well-chosen physical examples of the theory. Exercises. Index. Bibliography. 289pp. 5⅜ x 8½. 65582-2 Pa. $10.95

INTRODUCTION TO NUMERICAL ANALYSIS (2nd Edition), F.B. Hildebrand. Classic, fundamental treatment covers computation, approximation, interpolation, numerical differentiation and integration, other topics. 150 new problems. 669pp. 5⅜ x 8½. 65363-3 Pa. $16.95

INVESTIGATIONS ON THE THEORY OF THE BROWNIAN MOVEMENT, Albert Einstein. Five papers (1905–8) investigating dynamics of Brownian motion and evolving elementary theory. Notes by R. Fürth. 122pp. 5⅜ x 8½. 60304-0 Pa. $5.95

CATASTROPHE THEORY FOR SCIENTISTS AND ENGINEERS, Robert Gilmore. Advanced-level treatment describes mathematics of theory grounded in the work of Poincaré, R. Thom, other mathematicians. Also important applications to problems in mathematics, physics, chemistry and engineering. 1981 edition. References. 28 tables. 397 black-and-white illustrations. xvii + 666pp. 6⅛ x 9¼. 67539-4 Pa. $17.95

AN INTRODUCTION TO STATISTICAL THERMODYNAMICS, Terrell L. Hill. Excellent basic text offers wide-ranging coverage of quantum statistical mechanics, systems of interacting molecules, quantum statistics, more. 523pp. 5⅜ x 8½. 65242-4 Pa. $13.95

STATISTICAL PHYSICS, Gregory H. Wannier. Classic text combines thermodynamics, statistical mechanics and kinetic theory in one unified presentation of thermal physics. Problems with solutions. Bibliography. 532pp. 5⅜ x 8½. 65401-X Pa. $14.95

ORDINARY DIFFERENTIAL EQUATIONS, Morris Tenenbaum and Harry Pollard. Exhaustive survey of ordinary differential equations for undergraduates in mathematics, engineering, science. Thorough analysis of theorems. Diagrams. Bibliography. Index. 818pp. 5⅜ x 8½. 64940-7 Pa. $19.95

STATISTICAL MECHANICS: Principles and Applications, Terrell L. Hill. Standard text covers fundamentals of statistical mechanics, applications to fluctuation theory, imperfect gases, distribution functions, more. 448pp. 5⅜ x 8½. 65390-0 Pa. $14.95

ORDINARY DIFFERENTIAL EQUATIONS AND STABILITY THEORY: An Introduction, David A. Sánchez. Brief, modern treatment. Linear equation, stability theory for autonomous and nonautonomous systems, etc. 164pp. 5⅜ x 8¼. 63828-6 Pa. $6.95

THIRTY YEARS THAT SHOOK PHYSICS: The Story of Quantum Theory, George Gamow. Lucid, accessible introduction to influential theory of energy and matter. Careful explanations of Dirac's anti-particles, Bohr's model of the atom, much more. 12 plates. Numerous drawings. 240pp. 5⅜ x 8½. 24895-X Pa. $7.95

THEORY OF MATRICES, Sam Perlis. Outstanding text covering rank, nonsingularity and inverses in connection with the development of canonical matrices under the relation of equivalence, and without the intervention of determinants. Includes exercises. 237pp. 5⅜ x 8½. 66810-X Pa. $8.95

GREAT EXPERIMENTS IN PHYSICS: Firsthand Accounts from Galileo to Einstein, edited by Morris H. Shamos. 25 crucial discoveries: Newton's laws of motion, Chadwick's study of the neutron, Hertz on electromagnetic waves, more. Original accounts clearly annotated. 370pp. 5⅜ x 8½. 25346-5 Pa. $11.95

INTRODUCTION TO PARTIAL DIFFERENTIAL EQUATIONS WITH APPLICATIONS, E.C. Zachmanoglou and Dale W. Thoe. Essentials of partial differential equations applied to common problems in engineering and the physical sciences. Problems and answers. 416pp. 5⅜ x 8½. 65251-3 Pa. $11.95

BURNHAM'S CELESTIAL HANDBOOK, Robert Burnham, Jr. Thorough guide to the stars beyond our solar system. Exhaustive treatment. Alphabetical by constellation: Andromeda to Cetus in Vol. 1; Chamaeleon to Orion in Vol. 2; and Pavo to Vulpecula in Vol. 3. Hundreds of illustrations. Index in Vol. 3. 2,000pp. 6⅛ x 9¼. 23567-X, 23568-8, 23673-0 Pa., Three-vol. set $46.85

CHEMICAL MAGIC, Leonard A. Ford. Second Edition, Revised by E. Winston Grundmeier. Over 100 unusual stunts demonstrating cold fire, dust explosions, much more. Text explains scientific principles and stresses safety precautions. 128pp. 5⅜ x 8½. 67628-5 Pa. $5.95

AMATEUR ASTRONOMER'S HANDBOOK, J.B. Sidgwick. Timeless, comprehensive coverage of telescopes, mirrors, lenses, mountings, telescope drives, micrometers, spectroscopes, more. 189 illustrations. 576pp. 5⅜ x 8¼. (Available in U.S. only) 24034-7 Pa. $13.95

SPECIAL FUNCTIONS, N.N. Lebedev. Translated by Richard Silverman. Famous Russian work treating more important special functions, with applications to specific problems of physics and engineering. 38 figures. 308pp. 5⅜ x 8½. 60624-4 Pa. $9.95

THE EXTRATERRESTRIAL LIFE DEBATE, 1750–1900, Michael J. Crowe. First detailed, scholarly study in English of the many ideas that developed between 1750 and 1900 regarding the existence of intelligent extraterrestrial life. Examines ideas of Kant, Herschel, Voltaire, Percival Lowell, many other scientists and thinkers. 16 illustrations. 704pp. 5⅜ x 8½. 40675-X Pa. $19.95

INTEGRAL EQUATIONS, F.G. Tricomi. Authoritative, well-written treatment of extremely useful mathematical tool with wide applications. Volterra Equations, Fredholm Equations, much more. Advanced undergraduate to graduate level. Exercises. Bibliography. 238pp. 5⅜ x 8½. 64828-1 Pa. $8.95

POPULAR LECTURES ON MATHEMATICAL LOGIC, Hao Wang. Noted logician's lucid treatment of historical developments, set theory, model theory, recursion theory and constructivism, proof theory, more. 3 appendixes. Bibliography. 1981 edition. ix + 283pp. 5⅜ x 8½. 67632-3 Pa. $8.95

MODERN NONLINEAR EQUATIONS, Thomas L. Saaty. Emphasizes practical solution of problems; covers seven types of equations. ". . . a welcome contribution to the existing literature...."–*Math Reviews.* 490pp. 5⅜ x 8½. 64232-1 Pa. $13.95

FUNDAMENTALS OF ASTRODYNAMICS, Roger Bate et al. Modern approach developed by U.S. Air Force Academy. Designed as a first course. Problems, exercises. Numerous illustrations. 455pp. 5⅜ x 8½. 60061-0 Pa. $12.95

INTRODUCTION TO LINEAR ALGEBRA AND DIFFERENTIAL EQUATIONS, John W. Dettman. Excellent text covers complex numbers, determinants, orthonormal bases, Laplace transforms, much more. Exercises with solutions. Undergraduate level. 416pp. 5⅜ x 8½. 65191-6 Pa. $11.95

INCOMPRESSIBLE AERODYNAMICS, edited by Bryan Thwaites. Covers theoretical and experimental treatment of the uniform flow of air and viscous fluids past two-dimensional aerofoils and three-dimensional wings; many other topics. 654pp. 5⅜ x 8½. 65465-6 Pa. $16.95

INTRODUCTION TO DIFFERENCE EQUATIONS, Samuel Goldberg. Exceptionally clear exposition of important discipline with applications to sociology, psychology, economics. Many illustrative examples; over 250 problems. 260pp. 5⅜ x 8½. 65084-7 Pa. $8.95

THREE PEARLS OF NUMBER THEORY, A. Y. Khinchin. Three compelling puzzles require proof of a basic law governing the world of numbers. Challenges concern van der Waerden's theorem, the Landau-Schnirelmann hypothesis and Mann's theorem, and a solution to Waring's problem. Solutions included. 64pp. 5⅜ x 8½. 40026-3 Pa. $4.95

LECTURES ON CLASSICAL DIFFERENTIAL GEOMETRY, Second Edition, Dirk J. Struik. Excellent brief introduction covers curves, theory of surfaces, fundamental equations, geometry on a surface, conformal mapping, other topics. Problems. 240pp. 5⅜ x 8½. 65609-8 Pa. $9.95

ROTARY-WING AERODYNAMICS, W.Z. Stepniewski. Clear, concise text covers aerodynamic phenomena of the rotor and offers guidelines for helicopter performance evaluation. Originally prepared for NASA. 537 figures. 640pp. 6⅛ x 9¼.
64647-5 Pa. $16.95

DIFFERENTIAL GEOMETRY, Heinrich W. Guggenheimer. Local differential geometry as an application of advanced calculus and linear algebra. Curvature, transformation groups, surfaces, more. Exercises. 62 figures. 378pp. 5⅜ x 8½.
63433-7 Pa. $11.95

INTRODUCTION TO SPACE DYNAMICS, William Tyrrell Thomson. Comprehensive, classic introduction to space-flight engineering for advanced undergraduate and graduate students. Includes vector algebra, kinematics, transformation of coordinates. Bibliography. Index. 352pp. 5⅜ x 8½.
65113-4 Pa. $10.95

A SURVEY OF MINIMAL SURFACES, Robert Osserman. Up-to-date, in-depth discussion of the field for advanced students. Corrected and enlarged edition covers new developments. Includes numerous problems. 192pp. 5⅜ x 8½.
64998-9 Pa. $8.95

ANALYTICAL MECHANICS OF GEARS, Earle Buckingham. Indispensable reference for modern gear manufacture covers conjugate gear-tooth action, gear-tooth profiles of various gears, many other topics. 263 figures. 102 tables. 546pp. 5⅜ x 8½.
65712-4 Pa. $16.95

SET THEORY AND LOGIC, Robert R. Stoll. Lucid introduction to unified theory of mathematical concepts. Set theory and logic seen as tools for conceptual understanding of real number system. 496pp. 5⅜ x 8¼.
63829-4 Pa. $14.95

A HISTORY OF MECHANICS, René Dugas. Monumental study of mechanical principles from antiquity to quantum mechanics. Contributions of ancient Greeks, Galileo, Leonardo, Kepler, Lagrange, many others. 671pp. 5⅜ x 8½.
65632-2 Pa. $18.95

FAMOUS PROBLEMS OF GEOMETRY AND HOW TO SOLVE THEM, Benjamin Bold. Squaring the circle, trisecting the angle, duplicating the cube: learn their history, why they are impossible to solve, then solve them yourself. 128pp. 5⅜ x 8½.
24297-8 Pa. $5.95

MECHANICAL VIBRATIONS, J.P. Den Hartog. Classic textbook offers lucid explanations and illustrative models, applying theories of vibrations to a variety of practical industrial engineering problems. Numerous figures. 233 problems, solutions. Appendix. Index. Preface. 436pp. 5⅜ x 8½.
64785-4 Pa. $13.95

CURVATURE AND HOMOLOGY: Enlarged Edition, Samuel I. Goldberg. Revised edition examines topology of differentiable manifolds; curvature, homology of Riemannian manifolds; compact Lie groups; complex manifolds; curvature, homology of Kaehler manifolds. New Preface. Four new appendixes. 416pp. 5⅜ x 8½.
40207-X Pa. $14.95

HISTORY OF STRENGTH OF MATERIALS, Stephen P. Timoshenko. Excellent historical survey of the strength of materials with many references to the theories of elasticity and structure. 245 figures. 452pp. 5⅜ x 8½.
61187-6 Pa. $14.95

GEOMETRY OF COMPLEX NUMBERS, Hans Schwerdtfeger. Illuminating, widely praised book on analytic geometry of circles, the Moebius transformation, and two-dimensional non-Euclidean geometries. 200pp. 5⅝ x 8¼. 63830-8 Pa. $8.95

MECHANICS, J.P. Den Hartog. A classic introductory text or refresher. Hundreds of applications and design problems illuminate fundamentals of trusses, loaded beams and cables, etc. 334 answered problems. 462pp. 5⅜ x 8½. 60754-2 Pa. $12.95

TOPOLOGY, John G. Hocking and Gail S. Young. Superb one-year course in classical topology. Topological spaces and functions, point-set topology, much more. Examples and problems. Bibliography. Index. 384pp. 5⅝ x 8¼. 65676-4 Pa. $11.95

STRENGTH OF MATERIALS, J.P. Den Hartog. Full, clear treatment of basic material (tension, torsion, bending, etc.) plus advanced material on engineering methods, applications. 350 answered problems. 323pp. 5⅜ x 8½. 60755-0 Pa. $9.95

ELEMENTARY CONCEPTS OF TOPOLOGY, Paul Alexandroff. Elegant, intuitive approach to topology from set-theoretic topology to Betti groups; how concepts of topology are useful in math and physics. 25 figures. 57pp. 5⅜ x 8½. 60747-X Pa. $4.95

ADVANCED STRENGTH OF MATERIALS, J.P. Den Hartog. Superbly written advanced text covers torsion, rotating disks, membrane stresses in shells, much more. Many problems and answers. 388pp. 5⅜ x 8½. 65407-9 Pa. $11.95

COMPUTABILITY AND UNSOLVABILITY, Martin Davis. Classic graduate-level introduction to theory of computability, usually referred to as theory of recurrent functions. New preface and appendix. 288pp. 5⅜ x 8½. 61471-9 Pa. $8.95

GENERAL CHEMISTRY, Linus Pauling. Revised 3rd edition of classic first-year text by Nobel laureate. Atomic and molecular structure, quantum mechanics, statistical mechanics, thermodynamics correlated with descriptive chemistry. Problems. 992pp. 5⅜ x 8½. 65622-5 Pa. $19.95

AN INTRODUCTION TO MATRICES, SETS AND GROUPS FOR SCIENCE STUDENTS, G. Stephenson. Concise, readable text introduces sets, groups, and most importantly, matrices to undergraduate students of physics, chemistry, and engineering. Problems. 164pp. 5⅜ x 8½. 65077-4 Pa. $7.95

THE HISTORICAL BACKGROUND OF CHEMISTRY, Henry M. Leicester. Evolution of ideas, not individual biography. Concentrates on formulation of a coherent set of chemical laws. 260pp. 5⅜ x 8½. 61053-5 Pa. $8.95

THE PHILOSOPHY OF MATHEMATICS: An Introductory Essay, Stephan Körner. Surveys the views of Plato, Aristotle, Leibniz & Kant concerning propositions and theories of applied and pure mathematics. Introduction. Two appendices. Index. 198pp. 5⅜ x 8½. 25048-2 Pa. $8.95

THE DEVELOPMENT OF MODERN CHEMISTRY, Aaron J. Ihde. Authoritative history of chemistry from ancient Greek theory to 20th-century innovation. Covers major chemists and their discoveries. 209 illustrations. 14 tables. Bibliographies. Indices. Appendices. 851pp. 5⅜ x 8½. 64235-6 Pa. $18.95

DE RE METALLICA, Georgius Agricola. The famous Hoover translation of greatest treatise on technological chemistry, engineering, geology, mining of early modern times (1556). All 289 original woodcuts. 638pp. 6¾ x 11. 60006-8 Pa. $21.95

SOME THEORY OF SAMPLING, William Edwards Deming. Analysis of the problems, theory and design of sampling techniques for social scientists, industrial managers and others who find statistics increasingly important in their work. 61 tables. 90 figures. xvii + 602pp. 5⅜ x 8½. 64684-X Pa. $16.95

THE VARIOUS AND INGENIOUS MACHINES OF AGOSTINO RAMELLI: A Classic Sixteenth-Century Illustrated Treatise on Technology, Agostino Ramelli. One of the most widely known and copied works on machinery in the 16th century. 194 detailed plates of water pumps, grain mills, cranes, more. 608pp. 9 x 12. 28180-9 Pa. $24.95

LINEAR PROGRAMMING AND ECONOMIC ANALYSIS, Robert Dorfman, Paul A. Samuelson and Robert M. Solow. First comprehensive treatment of linear programming in standard economic analysis. Game theory, modern welfare economics, Leontief input-output, more. 525pp. 5⅜ x 8½. 65491-5 Pa. $17.95

ELEMENTARY DECISION THEORY, Herman Chernoff and Lincoln E. Moses. Clear introduction to statistics and statistical theory covers data processing, probability and random variables, testing hypotheses, much more. Exercises. 364pp. 5⅜ x 8½. 65218-1 Pa. $10.95

THE COMPLEAT STRATEGYST: Being a Primer on the Theory of Games of Strategy, J.D. Williams. Highly entertaining classic describes, with many illustrated examples, how to select best strategies in conflict situations. Prefaces. Appendices. 268pp. 5⅜ x 8½. 25101-2 Pa. $8.95

CONSTRUCTIONS AND COMBINATORIAL PROBLEMS IN DESIGN OF EXPERIMENTS, Damaraju Raghavarao. In-depth reference work examines orthogonal Latin squares, incomplete block designs, tactical configuration, partial geometry, much more. Abundant explanations, examples. 416pp. 5⅜ x 8¼. 65685-3 Pa. $10.95

THE ABSOLUTE DIFFERENTIAL CALCULUS (CALCULUS OF TENSORS), Tullio Levi-Civita. Great 20th-century mathematician's classic work on material necessary for mathematical grasp of theory of relativity. 452pp. 5⅜ x 8½. 63401-9 Pa. $11.95

VECTOR AND TENSOR ANALYSIS WITH APPLICATIONS, A.I. Borisenko and I.E. Tarapov. Concise introduction. Worked-out problems, solutions, exercises. 257pp. 5⅝ x 8¼. 63833-2 Pa. $9.95

THE FOUR-COLOR PROBLEM: Assaults and Conquest, Thomas L. Saaty and Paul G. Kainen. Engrossing, comprehensive account of the century-old combinatorial topological problem, its history and solution. Bibliographies. Index. 110 figures. 228pp. 5⅜ x 8½. 65092-8 Pa. $7.95

CATALOG OF DOVER BOOKS

CATALYSIS IN CHEMISTRY AND ENZYMOLOGY, William P. Jencks. Exceptionally clear coverage of mechanisms for catalysis, forces in aqueous solution, carbonyl- and acyl-group reactions, practical kinetics, more. 864pp. 5⅜ x 8½.
65460-5 Pa. $19.95

PROBABILITY: An Introduction, Samuel Goldberg. Excellent basic text covers set theory, probability theory for finite sample spaces, binomial theorem, much more. 360 problems. Bibliographies. 322pp. 5⅜ x 8½.
65252-1 Pa. $10.95

LIGHTNING, Martin A. Uman. Revised, updated edition of classic work on the physics of lightning. Phenomena, terminology, measurement, photography, spectroscopy, thunder, more. Reviews recent research. Bibliography. Indices. 320pp. 5⅜ x 8¼.
64575-4 Pa. $8.95

PROBABILITY THEORY: A Concise Course, Y.A. Rozanov. Highly readable, self-contained introduction covers combination of events, dependent events, Bernoulli trials, etc. Translation by Richard Silverman. 148pp. 5⅜ x 8¼.
63544-9 Pa. $7.95

AN INTRODUCTION TO HAMILTONIAN OPTICS, H. A. Buchdahl. Detailed account of the Hamiltonian treatment of aberration theory in geometrical optics. Many classes of optical systems defined in terms of the symmetries they possess. Problems with detailed solutions. 1970 edition. xv + 360pp. 5⅜ x 8½.
67597-1 Pa. $10.95

STATISTICS MANUAL, Edwin L. Crow, et al. Comprehensive, practical collection of classical and modern methods prepared by U.S. Naval Ordnance Test Station. Stress on use. Basics of statistics assumed. 288pp. 5⅜ x 8½.
60599-X Pa. $8.95

DICTIONARY/OUTLINE OF BASIC STATISTICS, John E. Freund and Frank J. Williams. A clear concise dictionary of over 1,000 statistical terms and an outline of statistical formulas covering probability, nonparametric tests, much more. 208pp. 5⅜ x 8½.
66796-0 Pa. $7.95

STATISTICAL METHOD FROM THE VIEWPOINT OF QUALITY CONTROL, Walter A. Shewhart. Important text explains regulation of variables, uses of statistical control to achieve quality control in industry, agriculture, other areas. 192pp. 5⅜ x 8½.
65232-7 Pa. $8.95

METHODS OF THERMODYNAMICS, Howard Reiss. Outstanding text focuses on physical technique of thermodynamics, typical problem areas of understanding, and significance and use of thermodynamic potential. 1965 edition. 238pp. 5⅜ x 8½.
69445-3 Pa. $8.95

STATISTICAL ADJUSTMENT OF DATA, W. Edwards Deming. Introduction to basic concepts of statistics, curve fitting, least squares solution, conditions without parameter, conditions containing parameters. 26 exercises worked out. 271pp. 5⅜ x 8½.
64685-8 Pa. $9.95

TENSOR CALCULUS, J.L. Synge and A. Schild. Widely used introductory text covers spaces and tensors, basic operations in Riemannian space, non-Riemannian spaces, etc. 324pp. 5⅜ x 8¼.
63612-7 Pa. $11.95

A CONCISE HISTORY OF MATHEMATICS, Dirk J. Struik. The best brief history of mathematics. Stresses origins and covers every major figure from ancient Near East to 19th century. 41 illustrations. 195pp. 5⅜ x 8½. 60255-9 Pa. $8.95

A SHORT ACCOUNT OF THE HISTORY OF MATHEMATICS, W.W. Rouse Ball. One of clearest, most authoritative surveys from the Egyptians and Phoenicians through 19th-century figures such as Grassman, Galois, Riemann. Fourth edition. 522pp. 5⅜ x 8½. 20630-0 Pa. $13.95

HISTORY OF MATHEMATICS, David E. Smith. Nontechnical survey from ancient Greece and Orient to late 19th century; evolution of arithmetic, geometry, trigonometry, calculating devices, algebra, the calculus. 362 illustrations. 1,355pp. 5⅜ x 8½. 20429-4, 20430-8 Pa., Two-vol. set $27.90

THE GEOMETRY OF RENÉ DESCARTES, René Descartes. The great work founded analytical geometry. Original French text, Descartes' own diagrams, together with definitive Smith-Latham translation. 244pp. 5⅜ x 8½. 60068-8 Pa. $8.95

GAMES, GODS & GAMBLING: A History of Probability and Statistical Ideas, F. N. David. Episodes from the lives of Galileo, Fermat, Pascal, and others illustrate this fascinating account of the roots of mathematics. Features thought-provoking references to classics, archaeology, biography, poetry. 1962 edition. 304pp. 5⅜ x 8½. (USO) 40023-9 Pa. $9.95

THE HISTORY OF THE CALCULUS AND ITS CONCEPTUAL DEVELOPMENT, Carl B. Boyer. Origins in antiquity, medieval contributions, work of Newton, Leibniz, rigorous formulation. Treatment is verbal. 346pp. 5⅜ x 8½. 60509-4 Pa. $9.95

THE THIRTEEN BOOKS OF EUCLID'S ELEMENTS, translated with introduction and commentary by Sir Thomas L. Heath. Definitive edition. Textual and linguistic notes, mathematical analysis. 2,500 years of critical commentary. Not abridged. 1,414pp. 5⅜ x 8½. 60088-2, 60089-0, 60090-4 Pa., Three-vol. set $34.85

GAMES AND DECISIONS: Introduction and Critical Survey, R. Duncan Luce and Howard Raiffa. Superb nontechnical introduction to game theory, primarily applied to social sciences. Utility theory, zero-sum games, n-person games, decision-making, much more. Bibliography. 509pp. 5⅜ x 8½. 65943-7 Pa. $14.95

THE HISTORICAL ROOTS OF ELEMENTARY MATHEMATICS, Lucas N.H. Bunt, Phillip S. Jones, and Jack D. Bedient. Fundamental underpinnings of modern arithmetic, algebra, geometry and number systems derived from ancient civilizations. 320pp. 5⅜ x 8½. 25563-8 Pa. $8.95

CALCULUS REFRESHER FOR TECHNICAL PEOPLE, A. Albert Klaf. Covers important aspects of integral and differential calculus via 756 questions. 566 problems, most answered. 431pp. 5⅜ x 8½. 20370-0 Pa. $9.95

CATALOG OF DOVER BOOKS

CHALLENGING MATHEMATICAL PROBLEMS WITH ELEMENTARY SOLUTIONS, A.M. Yaglom and I.M. Yaglom. Over 170 challenging problems on probability theory, combinatorial analysis, points and lines, topology, convex polygons, many other topics. Solutions. Total of 445pp. 5⅜ x 8½. Two-vol. set.

Vol. I: 65536-9 Pa. $8.95
Vol. II: 65537-7 Pa. $7.95

FIFTY CHALLENGING PROBLEMS IN PROBABILITY WITH SOLUTIONS, Frederick Mosteller. Remarkable puzzlers, graded in difficulty, illustrate elementary and advanced aspects of probability. Detailed solutions. 88pp. 5⅜ x 8½.

65355-2 Pa. $4.95

EXPERIMENTS IN TOPOLOGY, Stephen Barr. Classic, lively explanation of one of the byways of mathematics. Klein bottles, Moebius strips, projective planes, map coloring, problem of the Koenigsberg bridges, much more, described with clarity and wit. 43 figures. 210pp. 5⅜ x 8½.   25933-1 Pa. $6.95

RELATIVITY IN ILLUSTRATIONS, Jacob T. Schwartz. Clear nontechnical treatment makes relativity more accessible than ever before. Over 60 drawings illustrate concepts more clearly than text alone. Only high school geometry needed. Bibliography. 128pp. 6⅛ x 9¼.   25965-X Pa. $7.95

AN INTRODUCTION TO ORDINARY DIFFERENTIAL EQUATIONS, Earl A. Coddington. A thorough and systematic first course in elementary differential equations for undergraduates in mathematics and science, with many exercises and problems (with answers). Index. 304pp. 5⅜ x 8½.   65942-9 Pa. $9.95

FOURIER SERIES AND ORTHOGONAL FUNCTIONS, Harry F. Davis. An incisive text combining theory and practical example to introduce Fourier series, orthogonal functions and applications of the Fourier method to boundary-value problems. 570 exercises. Answers and notes. 416pp. 5⅜ x 8½.   65973-9 Pa. $13.95

AN INTRODUCTION TO ALGEBRAIC STRUCTURES, Joseph Landin. Superb self-contained text covers "abstract algebra": sets and numbers, theory of groups, theory of rings, much more. Numerous well-chosen examples, exercises. 247pp. 5⅜ x 8½.

65940-2 Pa. $8.95

STARS AND RELATIVITY, Ya. B. Zel'dovich and I. D. Novikov. Vol. 1 of *Relativistic Astrophysics* by famed Russian scientists. General relativity, properties of matter under astrophysical conditions, stars and stellar systems. Deep physical insights, clear presentation. 1971 edition. References. 544pp. 5⅜ x 8½.

69424-0 Pa. $14.95
